T0329130

BIOFLUID MECHANICS

SECOND EDITION

BIOFLUID MECHANICS

AN INTRODUCTION TO FLUID MECHANICS, MACROCIRCULATION, AND MICROCIRCULATION

SECOND EDITION

DAVID A. RUBENSTEIN, WEI YIN AND MARY D. FRAME
The Department of Biomedical Engineering
Stony Brook University
Stony Brook, NY, USA

AMSTERDAM • BOSTON • HEIDELBERG • LONDON
NEW YORK • OXFORD • PARIS • SAN DIEGO
SAN FRANCISCO • SINGAPORE • SYDNEY • TOKYO
Academic Press is an imprint of Elsevier

Academic Press is an imprint of Elsevier
125, London Wall, EC2Y 5AS.
525 B Street, Suite 1800, San Diego, CA 92101-4495, USA
225 Wyman Street, Waltham, MA 02451, USA
The Boulevard, Langford Lane, Kidlington, Oxford OX5 1GB, UK

Second edition 2015

Copyright © 2015, 2012 Elsevier Inc. All rights reserved.

No part of this publication may be reproduced or transmitted in any form or by any means, electronic or mechanical, including photocopying, recording, or any information storage and retrieval system, without permission in writing from the publisher. Details on how to seek permission, further information about the Publisher's permissions policies and our arrangements with organizations such as the Copyright Clearance Center and the Copyright Licensing Agency, can be found at our website: www.elsevier.com/permissions.

This book and the individual contributions contained in it are protected under copyright by the Publisher (other than as may be noted herein).

Notices
Knowledge and best practice in this field are constantly changing. As new research and experience broaden our understanding, changes in research methods, professional practices, or medical treatment may become necessary.

Practitioners and researchers must always rely on their own experience and knowledge in evaluating and using any information, methods, compounds, or experiments described herein. In using such information or methods they should be mindful of their own safety and the safety of others, including parties for whom they have a professional responsibility.

To the fullest extent of the law, neither the Publisher nor the authors, contributors, or editors, assume any liability for any injury and/or damage to persons or property as a matter of products liability, negligence or otherwise, or from any use or operation of any methods, products, instructions, or ideas contained in the material herein.

ISBN: 978-0-12-800944-4

British Library Cataloguing-in-Publication Data
A catalogue record for this book is available from the British Library.

Library of Congress Cataloging-in-Publication Data
A catalog record for this book is available from the Library of Congress.

For Information on all Academic Press publications
visit our website at http://store.elsevier.com/

Quotes on Engineering, Science, Research, and Related Matters

The Sun, with all the planets revolving around it and dependent on it, can still ripen a bunch of grapes as though it had nothing else in the Universe to do. ***Galileo Galilei***

The scientists of today think deeply instead of clearly. One must be sane to think clearly, but one can think deeply and be quite insane. ***Nikola Tesla***

Life is pretty simple: You do some stuff. Most fails. Some works. You do more of what works. If it works big, others quickly copy it. Then you do something else. The trick is the doing something else. ***Leonardo da Vinci***

It is not the strongest of the species that survives, nor the most intelligent that survives. It is the one that is the most adaptable to change. ***Charles Darwin***

Equipped with his five senses, man explores the universe around him and calls the adventure Science. ***Edwin Hubble***

It would be nice if all of the data which sociologists require could be enumerated because then we could run them through IBM machines and draw charts as the economists do. However, not everything that can be counted counts, and not everything that counts can be counted. ***William Bruce Cameron***

Contents

5. Blood Flow in Arteries and Veins

III MICROCIRCULATION

6. Microvascular Beds

7. Mass Transport and Heat Transfer in the Microcirculation

8. The Lymphatic System

IV SPECIALITY CIRCULATIONS AND OTHER BIOLOGICAL FLOWS

9. Flow in the Lungs

V

MODELING AND EXPERIMENTAL TECHNIQUES

14. *In Silico* Biofluid Mechanics

15. *In Vitro* Biofluid Mechanics

16. *In Vivo* Biofluid Mechanics

Preface

The purpose of this textbook is to serve as an introduction to biofluid mechanics with emphasis on the macrocirculation, microcirculation, and other important biological flows in the human body. As the reader will see, an integral component of the human body is the cardiovascular system, and consequently it may be one of the most important biofluid scenarios to study. Furthermore, most implantable devices, therapeutic agents, and any other device that can come into contact with the body may augment biofluid properties. With this textbook, the authors hope to provide a systematic teaching and learning tool that will motivate students to continue their education in biofluid mechanics, as well as provide an effective educational structure to aid in biofluid mechanics instruction. To complete these aims, the authors have provided first a rigorous review of the salient fluid mechanics principles and have related these principles to biological systems. This is followed by two sections on the application of these principles to the macrocirculation and the microcirculation, the two most commonly studied systems in biofluid mechanics. Next, sections on biological flows within other systems and experimental techniques for biofluid mechanics are included to illustrate that (i) fluid mechanics principles are not restricted to the cardiovascular system and (ii) advanced techniques are needed to tease out biofluid mechanics properties. Finally, the authors relate this material to many principles that the readers should be familiar with to provide concrete and relevant examples to supplement learning. The authors believe that the combination of all of these features makes the book novel and unique, and they hope that this will facilitate the learning of biofluid mechanics at all levels.

This textbook has been put together to fill a need in the evolving biofluid mechanics community. As more undergraduate biomedical engineering departments become established, textbooks tailored to undergraduate education are needed in the core biomedical engineering courses. This book has made use of a problem-based approach and modern pedagogy in its development. The problem-based approach is used to capture the reader's interest and show insight into biofluid properties. Furthermore, there are extensive problems worked out in detail in each section to provide examples of how to approach biofluid mechanics problems, typical pitfalls, and a structured solution methodology that students can follow when studying at home. Numerous pedagogical tools have been implemented in this book to aid student understanding. Each chapter starts with a learning outcomes section, so that the reader can anticipate what the key concepts are that will be covered in the particular chapter. Furthermore, each chapter concludes with a summary section that reiterates the salient points and key equations. Further readings and classical biofluid mechanics papers are cited for supplemental reading. These sections will hopefully be used as a springboard for learning and act to reinforce key concepts.

ANCILLARIES

For instructors using this text in a course, a solutions manual and a set of electronic images are available by registering at: http://textbooks.elsevier.com/9780128009444

Acknowledgments

The authors would like to thank the following reviewers of this project, including those who reviewed at various stages for organization, content, and accuracy:

Iskander S. Akhatov, North Dakota State University
Naomi C. Chesler, University of Wisconsin–Madison
Dr. Michael C. Christie, Florida International University
Prof. Zhixiong Guo, Rutgers University
Julie Y. Ji, Indiana University-Purdue University Indianapolis
Dr. Carola König, Brunel University
Jani Macari Pallis, University of Bridgeport
Keefe B. Manning, Ph.D., The Pennsylvania State University
Dr. Manosh C. Paul, University of Glasgow
David B. Reynolds, Ph.D., Wright State University
Philippe Sucosky, University of Notre Dame
Lucy T. Zhang, Rensselaer Polytechnic Institute
Anonymous Reviewer, Oregon Health & Science University
Anonymous Reviewer, Purdue University
Anonymous Reviewer, Worcester Polytechnic Institute

This book has also been class-tested by students in manuscript form, and their feedback has proved valuable in improving the usefulness of the book. We appreciate all the efforts of these individuals in helping improve the quality of the book, but recognize that any errors that might still exist despite our best efforts are the responsibility of the authors themselves.

We would also like to thank the efforts of the publisher in helping bring this book about:

Joe Hayton (Publisher)
Stephen Merken (Sr. Acquisitions Editor)
Nate McFadden (Sr. Development Editor)
Kiruthika Govindaraju (Project Manager)

PART I

FLUID MECHANICS BASICS

CHAPTER

1

Introduction

LEARNING OUTCOMES

1. Identify basic engineering skills that will be used in this course
2. Describe the fields and the importance of biomedical engineering, fluid mechanics, and biofluid mechanics
3. Review concepts of dimensions and units
4. Discuss two of the salient dimensionless numbers in biofluid mechanics

1.1 NOTE TO STUDENTS ABOUT THE TEXTBOOK

The goal of this textbook is to clearly describe how fluid mechanics principles can be applied to different biological systems and, in parallel, discuss current research avenues in biofluid mechanics and common pathological conditions that are associated with altered biofluids, biofluid flows, and/or biofluid organs. Classic fluid mechanics laws, which the reader may be familiar with from a previous course in fluid mechanics (but will be reviewed in Part 1 of this textbook), have been used extensively to describe blood flow through the vascular system for decades. One major goal of this textbook is to discuss how these laws apply to the vascular system, but we also aim to highlight some of the specialty flows that can be described using the same fluid mechanics principles. Part 2, Macrocirculation, and Part 3, Microcirculation, focus on the application of these classic principles to the vascular system and develop mathematical formulas and relationships to help the reader understand the fluid mechanics associated with blood flow through blood vessels of various sizes. Part 4, Specialty Circulations, describes fluid flows through the lungs, eyes, diarthroses joints, kidneys, and the splanchnic circulation, which are not traditionally covered in biofluid mechanics courses but are very important biological flows in the human body. Note that there are other important specialty circulations, which can be

 © 2015 Elsevier Inc. All rights reserved.

modeled using the fluid mechanics principles that will be developed in Part 1 of this textbook. We may touch on some of those circulations, but we will not fully develop an analysis of these flows. For the most part, similar fluid mechanics principles can be used to describe the specialty circulations with some slight modifications to accurately depict the particular special conditions associated with the circulation. Part 5, Experimental Techniques, briefly highlights different procedures that are currently being used in biofluid mechanics laboratories to elucidate flow characteristics. At the same time, we will highlight some of the current innovative work that is being conducted to elucidate biofluid mechanics phenomena. The overarching goal of this textbook is to establish a foundation for students' future studies in biofluid mechanics, whether in a more advanced course or in a research environment.

In writing this textbook, we hope to meet the needs of both the students and the instructors who may use this textbook. We believe that this textbook is written in a way that instructors can use the material presented either as the sole course material (introductory biofluid mechanics course) or as the foundation for more in-depth discussions of biofluid mechanics (upper-division/graduate courses). However, an introductory textbook, such as this one, cannot include every detail of importance to biofluid mechanics. There are multiple exceptional references that can be used in conjunction with this textbook (some of which are highlighted in the Further Reading section). Therefore, we encourage you to visit your local libraries or to search the Internet for more in-depth details that are not included in this textbook. This textbook cannot and does not aim to replace traditional fluid mechanics or physiology textbooks, but it will provide the information necessary to (i) set a foundation for a broad biofluid mechanics discussion, (ii) analyze some of the particular biofluid mechanics principles, (iii) quantify some of the salient biofluid mechanics flows, and (iv) serve as a springboard for future more detailed and in-depth discussion. At the end of most of the chapters, we provide suggested references for the students and instructors, if more information is desired.

Your instructor and other students in your class are other good resources to learn about biofluid mechanics. However, we believe that you will learn these principles best by working example problems at home. We included extensive examples within the text, all completely worked out, so you can see the level of detail needed to solve biofluid mechanics problems. We also include various levels of homework problems at the end of each chapter for you to practice on your own time. Your success in this course will depend not only on the material presented within this textbook and from your instructor's notes, but also on your willingness to comprehend the material and work biofluid mechanics problems yourself. We hope that this textbook can serve as a stepping-stone on your way to becoming experts in biofluid mechanics principles and applications. If you feel there are shortcomings or omissions in this textbook, please let us know so that the situation can be remedied in future editions.

1.2 BIOMEDICAL ENGINEERING

One of the first questions that should arise when studying biomedical engineering (in this textbook the term biomedical engineering will be used interchangeably with bioengineering) is, What is biomedical engineering? The National Institutes of Health working definition (as of July 24, 1997) of biomedical engineering is

> Biomedical engineering integrates physical, chemical, mathematical, and computational sciences and engineering principles to study biology, medicine, behavior, and health. It advances fundamental concepts; creates knowledge from the molecular to the organ systems level; and develops innovative biologics, materials, processes, implants, devices and informatics approaches for the prevention, diagnosis, and treatment of disease, for patient rehabilitation, and for improving health.

This definition is broad and can encompass many different engineering disciplines and, in fact, biomedical engineers can apply electrical, mechanical, chemical, and materials engineering principles to the study of biological tissue and to how these tissues function and respond to different conditions. Biomedical engineers also focus on many other fundamental engineering disciplines, such as systems and controls problems through the design of new devices for medical imaging, rehabilitation, and disease diagnosis, among others. The nature of biomedical engineering is thus interdisciplinary because of the need to understand both engineering principles and physiology and apply concepts from both disciplines to your area of investigation. The goal of biomedical engineering is to mold these disciplines together to describe biological systems or design and fabricate devices to be used in a biological or medical setting.

In this textbook, the focus is on mechanical engineering principles and how they are related to biofluid mechanics. This is not to say that other engineering principles are not or cannot be applied to biofluid mechanics. For this textbook, we will take the approach that starts from the fundamental engineering statics and dynamics laws to derive the fluid mechanics equations of state. In parallel, thermodynamics equations will be developed for the study of heat transfer within biological systems. Most of these equations should be familiar, but we will discuss and develop them in subsequent chapters where a review is needed. Most biofluid mechanics problems deal with describing the flow in a particular tissue, which can be considered an extension or a special case of the fluid mechanics problems that have been studied previously (if a fluid mechanics course was taken prior to this course). For example, if we were interested in describing the blood flow through the coronary artery, we can use fluid mechanics principles, but we would also need to consider the mechanical properties of the blood vessel and how this may alter the fluid flow. Likewise, if we were interested in designing a new implantable cardiovascular device, we would need to understand and consider not only the mechanical flow principles, but also the material properties, the electrical components, and the physiological effects that the device may have on the cardiovascular system. This type of problem approaches the heart of what a biomedical engineer does: design a device to remedy a physiological problem and describe the effects of that device in physiologically relevant settings. Some biomedical engineers focus solely on the engineering design aspect, while others focus on physiological applications. Herein, we discuss both aspects of biofluid mechanics to solve multiple types of problems.

1.3 SCOPE OF FLUID MECHANICS

Now that we have defined what a biomedical engineer is and what a biomedical engineer does, the question should arise, "Why do we need to study biofluid mechanics?" We will first answer the question, "Why should we study fluid mechanics?" and return to the earlier question later. Any system that operates in a fluid medium can be analyzed using fluid mechanics principles. This includes anything that moves in a gas (e.g., airplanes, cars, trains, birds) or in a liquid (e.g., submarines, fish), or anything that is designed to have at least one boundary surface with a gas or a liquid (e.g., bridges, skyscrapers, boats, cells). Situations may arise in which objects can be described as "flowing" through a fluid medium (fish swimming). These types of situations need to be concerned with both the flow of the object (fish) and the flow of the medium (water surrounding the fish). These types of flow scenarios are common in biofluid mechanics in which cells, which exhibit fluid properties, are submerged within a moving fluid (e.g., red blood cells flowing through blood within the cardiovascular system). Stationary flow structures are concerned with the forces (drag, shear, pressure) that can be transmitted from a flowing fluid to the structure that is containing or bounding the fluid. This is a critical area of analysis for the design or evaluation of any fluid that flows through any "channel," which includes blood flowing through blood vessels, lymph flowing through lymph vessels, and air flowing through the respiratory tract. Fluid mechanics principles can be used to describe all of these highlighted systems, among many others.

Also integral to the study of fluid mechanics is the design of fluid machinery. Fluid machinery includes pumps, turbines, and anything that has a lubrication layer between two moving and usually solid parts. Typically, fluid machinery comes into play in the design of heating and ventilation systems, cars, airplanes, and a long list of other devices/systems, which include devices that interact with or within biological systems. Fluid mechanics is not an academic problem, but one that every engineer will face at some time during his or her career. This textbook is not going to replace a standard fluid mechanics text or a fluid mechanics course. Instead, we use fluid mechanics as a starting point to describe one particular application of fluid mechanics: biological fluid flows.

1.4 SCOPE OF BIOFLUID MECHANICS

So, why should we study biofluid mechanics? First, your body is composed of approximately 65% water. All cells have an intracellular water component and each cell is immersed within an extracellular water compartment. There are some forces that are distributed and transmitted through this water layer that act on each and every immersed cells. Also, some cells in your body are non-adherent cells under non-pathological conditions (i.e., red blood cells, white blood cells, platelets). These highlighted cells are convected through your blood stream (within the cardiovascular system) and experience many types of fluid forces (including shear forces and pressure forces), which can alter their functions. Gas movement (such as oxygen and carbon dioxide exchange within your lungs or air motion through the respiratory tree) can also be described by fluid mechanics principles. Joint lubrication, a major research area of biofluid mechanics, is critical to locomotion: with the degeneration of the lubrication layer, movement becomes difficult. Prosthetics cannot be designed without fluid mechanics. It is our hope that throughout this textbook, the readers will understand how fluid mechanics laws can be applied to biological systems, and the significance of fluid mechanics laws to the biological system as a whole.

Furthermore, the design of many implantable devices must consider fluid mechanics laws. Obvious examples are those devices that are directly implanted into the human body and are in contact with blood, such as stents and mechanical heart valves, among others. However, a total artificial heart is a pump that will have a biological fluid flowing within it and replaces a portion of the cardiovascular system. Other examples of implantable devices that involve biofluid mechanics include extracorporeal devices that must maintain steady flow without aggravating cells or introducing harmful chemicals (for dialysis) and contact lenses that must consider the wetting of the eye, as well as gas diffusion to the eye, to function properly. Indeed, nearly every device intended for biological use will have to consider fluid mechanics laws, which are critical for proper design and functioning of the device.

Why is biofluid mechanics so critical to study? According to the American Heart Association, in 2014, more than 83,600,000 people in the United States had at least one cardiovascular disease. The majority of these cases are associated with high blood pressure (approximately 78 million people, which is approximately one-third of the United States population; this number only includes people older than 20 years) or coronary heart disease (approximately 16 million people; note that some of these patients can overlap with the first group). Also, cardiovascular diseases are the cause for approximately one of every three deaths in the United States, which is approximately 1.5 times more deaths than caused by cancer. For the complete statistics, which are divided into groups based on risk factors, age, race, etc. (see Ref. [1]). Coronary heart disease accounts for nearly 50% of these deaths and remains as the leading cause of death in the United States. One of the most telling statistics associated with cardiovascular diseases is that approximately every 30 s, one American will experience a coronary event and that approximately every 1.5 min, one American will die due to a coronary event. Along with the American Heart Association, the Centers for Disease Control and Prevention generates national maps documenting the country-level heart disease death rates and the heart stroke death rates by county (see http://www.cdc.gov/dhdsp/maps/national_maps/index.htm). Both of these maps only include data for people in the United States who are older than 34 years, and both provide evidence for where the highest incident rates of heart disease occur in the United States. It is clear that states along the lower Mississippi River valley have a higher rate of death associated with cardiovascular diseases, than the majority of other states.

Biofluid mechanics is concerned not only with cardiovascular system and diseases but also with lung diseases. Lung cancer is probably the first disease that comes to mind, and you are probably also considering that the most likely origin of lung cancer is exposure to cigarette smoke (either first-hand or second-hand). Indeed, approximately 85% of deaths associated with lung cancer are associated with cigarette smoking. Burning cigarettes emit close to 4,000 different chemicals, with approximately 50 of these known carcinogens. Interestingly, smoking cigarettes that have a reduced amount of tar does not significantly decrease the chances of developing lung cancer because of the plethora of other carcinogens within the cigarette. As a fact, tobacco companies are allowed to add approximately 600 chemicals to their product, including ammonia, 1-butanol, caffeine, cocoa, ethanol, 1-octanol, raisin juice, sodium hydroxide, and urea. In excess, some of the mentioned additives can lead to death. Furthermore, smoking is not just associated with lung cancer, but a smoker can succumb to other cancers as well, including oral, pharynx, larynx, bladder, kidney, stomach, and pancreatic cancer, due to the chemical additives that compose a cigarette. Interestingly, the rate at which e-cigarettes, which may contain some of the same chemicals that are within tobacco-based cigarettes, induce various cancers are currently unknown.

A third disease that is commonly investigated by people interested in biofluid mechanics is chronic kidney disease. Chronic kidney diseases are a group of diseases that alter the function of the kidneys and typically reduces their ability to remove toxins from the blood properly. Kidney diseases are linked to cardiovascular diseases, such as high blood pressure. Current statistics show that approximately 26 million American adults have some form of chronic kidney diseases. Finally, the major cause of death for people with chronic kidney diseases is some form of cardiovascular event.

To bring us back to the cardiovascular system, we would like to highlight some of the major milestones that have transformed the study of cardiovascular diseases and the management of cardiovascular pathologies. This is clearly not an all-inclusive list, but just a list to describe some of the events that we find important.

- 1538: Andreas Vesalius published a new anatomy textbook, which contained two large and accurate anatomical charts dedicated to the heart.
- 1628: William Harvey, concluded that the heart is a pump and that veins contain valves to prevent the backflow of blood.
- 1715: Raymond de Vieussens, established that the heart consists of two chambers that isolate arterial and venous blood.
- 1733: Stephen Hales, made the first blood pressure measurements, in a horse.
- 1816: René Laennec invented the earliest stethoscope to listen to blood moving through the heart.
- 1891: cardiac chest compression was used to revive patients (reported by Friedrich Maass).
- 1902: Willem Einthoven invented the electrocardiograph and later received the Nobel Prize in physiology or medicine (1924) for this development.
- 1929: Werner Forssmann inserted a "catheter" into his own arm and passed it through his own cardiovascular system into his own right atrium. He then had colleagues take an x-ray of this procedure to document the catheterization. He later was awarded the Nobel Prize in physiology or medicine (1956) for his efforts.
- 1938: the first successful heart surgery was completed by Robert Gross.
- 1944: Alfred Blalock and Helen Taussig performed the first successful bypass surgery.
- 1947: the first successful defibrillation was completed.
- 1953: Charles Hufnagel implanted the first artificial heart valve.
- 1958: the first pacemaker and coronary angiography was completed.
- 1967: Christiaan Barnard successfully transplanted the first human heart.
- 1977: Andreas Grüntzig performed a balloon dilation of a stenosed coronary artery.
- 1970s–1980s: Robert Jarvik and Willem Kolff worked on and successfully designed a total artificial heart, which was first implanted in a patient in 1982.
- 1986: the first metal stent was implanted into an artery.
- 2000: tissue engineering and stem cells have been used to improve the function of diseased cardiac tissue.

Clearly, we have not discussed all of the critical studies and findings that have helped to progress cardiovascular disease management, but this discussion highlights some of the major advances.

1.5 DIMENSIONS AND UNITS

Dimensions and units are commonly confused, even though the solution to all engineering problems must include units. Dimensions are physical quantities that can be measured, whereas units are arbitrary names that correlate to particular dimensions to make it relative (e.g., a dimension is length, whereas a meter is a relative unit that describes length). All units for the same dimension are related to each other through a conversion factor (e.g., 2.54 cm is exactly equal to 1 in). There are seven base dimensions that can be combined to describe all of the other dimensions of interest in engineering and physics, among other disciplines. In fluid mechanics, we generally pick length, mass, time, and temperature as base dimensions. This makes force a function of length, mass, and time (i.e., force is equal to mass multiplied by length all divided by time squared). Others define force as one of their base dimensions and define mass by dividing force by the gravitational acceleration. This is one of the major differences between the standard English unit system and metric unit system. Those who choose to use metric units make use of the units kilogram, meter, and second to define the Newton. In contrast, those that use the English units use the units pound, foot, and second to define the slug.

Système International d'Unités (SI) units were the first international standard for units. English units followed later and are currently defined from the standard SI units. To define the seven base units using the SI system, scientists and engineers developed the following standards in order to quantify the dimension. The base unit for length is the meter (m). One meter is defined as the distance traveled by light in a vacuum during 1/299,792,458 of a second (as of 1983). One inch (the English unit counterpart) is defined as exactly 0.0254 m (1 in = 2.54 cm). Prior to the current definition, the meter was defined to the length of a pendulum with a half period of 1 s (1668), then one ten-millionth of the length of the Earth's meridian (1791), followed by approximately 1.6 million wavelengths of krypton-86 radiation in a vacuum (1960). The base unit for time is the second (s). One second is defined as the time for 9,192,631,770 periods of the radiation of a cesium-133 atom transitioning between two hyperfine ground states (1967). Prior to this definition, an interestingly calculated hypothetical year and time were used to define the second, as the fraction 1/31,556,925.9747 of the tropical year for 1900 January 0 at 12 h ephemeris time. The standard unit for mass is the kilogram (kg). A kilogram is defined by the mass of a platinum–iridium cylinder that is housed at the International Bureau of Weights and Measures (Paris, France). This mass of 1 g was originally defined as the mass of 1 cm^3 of water at 4°C, making a kilogram the mass of 1 L of water. However, the first prototype kilogram mass, which is what is currently in use today, has the mass of 1.000025 L of water. The base unit for temperature is the Kelvin (K). The Kelvin scale is defined from absolute zero (where no heat remains in an atom) and the triple point of water. From these four base units most of the parameters used in fluid mechanics can be derived. The three remaining base units are electric current (ampere (A)), amount of substance (mole (mol)), and luminous intensity (candela (cd)). The definition of ampere is currently under review by the International Committee for Weights and Measures, but will likely include the amount of elementary particles moving past a particular point in 1 s (at the time of writing, the definition appears to be approximately equal to 6.241×10^{18} elementary particles). The mole was defined when considerations on molecular mass, atomic mass, and Avogadro's number were under consideration. The candela is the luminous intensity of a source that emits a monochromatic radiation of frequency 540×10^{12} Hz and that has a radiant intensity of 1/683 watt/square radian in that same direction. The three last base units/dimensions are not as applicable to biofluid mechanics problems but may arise in problems throughout the textbook.

When converting between two different units, it is imperative to make sure that you track the units you are converting and to make sure that the quantities are being converted properly. For instance, if you are converting area, which is a length squared quantity, you must multiply by the conversion factor twice. If there is an addition or subtraction within your equation, you also need to make sure that the units are the same prior to the addition or subtraction operation because 3 meters minus 2 feet is not equal to 1 meter (or 1 foot). You would first need to convert 2 feet to x many meters to do this subtraction properly. This might seem trivial at this stage, but when your problem involves multiple dimensions and multiple quantities, you must make sure that your units are correct before you do the algebra.

1.6 SALIENT BIOFLUID MECHANICS DIMENSIONLESS NUMBERS

Dimensionless numbers tend to be very useful in characterizing many types of engineering systems. You may be familiar with this type of analysis from different classes. For instance, you may have encountered the engineering parameter of strain, which is a dimensionless number that relates the percent of stretch that a material experiences to the resting length of that same material (note that some prefer to report strain in dimensions of length/length, e.g., mm/mm, but these dimensions do cancel out and thus you would remain with a dimensionless quantity). This type of dimensionless number helps us to scale a parameter across multiple types of scenarios that engineers may come across. In fluid mechanics, you may also encounter this type of dimensionless number to simplify the analysis. Some fluid mechanics engineers will report variables divided by some characteristics or constant value. For instance, some will divide the velocity (which can be a variable of space and time) by the inflow velocity (if it is uniform) or the centerline velocity at a given point of interest. Therefore, all of the velocity values become related to this value and the velocity profile will be scaled by this constant value. Fundamentally, this does not change the fluid properties or analysis of the problem, however, it may be easier to report data in this manner or analyze the problem under these conditions.

There is a second type of dimensionless number that exists in engineering fields. This second type of dimensionless number provides a measure of the importance of phenomenon that plays a role in dictating how an event occurs. This second type of dimensionless number also helps us to rescale problems as needed. Since, this type of dimensionless number provides important information about the flow conditions, many different dimensionless quantities have been developed. We will discuss many of these dimensionless numbers and one method to derive these dimensionless numbers in Section 14.3. However, due to the importance of two dimensionless numbers in biofluids mechanics phenomena, we will briefly discuss them here, and leave the more thorough discussion for Section 14.3.

The Reynolds number (Re) is the first dimensionless number that is important to nearly all biofluid mechanics flows. As stated above, dimensionless numbers relate two (or more) important phenomenon that play a role in the flow that is being analyzed. The Reynolds number relates the overall inertial forces that govern the flow to the viscous forces that will impede the flow. This is important because for any flow to occur, enough force must be present to overcome the fluids resistance to flow. This does not mean that the Reynolds number must always be greater than one, because the Reynolds number does not relate the driving forces to the resistance to flow; it only relates some of the inertial forces to some of the viscous forces (e.g., adhesion of the fluid to do the bounding surface is not in this quantification).

Recall that inertia is defined as the objects resistance to change its velocity. Objects with a very high inertia resist changes to velocity very strongly, whereas objects with a low inertia do not resist changes to velocity very strongly. A simple experiment to observe inertia is to take a pen and hold it at its center between your forefinger and thumb. Now move your fingers to cause the ends of the pen to wiggle up and down. With a typical pen, these constant changes in motion are relatively easy to accomplish. Now conduct the same experiment, with the same pen, however, hold the pen at one of the ends. The pen resists the changes in motion much greater under the second conditions. The moment of inertia of the pen in the second scenario is much larger than the first scenario. Fluid viscosity is simplistically defined as the internal resistance of a fluid to deform under shear loading conditions (we will discuss the more stringent definition of viscosity in later sections). A fluid with a large viscosity, requires a large shearing force to deform the sample (or a small force applied for a very lengthy time), whereas a fluid with a lower viscosity, requires a smaller shearing force to deform the sample. Another simple experiment can illustrate viscosity. Hold a glass of water and tip it at an angle (make sure not to spill the water!). Did the water deform? It should have and the force was very low on the water. Now conduct the same experiment with syrup, toothpaste or honey. Did the second fluid deform as easily; probably not. In this case, water has a lower viscosity as compared with the second fluid. A fluid that will experience a change in its velocity will have to balance both the inertial forces and the viscous forces, which resist changes to velocity or deformation, respectively, since when a fluid proceeds to a new velocity, the inertia will play a role in resisting this change and the viscous interactions will play a role in the deformation changes. Thus, the Reynolds number provides a measure of which forces dominate changes to a fluids velocity. To relate these important parameters, the Reynolds number is defined as

$$\mathrm{Re} = \frac{\text{inertial forces}}{\text{viscous forces}} = \frac{\rho v d}{\mu} \tag{1.1}$$

where ρ is the density of the fluid, v is some characteristic velocity (e.g., centerline velocity, max velocity, or other), d is a characteristic length of the flow (e.g., channel length, radius, diameter, or other), and μ is the dynamic viscosity of the fluid. Even though there are choices inherent within the Reynolds number formulation, there are some conventional choices for the characteristic velocity and length. If your flow can be approximated to be passing through a perfect tube with a constant cross-section, v is conventionally chosen as the spatial mean flow velocity over the circular cross-section and d is conventionally chosen as the diameter of the tube.

The Reynolds number also provides a measure of flow characteristics. For instance, low Reynolds number flows tend to be laminar whereas high Reynolds number flows tend to be turbulent. The transition between laminar and turbulent flows tends to be hard to strictly define, since there are many properties that affect the overall laminar versus turbulent flow properties. However, for perfect tubes, the flow begins to transit to turbulence when the flow exceeds a Reynolds number of approximately 2,300. However, true turbulence (as defined in Section 2.12), will only be found in flows with a Reynolds number of approximately 10,000. Flows in-between these two values are said to be transitioning and exhibit both laminar and turbulent flow properties. If the geometry of the flow changes, these transitioning values will also change; typical Reynolds number values, to describe laminar or turbulent flows can be found in fluid mechanics textbooks.

A second salient dimensionless number for biofluid mechanics flows is the Womersley number, which is related to the pulsatility of the flow. Flows that have regular oscillating time-dependent components (e.g., are not steady), are said to have pulsatility. The Womersley number is a ratio of the fluids oscillatory inertia to the viscous momentum. The oscillatory inertia is a measure of the forces that are governing the pulsatile flow, whereas the viscous forces are again a measure of the fluids' overall resistance to changes in its velocity. The Womersley parameter can be defined as

$$\alpha = \frac{\text{oscillatory inertia}}{\text{viscous forces}} = \sqrt{\frac{\rho v \omega}{\mu\left(\frac{v}{d^2}\right)}} = d\left(\frac{\omega}{\nu}\right)^{\frac{1}{2}} \tag{1.2}$$

where ρ is the fluid density, v is a characteristic velocity, ω is the angular frequency associated with the oscillation, μ is the dynamic viscosity, d is a characteristic length, and ν is the kinematic viscosity. A fluid with a low Womersley number is characterized by having essentially no phase difference between the pulsatile pressure waveform that is driving the flow and the pulsatile velocity waveform associated with the flow. With increasing Womersley numbers, a phase difference between the pressure and velocity waveform can be observed; which is due to the inertia of the fluid resisting changes governed by the pressure waveform and that the pulse frequency is relatively high. Other important dimensionless numbers and how dimensionless numbers can be used in various biofluid mechanics applications will be discussed in Section 14.3.

END OF CHAPTER SUMMARY

1.1 This textbook will discuss basic fluid mechanics principles, flows within the macrocirculation, flows within the microcirculation, specialty circulations, and experimental techniques.

1.2 The National Institutes of Health working definition of biomedical engineering is "Biomedical engineering integrates physical, chemical, mathematical, and computational sciences and engineering principles to study biology, medicine, behavior, and health. It advances fundamental concepts; creates knowledge from the molecular to the organ systems level; and develops innovative biologics, materials, processes, implants, devices and informatics approaches for the prevention, diagnosis, and treatment of disease, for patient rehabilitation, and for improving health."

1.3 Fluid mechanics is useful for the analysis of anything that includes an interaction with a liquid or gas. This includes traditional engineering applications, as well as many biological applications.

1.4 Biofluid mechanics is focused on how biological systems interact with and/or use liquids/gases. For humans, this includes obtaining and transporting oxygen, maintaining body temperature, and regulating homeostasis. Cardiovascular diseases account for nearly one of three deaths in the United States; the highest prevalence regions of cardiovascular diseases are along the lower Mississippi River valley.

1.5 Dimensions are physical properties, whereas units are arbitrary names that correlate to a measurement. There are seven base dimensions, including time, length, mass, temperature, electric current, amount of substance, and luminous intensity. All other physical parameters can be related to these base units. Depending on which system of units are chosen, the base units can change, but again all dimensions/units can be defined from one another.

1.6 Dimensionless numbers are important for either scaling fluid properties, relating important parameters that govern fluid flows or both. Two of the most widely used biofluid mechanics dimensionless numbers are the Reynolds number

$$\mathrm{Re} = \frac{\rho v d}{\mu}$$

and the Womersley number

$$\alpha = d\left(\frac{\omega}{\nu}\right)^{\frac{1}{2}}$$

which relate the inertial forces to the viscous forces and the pulsatility to the viscous forces, respectively.

Reference

[1] A.S. Go, D. Mozaffarian, V.L. Roger, E.J. Benjamin, J.D. Berry, M.J. Blaha, et al., Heart Disease and Stroke Statistics – 2014 update: a report from the American Heart Association, Circulation 129 (2014) e28–e292.

CHAPTER

2

Fundamentals of Fluid Mechanics

LEARNING OUTCOMES

1. Describe fluid mechanics principles
2. Identify the principle relationships and laws that govern fluid flow
3. Specify different analysis techniques that can be used to solve fluid mechanics problems
4. Describe the continuum principle
5. Evaluate the pressures and stresses that act on differential fluid elements
6. Define kinematic relationships for fluid flows
7. Explain the relationships among shear rate, shear stress, and viscosity
8. Define different classifications for fluids
9. Describe fundamental changes at the microscale level
10. Describe the relevance of turbulence in biofluid mechanics
11. Identify the common features of turbulent flow and explain mathematical principles that can be used to describe turbulent flow

2.1 FLUID MECHANICS INTRODUCTION

In the broadest sense, *fluid mechanics* is the study of fluids at rest and in motion. A *fluid* is defined as any material that deforms continually under the application of a *shear stress*, which is a stress directed tangentially to the surface of the material. In other words, no matter how small the applied shear stress acting on a fluid surface, that fluid will flow under the applied stress (for some fluids a yield stress must be met before the fluid flows; however, once this yield stress is met, the previous definition applies. See Section 2.8 for more information about these and other special types of fluids.) Some define a fluid as any material that takes the shape of the container in which it is held. It is easily seen that any substance in the liquid or the gas phase would fall under all of these definitions and are, therefore, fluids. The distinction between a fluid and a solid is also clear from these definitions because solids do not take the shape of their containers; for instance, a shoe in a

© 2015 Elsevier Inc. All rights reserved.

shoebox is not a rectangular cuboid that only takes the shape of a foot when worn, whereas if you place a fluid in a shoebox it would take the shape of a rectangular cuboid. Also, if you compare the action of solids and fluids under a shear stress loading condition, the distinction is clear: a solid will deform under shear, but this deformation does not continue when the shearing force is removed or held constant. However, as a caveat to that statement, some materials exhibit solid and fluid properties both. These materials are termed "viscoelastic" materials, and they deform continually under shear loading until some threshold deformation has been reached. In fact, all real solid materials must exhibit some fluid properties, but the fluid-like properties can be neglected in most practical situations.

We will begin our discussion with a review of shear loading, which should be familiar from a solid mechanics course. A typical manner to induce a shear loading condition on a solid material is through torsion. *Torsion* is defined as the twisting of a material when it is loaded by moments (or torques) that produce a rotation about an axis through the material. In solid mechanics courses, torsion analysis is typically applied to a solid bar, which is fixed on one end and has a moment (M) applied at some location along the length of the bar (see Figure 2.1, which illustrates the loading at the free end of the bar). We will not define the appropriate equations of state for this loading condition. However, refer to a textbook on solid mechanics for an appropriate review. (We suggest some textbooks in the *Further Readings* section.) The same analysis that is applied to a fixed solid bar can be applied to a bone subjected to torsion. Imagine an athlete who plants his or her foot on a surface while making a quick rotation about that foot. This would generate a moment throughout the bones in the leg. To solve for the shear forces/stresses within the bone, we can assume that the foot is fixed to the surface and that the bone is modeled as a hollow cylinder (with a taper, if necessary, to make the solution more accurate) (see Figure 2.1, which illustrates different modeling methods used in biomedical engineering). Clearly, as was the case in many introductory solid mechanics courses, this simplifying assumption ignores inhomogeneities (if any) in the bone material properties. We will see that this, or a similar, assumption is made in many biofluid mechanics examples that we will discuss

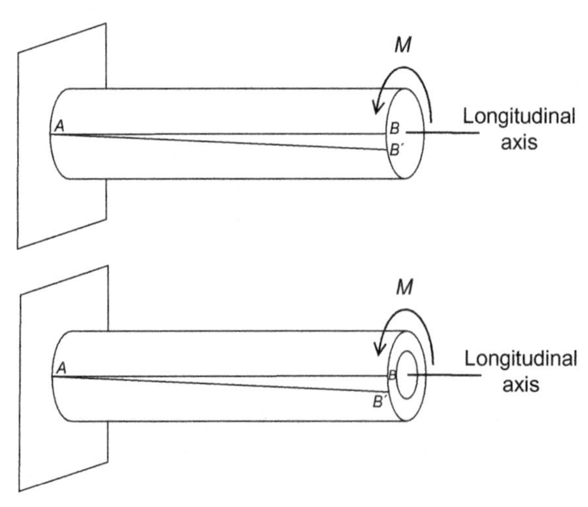

FIGURE 2.1 A depiction of a method to model torsion throughout a solid material. Through the application of a moment at a location on the bar, a line AB would deform to the arc AB'. With removal of the moment, point B' would move back to B, assuming the induced stress did not exceed the materials yield stress. The second figure depicts a cylinder, which may be used for torsion analysis in bones.

because it is very difficult to mathematically quantify inhomogeneities within a fluid (we typically assume a uniform distribution of molecules and cells throughout the fluid although it has been shown that the distribution of these molecules is not homogeneous).

To bring us back to the case of torsional analysis, under pure torsion loading, the shear forces throughout the material are dependent on the applied moment, the geometry of the material, and the load application position (radial and axial) within the material. Assuming that the elastic limit (in shear) of the material is not reached and that the applied moment is constant, the material will deform to a certain extent (this would be seen as a twist of the bar and is illustrated in Figure 2.1 where line *AB* deforms to arc *AB′*). With removal of the moment, a solid material will return to its original conformation (e.g., the arc *AB′* will return to line *AB*, if the material does not experience any plastic deformations). We will learn that the shear stresses that are generated in a bar subjected to torsional loading are similar to what develops in a fluid subjected to shear loading; however, the mechanisms of the loading conditions and stress development are fundamentally different. We use the example of torsional loading on a bar because it should be familiar from previous courses and should highlight that some assumptions, such as geometry or homogeneity, may need to be relaxed in physiological environments.

In contrast to the analysis just described for solid materials, any fluid subjected to a shearing force will deform continually assuming that the applied force exceeds the fluids yield stress, which may be zero. A typical example (from a fluid dynamics course) to depict a fluid under a shear loading condition is with a parallel plate design with a thin layer of fluid between the plates (Figure 2.2). Imagine that a constant force with a horizontal component is applied to the top plate, which causes the top plate to move with some velocity in the positive *X*-direction, to one particular fixed location. The fluid that is in contact with the top plate will move with the same velocity and move to the same displacement as the top plate because of the no-slip boundary condition. The no-slip boundary condition dictates that a fluid layer in contact with a solid boundary must have the same velocity as that boundary. We will see later that the no-slip boundary condition can lead to the fluid along the wall having a zero or negligible velocity. Due to the constant horizontal force that was applied to the top plate, a fluid element denoted by *ABCD* will deform to *ABC′D′* during the instantaneous application of the force. However, if one looks at a later time, the fluid element will continue to deform, even though the applied force has moved through its physical line of action, is no longer actively acting on the plate and the top plate is no longer moving as a result of the applied force *F*. At the later

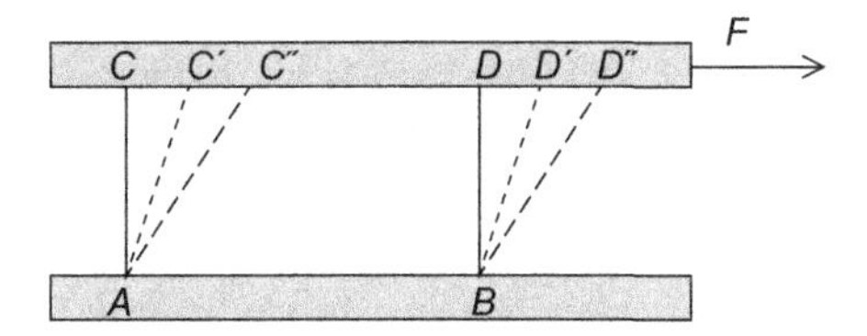

FIGURE 2.2 **Under a constant shearing force (*F*), a fluid will continually deform.** Immediately after the application of a constant force to the top plate, a fluid element *ABCD*, deforms to *ABC′D′*. At a later time, the element will have deformed to *ABC″D″* even though the force *F* is held constant and no additional force has been added to the system.

time, the fluid element can then be depicted as the element *ABC″D″*. In biology, a condition depicted in Figure 2.2 can occur in the lubrication of joints, where one solid boundary (a bone) moves in relation to another solid boundary (a second bone). One of these boundaries could be considered stationary through the use of a translating coordinate system fixed to that boundary. (This simplifies our analysis instead of having two moving boundaries.) The other boundary will then move with some velocity relative to the stationary boundary. Additionally, blood flowing through a vessel that is being deformed by some external force can be simplified to this type of analysis. The purpose of the discussion up to this point is to highlight one of the fundamental differences between solid and fluid materials: solids will deform to one particular conformation under a constant shearing force, whereas fluids continually deform under a constant shear force. Recall that an exception to this definition exists. There are viscoelastic materials that exhibit properties of solids and fluids both and, therefore, continually deform under constant loading (like a fluid) to some threshold deformation state (like a solid).

Fluid mechanics can be divided into two main categories: static fluid mechanics and dynamic fluid mechanics. In static fluid mechanics, the fluid is either at rest or is undergoing rigid body motion. Therefore, the fluid elements are in the same arrangement at all times. This also suggests that the fluid elements do not experience any type of deformation (linear or angular) during their motion. Since a fluid element deforms continually under a shear force and there are no deformations in static fluid mechanics, the implication is that no shear stresses are acting on the fluid. Unfortunately, no true fluid can experience only rigid body motion and at the microscopic level, fluid molecules are continually in motion (unless the fluid is at absolute zero). In general, the salient aspect of static fluid mechanics is the pressure distribution throughout the fluid. For dynamic fluid mechanics, the fluid may have an acceleration term (i.e., nonconstant velocity) and can undergo deformations. In the case of dynamic fluid mechanics, Newton's second law of motion can be used to evaluate the forces acting on the fluid. Generally, for this type of analysis, the pressure distribution and the velocity distribution throughout the fluid are of interest. From these calculated fluid parameters, any other parameter of interest, such as acceleration, wall shear stress or shear rate, can be obtained.

2.2 FUNDAMENTAL FLUID MECHANICS EQUATIONS

Determining the solution of any engineering problem should begin by writing down the known quantities (also termed "the givens"), including the equations that govern the system being addressed. In general, there are five fundamental relationships (Table 2.1) of interest in fluid mechanics. In no way, do these relationships limit a person to five solution methods or five problem types. Instead, they provide a foundation for solving a variety of complex problems. These possible solution routes also use a variety of different computational analysis methods. For instance, if one was interested in the velocity profile of the fluid, then one would probably focus on kinematic relationships throughout the fluid. Yet, conservation laws may be necessary to help with the calculations. However, if the stress distribution was of interest, one might start with kinetic relationships and use kinematics/conservation laws to help solve the problem. Algebra, trigonometry, and calculus techniques are some computational methods that may be needed to solve biofluid mechanics problems. Although in

TABLE 2.1 Relationships That Are Useful in Fluid Mechanics Problems

Kinematic	Velocities, accelerations, shear rates—Fluid properties
Stresses	Force intensities acting over particular area—Flow properties
Conservation	Mass, momentum, and energy—Physical properties
Regulating	Initial and/or boundary conditions—Geometric and/or limiting properties
Constitutive	Mathematically describes the fluid—Laws to describe flow

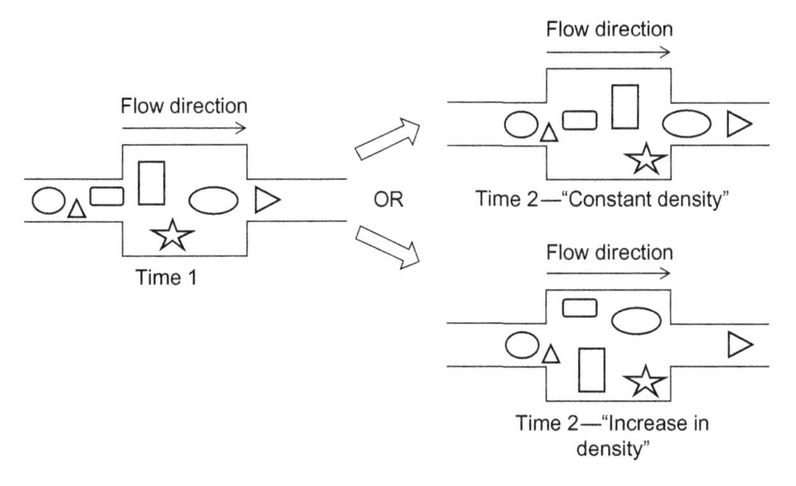

FIGURE 2.3 **Pictorial representation of the conservation of mass principle.** At time 1, the density of the fluid within the square chamber is represented with approximately three shape elements within the square chamber. At some time later, the density of the fluid can either remain the same ("constant density" case) because the same quantity of elements is present or increase because one of the shape elements did not leave the volume of interest ("increase in density" case).

practice, in order to solve more physiological and more accurate flow scenarios, computational methods and software might need to be employed, because simpler hand calculations may not be possible. Known variables in most fluid mechanics problems typically include geometric constraints, fluid material properties (density/viscosity), and temperature of the system, among other inflow and outflow boundary conditions (e.g., the inflow velocity profile and the outflow pressure). Also, material properties of the bounding container and the temporal variations in fluid/flow properties (if any) are typically known.

The laws that govern fluid flow are common among other engineering disciplines and they should not be encountered for the first time in this textbook. These laws are described in detail in the chapters that follow, and the derivation of common fluid mechanics parameters and equations of state will be detailed in those chapters. Here we will briefly summarize the laws that fluid flow must obey.

1. *Conservation of Mass*: Stated simply, whatever mass flows into a system must flow out of the system or be accounted for by a change in the mass of the system or via a change in density (ρ) of the fluid (Figure 2.3) (Remember that *density* is defined as mass

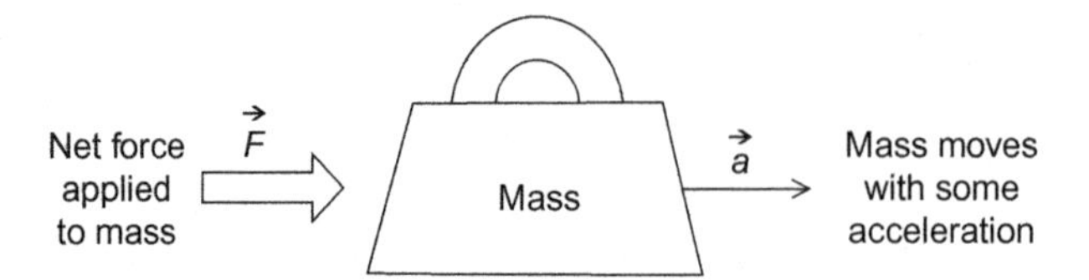

FIGURE 2.4 **When a force is applied to a mass, the mass will move with a velocity that is proportional to the applied force and in the same direction as the net force.** In this simple representation, we are not depicting all of the forces (such as friction or drag), just the net force that is acting on the mass.

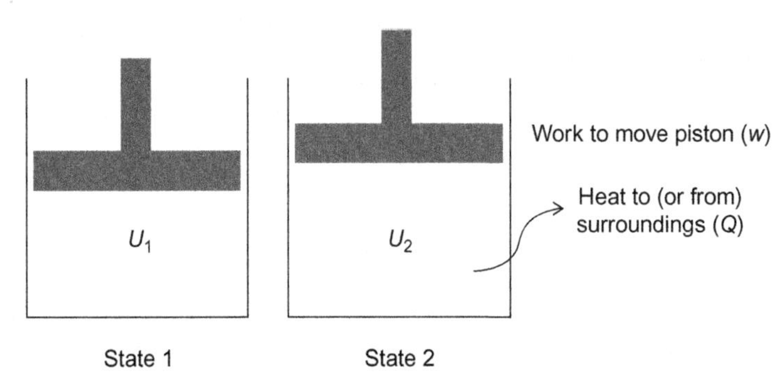

FIGURE 2.5 **Representation of the first law of thermodynamics where the internal energy of a system is equal to the heat added to the system from the surroundings minus the work that the system conducts on its surroundings.** In this example, a piston is moved up by the material within the system as heat is being lost to the surroundings.

divided by volume [*V*].) This law is similar to Kirchhoff's current law, which is common in circuit analysis.

$$m_{system} = m_{in} - m_{out} - \frac{d(\rho V)}{dt} \tag{2.1}$$

2. *Conservation of Linear Momentum*: The time rate of change of momentum (mass multiplied by velocity) of a body is proportional to the net force acting on that body. This is a vector equation that can be divided into component directions as needed. If the mass of the body remains constant with time, then the force is equal to the mass multiplied by the acceleration of the body (Newton's second law of motion, in simplified form, Figure 2.4). Also, the acceleration of the body will proceed in the same direction as the direction of the net force. This is a common equation used in solid mechanics, engineering statics, and engineering dynamics.

$$\vec{F} = \frac{d(m\vec{v})}{dt} \tag{2.2}$$

3. *The First Law of Thermodynamics*: The change of the internal energy (δU) of a system is equal to the change of the amount of energy added to a system (through heat, δQ) minus the change in the amount of energy lost by the system due to the work (δw) done by the system on the surroundings (Figure 2.5). Another relationship from this law states that an isolated system can hold or lose energy, but the total energy of the entire system remains constant. This law is dependent on the particular pathway in which energy is added/removed to/from the system and not just the difference between the beginning

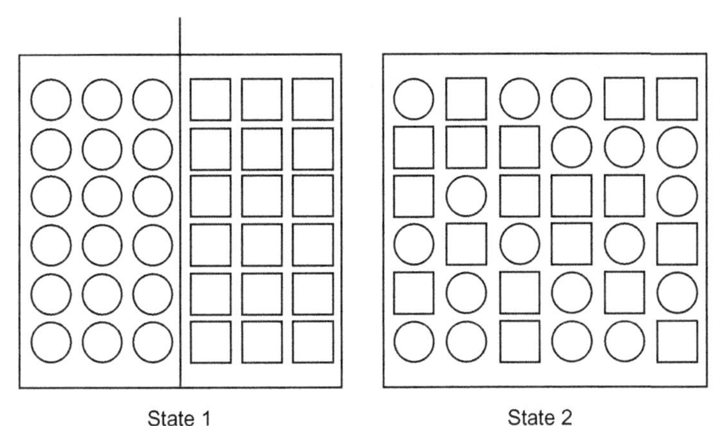

FIGURE 2.6 **The second law of thermodynamics states that the disorder of a system will increase with time.** In this case, there is a boundary between circular and square objects. Once this boundary has been removed, the system will tend to move from an orderly system to a disorderly system.

and end points (recall that for some energy changes, such as gravitational potential energy, it is typically assumed that the objects path does not play a role in the energy change; only the starting point of the material and the ending point of the material matter). This law should be familiar from a beginning thermodynamics course.

$$\delta U = \delta Q - \delta w \tag{2.3}$$

4. *The Second Law of Thermodynamics*: The entropy (S, disorder) of an isolated system will increase over time, until the system reaches equilibrium (Figure 2.6). In physics, this law is used to determine the direction in which time moves (is time travel possible?) and the total energy or efficiency of a system. The classic example of this law is that a mug that falls off a table and breaks will not mend, unless external energy is put into the system. The same would be true in biological systems; if a bone breaks it does not heal without addition of energy into the system. This law should also be familiar from beginning thermodynamics courses

$$\frac{dS}{dt} \geq 0 \tag{2.4}$$

5. *Conservation of Angular Momentum*: Unless acted upon by an external torque, a body rotating about an axis will continue to rotate about that axis. The mass moment of inertia of a body is its resistance against changes in angular momentum (this is the second mass moment of inertia with units of mass multiplied by length squared). Moment of inertia is a state property of the material, which is related to the geometry of the object and the material(s) that compose the object. An example of the conservation of angular momentum is as follows: a figure skater spinning around his or her centerline with arms outstretched will increase his or her angular velocity by bringing his or her arms into the body. Conversely, by stretching out his or her arms, the angular velocity will decrease. This is due to changes in the body's moment of inertia (Figure 2.7). Assuming that friction from the ice and drag from the surrounding air are negligible, the angular momentum is conserved under these motions. *Angular momentum* ($\vec{L}$) is defined by the cross-product between the position vector of a particle relative to the axis of rotation ($\vec{r}$, i.e., hands relative to the centerline) and the linear

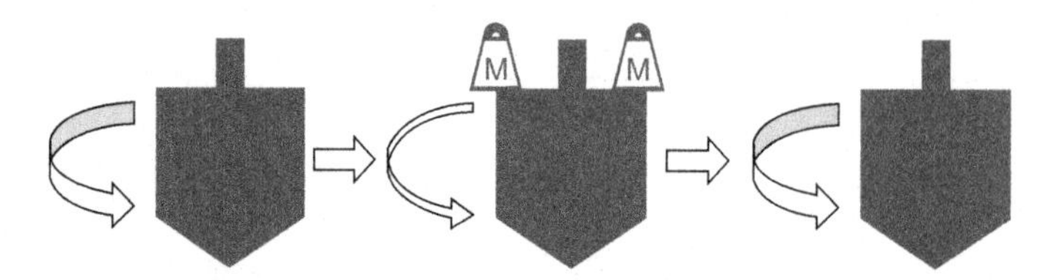

FIGURE 2.7 The conservation of angular momentum states that a spinning body will have a particular angular velocity. As more mass is added to the body at a larger radius from the axis of rotation, the angular velocity will decrease. If the added mass is removed, the angular velocity will return to the original angular velocity (the magnitude of angular velocity is represented by the width of the curved arrow).

momentum of that particle ($\vec{p}$). *Conservation of angular momentum* is defined as no change in angular momentum with time as long as the mass moment of inertia remains the same. When the mass moment of inertia increases, the momentum decreases, and vice versa. Equation (2.5) is the definition of angular momentum for discrete particles; for bodies, the angular momentum must be summated over the entire mass

$$\vec{L} = \vec{r} \times \vec{p} \tag{2.5}$$

In most of the examples we will discuss in this textbook, not all of these governing laws will be required to solve the problem. However, they will all be applied to some extent during the course of this textbook so that it can be realized how these laws affect different situations in biofluid mechanics. It is also true that problem solutions may require other relationships, such as constitutive or regulating rules, to solve the problem. However, the five laws listed previously are always a good starting point for many biofluid mechanics problems. For example, the ideal gas law (which is a constitutive rule), $p = \rho RT$, is useful in solving many gas flow problems that might occur in biological systems but would not be as applicable in purely liquid flow if there are no significant changes in temperature or pressure (which is typically the case in biological systems). Also, the ideal gas law assumes that the particles in the fluid have no interactions but can exchange energy during any of the infrequent particle collisions. Liquid molecules interact with each other constantly and, therefore, this law would not be useful for many liquid flow problems. In subsequent chapters, we will take the basic laws of mechanics and thermodynamics to derive constitutive equations that are useful for solving a variety of fluid problems. As a caveat, not all problems that we will encounter can be solved analytically. Chapter 14 will discuss some of the numerical methods that can be used to solve computationally complex problems, along with some of the current research topics that use computational approaches to understand a particular system of interest.

2.3 ANALYSIS METHODS

The first step in solving a fluid mechanics problem is understanding the system that the problem addresses. In engineering mechanics, the laws that govern a system, normally

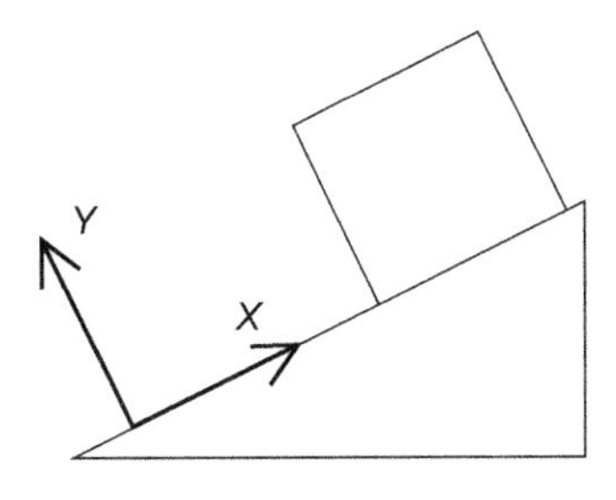

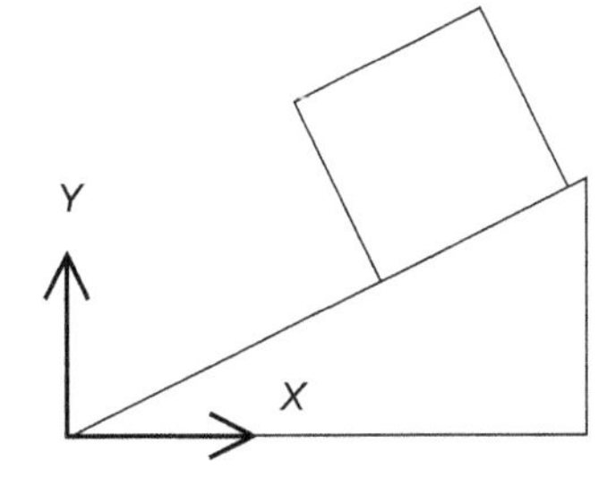

FIGURE 2.8 **Two common ways to set up the coordinate axes in a solid mechanics problem.** Either convention can produce a solution but the solution using the coordinate axis shown on the left would generally work out easier than the right coordinate axis because many variables are aligned with that axis system.

represented within a free body diagram, are used to analyze the problem. In thermodynamics, we generally discuss the system as being either open or closed to other neighboring systems. Whatever approach is taken, governing equations are then written based on our system convention choices. In fluid mechanics, we will discuss similar parameters that should be familiar from engineering mechanics and thermodynamics. We define them in this textbook as the *system of interest* and the *volume of interest*. By properly defining these parameters and by choosing the proper laws that govern the entire system, the solutions to fluid mechanics problems typically become significantly easier. By choosing these parameters in a different way, the fluid mechanics problems can still be solved, but this may become laborious. This is not to say that by choosing the system of interest and the volume of interest in the most ideal way, the solution will be computationally easy (e.g., algebra); the use of computational fluid dynamics may still be necessary to solve the problem (see Chapter 14). Also, not choosing the best system of interest and volume of interest does not mean that you cannot arrive at a solution by hand, but it will generally be more difficult than the most ideal choice. A good analogy to reinforce this concept is from engineering mechanics courses. When drawing a free body diagram, an assumption must be made for the Cartesian coordinates for the problem. If the problem consists of a mass on an inclined plane, there are two common ways to set up the coordinate system (Figure 2.8). In the first, the x-axis of the coordinate system is aligned with the slope of the inclined plane and, in the second, the x-axis is horizontal. Using either of these coordinate conventions, the problem could be solved; however, it is typically easier to solve this problem when the coordinate axis is aligned with the inclined plane because the frictional forces, velocity and acceleration terms all align with that coordinate system, whereas in the other configuration you must be careful regarding the decomposition of those same forces into the coordinate directions. The solution to fluid mechanics problems works in a similar manner; there is typically an easier way to arrange the system and volume of interest.

For fluid mechanics problems, a *system of interest* is defined here as a fixed quantity of mass and this typically encompasses the entire system of interest. The boundaries of this system separate the materials of interest and the surrounding materials that are not of interest. For instance, if our system was blood flowing through a blood vessel, the system of interest would be the blood vessel and everything within it, and the boundaries of our system of interest would be the vessel wall. These boundaries are movable with time, but mass cannot cross this boundary (this is a good approximation of flow within the macrocirculation

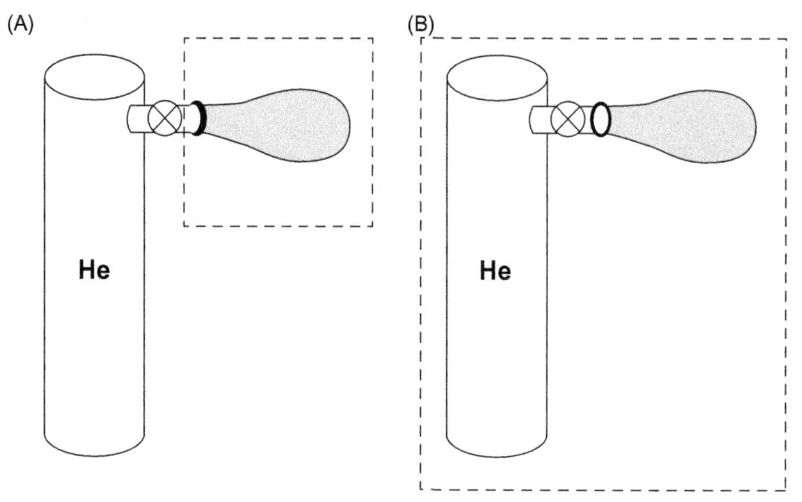

FIGURE 2.9 Two choices for the system of interest (dashed line). With the first choice (A) the balloon will remain uninflated for the entire problem because no mass can transfer through the system boundary. For the second choice (B) the balloon can be inflated and the boundary would follow the balloons wall in time.

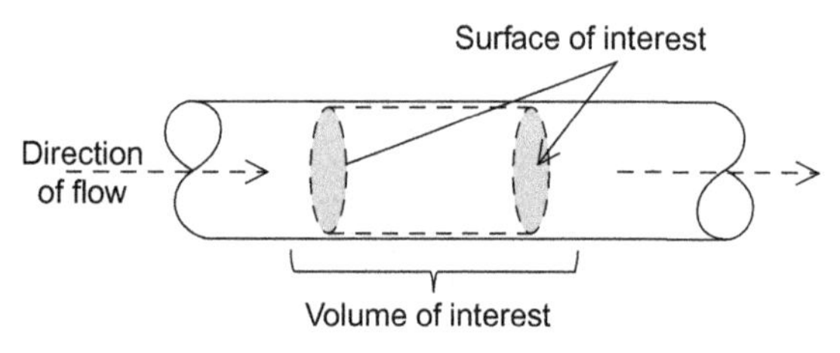

FIGURE 2.10 **Examples of a volume of interest and a surface of interest within a blood vessel.** The two highlighted surfaces of interest are imaginary. The boundary of the dashed cylinder and the tube wall is a real surface. By choosing the surfaces in this manner the solution of fluid mechanics problem would be easier, compared with other choices for the surfaces/volumes of interest.

but would not be valid for flows within the microcirculation). This is the key to defining the system of interest properly. For instance, if your system is an uninflated balloon, then no mass (e.g., helium gas) can cross into this boundary; hence, the balloon would remain uninflated for the entire duration of the problem (Figure 2.9A). However, if you choose your system to be the balloon plus the nearby helium tank, the balloon can be inflated during the course of the example (Figure 2.9B) because mass can cross the imaginary system boundary (the dashed box in the figure). In this same situation, the system boundary will also be variable with time. When the valve to the helium tank is closed (time 0) the system boundary is the flattened uninflated wall of the balloon. However, as you open the tank valve, the balloon gains mass and volume and the system boundary will follow the balloon wall. The same is true for biofluid problems that include the lungs; it is wise to include the atmosphere in the system of interest, if the problem deals with inspiration and exhalation.

As stated previously, a free body approach is used extensively in mechanics of solids. This is an easy approach because a system can be chosen with a defined mass and this mass will generally not change with time (for general solid mechanics problems). However, in fluid mechanics, the flow of fluids through a system (pipe, turbine, blood vessel) is the most interesting problem, unless you are analyzing a static fluid dynamics problem. Therefore, it is not easy to define a finite mass that we are interested in because mass is exchanged with time as it flows. To circumvent this type of problem, a *volume of interest* is defined. This volume of interest is a volume in space with which a fluid flows through. The boundaries that define the volume of interest are the *surfaces of interest* (Figure 2.10). These surfaces can experience motion with time. For instance, large blood vessels on the arterial side and most air conduits in the lungs experience some form of translation with time, whereas smaller vessels, such as capillaries, do not experience much translation and are generally considered to be stationary. Figure 2.10 depicts a blood vessel with a fluid flowing through it. The dashed cylinder within the blood vessel is our choice of a volume of interest for a particular problem. The choice that was made in this figure depicts an example in which the volume of interest has both real and imaginary surfaces of interest. The real surface is the inside of the blood vessel wall, which surrounds our volume of interest. Real surfaces can be thought as the fluid boundaries, which prevent possible mass transfer. However, the vertical light grey circles, with which fluid will enter or exit our volume, are imaginary surfaces. These surfaces are solely chosen for convenience, so that we can solve the fluid mechanics problem more easily. Using these imaginary surfaces, we can impose boundary conditions or initial conditions with ease. Other choices for this surface of interest/volume of interest may make the boundary value and initial condition definitions more difficult (remember problems depicted in Figure 2.8 from engineering mechanics courses). It is of the upmost importance to choose these surfaces wisely because the methodology of fluid mechanics problems will change based on the choices made for the surface of interest and volume of interest. Some choices (as in Figure 2.10) can make the solution simple, whereas other choices can make the problem difficult to solve.

Depending on the particular application of the fluid mechanics problem, there are different approaches to take to reach a solution to the problem. First, if the point-by-point flow characteristics (an infinitesimal volume, i.e., one water molecule) are of interest within the fluid, a differential approach to solve the governing equations should be used. This solution would be based on the differential equations of motion. This approach is useful in biofluid mechanics if one is interested in two-phase flows and the location of individual proteins or cells, etc., within the blood stream and how these individual particles interact with other particles or the vessel wall. However, in some cases, this very detailed knowledge of the flow profile is unnecessary and too labor intensive to calculate (by hand and/or computationally). In this case, one is most likely interested in the bulk properties of the fluid. Quantities of interest are most likely the velocity profile at a certain location and the pressure gradient across a particular length, among others. The solution to these types of problems takes the form of the integral equations of motion, and the volumes of interest are finite. An example of this approach is by looking at one particular section of a blood vessel (see the volume of interest in Figure 2.10) and obtaining the overall velocity profile or other fluid property through the entire volume of interest. This method is generally easier to solve analytically and will be developed first in this textbook.

In general engineering mechanics, a number of methods can be used to describe the system that is being illustrated. The most common methods that are implemented in fluid mechanics are the Lagrangian method and the Eulerian method. When it is easy to record and solve for identifiable particles with a real mass and volume (e.g., cells, molecules, and other identifiable objects in the flow field) the Lagrangian method is the most applicable. This method of solution tracks each particle and continually solves the governing equations for each particle in the flow stream. This method is the point-by-point differential approach that we described previously. By using this method, Newton's second law of motion becomes a summation over all of the identified particles within the fluid. This leads to why the solution may become difficult computationally in biofluid mechanics problems. For example, in the blood, cellular matter constitutes approximately 40% of the total blood volume. To solve a Lagrangian biofluids problem, every cell within this 40% must be solved individually. If your volume of interest is 1 μL of blood, your solution would consist of approximately 5×10^6 red blood cells, 6×10^3 white blood cells and 2×10^5 platelets, as well as all of the plasma proteins, sugars, ions, water molecules, and other compounds dissolved in blood. To monitor each of these identifiable particles with a mass and volume properly, a lot of bookkeeping is required! With the Lagrangian approach, the acceleration, velocity, or position of interest is associated with the center of mass of the particle/molecule of interest, and this may not make sense in biological systems, in which the cell shape can and does change with time. Because of this shape change, there would most likely be a redistribution of mass and, at each time point, a new center of mass must be found for each particle. The Lagrangian approach, however, is very useful in calculating particle trajectories in basic kinematic problems, if one can assume that the center of mass stays the same throughout the problem.

When considering a real fluid, which is composed of a large number of finite particles, the Eulerian method is more commonly used. This formulation is used because it is difficult (as described previously) to record all the information about each particle within the fluid at every instant in time. The Eulerian method focuses on the overall flow properties within a certain volume over a certain time interval. Formulations that describe the flow are, therefore, spatially and temporally dependent. For instance, if one was interested in solving for the velocity profile 5 cm downstream of the aortic valve for every 5 ms interval during one cardiac cycle, the Eulerian approach is much more helpful than particle tracking with the Lagrangian approach. This is because the solution for velocity, acceleration, among others is associated with this particular volume and time but not with individual particles within the flow field. Using the Eulerian method leads to the assumption that a fluid can be considered as continuous medium. If the fluid is a true continuum then the Lagrangian approach and the Eulerian approach will obtain the same solution. If the fluid is not a continuum then the solutions using these two methods will be different.

2.4 FLUID AS A CONTINUUM

In our previous discussions about fluid characteristics, we did not discuss the molecular nature of a fluid. A fluid is composed of molecules in constant motion, within the fluid itself (random motion of particles or diffusion) and possibly within the constraining volume (i.e., is the flow static or dynamic?). In most applications, the bulk properties of the fluid are what we are interested in because this is what we can easily measure; however, to fully characterize the fluid, a point-by-point analysis must be taken. The bulk properties can also be considered as the average of the particular property of all of the fluid particles within the fluid element of interest. If we can assume that each fluid particle has the same properties as the bulk fluid properties, our analysis will be simplified. Making this assumption allows us to consider a fluid as an infinitely divisible substance, which is the definition of a *continuum*. Therefore, the effects of individual molecules and the relative changes from the bulk fluids state properties induced by dividing the medium

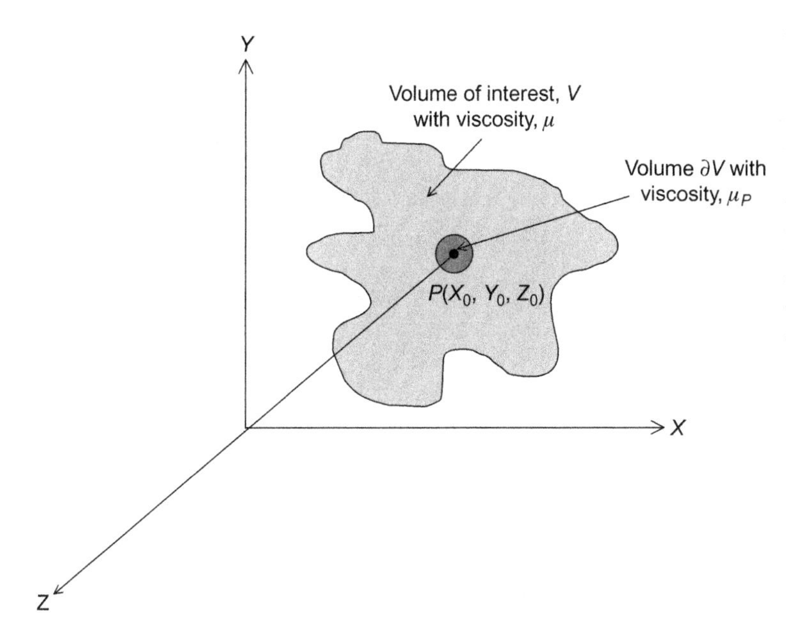

FIGURE 2.11 **The bulk viscosity (μ) of a fluid element can be described by the weighted average of the individual viscosities (μ_P) of differential elements (∂V).** The viscosity within each differential element can be greater or lower than the bulk viscosity, but if we ignore these changes, we would assume that the fluid is a continuum.

into very minute parts are ignored. If you have taken a course in mechanics of materials, the continuum principle should be familiar. In most mechanics examples when a deformation, stress, or strain is calculated within a certain material, the changes in the elastic modulus or deformities/inhomogeneities throughout the material are ignored. Under these cases, the material is assumed to be homogenous throughout the section of interest. In biofluid mechanics, the assumption that a fluid is a continuum is legitimate when we consider flows within the macrocirculation (see Part 2), but we must caveat that this is an assumption because blood is not a homogenous material. However, when discussing the microcirculation (see Part 3), molecular effects must be considered and the fluid should not be classified as a continuum, although we make that assumption here to simplify some of the calculations. As a consequence of using the assumption that a fluid is a continuum, properties such as density, viscosity, and temperature are continuous throughout the fluid. This is not necessarily a bad assumption on a large scale (i.e., bulk or average properties within the macrocirculation). Remember though these properties can be a function of time and/or spatial location, but within a particular instant in time and in one particular location, the medium is continuously divisible and the value of interest will not change based on molecular interactions. Once again, this is very useful for general fluid mechanics problems, but this assumption breaks down in some biological systems.

A good illustration of this phenomenon is with the viscosity of a fluid. If we look at a particular volume of fluid (Figure 2.11, light grey), which will be used to calculate the viscosity of the fluid, we obtain the average bulk viscosity for that volume. In general, the value for the bulk viscosity is similar to the individual viscosities of smaller sections within that same volume (i.e., dark grey section **P**). Imagine if we choose to divide the light grey section into 10 equal volumes. The viscosity of these 10 parts should be similar to the average viscosity for the entire volume (light grey) and, if this is true, we can state that the fluid

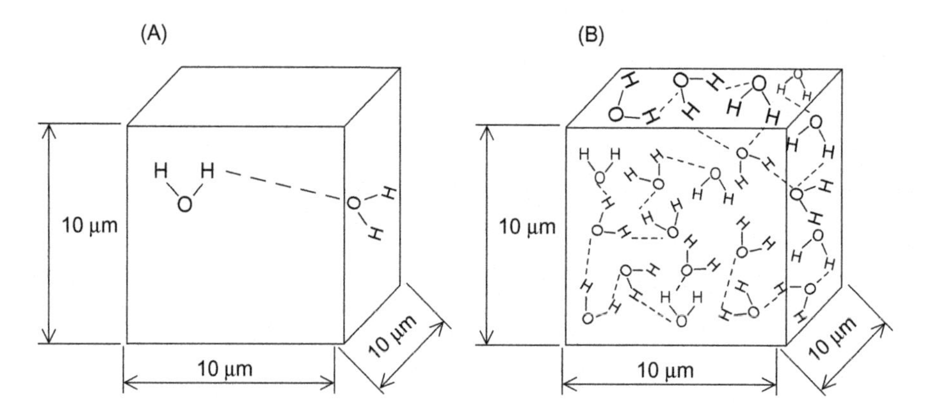

FIGURE 2.12 **If a noncontinuum fluid is divided into small volumes (e.g., 1 pL), the properties of the fluid may be different than the bulk property.** In this figure, panel A shows a volume with a smaller viscosity compared with panel B because there are less intermolecular interactions (dashed lines) between the molecules in panel A than in B.

behaves as a continuum. However, if we choose our section to be on the order of a few picoliters (10^{-12} L) of a fluid taken from a 1-L volume of interest, the viscosity of the small section would most likely be different from the average bulk viscosity for the entire section. This is because viscosity itself is dependent on molecular interactions within the fluid (see Section 2.7). In our small volume, there may be two fluid molecules that are interacting with each other to govern the viscosity of the solution within that volume (Figure 2.12). However, if there were 20 fluid molecules within the same picoliter volume, there would be more interactions, which alter the viscosity of that differential volume of fluid. The quantity of molecules per small volume differs as a function of time and space based on random molecular fluctuations within the fluid. Regardless of what the fluid viscosity is on this small scale, if the bulk viscosity is of interest, we can average each small sections viscosity to obtain the bulk viscosity. Therefore, if we can (or purposefully do) ignore the changes in viscosity between each of these small sections, we would then consider the fluid a continuum. Note that this same discussion can be used for any fluid property of interest.

One way to think about the continuum principle is by considering the average height of adults in the United States. If you take the average height of the students in your class, will this be the same as the average US height? Will this be close enough for an engineering approximation? The students in your class can be considered as differential volume for the US population. Can the average height of people living in the United States be obtained? Similar to the problem of the average height, can the true bulk viscosity (or any other fluid property) be quantified? It would be very difficult to measure the true viscosity because it is nearly impossible to quantify and average the true molecular viscosity of individual fluid volumes that are on the order of a few molecules in size. This problem is compounded when the properties can change with time. Imagine trying to obtain the viscosity of a differential element of fluid, to only have the fluid molecules continually move into and out of the differential element. If one single static snapshot of the bulk fluid can be obtained and used to quantify the viscosity of every differential element within that fluid, then this would not be a problem, however, with current technology this is not practical. Partially due to the difficulty of quantifying fluid properties and the observation that fluid properties change with time, a quantity termed the "apparent viscosity" has become popular (see Section 6.7) for flows within the microcirculation. The apparent viscosity relates the bulk viscosity to the viscosity that we can calculate by measuring other flow parameters. Clearly, depending on the system that we are interested in, the bulk viscosity may be a good approximation; however, it may be more realistic to discuss the apparent viscosity. It is also true that viscosity is not only a function of space but also of time and temperature. Think about the oil in your car engine. On a cold winter morning when you start your car the oil will initially be very viscous. However, after the car has been running for 5–10 min, the oil heats up and the viscosity reduces. Over that 5-min period, the viscosity decreased due to an increase in engine temperature. Most biological fluids in humans see an approximate constant temperature (approximately 37°C), but remember that physical property numeric values can change in time based on other physical or mechanical properties. For instance, an increase in shear rate between particular joints can decrease the viscosity of the fluid in the joint space (see Chapter 11). Sometimes, temperature can be critical to a particular problem, such that if you were to analyze the blood flow throughout your arm/hand when throwing snowballs barehanded at your classmates, you might want to include temperature changes. A second case could be taking a breath of air on a cold winter morning. The heat transfer properties combined with the fluid flow properties would need to be considered. Be cautioned that temperature fluctuations can arise in biological problems, and these fluctuations can alter the fluid properties.

2.5 ELEMENTAL STRESS AND PRESSURE

You may recall from engineering mechanics that there are a number of different ways to define stress, including the Cauchy stress, the first Piola–Kirchhoff stress, and the second Piola–Kirchhoff stress. The most useful definition for stress in fluid mechanics is the Cauchy stress (σ), which is a measure of all of the forces acting on a differential volume oriented in space in the current configuration. This is in contrast to the Piola–Kirchhoff stresses, which measures the forces on a differential element and relates them to a reference configuration, which may be different from the current configuration. The reference configuration is normally a cube in space oriented along an assumed Cartesian coordinate axes, whereas the current configuration could be that same cube rotated to any degree off any of the three Cartesian axes. In general, we can resolve the forces that act on a cube in three-dimensional space in terms of nine stress components, otherwise known as the stress tensor (Eq. 2.6). The stress tensor is represented as a 3×3 matrix (σ), with subscript entries representing first the face the force is acting upon and second the direction that the force acts. As an example, the first stress component σ_{xx} is the stress that acts on the x-face of the cube (the x-face is parallel to the yz-plane) and acts in the x-direction. The stress σ_{xz} acts on the x-face in the z-direction. Only six of the nine components in the stress tensor are independent because angular momentum must be conserved on each differential volume element. The six independent values of stress are σ_{xx}, σ_{xy}, σ_{xz}, σ_{yy}, σ_{yz}, and σ_{zz}. To conserve angular momentum, σ_{yx} must be equal to σ_{xy}; σ_{zx} must be equal to σ_{xz}; and σ_{zy} must be equal to σ_{yz}.

$$\sigma = \begin{bmatrix} \sigma_{xx} & \sigma_{xy} & \sigma_{xz} \\ \sigma_{yx} & \sigma_{yy} & \sigma_{yz} \\ \sigma_{zx} & \sigma_{zy} & \sigma_{zz} \end{bmatrix} \tag{2.6}$$

$$\sigma_{(\text{face})(\text{direction})}$$

Let us develop how to quantify stress on a surface. Any flowing surface is in contact with either another flowing surface (termed a "lamina of fluid") or the bounding surface of the fluid volume. The movement of this fluid surface will generate a force on the neighboring surface (either fluid or boundary), if it moves at a different velocity than that neighboring surface. The force that is generated on the two laminae of fluid must be in balance with each other so that conservation of momentum (Newton's laws) is satisfied (Figure 2.13). This figure illustrates that the areas in contact with each other from two adjacent laminas of fluid will have the same forces acting upon them (i.e., $\delta\vec{F}_2$), whereas the surfaces not in contact do not necessarily have the same forces acting upon them ($\delta\vec{F}_1$ and $\delta\vec{F}_3$). Also, the actual forces on opposing lamina surfaces, within one lamina, do not need to balance (e.g., $\delta\vec{F}_1$ and $\delta\vec{F}_2$) because elements within a dynamic fluid deform and the elements do not experience a uniform velocity profile across the differential height, h. Due to these deformations, the surfaces that are experiencing the forces we are describing do not necessarily have the same area (as shown on the right-hand side of Figure 2.13), although the area would equate to the surface on the neighboring lamina. Recall that forces of different magnitudes acting on areas of difference sizes can still satisfy the conservation laws. It is likely that these different stresses acting on different areas on opposing surfaces of a single lamina would end up balancing over the entire differential element. However, as the height of these laminae approaches the differential element size ($\delta h \to 0$), the forces on opposing surfaces of the same lamina will nearly be balanced, as described previously. Let us define the area of the original lamina (lamina 2) in contact with a fluid lamina above it (lamina 1) as δA_2. Lamina 1 is in contact with lamina 2 and the blood vessel wall in this example. It has the same area in contact with lamina 2 as lamina 2 has with lamina 1 (i.e., δA_2). Because of this defined contact area and the motion of the fluid, there is some force distributed over the area of each element. Consider a small element of area, $\delta\vec{A}_2$, which has a force component acting on it ($\delta\vec{F}_2$). This force can be resolved into two elements; one that is normal to $\delta\vec{A}_2$ and one that is tangential to $\delta\vec{A}_2$. These forces are denoted as $\delta\vec{F}_n$ and $\delta\vec{F}_t$, respectively. The normal stress, σ_n (Eq. (2.7)), and the shear stress, τ_n (Eq. (2.8)), can then be defined as

$$\sigma_n = \lim_{\delta\vec{A}_n \to 0} \frac{\delta\vec{F}_n}{\delta\vec{A}_n} \tag{2.7}$$

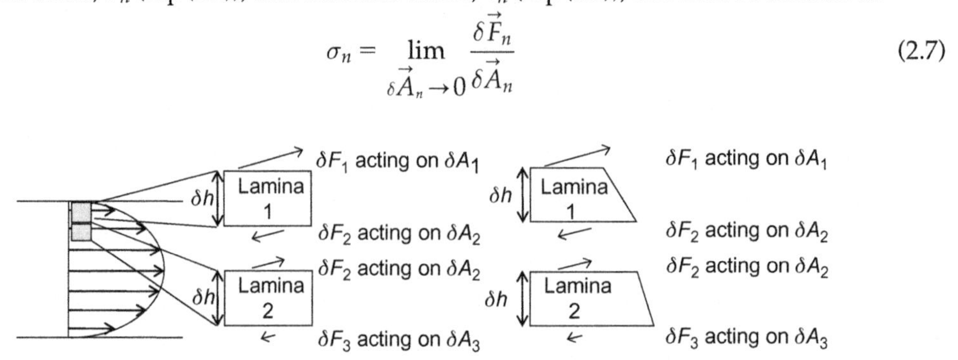

FIGURE 2.13 **Internal laminae of fluid exert forces on neighboring laminae.** These forces must be balanced within the fluid. Areas in contact of the individual laminae are the same and are represented as δA_2 as in the text.

$$\tau_n = \lim_{\delta\vec{A}_n \to 0} \frac{\delta\vec{F}_t}{\delta\vec{A}_n} \tag{2.8}$$

where n denotes the unit normal vector with regard to the differential area element. As the area approaches zero, the normal stress equation (Eq. (2.9)) and the shear stress equation (Eq. (2.10)) can be represented in differential form.

$$\sigma_n = \frac{d\vec{F}_n}{d\vec{A}_n} \tag{2.9}$$

$$\tau_n = \frac{d\vec{F}_t}{d\vec{A}_n} \tag{2.10}$$

To relate these equations to the stress tensor (Eq. (2.6)), you must define a coordinate system and define how the areas and forces relate to this system. In the most simplistic form, when a Cartesian coordinate system is set up in which the axes of this system are directly aligned with the fluid element, the normal stress σ_{xx} can be represented as Eq. (2.11) from Eq. (2.9) and the shear stress σ_{xy} can be represented as Eq. (2.12) from Eq. (2.10) (this is for a simplified two-dimensional case):

$$\sigma_{xx} = \frac{d\vec{F}_x}{d\vec{A}_x} \tag{2.11}$$

$$\tau_{xy} = \sigma_{xy} = \frac{d\vec{F}_y}{d\vec{A}_x} \tag{2.12}$$

For our analysis, we will normally use an orthogonal coordinate system. In Cartesian coordinates, this is the standard x, y, and z directions, which you should be familiar with. When we begin to discuss flow through blood vessels, we will begin to adopt the cylindrical coordinate system for ease. Hopefully, you have some familiarity with the cylindrical coordinate system, in which each point is defined with a radial distance (r), an angular coordinate (θ), and a height (z). Although, if you are asked to solve a problem involving two-dimensional blood flow through a vessel, you may choose to solve the problem using the Cartesian system, if you assume that the flow is symmetrical about the centerline in the third direction (although this would mimic an infinitely wide rectangular channel solution instead of a circular cylinder).

If you consider a differential rectangular element in three dimensions, we can see how the components of the stress tensor fall on this element (Figure 2.14) giving us the standard nomenclature used within the Cartesian stress tensor. The components take the form of normal stresses and shear stress (Eq. (2.13)), which is an extension of Eq. (2.6). Using this differential element and the Cartesian coordinate system, the stress tensor can be written as

$$\sigma = \begin{bmatrix} \sigma_{xx} & \tau_{xy} & \tau_{xz} \\ \tau_{xy} & \sigma_{yy} & \tau_{yz} \\ \tau_{xz} & \tau_{yz} & \sigma_{zz} \end{bmatrix} \tag{2.13}$$

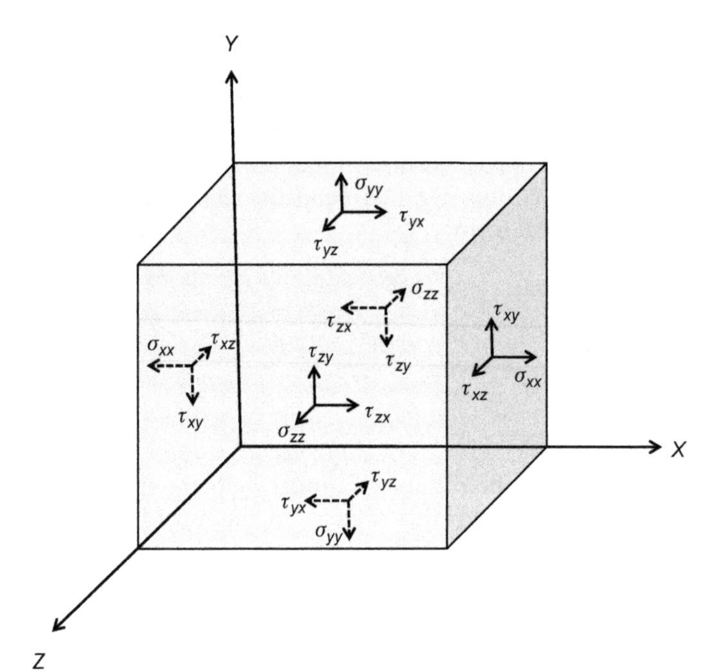

FIGURE 2.14 The nine components of the stress tensor on a differential fluid volume. For momentum balance, only six of these values are independent.

which has been reduced to its six independent values. As discussed previously, the shear stress values on different adjoining faces on a differential element must be equal so that momentum is conserved across the element. However, if one is evaluating a nondifferential element the stresses on opposing faces may not equate. There is a sign convention for stress values. If the force acts in the positive direction on a positive face or a negative direction on a negative face, then the stress value is positive. If the force acts in a negative direction on a positive face or a positive direction on a negative face, then the stress is negative. All stresses drawn on Figure 2.14 are positive stresses. The stresses represented as dashed lines are negative stresses acting on the negative faces. The stresses represented as solid lines are positive stresses acting on the positive faces. Note that positive faces and negative faces are defined by the direction of the unit normal vector.

Two different types of forces can account for stresses within the stress tensor: surface forces and body forces. Surface forces act directly on the boundaries of the volume of interest, not internally within the volume. These forces also act by direct contact of the substance that imparts the force with our fluid element. These forces are normally constant (they can vary with time) at the point of contact. Contrary to this, body forces are developed without direct contact with the fluid element. These forces are distributed throughout the entire fluid and are normally variable depending on where you look within the volume of interest or even within the same fluid element. Examples of surface forces are the shear force induced by the fluid flow of a neighboring volume (i.e., neighboring fluid lamina) or the mechanical force that the elastic blood vessel applies onto the fluid. An example of a body force is the force due to gravity, which is dependent on the spatial location within the fluid.

Pressure has many functions in fluid mechanics and is a major constituent of forces that arises in the stress tensor. First, it can be the driving force for fluid motion. This is discussed as the inflow pressure head or the pressure gradient (change in pressure) across the volume of interest. However, at rest, fluids still exert an internal pressure. This is the *hydrostatic pressure*, which is defined partially by the weight of the fluid above a particular plane within the fluid. This is an example of a body force because it varies depending on the location of the plane within the fluid. For instance, the pressure on a free fluid surface (open to any atmosphere) must be equal to the atmospheric pressure at the free surface. However, a differential element of fluid below a free surface must be able to support the weight of the fluid above it as well as the atmospheric pressure. To think of this, consider walking around your campus grounds with 1 atmosphere of pressure on your shoulders. If, however, you are carrying a heavy book bag while walking to your biofluid mechanics final examination, it might feel like the weight of the world is on your shoulders. However, in fact, the only added weight is that of the book bag; you still are supporting the weight of the 1 atmosphere of pressure. Also, the pressure from the book bag is not added to your head (although again the final examination might seem weighty), but it is added to everything from your shoulders down (even the floor you are standing on). One useful relationship between stress and pressure is that the hydrostatic pressure is equal to the average of the normal stress components (Eq. (2.14)). The negative sign is added to the formulation to make sure that the forces balance each other. For any other coordinate system (e.g., cylindrical) the hydrostatic pressure is equal to the average of the stresses along the main diagonal of the Cauchy stress tensor. (Remember to include the negative sign so that the forces can balance each other.)

$$p_{\text{hydrostatic}} = -\frac{1}{3}(\sigma_{xx} + \sigma_{yy} + \sigma_{zz}) \tag{2.14}$$

2.6 KINEMATICS: VELOCITY, ACCELERATION, ROTATION, AND DEFORMATION

Kinematics is the generalized study of motion. Using kinematic relationships you can describe the motion of any particle in space and time. In this section, we will develop the formulations for the motion of an infinitesimal area of fluid, $\delta x \delta y$, with a fixed identifiable mass, δm. (This derivation can be extended to three dimensions easily, by extending the area into a differential volume, denoted as $\delta x \delta y \delta z$.) As this mass moves throughout the fluid, it may experience translation, rotation, extension, pure shear, or any combination of these four movement classifications (Figure 2.15). Translation and rotation are rigid body motions and should be familiar from engineering mechanics courses. *Translation* is defined when all of the particles within the mass move with the same displacement and there is maintenance of the orientation of these particles. *Rotation* is defined when all the particles rotate about some point, maintaining the orientation between all particles. As depicted in Figure 2.15, not all of the particles experience the same displacement during rotation. Particles along the point of rotation (axis of rotation in three dimensions) have a zero displacement, while those at the periphery of the body have the largest displacement. In real fluid mechanics problems, fluid elements do not experience pure translation or rotation. However, in some cases, this can be used to

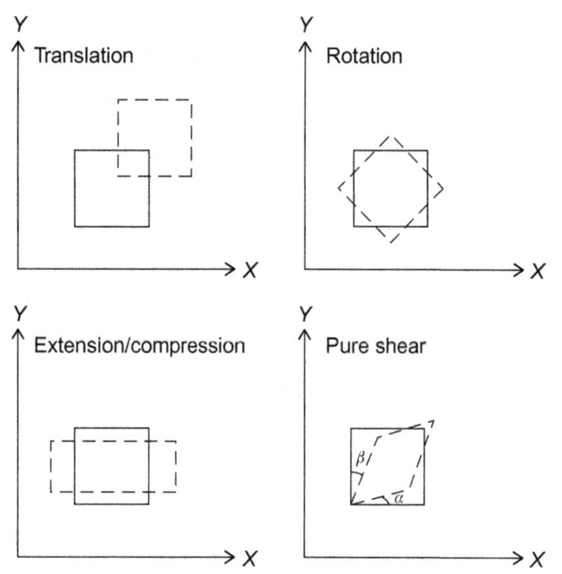

FIGURE 2.15 **Different motions that a fluid element can undergo.** These include translation, rotation, extension/compression, and pure shear. In general, these motions can occur all at the same time, which would be described as a general planar rotation/deformation.

approximate the flow scenario in order to make the calculations simpler. For most real fluid mechanics problems, when δm moves throughout the fluid, the internal organization of the particles is not conserved. If this is the case, the kinematic representation of the element will be a combination of extension/compression and pure shear (plus translation/rotation). If the element is subjected to pure extension or compression, there are changes in the boundary lengths of area δm, inducing some internal reorganization of the fluid particles to account for the changes in length. This is similar to compression or elongation in mechanics of materials problems. Pure shear is when only the internal angles along the boundary of δm are altered. This occurs if the element is under a pure shearing force and there are no normal forces acting on the element. Please note that α and β do not need to be equal to each other when the body is under pure shear. Figure 2.15 illustrates the kinematic movement of a mass in two dimension. These movements can be extended to three dimensions to describe any motion that a volume $\delta x \delta y \delta z$ would experience.

Similar to previous courses in mechanics, *velocity* is defined as the time rate of change of the position vector of any particle. Here we will define a particle's position vector in three-dimensional Cartesian coordinates as

$$\vec{x}(t) = u_x(t)\vec{i} + u_y(t)\vec{j} + u_z(t)\vec{k} \tag{2.15}$$

which is relative to the origin of a fixed coordinate system. This particle has a position of (x_i, y_i, z_i) at time t. Using this formulation, the velocity vector would be defined as

$$\vec{v}(t) = \frac{d\vec{x}(t)}{dt} = \frac{du_x(t)}{dt}\vec{i} + \frac{du_y(t)}{dt}\vec{j} + \frac{du_z(t)}{dt}\vec{k} = v_x(t)\vec{i} + v_y(t)\vec{j} + v_z(t)\vec{k} \tag{2.16}$$

As in mechanics, *acceleration* is defined as the time rate of change of the velocity vector of the particle in three-dimensional space. Similarly, it takes the form of

$$\begin{aligned}\vec{a}(t) &= \frac{d\vec{v}(t)}{dt} = \frac{d^2\vec{x}(t)}{dt^2} = \frac{dv_x(t)}{dt}\vec{i} + \frac{dv_y(t)}{dt}\vec{j} + \frac{dv_z(t)}{dt}\vec{k} \\ &= \frac{d^2u_x(t)}{dt^2}\vec{i} + \frac{d^2u_y(t)}{dt^2}\vec{j} + \frac{d^2u_z(t)}{dt^2}\vec{k} = a_x(t)\vec{i} + a_y(t)\vec{j} + a_z(t)\vec{k}\end{aligned} \tag{2.17}$$

Example

Given the following position vector, which describes the motion of a fluid element with respect to time, calculate the velocity and acceleration of the fluid particle as a function of time. This particle starts at the origin of the coordinate axis at time zero. Plot the fluid particles position and velocity for time 0, 1, 2, and 4 s.

$$\vec{x}(t) = (3\,ms^{-1})\left(t + \left(\frac{5}{2}s^{-1}\right)t^2\right)\vec{i} + (2\,ms^{-1})t\vec{j}$$

Solution

The velocity of an element is calculated as the time derivative of the position vector.

$$\begin{aligned}\vec{v}(t) &= \frac{d\vec{x}(t)}{dt} = \frac{d}{dt}\left[(3\,ms^{-1})\left(t + \left(\frac{5}{2}s^{-1}\right)t^2\right)\right]\vec{i} + \frac{d}{dt}\left[(2\,ms^{-1})\,t\right]\vec{j} \\ &= \left[(3\,ms^{-1})\left(1 + (5\,s^{-1})t\right)\right]\vec{i} + \left[(2\,ms^{1})\right]\vec{j}\end{aligned}$$

Acceleration is defined as the time rate of change of the velocity vector.

$$\vec{a}(t) = \frac{d\vec{v}(t)}{dt} = \frac{d}{dt}\left[(3\,ms^{-1})(1 + (5\,s^{-1})t)\right]\vec{i} + \frac{d}{dt}\left[(2\,ms^{-1})\right]\vec{j} = \left[(3\,ms^{-1})(5\,s^{-1})\right]\vec{i} = 15\,ms^{-2}\vec{i}$$

Figure 2.16 and Table 2.2 depict the motion of the particle investigated.

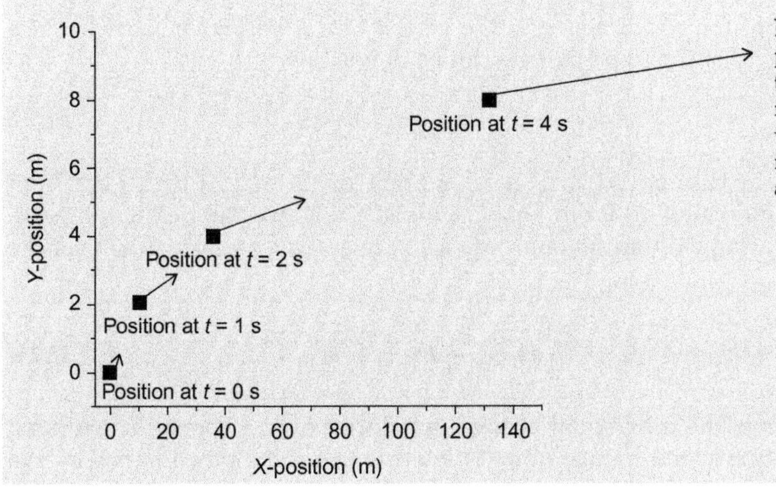

FIGURE 2.16 The fluid particles position as a function of time. The arrows represent the magnitude of velocity at each time. Please note that the x and y axes are not on the same scale.

TABLE 2.2 Kinematic Properties of the Particle as a Function of Time

Time (s)	*X*-Position (m)	*Y*-Position (m)	*X*-Velocity (m/s)	*Y*-Velocity (m/s)	Velocity, Magnitude, and Direction	*X*-Acceleration (m/s^2)
0	0	0	3	2	3.61 m/s @ 33.7°	15
1	10.5	2	18	2	18.11 m/s @ 6.34°	15
2	36	4	33	2	33.06 m/s @ 3.47°	15
4	132	8	63	2	63.03 m/s @ 1.82°	15

The formulations for velocity and acceleration are based on the Lagrangian assumption that you can define discrete particles within the fluid. It is a requirement that each particle has a known position as a function of time; therefore, velocity and acceleration can be defined from this position vector (see previous example). However, in fluid mechanics, it is sometimes more accurate to discuss the velocity as a property of the fluid as a whole, instead of the position of each fluid particle as a function of time. In this case, the velocity vector of the fluid is defined first because it is probably difficult to accurately describe each particle's position within the fluid (recall our earlier discussion on how many cells, proteins, molecules this would include for a small volume of blood). In this Eulerian type approach, the velocity of the fluid becomes

$$\vec{v} = v(x, y, z, t) = v_x + v_y + v_z + v_t \tag{2.18}$$

The acceleration, which is the time derivative of the velocity, is defined by considering that a particle at time t has a velocity of

$$v_{p|t} = v_{x|t} + v_{y|t} + v_{z|t} + v_{t|t} \tag{2.19}$$

where the added subscript for p indicates it is the velocity for a particular particle. At a later time point, the velocity of the same particle can be defined as

$$v_{p|t+dt} = v_{x+dx|t+dt} + v_{y+dy|t+dt} + v_{z+dz|t+dt} + v_{t+dt|t+dt} \tag{2.20}$$

assuming that the particle has moved to a new location of $(x + dx, y + dy, z + dz)$ from (x, y, z) during the time interval of dt. From Eqs. (2.19) and (2.20), we can define the particles change in velocity during the translation from (x, y, z) to $(x + dx, y + dy, z + dz)$ over the dt time interval as

$$d\vec{v}_p = \frac{\partial \vec{v}}{\partial x} dx_p + \frac{\partial \vec{v}}{\partial y} dy_p + \frac{\partial \vec{v}}{\partial z} dz_p + \frac{\partial \vec{v}}{\partial t} dt \tag{2.21}$$

where each term represents the differential change in velocity over a spatial or temporal dimension multiplied by the actual change in that dimension over the time interval dt. We can obtain the acceleration of this particle over this particular time interval by taking the time derivative of Eq. (2.21). The acceleration can then be represented by

$$\vec{a}_p = \frac{d\vec{v}_p}{dt} = \frac{\partial \vec{v}}{\partial x}\frac{dx_p}{dt} + \frac{\partial \vec{v}}{\partial y}\frac{dy_p}{dt} + \frac{\partial \vec{v}}{\partial z}\frac{dz_p}{dt} + \frac{\partial \vec{v}}{\partial t}\frac{dt}{dt} = \frac{\partial \vec{v}}{\partial t} + v_x\frac{\partial \vec{v}}{\partial x} + v_y\frac{\partial \vec{v}}{\partial y} + v_z\frac{\partial \vec{v}}{\partial z} \tag{2.22}$$

$\frac{\partial \vec{v}}{\partial t}$ is termed the local acceleration of the fluid and the remaining terms are the convective acceleration of the fluid. Equation (2.22) is typically termed as the total acceleration of a particle and is typically represented with the substantial derivative as follows:

$$\frac{D\vec{v}}{Dt} = \vec{a}_p = \frac{\partial \vec{v}}{\partial t} + v_x\frac{\partial \vec{v}}{\partial x} + v_y\frac{\partial \vec{v}}{\partial y} + v_z\frac{\partial \vec{v}}{\partial z} \tag{2.23}$$

We should recognize that there are two ways in which a fluid particle can experience an acceleration. First, due to spatial constrictions (such as a nozzle) or dilations the fluid elements may be convected into a region that has a higher velocity. This type of acceleration is due to spatial changes and is termed as the convective acceleration. If this flow is steady, meaning that the velocity profile at each spatial location is not a function of time, then this is the only acceleration that the particle will experience. However, if the flow is unsteady, meaning that the velocity is a function of time, then the particles experience an added acceleration term, which is due to the temporal changes in the velocity profile. This arises when the flow is driven by an unsteady pump, such as the heart. Different types of temporal driving forces and spatial driving forces are shown in Figure 2.17. We illustrate one cardiac cycle as a temporal driving force and different types of constrictions as spatial driving forces. It is important to note that velocity as well as acceleration can be altered by spatial and temporal driving forces. Fluid velocity and acceleration are field quantities, that is, they are functions of the fluid itself.

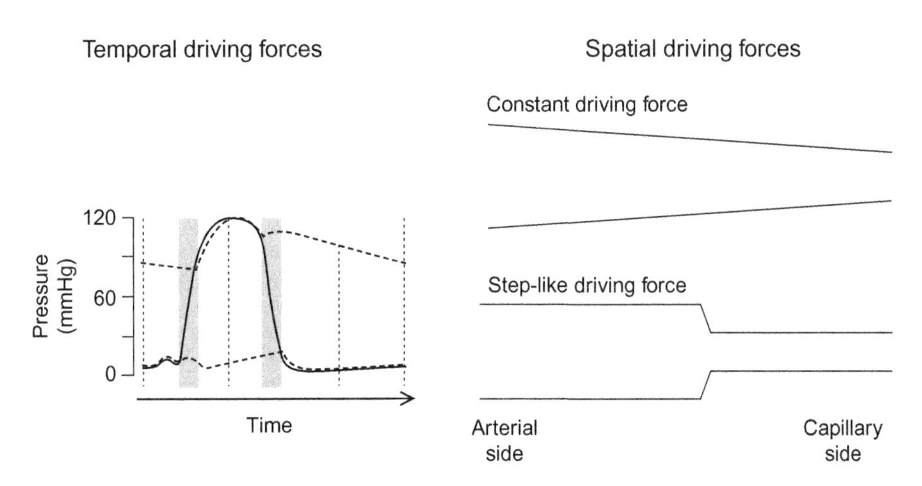

FIGURE 2.17 Examples of temporal and spatial driving forces that may arise in the vascular system. In general, temporal and spatial driving forces can be coupled within the same problem.

Since acceleration is itself a vector, it can be divided into the following three terms (in Cartesian coordinates)

$$\begin{aligned} a_x &= \frac{\partial v_x}{\partial t} + v_x \frac{\partial v_x}{\partial x} + v_y \frac{\partial v_x}{\partial y} + v_z \frac{\partial v_x}{\partial z} \\ a_y &= \frac{\partial v_y}{\partial t} + v_x \frac{\partial v_y}{\partial x} + v_y \frac{\partial v_y}{\partial y} + v_z \frac{\partial v_y}{\partial z} \\ a_z &= \frac{\partial v_z}{\partial t} + v_x \frac{\partial v_z}{\partial x} + v_y \frac{\partial v_z}{\partial y} + v_z \frac{\partial v_z}{\partial z} \end{aligned} \tag{2.24}$$

As you will see later in the text, it is this formulation that is critical to the foundation of many fluid mechanics problems. Using the Eulerian approach to calculate velocity and acceleration, the main quantity of interest is a particular volume within the system, not a particular particle within the fluid. Again, this may make the computations significantly easier because we solve for the entire fluid and not individual particles.

To define the rotation of fluid particles, we must take a slightly different approach. Figure 2.15 highlights the kinematic rotation of a fluid element at any time/space in the flow field. This can be described by rotation vectors, namely, the angular velocity vector (in three dimensions, Eq. (2.25))

$$\vec{\omega} = \omega_x \vec{i} + \omega_y \vec{j} + \omega_z \vec{k} \tag{2.25}$$

where each Cartesian component is the rotation of the element about that particular Cartesian axis. For instance, ω_x is the angular velocity of the element about the x-axis. Rotation components can be described by velocity components by realizing that the rotation of a small fluid element (Figure 2.18) can be represented as the mathematical limit (lim) as time approaches zero of the time rate of change of the angles (i.e., α, Figure 2.18) that the fluid element makes with the particular fixed coordinate system. From Figure 2.18, ω_z would be defined as

$$\omega_z = \lim_{\Delta t \to 0} \frac{\Delta \alpha}{\Delta t} \tag{2.26}$$

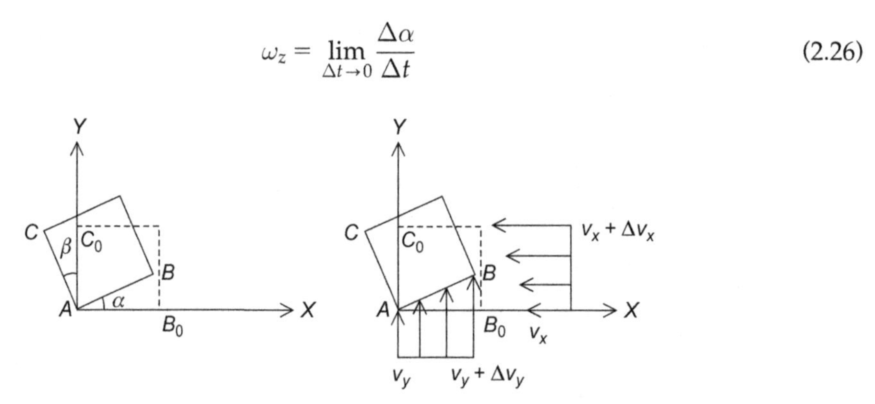

FIGURE 2.18 Rotation of a rigid body used to calculate angular velocity, vorticity, and shear rate. The rate of change of the displacement of point B is defined from the change in velocity over time across line segment AB. The same analysis conducted on all other line segments in three dimensions would yield the angular velocity formula. Remember that α and β do not have to be the same.

Using this approach, the angular velocity about the z axis can also be defined mathematically as the average of the angular velocity of two perpendicular line segments in the normal plane ($x-y$ plane) of that axis system. This formulation will end up with fluid quantities that are more easily quantified and are typically used because the angle α does not typically equal the angle β (Figure 2.18). Additionally, the time rate of change of these two angles does not need to be equal. The perpendicular line segments that will be used to define the changes in angular velocity are $\overline{AB_0}$ and $\overline{AC_0}$ (Figure 2.18). At some later time, B_0 and C_0 will translate to B and C, respectively (Figure 2.18), forming the angle α/β that was discussed previously (recall that these angles are not necessarily equal). To quantify the angular velocity, the difference in linear velocity at points A and B must first be determined. Since this element is a differential element, the velocity difference between A and B is Δv_y. (This velocity acts in the y-direction even though the original line segment was in the x-direction because over a small time interval the velocity needed to initiate a rotation of this fluid element will only be in the y-direction.) Similarly, the difference in velocity from point A to point C would be Δv_x, making the same assumption regarding a small time interval. This also makes the assumption that the x-velocity at A is defined as v_x and the y-velocity at A is defined as v_y. Using these definitions for velocity, the change in displacement for points A, B_0, and C_0 to points A, B, and C would be 0, $\left(v_y + \frac{\partial v_y}{\partial x}\Delta x - v_y\right)\Delta t$, and $-\left(v_x + \frac{\partial v_x}{\partial y}\Delta y - v_x\right)\Delta t$, respectively. To formulate these displacements (denoted as s), a change in velocity is multiplied by a change in time (i.e., $\Delta s = \Delta v \Delta t$). Also, the change in velocity is defined by the velocity at B $\left(v_y + \frac{\partial v_y}{\partial x}\Delta x\right)$ minus the y-velocity at A (v_y) or the velocity at C $-\left(v_x + \frac{\partial v_x}{\partial y}\Delta y\right)$ minus the x-velocity at A $-(v_x)$. The displacement of C is negative because the point moves in the negative x-axis direction and thus the velocity must be negative to induce this motion. Additionally, the partial derivative term for the velocity at B arises from the original Δv_y, but the assumption is made that $\Delta x \rightarrow 0$ for a differential element and thus the velocity changes can be represented with partial differential changes. Now take the limit as time approaches zero of the change in the angle (α or β) (from Eq. (2.26) to obtain Eq. (2.27)), to obtain the elements angular velocity

$$\lim_{\Delta t \to 0} \frac{\Delta \alpha}{\Delta t} \cong \lim_{\Delta t \to 0} \frac{\overline{BB_0}/\overline{AB_0}}{\Delta t} = \lim_{\Delta t \to 0} \frac{\left(v_y + \frac{\partial v_y}{\partial x}\Delta x - v_y\right)\Delta t/\Delta x}{\Delta t} = \lim_{\Delta t \to 0} \frac{1}{\Delta t}\left(\frac{\frac{\partial v_y}{\partial x}\Delta x \Delta t}{\Delta x}\right) = \frac{\partial v_y}{\partial x}$$

$$\lim_{\Delta t \to 0} \frac{\Delta \beta}{\Delta t} \cong \lim_{\Delta t \to 0} \frac{\overline{CC_0}/\overline{AC_0}}{\Delta t} = \lim_{\Delta t \to 0} \frac{-\left(v_x + \frac{\partial v_x}{\partial y}\Delta y - v_x\right)\Delta t/\Delta y}{\Delta t} = \lim_{\Delta t \to 0} \frac{1}{\Delta t}\left(-\frac{\frac{\partial v_x}{\partial y}\Delta y \Delta t}{\Delta y}\right) = -\frac{\partial v_x}{\partial y} \tag{2.27}$$

Therefore, angular velocity about the z axis will equal the average of these two quantities for a differential area element and can be represented as

$$\omega_z = \frac{1}{2}\left(\frac{\partial v_y}{\partial x} - \frac{\partial v_x}{\partial y}\right) \tag{2.28}$$

A similar analysis conducted about the other two Cartesian axes would yield

$$\omega_x = \frac{1}{2}\left(\frac{\partial v_z}{\partial y} - \frac{\partial v_y}{\partial z}\right) \text{ and } \omega_y = \frac{1}{2}\left(\frac{\partial v_x}{\partial z} - \frac{\partial v_z}{\partial x}\right) \tag{2.29}$$

Combining the terms for the angular velocity about each Cartesian direction into one angular velocity vector allows us to write an expression for the angular velocity of a fluid element as

$$\vec{\omega} = \frac{1}{2}\left[\left(\frac{\partial v_z}{\partial y} - \frac{\partial v_y}{\partial z}\right)\vec{i} + \left(\frac{\partial v_x}{\partial z} - \frac{\partial v_z}{\partial x}\right)\vec{j} + \left(\frac{\partial v_y}{\partial x} - \frac{\partial v_x}{\partial y}\right)\vec{k}\right] = \frac{1}{2}\text{curl}\,\vec{v} \tag{2.30}$$

Vorticity (ξ) is a measure of the rotation of fluid elements and is defined as $2\vec{\omega}$. An irrotational fluid has a vorticity of zero. This simplifies the governing equations of the fluid (to be derived later) and makes the computational analysis simpler. In most examples in this text, the vorticity will be zero. However, in biofluid mechanics, due to the large pressure and velocity gradients, recirculation zones, turbulence and shed vortices can be generated which can cause a fluid to have a non-zero vorticity; hence, the fluid would be rotational. Some people believe that an irrotational fluid is also inviscid; this is not necessarily true. However, an inviscid fluid that is initially irrotational will remain irrotational unless external forces are applied to the fluid element to induce rotation.

Example

Calculate the vorticity of the following velocity field. Is the flow rotational or not?

$$\vec{v} = [Cz\sin(xy) + Cxyz\cos(xy)]\vec{i} + Cx^2z\cos(xy)\vec{j} + Cx\sin(xy)\vec{k}$$

Solution

$$\xi = \left(\frac{\partial v_z}{\partial y} - \frac{\partial v_y}{\partial z}\right)\vec{i} + \left(\frac{\partial v_x}{\partial z} - \frac{\partial v_z}{\partial x}\right)\vec{j} + \left(\frac{\partial v_y}{\partial x} - \frac{\partial v_x}{\partial y}\right)\vec{k}$$

$$= \left(\frac{\partial(Cx\sin(xy))}{\partial y} - \frac{\partial(Cx^2z\cos(xy))}{\partial z}\right)\vec{i} + \left(\frac{\partial([Cz\sin(xy) + Cxyz\cos(xy)])}{\partial z} - \frac{\partial(Cx\sin(xy))}{\partial x}\right)\vec{j}$$

$$+ \left(\frac{\partial(Cx^2z\cos(xy))}{\partial x} - \frac{\partial[Cz\sin(xy) + Cxyz\cos(xy)]}{\partial y}\right)\vec{k}$$

$$(Cx^2\cos(xy) - Cx^2\cos(xy))\vec{i} + (C\sin(xy) + Cxy\cos(xy) - C\sin(xy) - Cxy\cos(xy))\vec{j}$$
$$+ (2Cxz\cos(xy) - Cx^2yz\sin(xy) - Cxz\cos(xy) - Cxz\cos(xy) + Cx^2yz\sin(xy))\vec{k}$$
$$= 0\vec{i} + 0\vec{j} + 0\vec{k}$$

The fluid is not rotational.

In solid mechanics, deformations were described by the change in length of a small element (i.e., strain). Here we will discuss the rate of change of the extension/compression of fluid elements. Strain has nine components that comprise the strain tensor (similar to stress components that comprise the stress tensor). Six of these components are independent of each other using the same assumptions that we made in developing the stress tensor. The rate of extension is defined by velocity changes through the flow field. If at some location in time, the velocity of a fluid is v_x and this changes to $v_x + \Delta v_x$ at time $t + \Delta t$ then the fluid elongates by $\Delta x + \Delta v_x \Delta t$. The first term in the elongation formulation arises from the original length between two points within the fluid. The second term is derived from the velocity component that may cause a change in the fluid element length. Similar to the angular velocity formulations, deformation in one dimension becomes

$$d_{xx} = \lim_{\Delta t \to 0} \frac{1}{\Delta t} \left(\frac{\Delta x + \Delta v_x \Delta t - \Delta x}{\Delta x} \right) = \frac{\partial v_x}{\partial x} \tag{2.31}$$

The other two normal deformations are represented as

$$d_{yy} = \frac{\partial v_y}{\partial y} \text{ and } d_{zz} = \frac{\partial v_z}{\partial z} \tag{2.32}$$

The other six strain tensor components are derived in a similar manner using the relationship that these stress components arise from shear stress (or shear forces) on the fluid elements and not from velocity changes. These shear stresses cause a change in the angle of our fluid element (which is normally represented as a cube, in our assumed Cartesian coordinate system). Similar to our angular velocity calculations, the strain will be calculated as the average of the time rate of change of the two angles α and β that the element makes with the fixed coordinate axis (these angles are shown in Figure 2.15).

$$\Delta\alpha \cong \frac{\{v_y + (\partial v_y/\partial x)\Delta x - v_y\}\Delta t}{\Delta x} \text{ and } \Delta\beta \cong \frac{\{v_x + (\partial v_x/\partial y)\Delta y - v_x\}\Delta t}{\Delta y} \tag{2.33}$$

Equation (2.33) assumes that the time interval is short enough that the angles do not change significantly and that the tangent of these angles is approximately equal to the angles themselves (small angle approximation). Additionally, we want to point out that the angles are not the same as in Figure 2.18, which illustrated a rigid body rotation about the z axis and that these angles are assumed to get smaller with time. Using these formulas, the rate of deformation due to shear simplifies to

$$d_{xy} = d_{yx} = \lim_{\Delta t \to 0} \frac{1}{\Delta t} \left(\lim_{\substack{\Delta x \to 0 \\ \Delta y \to 0}} \frac{1}{2} (\Delta\alpha + \Delta\beta) \right) = \frac{1}{2} \left(\frac{\partial v_y}{\partial x} + \frac{\partial v_x}{\partial y} \right) \tag{2.34}$$

The remaining components of deformation due to shear are defined in a similar manner, in Cartesian coordinates, making the deformation tensor

$$d_{xz} = d_{zx} = \frac{1}{2} \left(\frac{\partial v_z}{\partial x} + \frac{\partial v_x}{\partial z} \right) \text{ and } d_{yz} = d_{zy} = \frac{1}{2} \left(\frac{\partial v_z}{\partial y} + \frac{\partial v_y}{\partial z} \right) \tag{2.35}$$

$$\boldsymbol{d}=\begin{bmatrix} d_{xx} & d_{xy} & d_{xz} \\ d_{yx} & d_{yy} & d_{yz} \\ d_{zx} & d_{zy} & d_{zz} \end{bmatrix}=\frac{1}{2}\begin{bmatrix} 2\frac{\partial v_x}{\partial x} & \frac{\partial v_y}{\partial x}+\frac{\partial v_x}{\partial y} & \frac{\partial v_z}{\partial x}+\frac{\partial v_x}{\partial z} \\ \frac{\partial v_y}{\partial x}+\frac{\partial v_x}{\partial y} & 2\frac{\partial v_y}{\partial y} & \frac{\partial v_z}{\partial y}+\frac{\partial v_y}{\partial z} \\ \frac{\partial v_z}{\partial x}+\frac{\partial v_x}{\partial z} & \frac{\partial v_z}{\partial y}+\frac{\partial v_y}{\partial z} & 2\frac{\partial v_z}{\partial z} \end{bmatrix} \tag{2.36}$$

In fluid mechanics, the deformation tensor is known as the shear rate (similar to strain in solid mechanics). Again, this is the rate of change of the angles or changes in the element's geometry within a fluid and can be calculated from the velocity changes within the fluid.

Example

Calculate the shear rate of a fluid with the following velocity profile:

$$\vec{v}=Cxy\vec{i}+\sin(y)\vec{j}+Cz\vec{k}$$

Solution

$$d_{xx}=\frac{\partial v_x}{\partial x}=\frac{\partial(Cxy)}{\partial x}=Cy$$

$$d_{yy}=\frac{\partial v_y}{\partial y}=\frac{\partial(\sin(y))}{\partial y}=\cos(y)$$

$$d_{zz}=\frac{\partial v_z}{\partial z}=\frac{\partial(Cz)}{\partial z}=C$$

$$d_{xy}=d_{yx}=\frac{1}{2}\left(\frac{\partial v_x}{\partial y}+\frac{\partial v_y}{\partial x}\right)=\frac{1}{2}\left(\frac{\partial(Cxy)}{\partial y}+\frac{\partial(\sin(y))}{\partial x}\right)=\frac{Cx}{2}$$

$$d_{xz}=d_{zx}=\frac{1}{2}\left(\frac{\partial v_x}{\partial z}+\frac{\partial v_z}{\partial x}\right)=\frac{1}{2}\left(\frac{\partial(Cxy)}{\partial z}+\frac{\partial(Cz)}{\partial x}\right)=0$$

$$d_{yz}=d_{zy}=\frac{1}{2}\left(\frac{\partial v_y}{\partial z}+\frac{\partial v_z}{\partial y}\right)=\frac{1}{2}\left(\frac{\partial(\sin(y))}{\partial z}+\frac{\partial(Cz)}{\partial x}\right)=0$$

$$\boldsymbol{d}=\begin{bmatrix} Cy & \frac{Cx}{2} & 0 \\ \frac{Cx}{2} & \cos(y) & 0 \\ 0 & 0 & C \end{bmatrix}$$

2.7 VISCOSITY

Viscosity is a property of fluids that relates the shear stress acting on a fluid to the shear rate experienced by the fluid. Fundamentally, viscosity arises because of intermolecular interactions within the fluid. For instance, the intermolecular interactions in water are mostly hydrogen bonds (i.e., adhesion and/or cohesion; see Figure 2.12). These bonds are relatively weak; therefore, water does not have a high viscosity. Fluids that possess stronger intermolecular bonds will have a larger viscosity (e.g., syrup). Another way to consider viscosity is as the fluid's internal resistance to changes in motion that will result in deformations. (Recall that a body's resistance to acceleration changes is its mass, and the body's resistance to rotational acceleration changes is its mass moment of inertia.) Fluids with a larger viscosity will resist motions that enduce deformation more easily than those with a lower viscosity.

In solid mechanics, stress and strain are related by the modulus of elasticity. Fundamentally, viscosity is defined in the same way, as a value that completes the following proportionality:

$$\tau_{xy} \propto 2d_{xy} = \frac{\partial v_x}{\partial y} + \frac{\partial v_y}{\partial x} \rightarrow \tau_{xy} = \mu\left(\frac{\partial v_x}{\partial y} + \frac{\partial v_y}{\partial x}\right) \tag{2.37}$$

In this formulation, the viscosity (μ) is the dynamic viscosity of the fluid. This is the internal resistance to motions that cause deformation that was discussed previously. The common unit for viscosity is Poise (where $1\ \text{P} \equiv 1\ g/\text{cm} * s$). In fluid mechanics, the kinematic viscosity (ν) is a standardized version of the viscosity. The kinematic viscosity is defined as the dynamic viscosity divided by the fluid density. The unit for kinematic viscosity is Stoke (where $1\ \text{St} \equiv 1\ \text{cm}^2/s$).

Example

Calculate the shear stress on the upper and lower boundaries of the following fluid, given that the viscosity of the fluid is 3.5 cP. The distance between the two plates is 10 mm and the velocity is 0 mm/s at the lower plate and 30 mm/s at the upper plate (Figure 2.19).

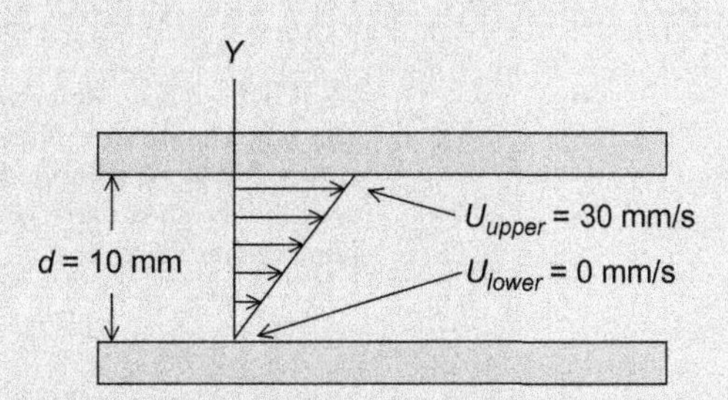

FIGURE 2.19 **Figure for the example problem.**

Solution

$$\tau_{upper} = \mu\left(\frac{\partial v_x}{\partial y} + \frac{\partial v_y}{\partial x}\right) = \mu\frac{\partial v_x}{\partial y}$$

There is no velocity in the y-direction so the second term within the parenthesis drops out of the relationship. Since the velocity profile is linear in the y-direction, the shear stress equation simplifies to

$$\tau_{upper} = \mu\frac{\Delta v_x}{\Delta y} = \mu\frac{U_{upper} - U_{lower}}{d_{upper} - d_{lower}} = \mu\frac{U_{upper}}{d_{upper}}$$

$$\tau_{upper} = 3.5\times10^{-2}P\left(\frac{1\ g/cm*s}{1\ P}\right)*\frac{30\ mm/s}{10\ mm} = 0.0105\ Pa = 0.105\ dyne/cm^2$$

Similarly, the shear stress on the lower plate is

$$\tau_{lower} = \mu\frac{\Delta v_x}{\Delta y} = \mu\frac{U_{lower} - U_{upper}}{d_{lower} - d_{upper}} = \mu\frac{U_{upper}}{d_{upper}}$$

$$\tau_{lower} = 3.5\times10^{-2}\ P\left(\frac{1\ g/cm*s}{1\ P}\right)*\frac{-30\ mm/s}{-10\ mm} = 0.0105\ Pa = 0.105\ dyne/cm^2$$

Although the shear forces have the same magnitude, their directions are opposite. The force on the top plate is acting in the negative x-direction, while the force on the bottom plate is acting in the positive x-direction. Remember the sign convention for stress that was discussed in Section 2.5, where a positive shear stress can occur on a positive face in the positive direction or on a negative face in the negative direction.

2.8 FLUID MOTIONS

Most fluids under normal conditions can be considered Newtonian fluids. A Newtonian fluid is classified by a constant dynamic viscosity under any shear rate. Similar to a purely elastic material, these fluids have a linear relationship between shear stress and shear rate. However, many fluids do not exhibit Newtonian properties and, therefore, are termed "Non-Newtonian" fluids (Figure 2.20). For example, ketchup will flow out of

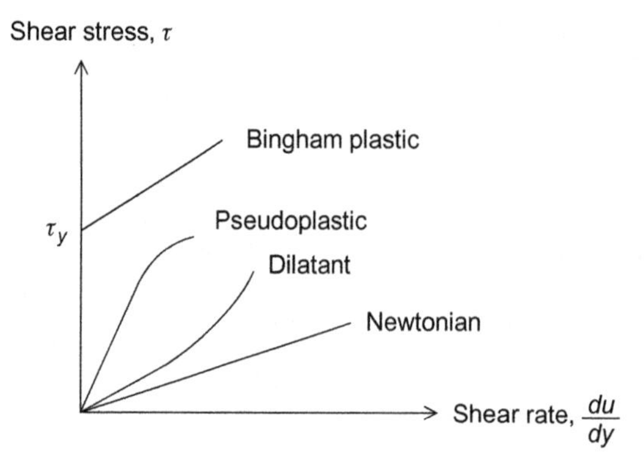

FIGURE 2.20 Relationship between shear stress and shear rate for Newtonian and non-Newtonian fluids. Newtonian fluids have a constant viscosity as a function of shear rate. Non-Newtonian fluids have a nonconstant viscosity.

the bottle very slowly with only the gravitational force pulling it. However, if you squeeze the bottle slightly, the ketchup will flow out of the bottle more quickly. This is an example of a shear-thinning fluid (also termed "pseudoplastic"). These fluids have a larger dynamic viscosity at low shear rates and a lower dynamic viscosity at high shear rates. Another non-Newtonian fluid is a sand–water mixture. When there is a large volume of water with a small amount of sand, this fluid will have a lower dynamic viscosity. Under a low shear rate, this type of mixture (large amounts of water to small amounts of sand, e.g., water:sand, 100:1) behaves as water. As the shear rate increases, water will flow out of the mixture more easily than the sand, leaving a similar quantity of sand and water (water:sand = 1:1). This 1:1 mixture will have a larger viscosity, at the higher shear rate, as compared with the 100:1 water:sand mixture. At some time, most of the water will be removed from the mixture (water:sand = 1:100), and it will become very difficult to make this fluid flow, because the internal resistance to deformation will be high. Although, under very large stresses, the fluid will deform because the resistance against motion is not infinite. This type of fluid is a shear-thickening fluid (also termed "dilatant"). To describe these fluids, the shear stress–shear rate relationship becomes (for three-dimensional volume in a Cartesian coordinate system, as derived above)

$$\tau_{xy} = k\left(\frac{\partial v_x}{\partial y} + \frac{\partial v_y}{\partial x}\right)^n \tag{2.38}$$

where k is termed the consistency index and n is the flow behavior index. To maintain the sign convention for shear stress, this expression may also be written as

$$\tau_{xy} = k\left|\frac{\partial v_x}{\partial y} + \frac{\partial v_y}{\partial x}\right|^{n-1}\left(\frac{\partial v_x}{\partial y} + \frac{\partial v_y}{\partial x}\right) = \eta\left(\frac{\partial v_x}{\partial y} + \frac{\partial v_y}{\partial x}\right) \tag{2.39}$$

where η is the apparent viscosity of the fluid. For Newtonian fluids, the apparent viscosity is the same as the dynamic viscosity because n is equal to 1. Pseudoplastic fluids have an $n < 1$ and dilatant fluids have an $n > 1$. One more classification of fluids is a Bingham plastic fluid. These types of fluids do not flow when small stresses are applied to them. Under these scenarios, some threshold shear stress level must be applied to the fluid to get the fluid to begin to flow. An example of this type of fluid is toothpaste. When in the tube, small forces will not cause the toothpaste to flow out of the tube (e.g., gravity when the tube is held upside down). A certain yield stress (τ_y) must be applied to the tube for the toothpaste to come out. In the case of toothpaste, the yield stress must exceed the atmospheric pressure plus any additional internal adhesive forces. Bingham plastic fluids are modeled as

$$\tau_{xy} = \tau_y + \mu\left(\frac{\partial v_x}{\partial y} + \frac{\partial v_y}{\partial x}\right) \tag{2.40}$$

where τ_y is the yield shear stress, which must be overcome to induce fluid motion. All real fluids will have a minimal yield stress and will like have a flow behavior index that is not equal to 1 and likely varies with the applied shear rate. However, for some fluids, like water, the yield stress is so small and the variation of the flow behavior index away from 1 is negligible.

Blood, however, is a Non-Newtonian fluid that is a combination of a pseudoplastic fluid and a Bingham plastic fluid. This means that blood must surpass an initial yield stress to begin to flow and that, as the shear rate increases, once blood is flowing, the dynamic viscosity of blood decreases. Through numerical and experimental studies, it has been shown that using the assumption that blood behaves as a Newtonian fluid overestimates the actual shear stress values that can be found in the fluid flow field. This is significant because, as we will see in later chapters, shear stress is a significant mediator of cardiovascular diseases and can elicit many physiological/pathological cellular responses. Therefore, an accurate prediction of shear stress must be made by using the appropriate fluid relationships for blood's properties.

Another classification can be made for fluids that exhibit a time-dependent change in viscosity. If the apparent viscosity decreases under a constant shear stress, this fluid is thixotropic. Biological examples of thixotropic fluids include synovial fluid (found in joints), cytoplasm, and ground substance (extracellular matrix found in connective tissue). Similar, rheopectic fluids have a time-dependent increase in apparent viscosity, under a constant shear stress. Currently, there are no known rheopectic fluids found natural in the human body, however, some printer inks can exhibit rheopectic behavior. The relationship between shear stress and shear rate for these fluids cannot be plotted on Figure 2.20 because the viscosity is also dependent on time, which is not a variable on this figure.

Inviscid fluids have a zero (or negligible viscosity). While there is no such real fluid that exhibits these properties, some fluid mechanics analysis can be done using this assumption and the obtained results are still meaningful (e.g., gas flows do not typically involve viscosity). Recall, that under these conditions, the fluid elements can only undergo rigid body motion. In most of the following examples in this textbook, we will use the no-slip boundary assumption. This means that the fluid at a boundary must have the same velocity as the boundary. (Refer to the previous parallel plate example in this textbook.) In most cases, when the boundary is stationary, the first lamina of fluid along the boundary will have a zero velocity. In some cases, the boundary can be moving and that first lamina will match the velocity of the boundary. However, this only holds for viscous fluids. Inviscid fluids do not resist flow internally and, therefore, all of the fluid layers must have the same velocity, regardless of the boundary velocity. If you look at a velocity profile for blood modeled as an inviscid fluid versus that of a viscous fluid, it could be depicted as in Figure 2.21, if all of the boundaries have a zero velocity component.

Flows can also be characterized as compressible or incompressible. If the density can vary throughout the fluid (because of geometry considerations or thermal consideration, among others) the fluid is compressible. Gases, for instance, are highly compressible. Incompressible fluids have negligible changes in density throughout the flow; suggesting that pressure and temperature (primarily) do not affect the fluid density. Most liquids exhibit this property to some extent and, in this textbook, we will normally make the assumption that liquids are incompressible. For most liquids, density can change with changes in temperature (for instance

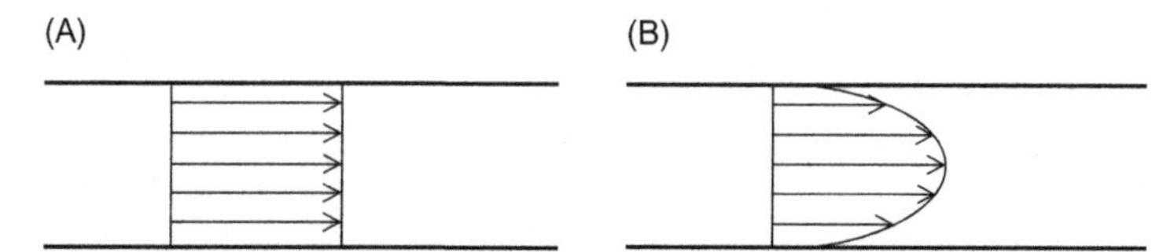

FIGURE 2.21 The velocity profile for blood flowing through an artery that is modeled as an inviscid fluid (A) or a viscous fluid (B). An inviscid fluid has a uniform velocity profile because there is no internal resistance to motion. A viscous fluid has a nonuniform velocity profile, which can be idealized to a parabolic velocity profile for many conditions.

the density of water at 5°C is 999.9 kg/m^3, at 35°C is 994 kg/m^3, at 55°C is 985.2 kg/m^3, and at 95°C is 961.5 kg/m^3). If temperature fluctuations are a property of your flow then you must use compressible flow formulations to describe your flow conditions or approximate the fluid density changes, with a manner of your choosing. However, in the human body, core body temperature is fairly constant at 37°C, so flow through internal organs may not need to account for blood density changes. The temperature at your extremities can vary significantly from this, depending on the atmospheric temperature, so blood density changes may become a significant property of your particular problem. During problem solution, it is imperative to determine whether or not this possible density change is significant. Furthermore, cavitation is a phenomenon that only occurs in compressible fluids (and this phenomenon can occur behind implanted mechanical heart valves). *Cavitation* occurs when the fluid pressure is below the fluid vapor pressure and air pockets form within the fluid. These pockets can then burst when the pressure rises again.

Viscous fluids can also be characterized by laminar or turbulent flows. Laminar regions are characterized by smooth transitions in flow. Turbulent regions have time-dependent fluctuations in fluid properties such as velocity, acceleration, and viscosity. These fluctuations are random and cannot be quantified easily. In laminar flow, there is generally no mixing of fluid laminae. Because of the random fluctuations in turbulent flows, there is mixing of fluid laminae and, generally, there is a large increase in shear stress because of the transfer of fluid particles (mass and energy is transferred) between laminae. Turbulent flows typically have a non-zero vorticity.

2.9 TWO-PHASE FLOWS

A two-phase flow is a fluid that is composed of either one fluid that is present in two discrete phases (e.g., water and steam in a turbine) or a mixture of fluids (e.g., water and oil). In principal, the two different fluid phases will have a different density, viscosity, among other parameters. To describe a two-phase flow, the same laws that will be developed in subsequent chapters can be applied, but they must be applied separately to each phase. In solid mechanics, if you were to analyze an axial loaded solid bar composed of one material on the outside and a separate material on the inside, this would introduce a statically indeterminate problem. However, by writing the equations independently and then equating variables that must be the same (e.g., displacement), this type of problem could be solved. In two-phase flow you must take a similar approach. In general, you must equate the pressure, velocity, and/or shear force at the phase boundary so that a solution can be reached. The difficulty with this type of fluid problem is whether or not the boundary between the two phases can be defined easily or not. For instance, a water and oil boundary may be easily defined but if you have a fluid that you are bubbling a gas through, there would be multiple boundaries that are transient with time.

One common two-phase (or multiphase) fluid is blood. The first phase to consider is the plasma, which contains water, salts, and proteins. The second phase is composed of the blood cells. These cells can be considered as fluid because they are mostly composed of water surrounded by an elastic membrane (cell membrane), which itself exhibits fluid properties (fluid-mosaic model). Currently, most blood simulations do not account for the different fluid properties that cells add to blood (note that each cell type may have different fluid properties); however, for a more accurate solution, these cells and their fluid properties should be accounted for because they can transfer energy to the fluid (see Chapter 14). Blood cells comprise approximately 40% of the blood volume and, in small blood vessels (see Part 3), the cells can group together, forming a type of plug flow region. In microvessels, pockets of blood that is composed only of plasma followed by pockets of blood which is composed mostly of cellular matter are found. Although we will not develop solutions for general two-phase flows, a few examples will be used to illustrate this phenomenon later in this textbook.

The most applicable two-phase flow modeling approach for biofluid mechanics scenarios is the dispersed particle flow, where each fluid is analyzed separately but the fluids interact over the boundary surface(s). This type of analysis can be applied to blood flow and air flow into the lungs with considerations on particulate matter. Under these scenarios, the following definitions are critical in order to be able to quantify the flow conditions. The number of particles per unit volume is necessary in order to determine the density of the particles and to determine which fluid should be considered as the continuous phase and which should be considered as the dispersed phase. The density of the dispersed phase can be represented as

$$n = \frac{\delta N}{\delta V} \tag{2.41}$$

where δN is the number of particles in a δV volume of interest. This density is limited by a value in which the continuous phase can no longer be considered as continuous. Using similar definitions, we can define the area that the particles occupy versus the area that the continuous phase occupies. The following relationships apply

$$\begin{aligned} A_p &= \frac{\delta V_p}{\delta V} \\ A_c &= \frac{\delta V_c}{\delta V} \\ A_p + A_c &= 1 \end{aligned} \tag{2.42}$$

where the subscript p denotes the particle phase and the subscript c denotes the continuous phase. Two other important definitions can be used to define the overall density (or other fluid parameter) of the two-phase flow and the loading ratio of the continuous phase and the particulate phase. In defining the overall density (and this would apply to any other fluid parameter), we make use of a simple weighted average formulation, which does not take into account any interactions that may occur between the phases. In this case, the overall density can be defined as

$$\rho_t = A_p \rho_p + A_c \rho_c \tag{2.43}$$

and the mass loading ration can be defined as

$$l_r = \frac{\dot{m}_p}{\dot{m}_c} \tag{2.44}$$

where $\dot{m}$ is the mass flow rate of a particular phase. Assuming that the particle density is high enough that collisions occur between particles, the time between collisions must be known, because momentum and energy can be transferred during collisions. Without deriving the entire equation, it is also important to highlight the momentum response time of particles, which is dependent on the fluid velocity and the particle velocity. This response time dictates how quickly changes in the fluid velocity will induce changes in the particle velocity. If we assume low Reynolds flow, we can obtain the following relationship for the momentum response time of the fluid particles

$$t_m = \frac{\rho_p d_p^2}{18\mu} \tag{2.45}$$

where d represents the particle diameter and μ represents the continuous phase viscosity. As briefly highlighted previously, the two different phases must be coupled in some manner in order to solve the two-phase flow scenario. There are two major cases that are currently under investigation in the literature. The first is when the particle density is relatively dilute compared with the total volume of the fluid. In this scenario, one would assume that the collision time is negligible and that no momentum is exchanged between particles. The flow around each particle (if the continuous fluid and the particle have different velocities) or the flow of the particles with the fluid could be analyzed in discrete regions. However, in the cases that the particle density is so large that collisions occur frequently, then momentum transfer between particles and the response time to momentum shifts becomes very important. Under these conditions, it is more likely that the entire flow field would be analyzed because the collision frequency would be so great that the particles typically would not have time to respond to changes in continuous fluid momentum.

There has been a recent significant effort on determining the governing equations and developing simplified governing equations for two-phase flows. However, it has been somewhat difficult since under most two-phase flow conditions, the particle density is not sparse enough to ensure that no collisions occur and is not dense enough to ensure that the momentum response time is smaller than the collision frequency. Under these "normal" conditions, the fluid behavior is much harder to predict and it appears to be a function of the fluid(s)/phase(s) involved.

2.10 CHANGES IN THE FUNDAMENTAL RELATIONSHIPS ON THE MICROSCALE

In the previous sections of this chapter, many relationships between various fluid mechanics parameters were described. These relationships still hold within the microcirculation; however, some of the parameters become more critical and can govern the blood flow throughout the vascular beds, more so than in the macrovasculature. Within the microcirculation, the viscosity of blood must be described by an apparent viscosity (η). This is because blood is no longer a homogenous fluid within the microcirculation and Newtonian flow rarely occurs. In the microcirculation, the blood vessel diameter approaches the size of red blood cells (approximately 8 μm) and this causes a redistribution of the cells and the plasma components into distinct regimes that will be discussed later. It is possible and likely to have a plasma only pocket followed by a cellular pocket

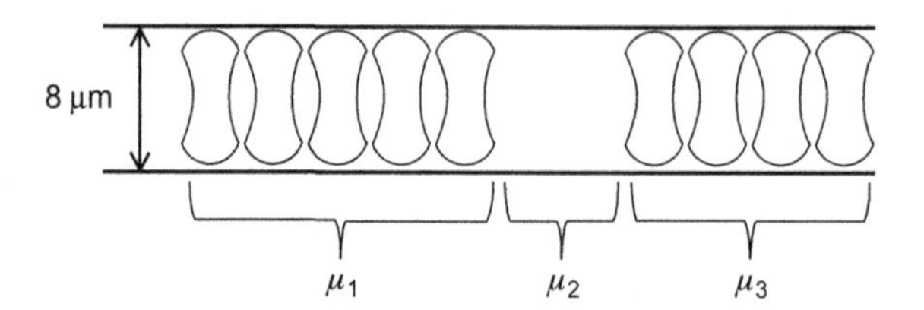

FIGURE 2.22 **The apparent viscosity within the microcirculation would represent the viscosity of the individual fluid and cell pockets, which may have different nominal values for viscosity.** The apparent viscosity could be considered a weighted average of these different viscosities but is realistically just an estimated value. Under most conditions, the apparent viscosity is not the same as the bulk viscosity.

(with negligible plasma) within the capillary and, therefore, different viscosities are associated with these two distinct phases (Figure 2.22). Additionally, it is unlikely that the cellular pockets will be composed of the same number of cells and therefore, each cellular pocket may have its own fluid parameters. The apparent viscosity would represent the overall viscosity of these distinct components. It is not an average, or a weighted average, of these components, but a best estimate of the viscosity.

As shown in Eq. (2.1), conservation of mass within the microcirculation must now include the movement of fluid out of the blood vessel (one cannot define the system of interest as only the blood vessel). In the macrocirculation, we will typically consider that whatever fluid enters the inlet must leave via the outlet (unless there is an increase or decrease in volume/density of the system). Within the microcirculation, this assumption is not valid because we must include mass transfer across the blood vessel wall within the solution. Furthermore, in Chapter 3, we will develop the Navier–Stokes formula, which can be used to determine the characteristics of any flow field. In the microcirculation, this formula simplifies to the Stokes equations because the viscous effects dominate the flow, compared with the inertial effects (recall that the relationship between inertial and viscous effects comprise the Reynolds number and thus flow within the microcirculation can be represented with low Reynolds flow equations). The Stokes equations will be described in detail in Parts 3 and 4. It is also important to note that small changes in blood vessel diameter have large effects on the pressure and velocity distribution within the blood vessel (see Sections 6.4 and 6.5). The take-home message from this discussion is that all of the fundamental relationships described for the macrocirculation will still be accurate within the microcirculation. However, some of the critical parameters that govern flow will change, which can either simplify (Stokes flow) or complicate (fluid exchange) the problem at hand. It is also critical to determine the appropriate value for viscosity within the microcirculation because we can no longer assume that the fluid is homogeneous described by one viscosity.

2.11 FLUID STRUCTURE INTERACTION

The interaction of a flowing fluid and a deformable boundary is defined as a *fluid structure interaction* (FSI). As you can imagine, this should be a major consideration when solving biofluids mechanics problems. Examples can be found in many biological flow conditions. Deformation of the trachea and bronchi can cause changes in air flow. A narrowed airway may cause snoring. All blood vessels are flexible to some degree. The expansion and contraction of the aorta helps to continuously drive blood forward in the systemic circulation, even when the ventricles are experiencing diastole. The cardiac muscle contraction during systole can stop coronary blood flow because the coronary arteries collapse during contraction. Proper muscular compression of leg veins is critical to return venous blood back to the heart. Disease conditions such as abdominal aortic aneurysm, coronary atherosclerosis, and venous valve dysfunction can change blood flow dynamics, which may potentially affect the functions of circulating blood cells and vascular wall endothelial cells. To mathematically describe the changes in structure (such as the trachea or a blood vessel), solid mechanics theories and modeling methods need to be used, to determine stress/strain distribution, deformation, and displacement of the structure. These types of models consider the materials/mechanical properties of the structure (e.g., density, stiffness, dimension, etc.) only. The Navier–Stokes equations, which will be developed in Chapter 3, can still be used to solve for the fluid movement within the fluid domain. However, as the structure deforms or moves, fluid boundaries change accordingly; and for every single change occurring in the structure, a new set of Navier–Stokes equations will need to be solved. However, these equations can only represent the fluid movement. Thus, an FSI model couples the solid mechanics equations as well as the fluid mechanics equations and usually imposes a large computational load, especially when the model is 3D and transient (multiple time steps). Results obtained from such models usually present different pressure-flow relationships compared to solid-wall fluid models. This type of FSI models has been widely used to solve biomedical problems such as wave (pressure and velocity) propagation in flexible arteries or bypass grafts, wall stress calculation in various arteries, tearing and rupturing of diseased blood vessels, movement of natural or artificial heart valves, and even pressure variations in the spinal cord.

Another definition for FSI is the interaction of components within the fluid (such as particles) and the surrounding containing medium. By solving these particle–fluid interaction problems, particle transport within the fluid, air bubble movement, particle collision, droplet formation, etc. can be mathematically described. This could be helpful in solving certain biofluids problems, in which cells, especially platelets and white blood cells (e.g., these cells would be defined as the particles), can interact with each other and adhere to vascular wall endothelial cells. Traditional numerical methods such as finite volume method and finite element method can be used to simulate particle movement and interaction with the flow. The immersed boundary method was reported to be more efficient to simulate flows with suspended particles. In this method, the outer surface of a particle could be defined as a deformable boundary, and the internal force induced by particle deformation and external forces induced by flow can be counted as body forces within the Navier–Stokes equations. However, when particles are considered, discrete phase (or multiphase) models are often used to describe phenomena such as particle movement, collision, binding, or dissociation mathematically. Note that although these methods are typically termed particle–fluid interactions, the "particle" has a definite mass and size with it that would potentially alter the flow field around the cell.

In general, FSI problems are typically too complex to solve analytically, due to the complex 3D geometry change (such as blood vessels), large computational load of mesh/grid regeneration, and the potential need to consider the suspending blood cells as a discrete phase. Just imagine the combined problem, when the blood vessel wall is deformable, which alters the fluid flow properties, which in turn alters the deformation/movement of suspended cells. This is a very complex but very "real" scenario in physiological and pathophysiological flows. A large component of biofluids computational fluid dynamic research is currently concerned with FSI modeling (see Section 14.2). In this textbook, we will not develop the equations needed to solve these types of problems, but we will describe some of the applications associated with FSI and when it would be critical to use this analysis to accurately solve biofluids problems.

2.12 INTRODUCTION TO TURBULENT FLOWS AND THE RELATIONSHIP OF TURBULENCE TO BIOLOGICAL SYSTEMS

We highlighted earlier that viscous flows can be classified as either laminar or turbulent. Laminar flows are characterized by smooth, orderly flow where no mixing exists between different lamina within the flow field. However, in some instances, this orderly flow breaks down and becomes unstable, disorderly and fluctuating. These types of flow can no longer be described by the classical fluid mechanics formula, but instead need to be represented by a more complex statistical analysis that is continually evolving. Turbulent flows begin to develop when the inertial forces far exceed the viscous forces. For internal flows in a smooth tube, flow remains laminar until the Reynolds number exceeds approximately 2000. When the Reynolds number exceeds 2000, the flow begins to transition to turbulence; however, enough viscous forces remain to impede the truly chaotic turbulent flow. Generally, although this varies with the surface properties of the tube, internal flows within a tube are considered transitional if the Reynolds number is between 2000 and 4000. Fully developed turbulent internal flows within a tube can be observed when the Reynolds number exceeds 10^4.

A number of ways exist to describe turbulent flows. We will discuss some of them for the remainder of this section, however, a more thorough discussion of these classifications can be found in some fluid mechanics textbooks. The primary classification for flows that have proceeded into the turbulent regime is time-dependent fluctuations in velocity, pressure, and temperature. Temperature fluctuations are only found if heat transfer is a component of the particular flow scenario (see Section 7.6). These fluctuations appear to be random and thus cannot be described well by mathematical formula. The random fluctuations in velocity and pressure cause mixing of fluid lamina, which is a second classification for turbulent flow. This mixing is much stronger than what is normally found due to the molecular motions associated with the fluid. The mixing occurs in all dimensions causing a rapid diffusion of mass, momentum, and energy throughout the fluid layers. Due to the mixing actions, the heat transfer and frictional forces associated with turbulent flows are significantly greater than with laminar flows. A third classification for turbulent flows is the presence of eddies, which are relatively small fluid sections that mix with each other. The size of eddies varies from the thickness of the shear layer (this is the layer that has significant shearing forces and is commonly discussed in nonfully developed flows and external flows and not internal flows) to the Kolmogorov length. The Kolmogorov length can be defined as

$$L_K = \left(\frac{\nu^3 \delta}{U^3}\right)^{\frac{1}{4}} \tag{2.46}$$

where ν is the kinematic viscosity of the fluid, δ is the shear layer thickness, and U is the mean inflow velocity. Finally, turbulent flows are self-sustaining. Increased frictional forces, due to mixing, will tend to reduce the random fluctuations, since extra energy would be needed to overcome these forces. The energy typically comes from the fluid itself and thus this energy usage will tend to reduce the overall inertia of the fluid. A transitioning or laminar flow which generates localized turbulence will dissipate these fluctuations rather quickly. In contrast, a truly turbulent flow is maintained because new eddies are continually formed, which replace those lost by the viscous frictional dissipation described above.

In the prior description, we have qualitatively described what can be observed during true fully developed turbulent flows. However, one of the goals of this textbook is to provide quantitative measurements for flow conditions. This is an immensely complicated situation because turbulent flows are random. Therefore, it is possible to observe and model turbulent flows but it is essentially unfeasible to predict turbulent flows. In the following discussion, we will highlight some of the important formula, which can be used to describe turbulent flows. To quantitatively analyze turbulent flows, the fluid properties of interest must be subdivided into a time averaged component and a fluctuating component. For example, it would not be possible to define a discrete function for velocity that is valid at locations and times within the flow field. However, through experimental observations of turbulent flow, one can plot the velocity at given spatial locations, as a function of time. Figure 2.23 illustrates that the turbulence and hence the velocity vary with the location within the fluid field. Figure 2.23A illustrates true self-sustaining turbulence that is typically found within the shear layer of an external flow. The average speed at this particular location is 500 cm/s, but the flow velocity varies significantly about this number over time. Figure 2.23B illustrates the fluid velocity at the boundary between the viscous flow regime and the free stream flow regime for an external turbulent flow. We can observe that the turbulent flow occurs for approximately 20% of the time, whereas for the remaining time the flow is largely steady (the average flow velocity was set at 500 cm/s as well for ease of understanding, but may vary from this number in a real turbulent flow scenario). The time associated with turbulent fluctuations can be quantified with an intermittency factor, which defines the percent of time that the flow is turbulent. The intermittency factor would be quantified by

$$\gamma = \frac{\text{turbulent time}}{\text{total time}} \tag{2.47}$$

Intermittency occurs because the viscous frictional force at the boundary between a free stream and the shear region may overcome the internal energy and dissipate the turbulent energy. When this occurs, the flow will appear laminar again. However, since the

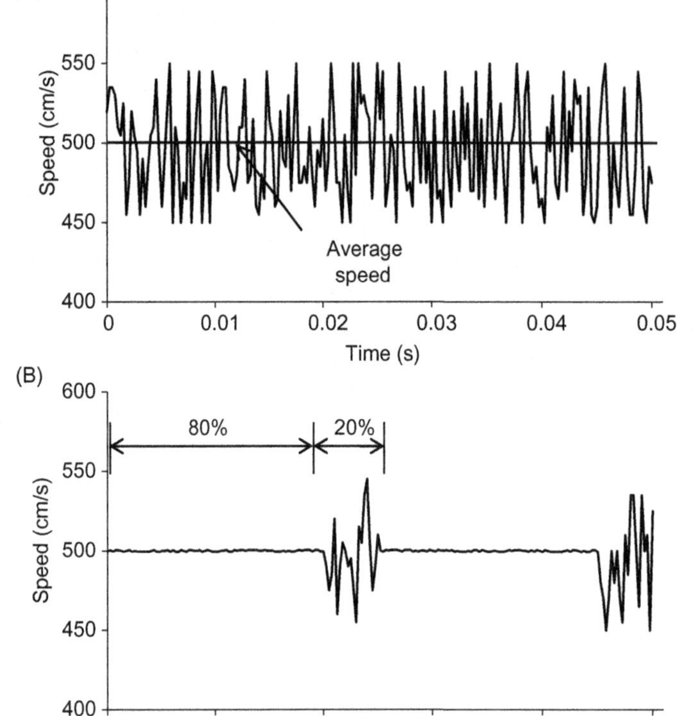

FIGURE 2.23 Turbulent absolute velocity fluctuations in the shear layer (A) and at the boundary (B) of an external flow. Velocity fluctuations, in both instances, have a mean speed of 500 cm/s, however, the turbulence in the shear layer is self-sustaining, whereas the turbulence at the boundary can dissipate and reform. The reformation of the turbulence at the boundary is due to the ambient fluid becoming entrained within the turbulent fluid. However, the viscous forces can overcome this entrainment for approximately 80% of the time.

underlying shear region is characterized by true turbulence, e.g., self-sustaining turbulence, for some times, the laminar boundary flow will become entangled with the turbulent flow. When this occurs, the flow will appear turbulent. If calculated, the intermittency factor for Figure 2.23A would be 1, whereas for Figure 2.23B it would be 0.2.

Since turbulent flows are characterized by time-dependent fluctuations, the fluid variables must be defined in a slightly different manner. Since we have been using the example of velocity, we will continue with velocity here, but note that any of the fluid properties of interest can be used for this type of quantification. The mean fluid velocity at a given location (x_0, y_0, z_0) in the flow field would be defined by

$$\overline{u} = \left(\frac{1}{T}\right) \int_{t_0}^{t_0+T} u dt \tag{2.48}$$

Therefore, the velocity fluctuation (u') at any time at the same location (x_0, y_0, z_0) would be defined as

$$u' = u - \overline{u} \tag{2.49}$$

Unfortunately, the mean fluctuation ($\overline{u'}$) must be equal to zero over a sufficient period of time, therefore, in order to describe the fluctuations, we must work with the root mean square value of the fluctuation terms (the root mean square valve is a common statistical method to describe varying fluctuations that summate to zero). The root mean square velocity fluctuation can be defined as

$$u'_{rms} = \sqrt{\overline{u'^2}} = \left(\frac{1}{T}\right)\int_{t_0}^{t_o+T} u'^2\, dt \tag{2.50}$$

The study of turbulent flows has been and will be studied for many years in the fluid mechanics fields. The question remains, how does turbulent flows relate to biofluid mechanics scenarios? Blood flow through the circulation and air flow through the respiratory system rarely enter the turbulence regime. Reynolds numbers for blood flow in the systemic circulation rarely exceed 100, whereas true fully developed self-sustaining turbulent flows have a Reynolds number in the range of 10,000 (if we assume that the blood flows through a smooth tube). However, there are some conditions when turbulent flows can develop transiently but dissipate nearly instantaneously in circulation. For instance, due to the complex geometry of the ascending aorta and the aortic arch, it is possible for random mixing to occur during the rapid jet deceleration phase during the end of systole. However, if this occurs, the turbulence is not fully developed and is also not self-sustaining. Thus, this is not true turbulence. Additionally, it is commonly to find some mixing under pathological conditions, however, this would dissipate quickly and would not be self-sustaining. Therefore, we would like to caution the reader that while some localized regions under very special conditions may experience turbulence, it is nearly impossible for true turbulence to develop in biological tissues. For completeness, we will discuss some of the mathematical formula that can be used to describe turbulence in Section 5.11, however, we will not make use of these formula in any of the example problems in the text. The reader can skip the turbulence portion of Section 5.11 without any loss of continuity to the textbook.

END OF CHAPTER SUMMARY

2.1 Fluid mechanics is the study of fluids at rest and in motion. A fluid is defined as a material that continuously deforms under a constant load.

2.2 Five relationships are useful in many fluid mechanics problems: kinematic, stresses, conservation, regulating, and constitutive. Five laws also govern most fluid flows

$$\text{Conservation of Mass: } m_{system} = m_{in} - m_{out} - \rho\frac{d(\rho V)}{dt}$$

$$\text{Conservation of Linear Momentum: } \vec{F} = \frac{(m\vec{v})}{dt}$$

$$\text{The First Law of Thermodynamics: } \delta U = \delta Q - \delta w$$

$$\text{The Second Law of Thermodynamics: } \frac{dS}{dt} \geq 0$$

$$\text{Conservation of Angular Momentum: } \vec{L} = \vec{r} \times \vec{p}$$

2.3 The analysis of fluid mechanics problems can be significantly altered depending on the choice of the system of interest and the volume of interest. Mass that is not originally within the system of interest cannot enter the volume of interest during problem solution.

2.4 By assuming a fluid is a continuum, we can suppose that there are no inhomogeneities within the smallest division of fluid that are made. Properties such as viscosity, density, and temperature, among others, must be constant throughout the entire fluid.

2.5 It is useful to use the stress tensor $\sigma = \begin{bmatrix} \sigma_{xx} & \tau_{xy} & \tau_{xz} \\ \tau_{yx} & \sigma_{yy} & \tau_{yz} \\ \tau_{zx} & \tau_{zy} & \sigma_{zz} \end{bmatrix}$ to define the various stresses on fluid elements. Normal stresses are defined as $\sigma_n \rightarrow \lim_{\delta \vec{A}_n \rightarrow 0} \frac{\delta \vec{F}_n}{\delta \vec{A}_n}$, whereas shear stresses are defined as $\tau_n = \lim_{\delta \vec{A}_n \rightarrow 0} \frac{\delta \vec{F}_t}{\delta \vec{A}_n}$. Only six of these stresses are independent of each other to conserve momentum throughout the fluid element. A useful relationship for the hydrostatic pressure as a function of the normal stresses is $p_{\text{hydrostatic}} = -\frac{1}{3}(\sigma_{xx} + \sigma_{yy} + \sigma_{zz})$.

2.6 A fluid can undergo four general motions. These are useful in defining the position of fluid packets, the velocity, and the acceleration of the fluid.

$$\vec{x}(t) = u_x(t)\vec{i} + u_y(t)\vec{j} + u_z(t)\vec{k}$$

$$\vec{v}(t) = \frac{d\vec{x}(t)}{dt} = \frac{du_x(t)}{dt}\vec{i} + \frac{du_y(t)}{dt}\vec{j} + \frac{du_z(t)}{dt}\vec{k} = v_x(t)\vec{i} + v_y(t)\vec{j} + v_z(t)\vec{k}$$

$$\vec{a}(t) = \frac{d\vec{v}(t)}{dt} = \frac{d^2\vec{x}(t)}{dt^2} = \frac{dv_x(t)}{dt}\vec{i} + \frac{dv_y(t)}{dt}\vec{j} + \frac{dv_z(t)}{dt}\vec{k}$$

$$= \frac{d^2u_x(t)}{dt^2}\vec{i} + \frac{d^2u_y(t)}{dt^2}\vec{j} + \frac{d^2u_z(t)}{dt^2}\vec{k} = a_x(t)\vec{i} + a_y(t)\vec{j} + a_z(t)\vec{k} \text{ or}$$

$$\vec{a} = \frac{d\vec{v}}{dt} = \frac{\partial \vec{v}}{\partial t}\frac{dt}{dt} + \frac{\partial \vec{v}}{\partial x}\frac{dx}{dt} + \frac{\partial \vec{v}}{\partial y}\frac{dy}{dt} + \frac{\partial \vec{v}}{\partial z}\frac{dz}{dt} = \frac{\partial \vec{v}}{\partial t} + v_x\frac{\partial \vec{v}}{\partial x} + v_y\frac{\partial \vec{v}}{\partial y} + v_z\frac{\partial \vec{v}}{\partial z}$$

Temporal and spatial driving forces can arise within the vascular system and appear in the substantial derivative equation for acceleration. The angular velocity of a fluid can be defined as

$$\vec{\omega} = \frac{1}{2}\left[\left(\frac{\partial v_z}{\partial y} - \frac{\partial v_y}{\partial z}\right)\vec{i} + \left(\frac{\partial v_x}{\partial z} - \frac{\partial v_z}{\partial x}\right)\vec{j} + \left(\frac{\partial v_y}{\partial x} - \frac{\partial v_x}{\partial y}\right)\vec{k}\right] = \frac{1}{2}\text{curl}\vec{v}$$

The deformation tensor takes the form of

$$\mathbf{d} = \frac{1}{2}\begin{bmatrix} 2\frac{\partial v_x}{\partial x} & \frac{\partial v_y}{\partial x} + \frac{\partial v_x}{\partial y} & \frac{\partial v_z}{\partial x} + \frac{\partial v_x}{\partial z} \\ \frac{\partial v_y}{\partial x} + \frac{\partial v_x}{\partial y} & 2\frac{\partial v_y}{\partial y} & \frac{\partial v_z}{\partial y} + \frac{\partial v_y}{\partial z} \\ \frac{\partial v_z}{\partial x} + \frac{\partial v_x}{\partial z} & \frac{\partial v_z}{\partial y} + \frac{\partial v_y}{\partial z} & 2\frac{\partial v_z}{\partial z} \end{bmatrix}$$

2.7 Viscosity is a quantity that relates the shear rate of a fluid to the shear stress. This relationship takes the general form of $\tau_{xy} = \mu\left(\frac{\partial v_x}{\partial y} + \frac{\partial v_y}{\partial x}\right)$.

2.8 The definition of a fluid as Newtonian depends on whether or not the viscosity is constant at various shear rates. Newtonian fluids are classified with constant viscosities for a range of shear rates, whereas non-Newtonian fluids have a nonconstant viscosity. For most biofluids applications, we can assume that the fluid behaves as a Newtonian fluid. An inviscid fluid has no viscosity, and this is useful in some formula derivations or simple assumptions of fluid flow.

2.9 A two-phase flow consists of a fluid in both a gas and liquid phase or two fluids (with different viscosities) within the same flow conditions. Blood is a two-phase flow because the cellular matter may have a different viscosity than the plasma component. To analyze two-phase flow conditions, we must consider some form of coupling between the different phases and this is normally a function of the particle seeding density within the continuous fluid.

2.10 Many fundamental fluid relationships change on the microscale. One of the most applicable for biofluids problems is the viscosity, which is described by an apparent viscosity within the microcirculation.

2.11 FSI modeling is important if the fluid can affect and cause a deformation on the flow boundary. This is important in the cardiovascular system, in which the blood vessel wall is deformable and cellular matter can interact with the wall.

2.12 Turbulent flows are qualitatively described by time varying, random fluctuations that cause mixing of the fluid lamina. Turbulent flows must be self-sustaining but can dissipate on the boundary between the turbulent flows and the nonturbulent flows. Turbulent flows must be characterized by time-dependent fluctuations

$$\overline{u} = \left(\frac{1}{T}\right)\int_{t_0}^{t_0+T} u dt$$

$$u' = u - \overline{u}$$

Additionally, statistical methods are needed to qualitatively describe the flow

$$u'_{rms} = \sqrt{\overline{u'^2}} = \left(\frac{1}{T}\right)\int_{t_0}^{t_o+T} u'^2 dt$$

In general, flows in biological systems never proceed into the turbulent regime, however, under very special circumstances, in very specific locations, turbulent flows may arise but they dissipate very quickly.

HOMEWORK PROBLEMS

2.1 For a particle with a velocity distribution of $v_x = 5x$, $v_y = -5y$, $v_z = 0$, determine the particles acceleration vector. Also, determine whether this velocity profile has a local and/or convective acceleration.

2.2 Consider a velocity vector $v = (xt^2 - y)\vec{i} + (xt - y^2)\vec{j}$. (i) Determine whether this flow is steady (*hint*: no changes with time). (ii) Determine whether this is an incompressible flow (*hint*: check if $\nabla \bullet v = 0$).

2.3 Given the velocity $v = (2x - y)\vec{i} + (x - 2y)\vec{j}$, determine whether it is irrotational.

2.4 The velocity vector for a steady incompressible flow in the xy plane is given by $\vec{v} = \frac{4}{x}\vec{i} + \frac{4y}{x^2}\vec{j}$, where the coordinates are measured in centimeters. Determine the time it takes for a particle to move from $x = 1$ cm to $x = 4$ cm for a particle that passes through the point $(x, y) = (1, 4)$.

2.5 Flow in a two-dimensional channel of width W has a velocity profile defined by

$$v_x(y) = k\left[\left(\frac{W}{2}\right)^2 - y^2\right]$$

where $y = 0$ is located at the center of the channel. Sketch the velocity distribution (with $k = 1$ and $W = 10$) and find the shear stress/unit width of the channel at the wall.

2.6 A velocity field is given by $v(x, y, z) = 20xy\vec{i} - 10y^2\vec{j}$. Calculate the acceleration, the angular velocity, and the vorticity vector at the point (−1,1,1), where the units of the velocity equation are millimeters per second.

2.7 Considering one-dimensional fluid (density, ρ; viscosity, μ) flow in a tube with an inlet pressure of p_i and outlet pressure of p_o and tube radius of r and length l, the density of the fluid can be represented as ρ. Express the wall shear stress as a function of these variables (*hint*: balance Newton's second law of motion).

2.8 Data obtained from one experiment are presented in Table 2.3. Plot the data and determine whether the fluid is pseudoplastic, Newtonian, or dilatant. Approximate the viscosity (or the apparent viscosity for this fluid).

TABLE 2.3 Shear Stress versus Shear Rate Data for Homework Problem 2.8

Shear stress (dyne/cm^2)	5	9	23	48	95
Shear rate (s^{-1})	0.1	1	10	100	1000

2.9 Consider a red blood cell that originates from the origin of our xy-coordinate system. The velocity of the fluid is unsteady and is described by

$$v = \begin{cases} v_x = 1\ cm/s & v_y = 0.5\ cm/s & 0\ \text{s} \le t \le 2\ \text{s} \\ v_x = 0.25\ cm/s & v_y = 1\ cm/s & 2\ \text{s} < t \le 5\ \text{s} \end{cases}$$

Plot the location of this red blood cell at time 0, 1, 2, 3, 4, and 5 s (enters from the origin at time 0). Also plot the location of the red blood cells that enter (from the origin) at time 1, 2, 3, and 4 s.

2.10 A flow is described by the velocity field

$$v = \left(\frac{y}{s}\right)\vec{i} + 0.75\frac{m}{s^2}t\vec{j}$$

At $t = 2$ s, what are the coordinates of the particles that passed through the point (1,5) at $t = 0$ s. What are the coordinates of the same particle at $t = 3$ s?

2.11 The velocity distribution for laminar flow between two parallel plates can be represented as

$$\frac{u}{u_{max}} = 1 - \left(\frac{2y}{h}\right)^2$$

where h is the separation distance between the two flat plates and the origin is located halfway between the plates. Consider the flow of blood at 37°C ($\mu = 3.5$ cP) with maximum velocity of 25 cm/s and a separation distance of 10 mm. Calculate the force on a 0.25 m^2 section of the lower plate.

2.12 A biofluid is flowing down an inclined plane. The velocity profile of this fluid can be described by

$$u = \frac{\rho g}{\mu}\left(hy - \frac{y^2}{2}\right)\sin\theta$$

if the coordinate axis is aligned with the inclined plane. Determine a function for the shear stress along this fluid. Plot the velocity profile and shear stress profile, if the fluid density is 900 kg/m^3 and viscosity is 2.8 cP. The fluid thickness h is equal to 10 mm and the plane is inclined at an angle of 40° ().

2.13 A block with mass M_1 is in contact with a thin fluid layer of thickness h. The block is connected to a second mass M_2 via a cable that goes over an ideal pulley. Determine a relationship for the velocity of the block when the second mass is released from rest (Figure 2.24).

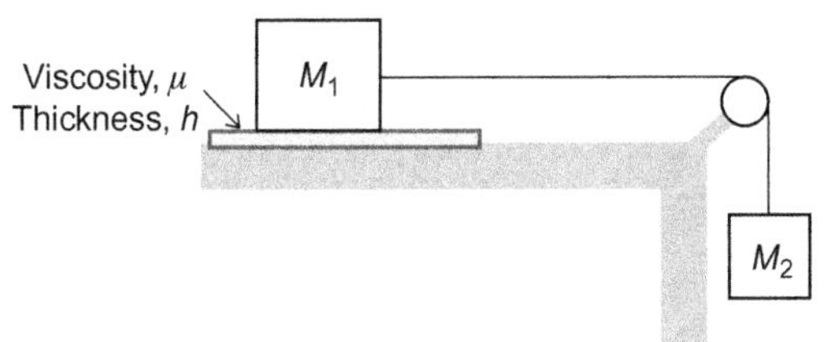

FIGURE 2.24 **Figure for Homework Problem 2.13.**

2.14 For the following velocity field:

$$v = (xt^2 + 2y - z^2)\vec{i} + (xt - 2yt)\vec{j} + (3xt + z)\vec{k}$$

calculate the shear rate tensor, the vorticity, the hydrostatic pressure (use a viscosity of μ), and the acceleration of the fluid.

2.15 A velocity field is given by

$$v = 0.2s^{-1}x\vec{i} - 0.35(mms)^{-1}y^2\vec{j}$$

where x and y are given in millimeters. Determine the velocity of a particle at point (2,4) and the position of the particle at $t = 6$ s of the particle that is located at (2,4) at $t = 0$ s.

2.16 A fluid is placed in the area between two parallel plates. The upper plate is movable and connected to a weight by a cable. Calculate the velocity of the plate for (i) the fluid between the plates is water; (ii) the fluid between the plates is blood (assumed to have no yield stress); and (iii) the fluid between the plates is blood with a yield stress of 50 mPa. For all cases take the mass to be 10 g, the height between the plates to be 10 mm, and the contact area to be 0.5 m^2 (Figure 2.25).

FIGURE 2.25 **Figure for Homework Problem 2.16.**

2.17 What type of fluid can be classified by the following shear stress–strain rate data? Plot the data and classify. Determine the viscosity of the fluid (Table 2.4).

TABLE 2.4 Shear Stress versus Shear Rate Data for Homework Problem 2.17

Shear stress (N/m^2)	0.4	0.82	2.50	5.44	8.80
Shear rate (s^{-1})	0	10	50	120	200

2.18 A plate is moving upwards through a channel filled with water as shown. The plate velocity is 10 cm/s, and the spacing is 50 mm on each side of the plate. The force required to move the plate is equal to 5 N. Determine the surface area of the plate (Figure 2.26).

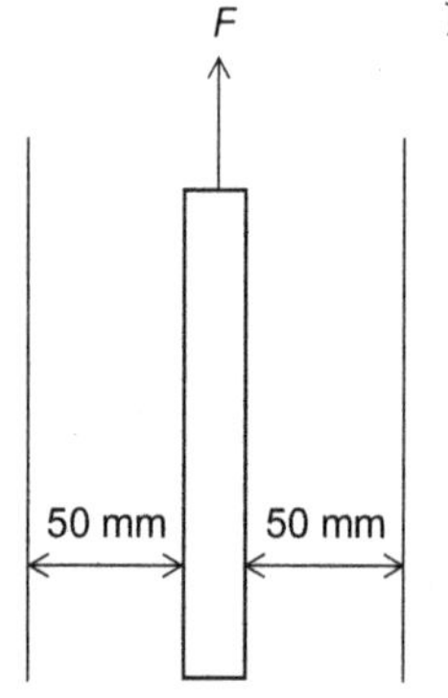

FIGURE 2.26 **Figure for Homework Problem 2.18/2.19.**

2.19 Use the same figure for Problem 2.18, except that the distance on the right-hand side is increased to 100 mm. The fluid on the right side is blood ($\mu = 3.5$ cP). Determine the viscosity of the fluid on the left side. The plate is moving with a constant velocity.

2.20 A conical shaft is turning within the conical bearing as shown below. The gap between the shaft and the bearing is filled with a fluid with a viscosity of 0.05 Ns/m^2. Determine a symbolic expression for the wall shear stress that is acting on the shaft and determine the wall shear stress with using the values given in the figure (Figure 2.27).

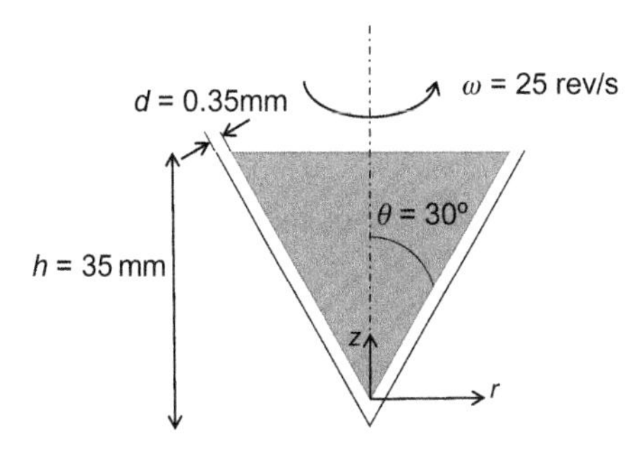

FIGURE 2.27 **Figure for Homework Problem 2.20.**

2.21 Two data points on the rheological diagram of a biofluid are provided. Determine the consistency index and the flow behavior index and the strain rate if the shear stress is increased to 3 dynes/cm^2. Assume that this is a two-dimensional flow.

$$\frac{dV}{dy} = 15\ rad/s, \quad \tau = 0.868\ dynes/cm^2$$

$$\frac{dV}{dy} = 30\ rad/s, \quad \tau = 0.355\ dynes/cm^2$$

2.22 Given that the volumetric flow rate for a fluid within a circular cross-section tube can be represented by

$$Q = \frac{\pi r^4 \Delta P}{8\mu L}$$

where r is the tube radius, μ is the fluid viscosity, P is the pressure drop across the tube, and L is the tube length, calculate the pressure drop across a tube of length 1 m and diameter of 23 mm. The fluid is blood ($\mu = 3.5$ cP) and has a volumetric flow rate of 4.5 L/min. Assuming the same conditions, what would the required pressure drop be for water ($\mu = 1$ cP) and chocolate syrup ($\mu = 15{,}000$ cP).

2.23 At a stenosis, the energy in the flowing fluid is skewed from potential energy to kinetic energy. Why? What effect may this have on locations downstream of the stenosis.

2.24 The no-slip boundary wall condition for Newtonian fluids has been well established. No known exceptions to this condition exist for Newtonian fluids. Blood is a non-Newtonian fluid. What should the boundary condition for blood and a solid surface be.

References

[1] Y. Alemu, D. Bluestein, Flow-induced platelet activation and damage accumulation in a mechanical heart valve: numerical studies, Artif. Organs. 31 (2007) 677.

[2] K.B. Chandran, H. Kim, Computational mitral valve evaluation and potential clinical applications, Ann. Biomed. Eng. (2014). Available from: http://dx.doi.org/10.1007/s10439-014-1094-5.

[3] Z.G. Feng, E.E. Michaelides, The immersed boundary-lattice Boltzmann method for solving fluid–particles interaction problems, J. Comput. Phys. 195 (2004) 602.

[4] A.L. Fogelson, C.S. Peskin, A Fast, Numerical method for solving the 3-dimensional stokes equations in the presence of suspended particles, J. Comput. Phys. 79 (1988) 50.

[5] M. Hasan, D.A. Rubenstein, W. Yin, Effects of cyclic motion on coronary blood flow, J. Biomech. Eng 135 (2013) 121002.
[6] X. Huang, C. Yang, G. Canton, M. Ferguson, C. Yuan, D. Tang, Quantifying effect of intraplaque hemorrhage on critical plaque wall stress in human atherosclerotic plaques using three-dimensional fluid–structure interaction models, J. Biomech. Eng. 134 (2012) 121004.
[7] J.B. Grotberg, O.E. Jensen, Biofluid mechanics in flexible tubes, Annu. Rev. Fluid Mech. 36 (2004) 121.
[8] R.D. Kamm, Cellular fluid mechanics, Annu. Rev. Fluid Mech. 34 (2002) 211.
[9] A.W. Khir, A. O'Brien, J.S. Gibbs, K.H. Parker, Determination of wave speed and wave separation in the arteries, J. Biomech. 34 (2001) 1145.
[10] B. Munson, D. Young, T Okiishi, W.W. Huebsch, Fundamentals of Fluid Mechanics, sixth ed., Wiley, Hoboken, NJ, 2009.
[11] D.A. Peter, Y. Alemu, M. Xenos, O. Weisberg, I. Avneri, M. Eshkol, et al., Fluid structure interaction with contact surface methodology for evaluation of endovascular carotid implants for drug-resistant hypertension treatment, J. Biomech. Eng. 134 (2012) 041001.
[12] P.J. Pritchard, Fox and McDonald's Introduction to Fluid Mechanics, eighth ed., Wiley, Hoboken, NJ, 2011.
[13] A.S. Tam, M.C. Sapp, M.R. Roach, The effect of tear depth on the propagation of aortic dissections in isolated porcine thoracic aorta, J. Biomech. 31 (1998) 673.
[14] V. Streeter, K. Bedford, E. Wylie, Fluid Mechanics, ninth ed., Mc-Graw Hill, New York, NY, 1998.
[15] D.M. Wang, J.M. Tarbell, Nonlinear analysis of oscillatory flow, with a nonzero mean, in an elastic tube (Artery), J. Biomech. Eng T ASME. 117 (1995) 127.
[16] F. White, Fluid Mechanics, seventh ed., Mc-Graw Hill, New York, NY, 2011.
[17] N.B. Wood, Aspects of fluid dynamics applied to the larger arteries, J. Theor. Biol. 199 (1999) 137–161.
[18] M. Xenos, Y. Alemu, D. Zamfir, S. Einav, J.J. Ricotta, N Labropoulos, et al., The effect of angulation in abdominal aortic aneurysms: fluid–structure interaction simulations of idealized geometries, Med. Bio. Eng. Comp. 48 (2010) 1175.

CHAPTER

3

Conservation Laws

LEARNING OUTCOMES

1. Develop equations that govern pressure variation within a static fluid
2. Determine the buoyancy forces that act on objects immersed within a fluid
3. Develop a general relationship for the time rate of change of any fluid system property
4. Apply the generalized formula for the time rate of change of a system property to the conservation of mass, conservation of momentum, and conservation of energy of a fluid system
5. Describe the conservation of momentum principle with acceleration
6. Derive and apply the Navier–Stokes equations
7. Explain the Bernoulli principle and the assumptions inherent in this principle

3.1 FLUID STATICS EQUATIONS

Fluid statics problems deal with fluids that either are at rest or are only undergoing constant velocity rigid body motions. This implies that the fluid is only subjected to normal stresses because by definition a fluid will continually deform under the application of a shear stress. Recall that shear stress acting on a fluid element would induce angular deformations within the fluid (see Figure 2.15) and therefore particular regions of the fluid element would experience accelerations. Another way to think about static fluid problems is that the relative position of all fluid elements remains the same after loading and the relative position of all components that make the larger fluid elements remains the same. Therefore, each fluid elements would only experience pure translation or pure rotation about the same axes of rotation. These types of problems fall under the class of hydrostatics and the analysis methods for these problems are typically simpler than fluid dynamics problems. Newton's second law of motion, simplified to the sum of the forces acting on

 © 2015 Elsevier Inc. All rights reserved.

the fluid is equal to zero ($\sum \vec{F} = 0$), is the primary governing equation that is used to solve these types of problems.

Although fluid statics problems make the assumption that the fluid elements are not undergoing deformation, it is still possible to gain important information and insight from this type of analysis on fluids that are undergoing some form of dynamic motion. Normal forces can be transmitted by fluids, and these forces can then be applied to devices within a biological system. For instance, by inserting a catheter into a patient, there will be some hydrostatic forces that the blood transmits onto the device. While moving the catheter throughout the cardiovascular system, the hydrostatic forces change, and it may be critical to determine these forces or the total force acting on the device. Under this condition, it is typical that the pressure forces are more critical than the shearing forces of the fluid on the catheter. Imagine undergoing balloon angioplasty (in which a small balloon attached to the end of a catheter is inflated within the cardiovascular system) and not knowing the hydrostatic pressure that is being applied to the end of the catheter from the fluid. If the physician does not overcome this pressure, the balloon will not inflate and the procedure will not be completed to remedy the patient. Therefore, it is critical to understand these principles (among others) to conduct balloon angioplasty. Hydrostatic pressure arises due to the weight of the fluid above the section of interest and the surrounding atmospheric pressure or other pressure that is induced by the containing surface (e.g., the elastic recoil of the blood vessels can locally increase the hydrostatic pressure). Therefore, the hydrostatic pressure is different at various locations throughout the body. When a person is standing upright, the hydrostatic pressure at the top of the head is lower than that at the heart, and the hydrostatic pressure at the feet is greater than that of the heart. One way to remember this principle is when you have been standing in the same position for a long time, without moving your legs, blood pools in your lower extremities. Typically, this would eventually lead to a "cramping" feeling followed by a "pins and needles" feeling when blood begins to move again through the vascular system. The reason that the blood pools in your lower limbs is that the blood in the legs cannot overcome the hydrostatic pressure to return back to the heart. Also, after sleeping, if you stand up too fast, the heart cannot overcome the new hydrostatic pressure difference and you may get light headed (while sleeping supine the hydrostatic pressure is effectively the same at the heart, head, and toes). When we discuss venous return and the heart mechanics, we will show how the body can compensate for rapid changes in pressure difference. Another way to recall this phenomenon is that after donating blood, the nurse will typically tell you to raise your arm. Why is this? This increases the pressure difference between your heart and your arm and will minimize the blood loss while a clot is forming at the venipuncture location. In this case, blood would have a hard time overcoming the new hydrostatic pressure gradient to enter the arterial circulation of the arm and the blood within the venous circulation of the arm will return to the heart.

As stated in the previous chapter, the primary quantity of interest within fluid statics problems is the pressure field throughout the fluid. Here we will develop the equations used in fluid statics analysis. To accomplish this, Newton's second law of motion will be applied to a differential element of fluid (Figure 3.1). Recall that Newton's second law of motion is the sum of all of the forces acting on an element (body forces and surface forces) is equal to the elements mass multiplied by the element's acceleration (if density is constant).

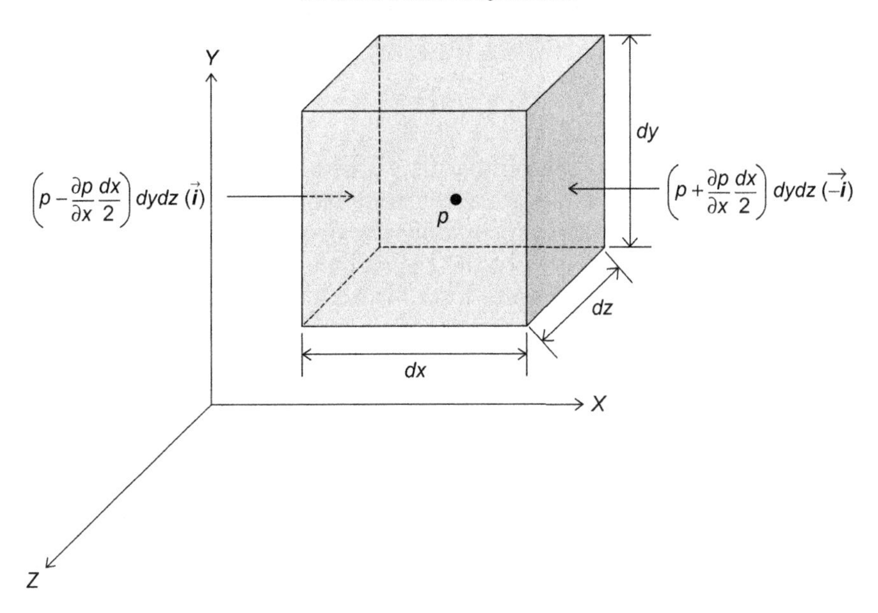

FIGURE 3.1 *X*-direction pressure forces that act on a differential fluid element. The same pressure forces can be derived for the other Cartesian directions.

We will assume here that the only body force acting on the element is due to gravity. In most biofluid mechanics problems in this textbook, this will be the only body force that is considered. However, be cautioned that other body forces can be applied via a magnetic field (blood flow of a patient within an MRI) or by an electric field, among other body forces. The mass of the differential element, dm, is equal to the fluid density multiplied by the differential volume of the element ($dm = \rho dV = \rho dxdydz$, in Cartesian coordinates; for other coordinate systems the analysis is similar, note that $V \equiv$ volume and $v \equiv$ velocity, for this textbook). Therefore, the force due to gravity becomes

$$d\vec{F}_b = \vec{g}dm = \vec{g}\rho dxdydz \tag{3.1}$$

where $\vec{g}$ is the gravitational constant.

Because the fluid is under static flow conditions, there are no shear forces applied to the fluid element. Therefore, the only surface force acting on the element is the pressure force. As we will see, pressure varies with position throughout the entire fluid, but the total pressure acting on a differential element can be represented by the summation of the pressure acting on each face of the differential element. Let us define the pressure at the center of the differential element to be p (Figure 3.1). The pressure at each face would be equal to p plus or minus the particular directional pressure gradient multiplied by the distance between the center of the element and the face. For instance, in the x-direction, the pressure on the right face of the differential element in the current orientation, shown in Figure 3.1, would be

$$p + \frac{\partial p}{\partial x}\frac{dx}{2}$$

acting in the negative x-direction, whereas the pressure on the left face would be

$$p - \frac{\partial p}{\partial x}\frac{dx}{2}$$

acting in the positive x-direction. Remember that pressure has the same unit as stress. In order to determine the force that the pressure exerts on each face, one must multiply the stress by the area over which it acts (again, only the two forces that act in the x-direction are shown in Figure 3.1). For the case of the x-direction pressures, the area can be represented as $dydz$. In this figure, the pressure force is also multiplied by a unit vector indicating the direction that the force acts within. Remember, for pressure there is a sign convention; a positive pressure is a compressive normal stress, and these are the forces that are shown on the differential element in Figure 3.1. To balance forces, the differential element would produce an equal and opposite force on the adjacent fluid element.

The four remaining pressure forces (that act in the y- and z-directions) can be obtained through a similar analysis method. By summing all of these six forces, the total surface force acting on the fluid element can be obtained. It would be represented as

$$\begin{aligned} d\vec{F}_s = &\left(p - \frac{\partial p}{\partial x}\frac{dx}{2}\right)(dydz)(\vec{i}) + \left(p + \frac{\partial p}{\partial x}\frac{dx}{2}\right)(dydz)(-\vec{i}) + \left(p - \frac{\partial p}{\partial y}\frac{dy}{2}\right)(dxdz)(\vec{j}) \\ &+ \left(p + \frac{\partial p}{\partial y}\frac{dy}{2}\right)(dxdz)(-\vec{j}) + \left(p - \frac{\partial p}{\partial z}\frac{dz}{2}\right)(dxdy)(\vec{k}) + \left(p + \frac{\partial p}{\partial z}\frac{dz}{2}\right)(dxdy)(-\vec{k}) \end{aligned} \tag{3.2}$$

Combining terms in the previous equations yields

$$\begin{aligned} d\vec{F}_s = &\left(p - \frac{\partial p}{\partial x}\frac{dx}{2}\right)(dydz)(\vec{i}) + \left(-p - \frac{\partial p}{\partial x}\frac{dx}{2}\right)(dydz)(\vec{i}) + \left(p - \frac{\partial p}{\partial y}\frac{dy}{2}\right)(dxdz)(\vec{j}) \\ &+ \left(-p - \frac{\partial p}{\partial y}\frac{dy}{2}\right)(dxdz)(\vec{j}) + \left(p - \frac{\partial p}{\partial z}\frac{dz}{2}\right)(dxdy)(\vec{k}) + \left(-p - \frac{\partial p}{\partial z}\frac{dz}{2}\right)(dxdy)(\vec{k}) \\ = &-2\left(\frac{\partial p}{\partial x}\frac{dx}{2}\right)(dydz)(\vec{i}) - 2\left(\frac{\partial p}{\partial y}\frac{dy}{2}\right)(dxdz)(\vec{j}) - 2\left(\frac{\partial p}{\partial z}\frac{dz}{2}\right)(dxdy)(\vec{k}) \\ = &-\left(\frac{\partial p}{\partial x}\vec{i} + \frac{\partial p}{\partial y}\vec{j} + \frac{\partial p}{\partial z}\vec{k}\right)dxdydz \end{aligned} \tag{3.3}$$

From a previous class in calculus, the final term in the parentheses (right-hand side of the equation) of Eq. (3.3) is the gradient (denoted as grad or ∇) of the pressure force in Cartesian coordinates. Therefore, the total surface forces acting on a differential fluid element can be mathematically represented as

$$d\vec{F}_s = -\nabla\vec{p}dxdydz \tag{3.4}$$

Returning to Newton's second law, the sum of the forces acting on a differential fluid element can then be represented as

$$d\vec{F} = d\vec{F}_b + d\vec{F}_s = \vec{g}\rho dxdydz - \nabla\vec{p}dxdydz = (\vec{g}\rho - \nabla\vec{p})dxdydz \tag{3.5}$$

If one divides the summation of the force acting on a differential element of fluid by the unit volume (Eq. (3.5)), then one gets a relationship that holds for fluid particles, and is in terms of density

$$\frac{d\vec{F}}{dxdydz} = \frac{d\vec{F}}{dV} = \vec{g}\rho - \nabla\vec{p} \tag{3.6}$$

For a static fluid flow case ($\vec{a} = 0$), Newton's second law of motion for a fluid element with a finite volume simplifies to

$$\frac{d\vec{F}}{dV} = \vec{g}\rho - \nabla\vec{p} = \rho\vec{a} = 0 \tag{3.7}$$

where density multiplied by the differential volume has been substituted for the differential mass, since the fluid density is typically known as compared with a differential mass. The significance of this equation is that the gravitational force must be balanced by the pressure force at each individual point within the fluid. Remember that this is only true if the fluid does not experience acceleration and the only body force is due to the gravitational acceleration. In terms of the vector component equations, which must independently summate to zero in each direction, for fluid static problems, Eq. (3.7) can be represented as

$$\begin{aligned} -\frac{\partial p}{\partial x} + \rho g_x = 0 \\ -\frac{\partial p}{\partial y} + \rho g_y = 0 \\ -\frac{\partial p}{\partial z} + \rho g_z = 0 \end{aligned} \tag{3.8}$$

It is conventional to choose a coordinate system in a particular way, so that the gravitational force acts in only one direction. For sake of a convention, we will typically align the gravitational force with the z-direction of the Cartesian coordinate system. With this definition, Eq. (3.8) simplifies to

$$\begin{aligned} \frac{\partial p}{\partial x} = 0 \\ \frac{\partial p}{\partial y} = 0 \\ \frac{\partial p}{\partial z} = -\rho g_z = \rho g \end{aligned} \tag{3.9}$$

because $g_x = g_y = 0$ and $g_z = -g$. Using these assumptions, the pressure is only a function of one coordinate variable (z) and it is therefore independent of the other two coordinate variables (x and y). Note that pressure can act in the x/y directions; however, the pressure must be constant in those directions. The assumptions made in this analysis are that the fluid is under static flow conditions (has no acceleration term), the only body force is the gravitational force, and that gravity is only aligned with the z-axis (using the Cartesian coordinate system). Combined, this allows the use of a total derivative instead of a partial derivative in Eq. (3.9). Therefore,

$$\frac{dp}{dz} = -\rho g_z \tag{3.10}$$

Equation (3.10) relates the pressure within a fluid to the vertical height of the fluid, if the assumptions made are valid or are within a reasonable estimate of the flow conditions. Equation (3.10) can be integrated to calculate the pressure distribution throughout a static fluid, if the correct boundary conditions are applied. In general, you would need to know if the fluid's density or if gravity varies with changes in vertical distances.

Example

Calculate the static fluid pressure in the cranium at the end of systole and at the end of diastole. Assume that the cranium is 30 cm above the aortic valve and that the pressure at the end of systole and end of diastole is 120 mmHg and 80 mmHg, respectively, at the aortic valve (Figure 3.2). The density of blood is 1050 kg/m^3.

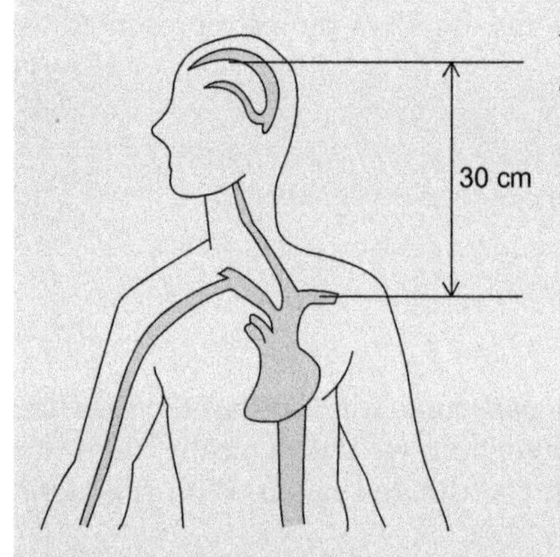

FIGURE 3.2 Difference in fluid static pressure between the aortic valve and the cranium based on height.

Solution

$$dp = -\rho g_z dz$$

$$\int_{p_0}^{p_1} dp = -\rho g_z \int_{z_0}^{z_1} dz$$

Since the fluid density and gravity is not a function of height, these variables can be pulled out of the integral.

$$p\Big|_{p_0}^{p_1} = -\rho g_z z\Big|_{z_0}^{z_1}$$

$$p_1 - p_0 = -\rho g_z (z_1 - z_0)$$

$$p_1 = p_0 - \rho g_z (z_1 - z_0)$$

End of Systole

$$p_1 = p_0 - \rho g_z(z_1 - z_0)$$

$$p_1 = 120\ mmHg - (1050\ kg/m^3)(9.81\ m/s^2)(30\ cm - 0\ cm)\left(\frac{1\ m}{100\ cm}\right)\left(\frac{1\ mmHg}{133.32\ Pa}\right)$$

$$p_1 = 96.83\ mmHg$$

End of Diastole

$$p_1 = p_0 - \rho g_z(z_1 - z_0)$$

$$p_1 = 80\ mmHg - (1050\ kg/m^3)(9.81\ m/s^2)(30\ cm - 0\ cm)\left(\frac{1\ m}{100\ cm}\right)\left(\frac{1\ mmHg}{133.32\ Pa}\right)$$

$$p_1 = 56.83\ mmHg$$

Notice that in the previous example, the absolute change in pressure remains the same at the end of systole and at the end of diastole ($\Delta p = -23.17$ mmHg). This change in pressure is constant because we assumed that the height and the blood density did not change. Again, this is only valid if the three assumptions we made to derive this formula can be applied to the particular example.

The previous example brings us to an important distinction in fluid statics situations. All pressures must be referenced to a specific reference value. For instance, we may have chosen to call the pressure at the aortic valve 0 mmHg, and this choice would make the cranium pressure exactly -23.17 mmHg for each of the two cases in the previous example. This is true even when the aortic pressure is variable, because the aortic pressure is our reference and is defined as 0 mmHg at all times. There are two types of pressures that we can discuss. The first is the absolute pressure, and the second is the gauge pressure. Absolute pressure is referenced to a vacuum; this would also be the absolute (or exact) pressure of the system at your particular point of interest. Gauge pressure is the pressure of the system related to some other reference pressure, which is conventionally local atmospheric pressure (which is variable). Therefore, gauge pressure is actually a pressure difference and is not the actual pressure of the system. In our example above, 120 mmHg and 80 mmHg are gauge pressures. This means that the actual pressure at the aortic valve would be 120 mmHg plus 1 atm (which would be equal to an absolute pressure of 880 mmHg). In this textbook, we will refer all gauge pressures to atmospheric pressure, so that

$$p_{\text{gauge}} = p_{\text{absolute}} - p_{\text{atmospheric}}$$

Equation (3.10) describes the pressure variation in any static fluid. Changes in the pressure force are only a function of density, gravity, and the height location, assuming that the gravitational force acts only in the z-direction. In the previous example, we made the unstated assumption that changes in gravity are negligible. For most practical biofluid mechanics problems, the variation in the gravitation force with height is insignificant. Assuming that gravity is 9.81 m/s^2 at sea level, for every kilometer above sea level gravity

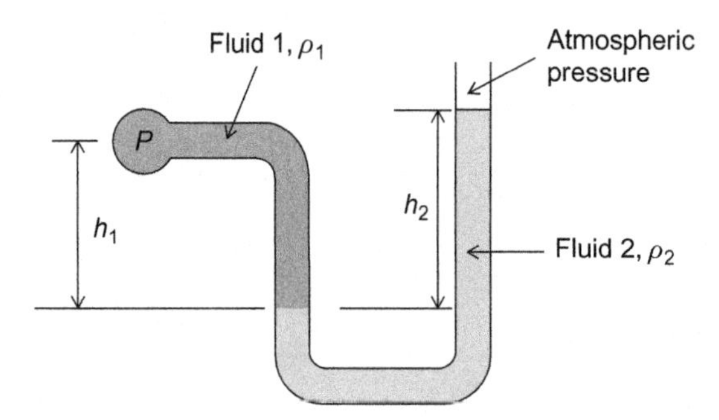

FIGURE 3.3 Schematic of a classic fluid mechanics manometer for measuring the pressure of a fluid at P. By measuring the differences in height, with a known open pressure, the hydrostatic pressure at P can be calculated.

reduces by approximately 0.002 m/s^2. Therefore, when the height changes that are being described are on the order of meters or centimeters (which will be typical in this textbook), the change in gravity with height is negligible. This is a reasonable assumption. However, in some biofluid situations, it may not be a good assumption that the density is constant, so be careful when applying the equation that we have been discussing.

Remember our definition for incompressible fluids; the fluid density is constant under all conditions. In this situation, it would be appropriate to use Eq. (3.10), in the form shown in the example problem. The pressure variation in a static incompressible fluid would then be

$$p = p_0 - \rho g_z(z - z_0) \tag{3.11}$$

A useful instrument to measure pressure variations solely based on height differences is a manometer. In classical fluid mechanics examples, manometers are used extensively to determine the pressure of a fluid compared to atmospheric pressure (Figure 3.3). Relating this to a biological example, manometers have been coupled to catheter systems in order to measure the intravascular pressure relative to atmospheric pressures. Even though the blood is flowing within the blood vessel, the blood that was diverted into the catheter system would not be flowing, but it would maintain the same pressure of the flowing blood within the vascular system. These systems however are not very accurate when quantifying pressure because of the effects that they may induce on the patient. Most likely, blood flow within the vessel would be shunted or the vessel would be ligated to insert the catheter. Therefore, the pressure that is being measured by the manometer system is not necessarily the exact physiological pressure, under normal conditions, but is the hydrostatic pressure under some modified conditions.

Example

Blood is flowing through point P (Figure 3.4), which is connected to a catheter tip manometer system. Blood enters the manometer and equilibrates the pressure of the various fluids within the system (as denoted in the figure). Calculate the pressure within the blood vessel assuming that blood density is 1050 kg/m^3 and that atmospheric pressure is 760 mmHg.

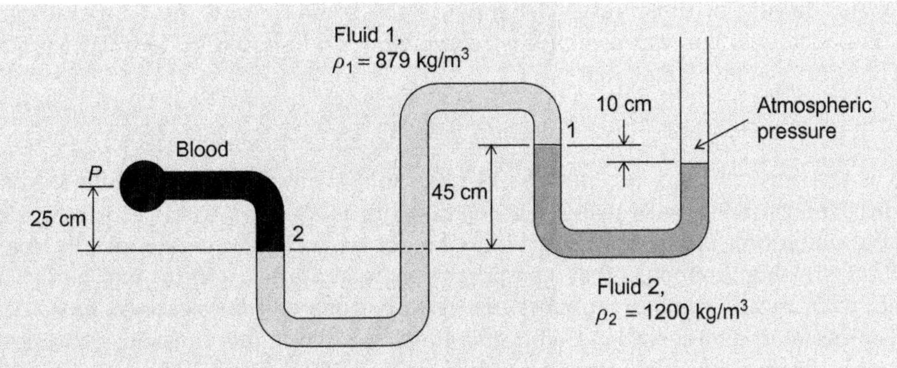

FIGURE 3.4 Schematic of a catheter tip manometer to measure intravascular blood pressure.

Solution

$$p_1 = p_{atm} - \rho_2 g(z_1 - z_0)$$

$$p_1 = 760\ mmHg - (1200\ kg/m^3)(9.81\ m/s^2)(10\ cm)\left(\frac{1\ m}{100\ cm}\right)\left(\frac{1\ mmHg}{133.32\ Pa}\right) = 751.17\ mmHg$$

$$p_2 = p_1 - \rho_1 g(z_2 - z_1)$$

$$p_2 = 751.17\ mmHg + (879\ kg/m^3)(9.81\ m/s^2)(45\ cm)\left(\frac{1\ m}{100\ cm}\right)\left(\frac{1\ mmHg}{133.32\ Pa}\right) = 780.28\ mmHg$$

$$p_{blood} = p_2 - \rho_{blood} g(z_p - z_2)$$

$$p_{blood} = 780.28\ mmHg - (1050\ kg/m^3)(9.81\ m/s^2)(25\ cm)\left(\frac{1\ m}{100\ cm}\right)\left(\frac{1\ mmHg}{133.32\ Pa}\right) = 760.96\ mmHg$$

The following example illustrates a few principles that should be remembered when working on fluid statics problems. The first is that the pressure at an interface between two different fluids is always the same. This is how we can equate the pressure at locations 1 and 2, 2 and *P*. If the fluid is continuous (same density), any location at the same height has the same pressure. Therefore, you can move around the bends without calculating each pressure change around those bends. Also, the dashed line through fluid one has the same pressure as location 2. Finally, pressure should increase as the elevation decreases and pressure should decrease as the elevation increases.

There are many biofluid problems in which the fluid density will vary. These types of fluids are compressible fluids and the density function would need to be stated within the problem. The density function would need to be given as a function of pressure and/or height. Once this function is known, then Eq. (3.10) can be used to solve for the pressure distribution throughout the fluid, assuming that gravity aligns with one direction. As an example, the density of most gases depends on the pressure and the temperature of the system. The ideal gas law represents this relationship and should be familiar to most students. The ideal gas law states that

$$p = \rho RT \tag{3.12}$$

where R is the universal gas constant (8.314 J/(g mol K)) and T is the absolute temperature (in kelvin). The problem with using this relationship is that it introduces a new variable, T, into the equation, which may vary with height as well. We will typically make the assumption in this textbook that temperature fluctuations within the body can be neglected. This means that for humans, the temperature will be assumed to be 310.15 K (37°C), unless stated otherwise. Using the ideal gas law, the pressure variation in a compressible fluid, with a constant temperature and with negligible changes in gravity with respect to height, is

$$\frac{dp}{dz} = -\rho g_z = -\frac{p}{RT} g_z$$

$$\frac{dp}{p} = -\frac{g_z}{RT} dz$$

$$\int_{p_0}^{p_1} \frac{dp}{p} = -\frac{g_z}{RT} \int_{z_0}^{z_1} dz$$

$$\ln(p)\Big|_{p_0}^{p_1} = -\frac{g_z}{RT} z\Big|_{z_0}^{z_1}$$

$$\ln(p_1) - \ln(p_0) = \ln\left(\frac{p_1}{p_0}\right) = -\frac{g_z}{RT}(z_1 - z_0)$$

$$\frac{p_1}{p_0} = e^{-\frac{g_z}{RT}(z_1 - z_0)}$$

$$p_1 = p_0 e^{-\frac{g_z}{RT}(z_1 - z_0)} \tag{3.13}$$

Depending on the particular application of the problem, the differential equation that relates pressure variations to height changes (Eq. (3.10)) can be solved for any fluid that has a density or temperature variation with height.

There is one important point to remember about hydrostatic pressure. Most students should be familiar with the concept that a pressure gradient acts as a driving force for fluid flow (e.g., the fluid will flow from high pressure to low pressure). Hydrostatic pressure is not this driving force; otherwise, it would be easier for blood to flow from the heart to the head then from the heart to the feet (when standing upright). Hydrostatic pressure is more similar to a friction concept; that is, to move an object, enough force must be applied to overcome the frictional forces. In fluids examples, there must be enough force applied to the fluid to overcome the hydrostatic pressure gradient in order to have the fluid accelerate in that direction. Also, in some instances, hydrostatic pressure can aid in fluid movement, whereas in other instances it can hinder movement.

3.2 BUOYANCY

Buoyancy is defined as the net vertical force acting on an object that is either floating on a fluid's surface or immersed within the fluid. To determine the net force acting on an immersed object, the same relationship for pressure variation within a static fluid can be applied. Starting from Eq. (3.10), the net pressure on a three-dimensional object would need to take into account the quantity of material that is in the z-direction (Figure 3.5). Again, by taking a differential element, the net force in the z-direction would be

$$dF_z = (p + \rho g h_1)dA - (p + \rho g h_2)dA = \rho g(h_1 - h_2)dA \tag{3.14}$$

The forces acting on the other sides of the cylinder would balance each other out (e.g., the forces in the radial and angular directions), so they do not need to be considered in the following formulation. This is assuming that gravity only acts in the z-direction for this case. Recall from a calculus course that

$$(h_1 - h_2)dA = dV$$

which is the volume of the element of interest (in Figure 3.5 this would be the shaded cylinder). Therefore, the summation of all of the forces that act on the immersed body would be

$$F_z = \int_V dF_z = \int_V \rho g dV = \rho g V$$

where V is the immersed volume of the element and gravity and density are not a function of the fluid volume of interest. This pressure force is equal to the force of gravity on the liquid displaced by the object. For biomedical applications, this is useful for designing any probe, which would be immersed within a biological fluid. Any cardiovascular implantable device would fall within this category as well.

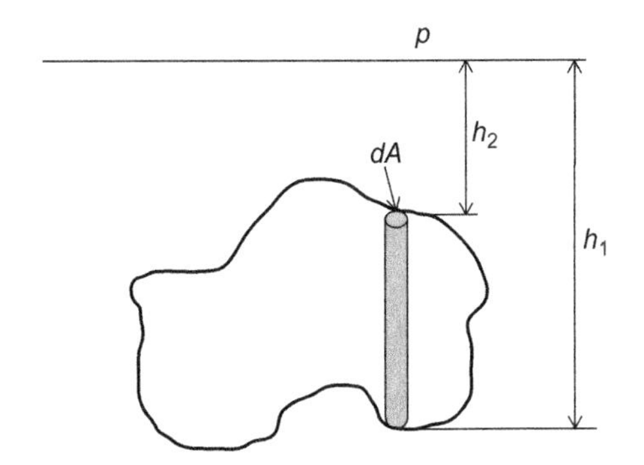

FIGURE 3.5 A body immersed in a static fluid. dA describes the cross-sectional area of the body at the location of h_1 and h_2 (which are measured in the z-axis). Various cross-sectional areas would be used to determine the buoyancy forces on an immersed object.

Example

Determine the maximum buoyancy of a catheter that is inserted into the femoral artery of a patient and is passed through the cardiovascular system to the coronary artery (Figure 3.6). The location where the catheter is inserted into the femoral artery is 50 cm below the aortic arch. The coronary artery is 5 cm below the aortic arch. Assume that the maximum buoyancy would occur at peak systole for a normal healthy individual (120 mmHg). Also assume that the catheter is perfectly cylindrical with a diameter of 2 mm.

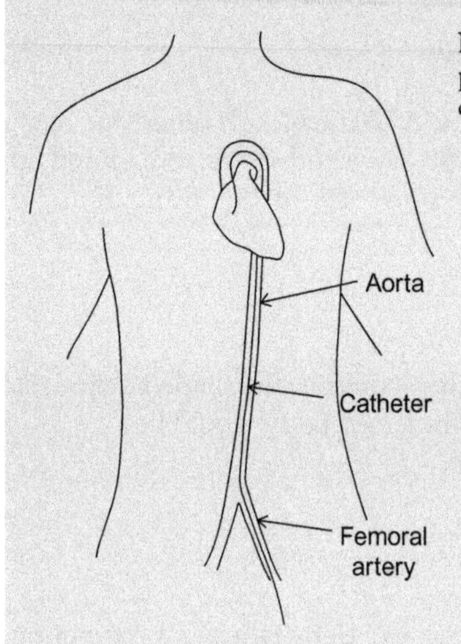

FIGURE 3.6 Catheter inserted at the femoral artery which is passed to the coronary artery. These catheters are commonly used during surgeries to remedy atherosclerotic lesions.

Solution

Pressure at incision:

$$p_1 = 120\ mmHg + (1050\ kg/m^3)(9.81\ m/s^2)(50\ cm)\left(\frac{1\ m}{100\ cm}\right)\left(\frac{1\ mmHg}{133.32\ Pa}\right) = 158.63\ mmHg$$

Pressure at coronary artery:

$$p_1 = 120\ mmHg + (1050\ kg/m^3)(9.81\ m/s^2)(5\ cm)\left(\frac{1\ m}{100\ cm}\right)\left(\frac{1\ mmHg}{133.32\ Pa}\right) = 123.86\ mmHg$$

Volume of catheter from femoral artery to aortic arch:

$$V = \pi\left(\frac{2\ mm}{2}\right)^2(50\ cm) = 1.57\ cm^3$$

Volume of catheter from aortic arch to coronary artery:

$$V = \pi\left(\frac{2\ mm}{2}\right)^2(5\ cm) = 0.157\ cm^3$$

Buoyancy force on catheter:

$$F = (1050\,kg/m^3)(9.81\,m/s^2)(1.57\,cm^3) + (1050\,kg/m^3)(9.81\,m/s^2)(0.157\,cm^3) = 0.0178N$$

Note that the force acting on the catheter was not affected by the absolute pressure in the system because these values cancel when adding the pressure terms (see Eq. (3.14)).

3.3 CONSERVATION OF MASS

The previous two sections described the pressure distribution in static fluids. However, in most biofluid mechanics problems, the fluid that we are interested in will be in motion (with an acceleration component), and therefore, the previous analysis may not be applicable or may not be the most accurate, although you may be able to obtain some meaningful information from static fluid analyses. In the following four sections, we will develop relationships that govern general fluid motion. Our analysis for each of these four sections will use a volume of interest (sometimes called a control volume) formulation, because it is normally quite difficult to identify the same mass of fluid throughout time. Remember that fluids under motion will deform, and therefore, some identifiable volume must be defined so that the laws of motion can be applied (Figure 3.7 illustrates different ways a fluid volume of interest may deform at later times, to illustrate that it may not be the most beneficial to follow one particular fluid volume but instead a general region). If you are interested in particular fluid elements and how those particular fluid elements move with time, then you must take a differential approach to Newton's laws. The laws that govern a system should be familiar from earlier courses in mechanics/thermodynamics. We will extend these principles to a volume in the following formulations.

First we will develop the conservation of mass principle for fluid mechanics. The basis for this principle is that mass can neither be created nor destroyed within the volume/system of interest. If the inflow mass flow rate is not balanced by the outflow mass flow rate, then there will be a change in volume or density within the volume of

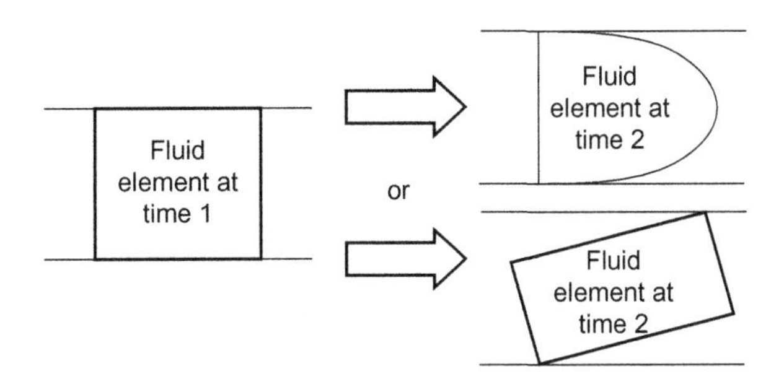

FIGURE 3.7 Two possible arrangements for a fluid element after the fluid experienced some motion. It is easier to maintain the control volume square of time 1 to analyze the fluid, instead of changing the volume of interest with time. This image shows that there are multiple possible arrangements for fluid elements after deformation, depending on the boundary conditions. It is critical during the analysis of biofluid mechanics problems to simplify these issues of deformability by choosing control volumes wisely.

interest. If the inflow mass flow rate exceeds the outflow mass flow rate, mass will accumulate within the system. If the reverse scenario is true, mass will be removed from the system. As an example, under normal conditions (e.g., not under strenuous activity), the blood volume within the heart remains constant from beat to beat. Stated in other words, the mass ejected from the aorta and the pulmonary arteries is recovered from the superior vena cava, the inferior vena cava, and the pulmonary veins. However, you can imagine a case where the residual fluid mass within the heart decreases from beat to beat. As an example, a patient that is experiencing severe blood loss due to a laceration, would initially continue to eject the normal amount of blood from the heart, but the venous return would not be equal to this ejection volume. Therefore, the blood volume in the heart would decrease. No matter what the case is regarding mass changes within the volume of interest, mass must be conserved within the system of interest.

Before we move forward into the derivation of the conservation of mass of a system, we will derive a general relationship for the time rate of change of a system property as a function of the same property per unit mass of the volume (inherent property). This is sometimes referred to as the Reynolds Transport Theorem (RTT) formulation. For mass balance, the system property is mass and the inherent property is 1 (i.e., mass divided by mass). For balance of linear momentum, the system property is momentum ($\vec{P}$) and the inherent property is velocity ($\vec{v}$) (i.e., momentum divided by mass). For energy balance, the system property is energy, E (or entropy, S), and the inherent property is energy per unit mass, e (or entropy per unit mass, s). The system and volume of interest used in this derivation will be a cube, but this same analysis technique can be applied to any geometry (Figure 3.8). We will also assume that the shape remains the same, but this analysis holds for deformation as well. The system and volume have been chosen so that there is a region that overlaps at some later time, Δt, during the motion (area 2 represents the region that overlaps at the two time points). Mass from area 1 enters the volume of interest during Δt and mass from area 3 exits the volume of interest during Δt.

The following derivation will relate the time rate of change of any general system property (W) to its inherent property (w). W and w are arbitrary properties that are only used for formula derivation (RTT). This formulation starts by using the definition of a derivative:

$$\frac{dW}{dt} = \lim_{\Delta t \to 0} \frac{W|_{t+\Delta t} - W|_t}{\Delta t}$$

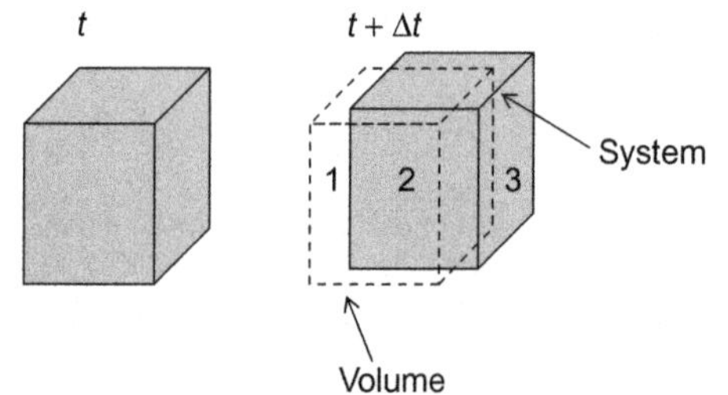

FIGURE 3.8 System and volume of interest used to derive the formula for conservation laws. The system of interest is shown by the gray shaded cube, and the volume of interest is the dashed cube. To use this formulation, one would need to know the change in time between the two states shown in this figure.

At any time t, the system is defined by the volume of interest, which keeps the same shape at all times. At time $t + \Delta t$, the system occupies areas 2 and 3, instead of areas 1 and 2 (for time t). Again, regardless of the area encompassed by the volume of interest, the volume (and its dimensions) remains constant at all time. Therefore, the following definitions apply for the system properties at time t and at time $t + \Delta t$:

$$W_t = W_{\text{volume of interest}} = W_{VI}$$
$$W_{t+\Delta t} = W_2 + W_3 = W_{VI} - W_1 + W_3$$

Using these definitions in the derivative formulation

$$\frac{dW}{dt} = \lim_{\Delta t \to 0} \frac{(W_{VI} - W_1 + W_3)_{t+\Delta t} - (W_{VI})_t}{\Delta t} \tag{3.15}$$

which is equal to

$$\frac{dW}{dt} = \lim_{\Delta t \to 0} \frac{(W_{VI})\Big|_{t+\Delta t} - (W_{VI})_t}{\Delta t} + \lim_{\Delta t \to 0} \frac{(W_3)_{t+\Delta t}}{\Delta t} - \lim_{\Delta t \to 0} \frac{(W_1)_{t+\Delta t}}{\Delta t} \tag{3.16}$$

The first term in Eq. (3.16) is equal to

$$\lim_{\Delta t \to 0} \frac{(W_{VI})\Big|_{t+\Delta t} - (W_{VI})_t}{\Delta t} = \frac{\partial W_{VI}}{\partial t} = \frac{\partial}{\partial t} \int_V w_{VI} \rho dV \tag{3.17}$$

Equation (3.17) is developed by recognizing that the left-hand side term and the middle term represent the definition of a derivative and an alternative way to represent a derivative, respectfully. The middle term is represented as a partial derivative since the general system property can be a function of space or time. The right-hand side term represents a third way to represent this derivative. The general inherent property is multiplied by density and volume (which is equal to mass) to obtain the time rate of change of the general system property. This formulation is typically more applicable since the inherent property and/or density can be a function of volume, which is the variable that we typically define in biofluid mechanics. For the remaining two terms in Equation (3.16), a similar analysis can be conducted to obtain

$$\begin{aligned} dW_3\Big|_{t+\Delta t} &= w_3 \rho dV\Big|_{t+\Delta t} = w_3 \rho \Delta x dA\Big|_{t+\Delta t} \\ dW_1\Big|_{t+\Delta t} &= w_1 \rho dV\Big|_{t+\Delta t} = w_1 \rho \Delta x(-dA)\Big|_{t+\Delta t} \end{aligned} \tag{3.18}$$

For this formulation, the variables cannot be described with the definition of a derivative. Also, remember that dV can be described as the change in length (i.e., from area 2 to area 3) multiplied by the differential area (in general, the cube can move in three-dimensional space). Also recall that a negative sign is included in the differential area term for the system property associated with area 1 formulation to take care of the direction that the normal area vector is facing. The mass is moving to the right (as shown in Figure 3.8), but the area vector for area 1 is oriented toward the left. Additionally, the change in length (Δx) can also be considered as the fluid path for any deformation that the fluid element experiences during the time interval that is being observed. If we integrate the two equations in (3.18), we get

$$\int_{area3} dW_3\Big|_{t+\Delta t} = W_3\Big|_{t+\Delta t} = \int_{area3} w_3\rho\Delta x dA\Big|_{t+\Delta t}$$
$$\int_{area1} dW_1\Big|_{t+\Delta t} = W_1\Big|_{t+\Delta t} = \int_{area1} w_1\rho\Delta x(-dA)\Big|_{t+\Delta t} \tag{3.19}$$

Substituting these values into Eq. (3.16)

$$\lim_{\Delta t\to 0}\frac{(W_3)\Big|_{t+\Delta t}}{\Delta t} = \frac{\int_{area3} w_3\rho\Delta x dA}{\Delta t} = \int_{area3} w_3\rho\vec{v}\cdot d\vec{A}$$
$$\lim_{\Delta t\to 0}\frac{(W_1)\Big|_{t+\Delta t}}{\Delta t} = \frac{\int_{area1} w_1\rho\Delta x(-dA)}{\Delta t} = -\int_{area1} w_1\rho\vec{v}\cdot d\vec{A} \tag{3.20}$$

The previous equation uses the equalities for

$$\lim_{\Delta t\to 0}\frac{\Delta x}{\Delta t} = \vec{v}$$
$$dA = d\vec{A}$$

to change the quantities between scalar and vector forms. Combining Eqs. (3.17) and (3.20) into Eq. (3.16)

$$\frac{dW}{dt} = \frac{\partial}{\partial t}\int_{VI} w_{VI}\rho dV + \int_{area3} w_3\rho\vec{v}\cdot d\vec{A} + \int_{area1} w_1\rho\vec{v}\cdot d\vec{A} \tag{3.21}$$

The entire system of interest consists of areas 1, 2, and 3, and we can make the assumption that there is no change in flow within region 2 during the time interval of Δt (this is why we choose to overlap the systems from the two time intervals). Therefore, $\vec{v}$ is zero for area 2, and we can combine the two area integrals in Eq. (3.21) into a general form, where "area" is equal to area 1 plus area 3.

$$\frac{dW}{dt} = \frac{\partial}{\partial t}\int_{V} w\rho dV + \int_{area} w\rho\vec{v}\cdot d\vec{A} \tag{3.22}$$

When developing the formulation for the time rate of change of a system property, we took the limit of the system as time approaches zero. This forces the relationship to be valid at the instant when the system and the control volume completely overlap, even though we developed the formulations as if the volume of interest has moved slightly. The first term of Eq. (3.22) is the time rate of change of any arbitrary general system property (W). The second term in Eq. (3.22) is the time rate of change of the inherent property within the volume of interest (w). The third term in Eq. (3.22) is the flux of the property out of the surface of interest or into the surface of interest, considering any possible flux from area 1 or 3. From this relationship, all of the conservation laws can be derived by substituting the appropriate system property and inherent property, which were described previously.

In Chapter 2, we defined conservation of mass as

$$m_{system} = m_{in} - m_{out} - \frac{d(\rho V)}{dt}$$

In a more concise form, mass balance can be stated as

$$\left.\frac{dm}{dt}\right|_{\text{system}} = 0 \tag{3.23}$$

The mass of a system can be defined as

$$m_{system} = \int_{\text{m-system}} dm = \int_{\text{V-system}} \rho dV$$

Substituting the appropriate values for mass into Eq. (3.22)

$$\left.\frac{dm}{dt}\right|_{\text{system}} = \frac{\partial}{\partial t}\int_{V} \rho dV + \int_{\text{area}} \rho \vec{v} \cdot d\vec{A} = 0 \tag{3.24}$$

The previous equation (Eq. (3.24)) describes the changes in mass within a system of interest. The first term (right-hand side of Eq. 3.24) describes the time rate of change of the mass within the volume of interest. This includes any possible change in density within the volume or changes of the volume itself. The second term (right-hand side) describes the mass flux into/out of the surfaces of interest. Mass that is entering into the volume of interest would be considered a negative flux (because the velocity vector acts in an opposite direction to the area normal vector), whereas mass leaving the volume of interest would be a positive flux (the velocity and the area vectors are acting in the same direction). By the conservation of mass principle, the time rate of change of mass within the volume of interest has to be balanced by the flux of mass into/out of the volume of interest.

Equation (3.24) can be simplified for specific fluid cases. For an incompressible flow, there is no change in density with time/space. This simplifies Eq. (3.24) to

$$\left.\frac{dm}{dt}\right|_{\text{system}} = \rho\frac{\partial V}{\partial t} + \rho\int_{\text{area}} \vec{v} \cdot d\vec{A} = 0 \tag{3.25}$$

because the volume integral of dV is simply the volume of interest. By canceling out the density terms and making a further assumption that the volume of interest does not change with time, Equation (3.25) becomes

$$\left.\frac{dm}{dt}\right|_{\text{system}} = \int_{\text{area}} \vec{v} \cdot d\vec{A} = 0 \tag{3.26}$$

A volume that does not change with time would be considered nondeformable or rigid. This is not always a good assumption in biofluids because blood vessels change shape when the heart's pressure pulse is passed through it, but the question would remain whether or not the change in the blood vessel shape significantly alters the fluid properties. If the change in volume does significantly change the fluid properties, then it should be included within the calculations, however, if the fluid properties are not altered significantly then volume changes may be ignored. Also, the lungs use a shape change to drive the flow of air into or out of the system (e.g., the volume of the lungs is altered and this translates into a change in pressure to drive air movement). In this textbook, we will assume that our volume of interest is nondeformable unless stated otherwise. Equation (3.26) does not make an assumption on the flow rate (i.e., is it steady or does it change with time), so this equation is valid for any incompressible flow through a nondeformable volume. Although, remember that by definition, steady flows can have no fluid property that changes with time. Therefore, in the case of a steady flow, the first integral term in Eq. (3.24) would be equal to zero. So for a general compressible steady flow situation, Eq. (3.24) would simplify to the mass flux equation:

$$\int_{area} \rho \vec{v} \cdot d\vec{A} = 0 \tag{3.27}$$

In fluid mechanics, the integral represented in Eq. (3.26) is commonly referred to as the volume (or volumetric) flow rate, Q. For an incompressible flow through a nondeformable volume, the volume flow rate into the volume must be balanced by the flow out of the volume. However, since the volume is nondeformable, the volume flow rate can be calculated at any one location at any time within the system of interest. Its definition would be

$$Q = \int_{area} \vec{v} \cdot d\vec{A} \tag{3.28}$$

The volume flow rate divided by area is defined as the average velocity at a particular section of interest:

$$v_{avg} = \frac{Q}{A} = \frac{1}{A} \int_{area} \vec{v} \cdot d\vec{A} \tag{3.29}$$

From the special cases that we have discussed, as well as the general formula, we can now use the conservation of mass to solve various fluid mechanics problems.

Example

Determine the velocity of blood at cross-section 4 of the aortic arch schematized in Figure 3.9. Assume that the diameter of the blood vessel is 3 cm, 1.5 cm, 0.8 cm, 1.1 cm, and 2.7 cm at cross-sections 1, 2, 3, 4, and 5, respectively. Branches 2, 3, and 4 make a 75°, a 85°, and a 70° angle with the horizontal direction, respectively. The velocity is 120, 85, 65, and 105 cm/s at 1, 2, 3, and 5, respectively. There is inflow at 1 and outflow at all of the remaining locations. Assume steady flow at this particular instant in time and that the volume of interest is nondeformable.

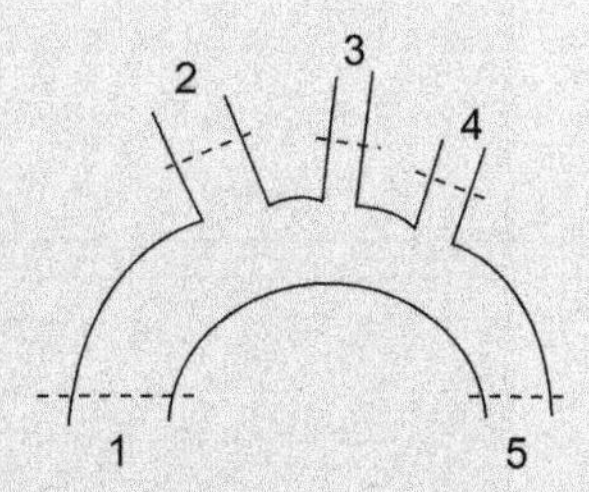

FIGURE 3.9 Schematic of the aortic arch.

Solution

Figure 3.10 highlights the given geometric constraints in this problem. The gray dashed box on this figure represents one of the possible choices for the volume of interest. We will also make the assumption that blood density does not change and has a value of 1050 kg/m^3.

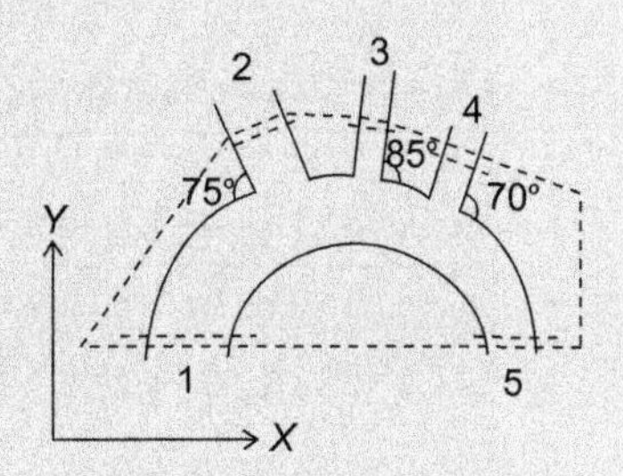

FIGURE 3.10 Figure depicting our choice for the control volume around the aortic arch. Notice that the area of interest at each location (1–5) are perpendicular to the mean velocity direction.

We can directly apply Eq. (3.26) because we have made the assumption that the density does not change with time, that the volume is not deformable, and that the flow is steady:

$$\int_{\text{area}} \vec{v} \cdot d\vec{A} = 0 \tag{3.26}$$

Equation (3.26) will further simplify to the volume flow rate at each location because of the assumptions that the velocity is constant with respect to area at the one particular time that we are interested in (e.g., the velocities that were provided in the prompt represent the peak velocity or the average velocity at each cross-section of interest). Therefore, Equation (3.26) can be represented as

$$-v_1A_1 + v_2A_2 + v_3A_3 + v_4A_4 + v_5A_5 = 0$$

because the velocity at location 1 opposes the area normal vector at 1 and at the remaining locations the velocity is in the same direction as the area normal vector (also, as a rule of thumb, inflow is negative and outflow is positive in these calculations).

The area at each location is

Location	Area
1	$\pi\left(\frac{3\,cm}{2}\right)^2 = 7.069\,cm^2$
2	$\pi\left(\frac{1.5\,cm}{2}\right)^2 = 1.767\,cm^2$
3	$\pi\left(\frac{0.8\,cm}{2}\right)^2 = 0.503\,cm^2$
4	$\pi\left(\frac{1.1\,cm}{2}\right)^2 = 0.950\,cm^2$
5	$\pi\left(\frac{2.7\,cm}{2}\right)^2 = 5.726\,cm^2$

Substituting the known values for area and velocity into the previous equation

$$v_4 = \frac{v_1A_1 - v_2A_2 - v_3A_3 - v_5A_5}{A_4}$$

$$v_4 = \frac{(120\,cm/s)(7.069\,cm^2) - (85\,cm/s)(1.767\,cm^2) - (65\,cm/s)(0.503\,cm^2) - (105\,cm/s)(5.726\,cm^2)}{0.950\,cm^2}$$

$$= 67.54\,cm/s$$

This velocity would flow at a 75° angle off the positive x-axis. Density was not used in any of the formulations because it would cancel out in each term, since density is assumed to be constant in this scenario. This problem also illustrates that it does not matter which direction (x or y) the velocity is acting, because all of the mass need to be conserved. We provided information regarding the branching angles of the vessels off of the aorta, but they were not used in this formulation.

Example

Calculate the time rate of change of air density during expiration. Assume that the lung (Figure 3.11) has a total volume of 6000 mL, the diameter of the trachea is 18 mm, the air flow velocity out of the trachea is 20 cm/s, and the density of air is 1.225 kg/m^3. Also assume that lung volume is decreasing at a rate of 100 mL/s.

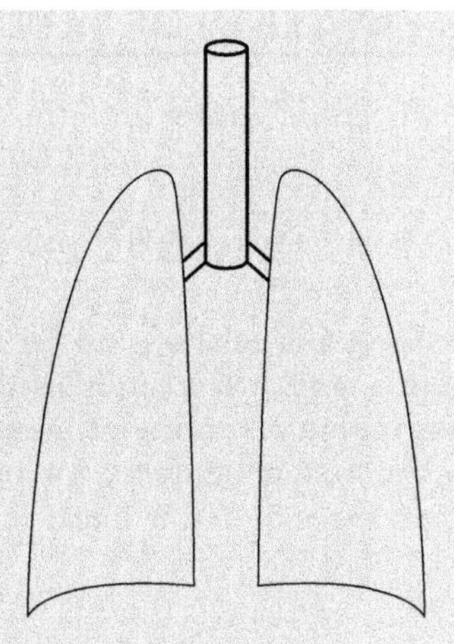

FIGURE 3.11 Schematic of the lung.

Solution

We will start from Eq. (3.24), since we are asked for the time rate of change of density. This suggests that the example is representing a nonsteady flow scenario. Also, we were told what the rate of change in the lung volume is during this procedure.

$$\frac{\partial}{\partial t}\int_V \rho dV + \int_{area} \rho\vec{v}\cdot d\vec{A} = 0 \tag{3.24}$$

Assume that at the instant in time that we are measuring the system, density is uniform within the volume of interest. This allows us to remove density from within the first integral:

$$\frac{\partial}{\partial t}\rho\int_V dV + \int_{area} \rho\vec{v}\cdot d\vec{A} = 0$$

The volume integral of the volume of interest is just the volume of interest. Therefore, Equation (3.24) can be represented as

$$\frac{\partial}{\partial t}(\rho V) + \rho\vec{v}A = 0$$

if we also assume that the density and velocity are not functions of the outflow area of interest (the second integral obtains a positive solution because it is an outflow velocity). Using the product rule

$$V\frac{\partial\rho}{\partial t} + \rho\frac{\partial V}{\partial t} = -\rho\vec{v}A$$

Solving this equation with the known values

$$\frac{\partial\rho}{\partial t} = \frac{-\rho\vec{v}A - \rho(\partial V/\partial t)}{V}$$

$$= \frac{-(1.225\ kg/m^3)(20\ cm/s)\left(\pi\left(\frac{18\ mm}{2}\right)^2\right) - (1.225\ kg/m^3)(-100\ mL/s)}{6000\ mL} = 0.01\ kg/m^3s$$

Note that the change in volume is negative since we were told that the lung volume is decreasing at the given rate.

3.4 CONSERVATION OF MOMENTUM

Newton's second law of motion can be written in terms of linear momentum, which is its most general form:

$$\vec{F} = \frac{d(\vec{P})}{dt} = \frac{d(m\vec{v})}{dt} \tag{3.30}$$

For this analysis, we want to develop a relationship for the linear momentum within a volume of interest. We will follow a similar technique as that used to develop a relationship for the conservation of mass within a volume of interest. As we alluded to, prior to developing Eq. (3.22), we know the system property for momentum ($\vec{P}$) and its inherent property ($\vec{v}$), but we will take some space to define linear momentum and how it relates to fluid mechanics.

Linear momentum is defined as

$$\vec{P} = \int_{\text{system mass}} \vec{v}dm = \int_V \vec{v}\rho dV \tag{3.31}$$

for a volume of interest. In writing Newton's law, we had to consider the summation of the forces that act on the fluid. Recall from our earlier discussion that the summation of the forces that act on a fluid element must include all body forces (denoted as $\vec{F}_b$) and all surface forces (denoted as $\vec{F}_s$), thus in considering linear momentum we may need to quantify all of the forces that act on the volume/surfaces of interest. Physically, linear momentum is a force of motion, which is conserved (and thus remains constant) unless other forces are applied to the system. By substituting the system property and the inherent property into Eq. (3.22), we can get the formulation for conservation of linear momentum using the RTT:

$$\frac{dP}{dt} = \frac{\partial}{\partial t}\int_V \vec{v}\rho dV + \int_{\text{area}} \vec{v}\rho\vec{v} \cdot d\vec{A} \tag{3.32}$$

Using Newton's relationship for momentum, Eq. (3.32) can be represented as

$$\frac{dP}{dt} = \vec{F} = \vec{F}_b + \vec{F}_s = \frac{\partial}{\partial t}\int_V \vec{v}\rho dV + \int_{\text{area}} \vec{v}\rho\vec{v} \cdot d\vec{A} \tag{3.33}$$

We make use of forces in lieu of the time rate of change of momentum, because it is typically easier to quantify forces acting on a fluid of interest. Equation (3.33) states that the summation of all forces acting on a volume of interest is equal to the time rate of change of momentum within the control volume and the summation of momentum flux through the surface of interest. To solve conservation of momentum problems, the first step will be to define the volume of interest and surfaces of interest and label all of the forces that are acting on this system (note though that some of the forces can cancel out and thus they may not be included in the formulation). This becomes especially critical when you define the coordinate system; for ease of quantification the coordinate system should align with the majority of the forces. For the other forces, you will need to know the trigonometric relationships between the forces and the chosen coordinate axes. If a standard Cartesian coordinate system is chosen, then gravity aligns with one of the axes, and typically gravity will be the only body force that acts on the system. Surface forces are due to externally applied loads and are normally denoted through a pressure acting on the system. In this case, the generalized surface force will be represented as

$$\vec{F}_s = \int_{\text{area}} -\vec{p} \cdot d\vec{A}$$

The negative sign in this formulation is added to maintain the sign convention for the forces acting on the system (Figure 3.12). In Figure 3.12, the pressure is positive, but because the pressure and the area vectors act in opposite directions, their vector product would be a negative force, which does not correspond with the positive directions chosen for the coordinate system, thus the inclusion of a negative sign.

Unlike the conservation of mass formula, the formulas derived for the conservation of linear momentum are vector equations (compare with aortic arch example for conservation of mass, where directionality did not matter). Equation (3.33) written in component form is

$$\begin{aligned} F_x = F_{bx} + F_{sx} &= \frac{\partial}{\partial t}\int_V u\rho dV + \int_{\text{area}} u\rho\vec{v} \cdot d\vec{A} \\ F_y = F_{by} + F_{sy} &= \frac{\partial}{\partial t}\int_V v\rho dV + \int_{\text{area}} v\rho\vec{v} \cdot d\vec{A} \\ F_z = F_{bz} + F_{sz} &= \frac{\partial}{\partial t}\int_V w\rho dV + \int_{\text{area}} w\rho\vec{v} \cdot d\vec{A} \end{aligned} \tag{3.34}$$

where u, v, and w are the velocity components in the x-, y-, and z-directions, respectively. As before, the product of $\rho\vec{v} \cdot d\vec{A}$ is a scalar whose sign depends on the directions of the normal area vector and the velocity vector (the sign of u, v, and w are defined by the velocity componets). If these two vectors act in the same direction, the product of the vectors is positive; if they act in opposite directions, then the product is negative. Remember that the velocity vector ($\vec{v}$) in this product is not a component of velocity but is the entire velocity vector. In scalar notation, the entire form of the product would be represented as $\pm|\rho v^2 A\cos\alpha|$,

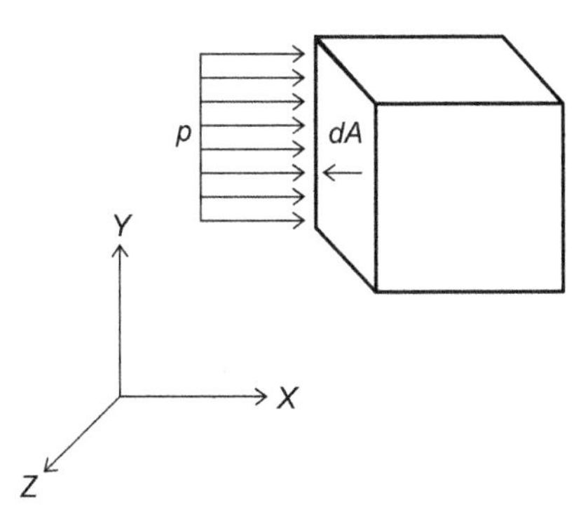

FIGURE 3.12 Pressure force acting on a surface of interest. Recall that the area vector for this surface would act in the negative x-direction, whereas the pressure forces are acting in the positive x-direction.

where α is defined by the coordinate system of choice and the positive or negative sign is defined through the velocity/normal area vectors relationship and the direction of the velocity component. This angle appears for the u, v, and w directional velocities. Some students find this type of overall formulation easier to understand, however, when applying this relationship, one must take significant care in determining the sign. However, regardless of how you choose to calculate the momentum flux, remember that the product of $\vec{v}\rho\vec{v} \cdot d\vec{A}$ is a vector, and the sign of this product depends on the coordinate system chosen (this defines the velocity vector sign) and the sign of the scalar product. The application of these sign conventions will become apparent in some of the example problems. To determine the sign of the momentum flux through a surface, first determine the sign associated with $|\rho v A cos\alpha|$, and then determine the sign of each velocity component (u, v, and w). By knowing the signs of these two parts, the sign of the overall product can be determined.

Example

Determine the force required to hold the brachial artery in place during peak systole (schematized in Figure 3.13). Assume at the inlet the pressure is 100 mmHg and at the outlet the pressure is 85 mmHg (these are gauge pressures). The diameter of the brachial artery is 18 mm at the inflow and 16 mm at the outflow. The blood flow velocity at the inlet is 65 cm/s. For simplicity, neglect the weight of the blood vessel and the weight of the blood within the vessel.

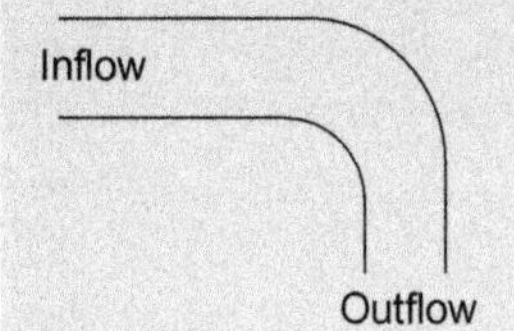

FIGURE 3.13 Brachial artery schematic for example problem.

Solution

Figures 3.13 and 3.14 depict what is known about this flow situation. The problem statement asks to solve for Fx and Fy. Remember that the weight of the blood vessel and the weight of the blood in the vessel can be incorporated as an additional force in the y-direction (the mass of blood would typically be obtained through the volume of the blood vessel and the density of blood).

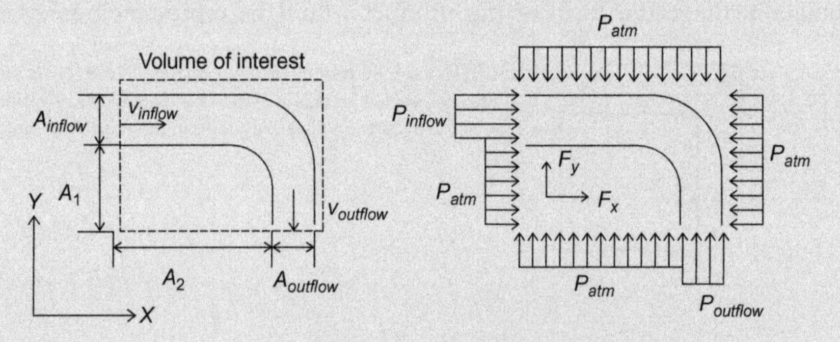

FIGURE 3.14 Free body diagram for the brachial artery example problem.

To solve this problem, assume that there is steady flow at the instant in time that we are interested in, that 1 atm = 760 mmHg, that the blood vessel does not move and is not deformable, and that the flow is incompressible. First, we will need to solve for the outflow velocity using the equations for conservation of mass:

$$\left.\frac{dm}{dt}\right|_{system} = \frac{\partial}{\partial t}\int_V \rho dV + \int_{\text{area}} \rho\vec{v}\cdot d\vec{A} = 0 \tag{3.24}$$

Since this is a steady flow problem, none of the variables change with time. Thus the conservation of mass can be rewritten as

$$\int_{\text{area}} \rho\vec{v}\cdot\left|d\vec{A}\right| = 0 \rightarrow \rho_1 v_1 A_1 = \rho_2 v_2 A_2$$

assuming that density and velocity are not a function of the inflow or outflow area.

$$\frac{(1050\ kg/m^3)(65\ cm/s)\left(\pi\left(\frac{18\ mm}{2}\right)^2\right)}{(1050\ kg/m^3)\left(\pi\left(\frac{16\ mm}{2}\right)^2\right)} = v_2 = 82.27\ cm/s(-\vec{j})$$

Note that the velocity accelerates due to the step down nature of the geometry. This is not representative of what occurs in the actual physiology, but this problem illustrates how to use conservation of mass and momentum together. Also, we added the direction vector so we can use it in the momentum calculation. Conservation of mass does not provide any information regarding the directionality; this information was gained from the figure and the choice of the "standard" Cartesian system.

First, let us solve for the x-component of the force needed to hold the brachial artery in place:

$$F_x = F_{bx} + F_{sx} = \frac{\partial}{\partial t}\int_V u\rho dV + \int_{\text{area}} u\rho\vec{v}\cdot d\vec{A}$$

$$F_{bx} = 0, u_2 = 0$$

$$F_{sx} = p_{\text{inflow}}A_{\text{inflow}} + p_{\text{atm}}A_1 - p_{\text{atm}}(A_{\text{inflow}} + A_1) + F_x = A_{\text{inflow}}(p_{\text{inflow}} - p_{\text{atm}}) + F_x$$

$$A_{\text{inflow}}(p_{\text{inflow}} - p_{\text{atm}}) + F_x = A_{\text{infow}}(p_{\text{inflow-gauge}}) + F_x = \int_{\text{area}} u\rho\vec{v}\cdot d\vec{A} = u_1 * (-\rho v_{\text{inflow}}A_{\text{inflow}})$$

Note that the u component of the velocity is positive, but the overall flux is negative because the velocity vector and the normal area vector act in opposite directions. Also, note that atmospheric pressure is acting on all of the surfaces of the blood vessel, but they cancel out and leave the gauge pressure at the inflow, which was provided in the prompt.

$$F_x = -u_1\rho v_{\text{inflow}}A_{\text{inflow}} - A_{\text{inflow}}(p_{\text{inflow-gauge}})$$

$$F_x = -(65\ cm/s)(1050\ kg/m^3)(65\ cm/s)\left(\pi\left(\frac{18\ mm}{2}\right)^2\right) - \left(\pi\left(\frac{18\ mm}{2}\right)^2\right)(100\ mmHg) = -3.5N$$

This means that this force acts toward the left because the flow/pressure is pushing toward the right. Also note that in this case, the total velocity and the x-velocity component are the same because (i) we assumed that the flow is only in the x-direction and (ii) the coordinate system aligns with the flow direction. Now solve for the y-component of force:

$$F_y = F_{by} + F_{sy} = \frac{\partial}{\partial t}\int_V v\rho dV + \int_{\text{area}} v\rho\vec{v} \cdot d\vec{A}$$

$$F_{by} = 0, v_1 = 0$$

$$F_{sy} = p_{\text{outflow}}A_{\text{outflow}} + p_{\text{atm}}A_2 - p_{\text{atm}}(A_{\text{outflow}} + A_2) + F_y = A_{\text{outflow}}(p_{\text{outflow}} - p_{\text{atm}}) + F_y$$

$$A_{\text{outflow}}(p_{\text{outflow}} - p_{\text{atm}}) + F_y = A_{\text{outflow}}(p_{\text{outflow-gauge}}) + F_y = \int_{\text{area}} v\rho\vec{v} \cdot d\vec{A} = v_2 * (\rho V_{\text{outflow}}A_{\text{outflow}})$$

Again, we are assuming that none of the flow variables change with time and that velocity and pressure are not functions of the outflow area of interest. Note that the v_2-velocity component is negative (this will be accounted for later) and the flux term is positive because the area and the velocity vectors act in the same direction.

$$F_y = v_2 * (\rho v_{\text{outflow}}A_{\text{outflow}}) - A_{\text{outflow}}(p_{outflow-gauge})$$

$$F_y = (-82.27\ cm/s)(1050\ kg/m^3)(82.27\ cm/s)\left(\pi\left(\frac{16\ mm}{2}\right)^2\right) - \left(\pi\left(\frac{16\ mm}{2}\right)^2\right)(85\ mmHg) = -2.42\ N$$

This means that this force acts downward. Note that since the chosen coordinate system aligned with the inflow in the x-direction and the outflow with the y-direction, we did not need to account for angles in either formulation. The overall net force is 4.25 N acting at an angle of 214° from the positive x-axis.

3.5 MOMENTUM EQUATION WITH ACCELERATION

In deriving Eq. (3.33), we made an assumption that the volume of interest (and therefore the fluid within the volume) had no acceleration at the instant in time that we were evaluating the system. The acceleration of interest here is not the possibility of the fluid to accelerate (or decelerate) within the containing volume due to volume changes, but this acceleration is associated with the volume itself accelerating. Therefore, Equation (3.33) does not hold for a volume of interest or system of interest that is accelerating during the time of interest. This is the case because typically when evaluating an accelerating system, you would use an inertial or fixed coordinate system (normally denoted as XYZ) instead of a reference (or relative) coordinate system, which follows the moving possibly accelerating volume (normally denoted as xyz). In the previous derivations, a relative coordinate system was used for simplicity and this relative system was either fixed or moving with a constant velocity. Furthermore, Eq. (3.33) does not hold for an inertial reference frame because the relative momentum for each system is not the same. That means that the sum of the forces that act on the system of interest equals the time rate of change of momentum of the reference system but it does not have to equal the time rate of change of the inertial system (if the volume is accelerating).

$$\vec{F} = \frac{d\vec{P}_{xyz}}{dt} \neq \frac{d\vec{P}_{XYZ}}{dt}$$

In order to develop an equivalent formulation as Eq. (3.33), for an accelerating control volume, a relationship between the inertial momentum ($\vec{P}_{XYZ}$) and the control volume momentum ($\vec{P}_{xyz}$) must be found. To start, let us define Newton's second law in terms of momentum and a system of interest:

$$\vec{F} = \frac{d\vec{P}_{XYZ}}{dt} = \frac{d}{dt}\int_{\text{system-mass}} \vec{v}_{XYZ} dm = \int_{\text{system-mass}} \frac{d\vec{v}_{XYZ}}{dt} dm \tag{3.35}$$

Equation (3.35) makes use of the standard momentum equations. To define the inertial velocity component in terms of the system velocity, use the following relationship:

$$\vec{v}_{XYZ} = \vec{v}_{xyz} + \vec{v}_r \tag{3.36}$$

where $\vec{v}_r$ is the velocity of the volume of interests reference frame relative to the inertial reference frame. Making the assumption that the fluid is irrotational

$$\frac{d\vec{v}_{XYZ}}{dt} = \vec{a}_{XYZ} = \frac{d\vec{v}_{xyz}}{dt} + \frac{d\vec{v}_r}{dt} = \vec{a}_{xyz} + \vec{a}_r \tag{3.37}$$

These equations should be familiar from a classical engineering dynamics course. In Equation (3.37), the first acceleration term ($\vec{a}_{XYZ}$) is the acceleration of the system relative to the inertial frame, the second acceleration term ($\vec{a}_{xyz}$) is the acceleration of the system relative to the system reference frame, and the third acceleration term ($\vec{a}_r$) is the acceleration of the system reference frame relative to the inertial reference frame. A rotational system would have multiple acceleration terms (see our brief discussion below about the added complexity that a rotational system adds). Substituting the acceleration terms into Equation (3.35)

$$\vec{F} = \int_{\text{system-mass}} \frac{d\vec{v}_{XYZ}}{dt} dm = \int_{\text{system-mass}} \frac{d\vec{v}_{xyz}}{dt} dm + \int_{\text{system-mass}} \vec{a}_r dm$$

$$\vec{F} - \int_{\text{system-mass}} \vec{a}_r dm = \int_{\text{system-mass}} \frac{d\vec{v}_{xyz}}{dt} dm = \frac{d\vec{P}_{xyz}}{dt} \tag{3.38}$$

In this formulation, we leave the reference coordinate system velocity as a velocity (instead of an acceleration) to have the equation resemble Newton's second law. Substituting Eq. (3.33) and converting the mass integral into a volume integral, Eq. (3.38) becomes

$$\vec{F}_b + \vec{F}_s - \int_V \vec{a}_r \rho dV = \frac{\partial}{\partial t}\int_V \vec{v}_{xyz}\rho dV + \int_{\text{area}} \vec{v}_{xyz}\rho\vec{v}_{xyz} \cdot d\vec{A} \tag{3.39}$$

To account for the acceleration of an inertial body, relative to the inertial reference frame, the conservation of linear momentum formulation requires one extra term. When the system is not accelerating relative to the inertial frame, $\vec{a}_r$ is zero and Eq. (3.39) simplifies to Eq. (3.33). To apply Eq. (3.39) to a system, it is required that there are two coordinate systems defined at the beginning of the problem; one is the inertial coordinate system (XYZ), and the other stays with the moving control volume (xyz) and note that the control volume variables are handled on the right-hand side of the equation. This formula is valid for one instant in time, similar to Eq. (3.33). However, it is possible in particular situations that the mass (i.e., $\vec{F}_b$,ρ) and the acceleration ($\vec{a}_r$) are known functions of time and can thus be incorporated into the equation relatively easily. Also, Eq. (3.39) is a vector equation, with all velocity components related to the noninertial reference frame (xyz). Written in component form, Eq. (3.39) becomes

$$
\begin{aligned}
\vec{F}_{bx} + \vec{F}_{sx} - \int_V \vec{a}_{rx}\rho dV &= \frac{\partial}{\partial t}\int_V u_{xyz}\rho dV + \int_{\text{area}} u_{xyz}\rho\vec{v}_{xyz} \cdot d\vec{A} \\
\vec{F}_{by} + \vec{F}_{sy} - \int_V \vec{a}_{ry}\rho dV &= \frac{\partial}{\partial t}\int_V v_{xyz}\rho dV + \int_{\text{area}} v_{xyz}\rho\vec{v}_{xyz} \cdot d\vec{A} \\
\vec{F}_{bz} + \vec{F}_{sz} - \int_V \vec{a}_{rz}\rho dV &= \frac{\partial}{\partial t}\int_V w_{xyz}\rho dV + \int_{\text{area}} w_{xyz}\rho\vec{v}_{xyz} \cdot d\vec{A}
\end{aligned}
\tag{3.40}
$$

Example

One of the first implantable mechanical heart valves was designed as a ball within a cage that acted as a check valve. Using the conservation of momentum (with acceleration), we will model the acceleration of the ball after it is hit by a jet of blood being ejected from the heart (Figure 3.15). The ball has a turning angle of 45° and a mass of 25 g. Blood is ejected from the heart at a velocity of 150 cm/s, through an opening with a diameter of 27 mm. Determine the velocity of the ball at 0.5 s. Neglect any resistance to motion (except mass).

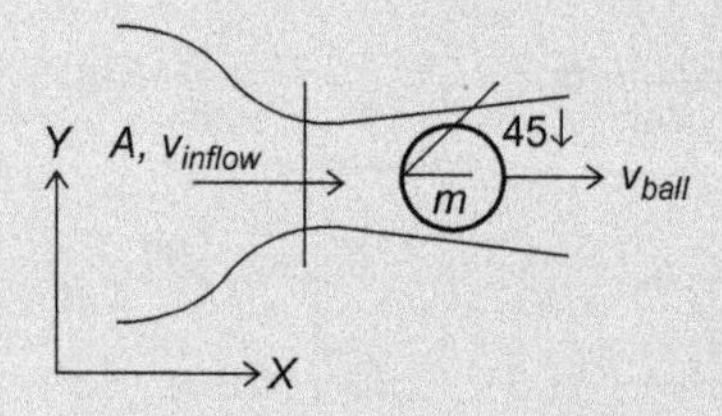

FIGURE 3.15 Acceleration of a ball and cage mechanical heart valve for the in-text problem.

Solution

The inertial reference frame is chosen at the aortic valve and the noninertial reference frame is chosen to coincide with the ball. With the volume of interest chosen to coincide with the flow direction around the ball, Figure 3.16 represents the flow situation. We will only

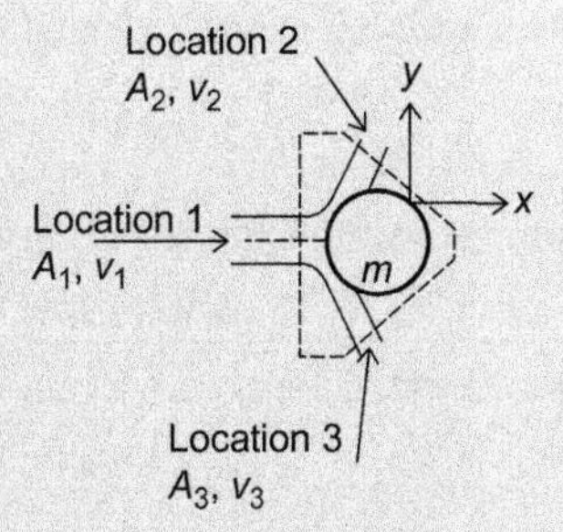

FIGURE 3.16 Free body diagram for ball and cage mechanical heart valve example problem.

analyze half of the equation, because the flow is assumed to be symmetrical around the uniform ball.

$$\vec{F}_{bx} + \vec{F}_{sx} - \int_V \vec{a}_{rx} \rho dV = \frac{\partial}{\partial t} \int_V u_{xyz} \rho dV + \int_{\text{area}} u_{xyz} \rho \vec{v}_{xyz} \cdot d\vec{A}$$

If we make the assumption that the blood flow is steady and uniform, the equation reduces to

$$-\int_V \vec{a}_{rx} \rho dV = \int_{\text{area}} u_{xyz} \rho \vec{v}_{xyz} \cdot d\vec{A}$$

because there are no external forces (e.g., pressure difference that was given in the prompt) acting on the system. Substituting known values into this equation, we get

$$-\int_V \vec{a}_{rx} \rho dV = u_1\left(-\frac{\rho v_1 A_1}{2}\right) + u_2(\rho v_2 A_2)$$

We will make the assumption that as blood flows around the ball, there is no loss of velocity (and the area does not change) due to friction between the ball and blood. We will also assume that there is no change in velocity between the aortic valve location and the location about the ball. This makes the magnitude of the following quantities:

$$\frac{v_1}{2} = u_1 = u_2 = v_{\text{inlet}} - v_{\text{ball}}$$

which is the relative velocity, the same. Furthermore, the area can be defined as

$$\frac{A_1}{2} = A_2 = A_3 = A$$

Simplifying each term of the momentum equation and only considering the top half of the ball:

$$\int_V \vec{a}_{rx} \rho dV = \vec{a}_{rx} \rho V = \vec{a}_{rx} m = \frac{dv_{\text{ball}}}{dt} m$$

Assuming that the relative acceleration and the density are not functions of the volume of interest

$$u_1(-\rho u_1 A) = (v_{\text{inlet}} - v_{\text{ball}})(-\rho(v_{\text{inlet}} - v_{\text{ball}})A) = -\rho A(v_{\text{inlet}} - v_{\text{ball}})^2$$

$$u_2(\rho u_2 A) = (v_{\text{inlet}} - v_{\text{ball}})\cos(45^\circ)(\rho(v_{\text{inlet}} - v_{\text{ball}})A) = \rho A(v_{\text{inlet}} - v_{\text{ball}})^2 \cos(45^\circ)$$

Substituting these values into the simplified conservation of momentum equation

$$-\frac{dv_{\text{ball}}}{dt}m = -\rho A(v_{\text{inlet}} - v_{\text{ball}})^2 + \rho A(v_{\text{inlet}} - v_{\text{ball}})^2\cos(45°) = (\cos(45°) - 1)\rho A(v_{\text{inlet}} - v_{\text{ball}})^2$$

To solve this differential equation, we must separate the variables as follows:

$$\frac{dv_{\text{ball}}}{(v_{\text{inlet}} - v_{\text{ball}})^2} = \frac{(1 - \cos(45°))\rho A}{m}dt$$

Integrate this equation as shown:

$$\int_0^{v_{\text{ball}}} \frac{dv_{\text{ball}}}{(v_{\text{inlet}} - v_{\text{ball}})^2} = \int_0^t \frac{(1 - \cos(45°))\rho A}{m}dt$$

$$\left.\frac{1}{(v_{\text{inlet}} - v_{\text{ball}})}\right|_0^{v_{\text{ball}}} = \left.\frac{(1 - \cos(45°))\rho At}{m}\right|_0^t = \frac{(1 - \cos(45°))\rho At}{m}$$

$$\frac{1}{(v_{\text{inlet}} - v_{\text{ball}})} - \frac{1}{v_{\text{inlet}}} = \frac{v_{\text{ball}}}{v_{\text{inlet}}(v_{\text{inlet}} - v_{\text{ball}})} = \frac{(1 - \cos(45°))\rho At}{m}$$

Solving this equation for $v_{\text{ball-max}}$

$$\frac{v_{\text{ball}}}{v_{\text{inlet}}^2 - v_{\text{inlet}}v_{\text{ball}}} = \frac{(1 - \cos(45°))\rho At}{m}$$

$$v_{\text{ball}} = (v_{\text{inlet}}^2 - v_{\text{inlet}}v_{\text{ball}})\left(\frac{(1 - \cos(45°))\rho At}{m}\right)$$

$$= v_{\text{inlet}}^2\left(\frac{(1 - \cos(45°))\rho At}{m}\right) - v_{\text{inlet}}v_{\text{ball}}\left(\frac{(1 - \cos(45°))\rho At}{m}\right)$$

$$v_{\text{ball}} + v_{\text{inlet}}v_{\text{ball}}\left(\frac{(1 - \cos(45°))\rho At}{m}\right) = v_{\text{ball}}\left(1 + v_{\text{inlet}}\left(\frac{(1 - \cos(45°))\rho At}{m}\right)\right)$$

$$= v_{\text{inlet}}^2\left(\frac{(1 - \cos(45°))\rho At}{m}\right)$$

$$v_{\text{ball}} = v_{\text{inlet}}^2\left(\frac{(1 - \cos(45°))\rho At}{m\left(1 + v_{\text{inlet}}\left(\frac{(1 - \cos(45°))\rho At}{m}\right)\right)}\right) = v_{\text{inlet}}^2\left(\frac{(1 - \cos(45°))At}{m + v_{\text{inlet}}((1 - \cos(45°))\rho At)}\right)$$

$$= (150\ cm/s)^2\left(\frac{(1 - \cos(45°))(1050\ kg/m^3)\frac{\pi}{2}\left(\frac{27\ mm}{2}\right)^2 t}{25g + (150\ cm/s)\left((1 - \cos(45°))(1050\ kg/m^3)\frac{\pi}{2}\left(\frac{27\ mm}{2}\right)^2 t\right)}\right)$$

$$= \frac{198\ g * m/s^2 * t}{25g + 132\ g/s * t}$$

To account for the bottom half of the flow

$$v_{\text{ball}} = 2\left(\frac{198\, g * m/s^2 * t}{25g + 132\, g/s * t}\right)$$

At $t = 0.5$ s

$$v_{\text{ball}} = 2\left(\frac{198\, g * m/s^2 * 0.5\, s}{25g + 132\, g/s * 0.5\, s}\right) = 218\, cm/s$$

This is consistent with a rapid opening of the valve, but the total length that the ball would traverse would only be approximately 4 cm (at most). Over time, the velocity of the ball would follow a logarithmic relationship (Figure 3.17), if there was no mechanism to stop the ball from moving (i.e., the cage).

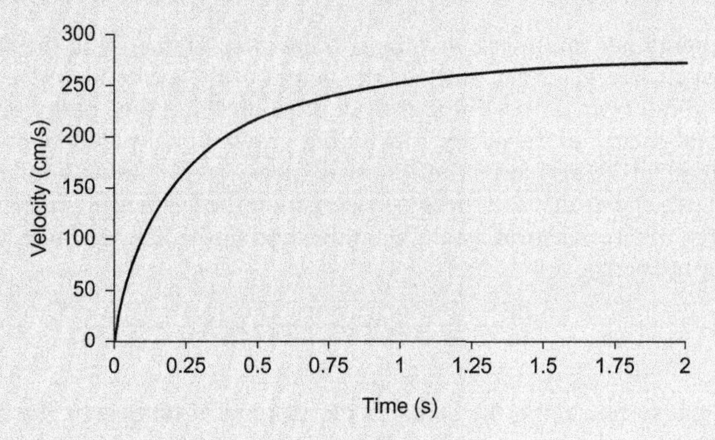

FIGURE 3.17 Velocity of the ball with respect to time.

We made the assumption in the derivation of Eq. (3.40) that the flow was irrotational, and therefore it only experienced pure translation. We will not show the derivation of the formula here, but the most general formula for the conservation of momentum must include all possible velocity components. This formula takes the form of

$$\begin{aligned}\vec{F}_b + \vec{F}_s - \int_V (\vec{a}_r + 2\vec{\omega} \times \vec{v}_{xyz} + \vec{\omega} \times (\vec{\omega} \times \vec{r}) + \vec{\omega} \times \vec{r}) \\ = \frac{\partial}{\partial t}\int_V \vec{v}_{xyz}\rho dV + \int_{\text{area}} \vec{v}_{xyz}\rho\vec{v}_{xyz} \cdot d\vec{A}\end{aligned} \tag{3.42}$$

where ω is the angular velocity (*note*: $\vec{\omega} \times \vec{v}_{xyz}$ is the Coriolis acceleration, $\vec{\omega} \times (\vec{\omega} \times \vec{r})$ is the centripetal acceleration, and $\vec{\omega} \times \vec{r}$ is the tangential acceleration due to angular velocity). This formula would be used if the fluid elements rotate, accelerate, and translate about each other or some reference coordinate axis (XYZ). Note again that this is a vector equation and can be written in terms of its component equation.

3.6 THE FIRST AND SECOND LAWS OF THERMODYNAMICS

The conservation of energy within a system is defined by the first law of thermodynamics, which is

$$\dot{Q} - \dot{W} = \frac{dE}{dt}$$

$\dot{Q}$ is the time rate of change of heat transfer and is positive when heat is added to the system. $\dot{W}$ is the time rate of change of work and is positive when work is done by the system. The energy of a system can be defined as

$$E = \int_V e\rho dV = \int_V \left(u + \frac{v^2}{2} + gz\right)\rho dV \tag{3.43}$$

where e is the energy per unit mass, u is the specific internal energy of the system, v is the speed of the system (not velocity), and z is the height of the system relative to a reference point. You may recall from a class in physics or dynamics that the energy per unit mass is related to the total energy of the system, including gravitational potential energy (e.g., gz), kinetic energy (e.g., $v^2/2$), and any other internal energy (represented here as u). In developing Eq. (3.22), we stated that for energy conservation the system property was E and the inherent property was e. Substituting these values into Eq. (3.22), we have a statement for the conservation of energy:

$$\frac{dE}{dt} = \frac{\partial}{\partial t}\int_V e\rho dV + \int_{\text{area}} e\rho\vec{v} \cdot d\vec{A} \tag{3.44}$$

If we define the system to be the same as the volume of interest at the instant in time that is of interest to us, we can make the following statement:

$$\dot{Q} - \dot{W}\Big|_{\text{system}} = \dot{Q} - \dot{W}\Big|_{\text{volume of interest}}$$

Substituting this into Eq. (3.44), we get

$$\begin{aligned} \frac{dE}{dt} &= \dot{Q} - \dot{W} = \frac{\partial}{\partial t}\int_V e\rho dV + \int_{\text{area}} e\rho\vec{v} \cdot \left|d\vec{A}\right| \\ &= \frac{\partial}{\partial t}\int_V \left(u + \frac{v^2}{2} + gz\right)\rho dV + \int_{\text{area}} \left(u + \frac{v^2}{2} + gz\right)\rho\vec{v} \cdot d\vec{A} \end{aligned} \tag{3.45}$$

In general, the rate of work is hard to quantify in fluid mechanics and is typically divided into four categories: work from normal stresses (W_n), work from shear stresses (W_{sh}), shaft work (W_s), and any other work (W_o).

From a physics course, you should remember that work is defined by force multiplied by the distance that the force acts over. When describing the work on a differential element

$$dW = \vec{F} \cdot d\vec{x}$$

The normal force acting on a differential element would be defined as

$$d\vec{F}_n = \sigma_n d\vec{A}$$

To define the time rate of change of work associated with the normal force

$$\dot{W}_n = \lim_{\Delta t \to 0} \frac{\partial W}{\Delta t} = \lim_{\Delta t \to 0} \frac{d\vec{F}_n \cdot d\vec{x}}{\Delta t} = \sigma_n d\vec{A} \cdot \vec{v}$$

because $\frac{d\vec{x}}{dt} = \vec{v}$. Therefore, the total work done by normal forces is

$$\dot{W}_n = -\int_{\text{area}} \sigma_n \vec{v} \cdot d\vec{A} \tag{3.46}$$

where the negative sign is needed to quantify the work done on the control volume instead of by the control volume.

The work of a shear force is defined in a similar way. Remember that shear force is defined as

$$d\vec{F} = \vec{\tau} dA$$

In this formulation, the shear stress is the vector quantity (to provide directionality of the stress which is different from the area vectors direction) not the area normal vector. Using the same process as above, the work of shear becomes

$$\dot{W}_{sh} = -\int_{\text{area}} \vec{\tau} \cdot \vec{v} dA \tag{3.47}$$

The work done by a shaft is not applicable to many biofluid mechanics problems, but would be defined as the negative of work input into the shaft to move the fluid. Similarly, other work would need to be defined by the type of work that is being done. For instance, if energy from an x-ray is being absorbed into the fluid, this can be considered as work being absorbed by the system. Using these definitions, Eq. (3.45) becomes

$$\dot{Q} + \int_{\text{area}} \sigma_n \vec{v} \cdot d\vec{A} + \int_{\text{area}} \vec{\tau} \cdot \vec{v} dA - \dot{W}_{\text{shaft}} - \dot{W}_{\text{other}}$$
$$= \frac{\partial}{\partial t} \int_V \left(u + \frac{v^2}{2} + gz \right) \rho dV + \int_{\text{area}} \left(u + \frac{v^2}{2} + gz \right) \rho \vec{v} \cdot d\vec{A}$$

by replacing e with various forms of internal energy that the system may contain. Rearranging we get.

$$\dot{Q} + \int_{\text{area}} \vec{\tau} \cdot \vec{v} dA - \dot{W}_{\text{shaft}} - \dot{W}_{\text{other}} = \frac{\partial}{\partial t} \int_V \left(u + \frac{v^2}{2} + gz \right) \rho dV$$
$$+ \int_{\text{area}} \left(u + \frac{v^2}{2} + gz \right) \rho \vec{v} \cdot d\vec{A} - \int_{\text{area}} \sigma_n \vec{v} \cdot d\vec{A}$$

We make use of the definition of specific volume (ν):

$$\rho = \frac{1}{\nu}$$

and then we combine like integrals to get

$$\dot{Q} + \int_{\text{area}} \vec{\tau} \cdot \vec{v} dA - \dot{W}_{\text{shaft}} - \dot{W}_{\text{other}}$$

$$= \frac{\partial}{\partial t}\int_V \left(u + \frac{v^2}{2} + gz\right)\rho dV + \int_{\text{area}} \left(u + \frac{v^2}{2} + gz\right)\rho \vec{v} \cdot d\vec{A} - \int_{\text{area}} \sigma_n \nu \rho \vec{v} \cdot d\vec{A}$$

$$= \frac{\partial}{\partial t}\int_V \left(u + \frac{v^2}{2} + gz\right)\rho dV + \int_{\text{area}} \left(u + \frac{v^2}{2} + gz - \sigma_n \nu\right)\rho \vec{v} \cdot d\vec{A}$$

From a previous discussion, we have defined that the normal stress is equal to the negative of the average hydrostatic pressure (in most cases without large viscous effects), therefore,

$$\dot{Q} + \int_{\text{area}} \tau \cdot \vec{v} dA - \dot{W}_{\text{shaft}} - \dot{W}_{\text{other}} = \frac{\partial}{\partial t}\int_V \left(u + \frac{v^2}{2} + gz\right)\rho dV + \int_{\text{area}} \left(u + \frac{v^2}{2} + gz + p\nu\right)\rho \vec{v} \cdot d\vec{A} \quad (3.48)$$

In most cases for this textbook, the pressure is acting in one dimension and we do not consider the hydrostatic pressure in the remaining directions.

Example

One of the functions of the cardiovascular system is to act as a heat exchanger, to maintain body temperature (Figure 3.18). Calculate the rate of heat transfer through a capillary bed, assuming that the blood velocity into the capillary is 100 mm/s and the flow velocity out of the capillary bed is 40 mm/s. The pressure on the arterial side is 20 mmHg, and the pressure on the venous side is 12 mmHg. Assume that the arteriole diameter is 75 μm and the venule diameter is 50 μm. The temperature on the arterial side is 35°C, and the temperature on the venous side is 33°C. Assume that the power put into the system throughout the muscular system is 15 μW.

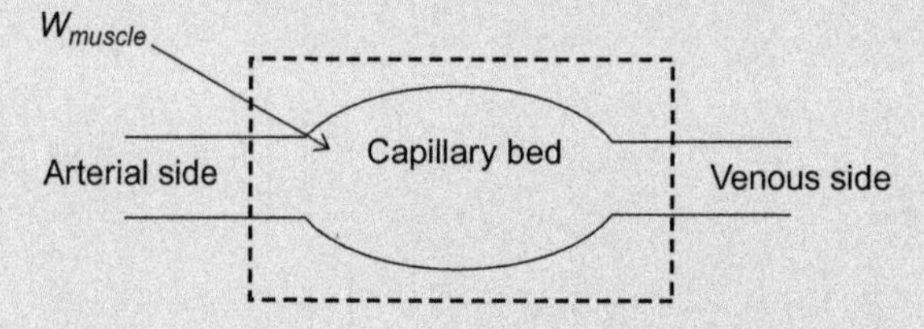

FIGURE 3.18 Schematic of a capillary heat exchanger for the example problem.

Solution

To solve this problem, we will make the assumptions that the flow is steady, uniform, the height difference between the arterial side and venous side is zero, and there is negligible internal energy or work done by stresses in the capillary bed. The conservation of mass does not hold across the capillary bed, because fluid is lost into the interstitial space:

$$\dot{Q} + \int_{area} \vec{\tau} \cdot \vec{v} dA - \dot{W}_{shaft} - \dot{W}_{other} = \frac{\partial}{\partial t}\int_{V}\left(u + \frac{v^2}{2} + gz\right)\rho dV + \int_{area}\left(u + \frac{v^2}{2} + gz + p\nu\right)\rho\vec{v} \cdot d\vec{A}$$

$$\dot{Q} = -\dot{W}_{muscle} + \int_{area}\left(u + \frac{v^2}{2} + gz + p\nu\right)\rho\vec{v} \cdot d\vec{A}$$

$$\dot{Q} = -\dot{W}_{muscle} + \left(\frac{v_1^2}{2} + gz_1 + p_1\nu_1\right)(-\rho v_1 A_1) + \left(\frac{v_2^2}{2} + gz_2 + p_2\nu_2\right)(\rho v_2 A_2)$$

Substituting known values into this equation

$$\dot{Q} = -15\ \mu W + \left(\frac{(100\ mm/s)^2}{2} + gz_1 + \frac{20\ mmHg}{1050\ kg/m^3}\right)\left(-(1050\ kg/m^3)(100\ mm/s)\pi\left(\frac{75\ \mu m}{2}\right)^2\right)$$

$$+\left(\frac{(40\ mm/s)^2}{2} + gz_2 + \frac{12\ mmHg}{1050\ kg/m^3}\right)\left((1050\ kg/m^3)(40\ mm/s)\pi\left(\frac{50\ \mu m}{2}\right)^2\right)$$

$$= -15\ \mu W + (-1\ \mu W - (4.64_E - 4\ g/s)gz_1) + (0.126\ \mu W + (8.25_E - 5\ g/s)gz_2)$$

$$= -15.874\ \mu W + g(0.825_E - 4(z_2) - 4.64_E - 4(z_1))\ g/s$$

We are making the assumption that there is no height difference between the arterial side and venous side ($z_2 = z_1$).

Therefore, the rate of heat transfer is $-15.874\ \mu W$. For every millimeter difference in height, where z_2 is lower than z_1, the rate of heat transfer would change by approximately -0.8 nW.

Example

Calculate the time rate of change of mass flow rate ($\rho v A$) of air entering the lungs. Assume that the lungs have a capacity of 6 L. The temperature of the lungs is 37°C. The air pressure inside of the lungs is 0.98 atm. At the instant that air enters the lungs, the temperature of the lungs raises by 0.0001°C/s. The height of the trachea is 20 cm. Assume that there is no work added to the system. Assume that air behaves as an ideal gas. Assume that the velocity is slow within the trachea.

Solution

$$\dot{Q} + \int_{\text{area}} \vec{\tau} \cdot \vec{v} dA - \dot{W}_{\text{shaft}} - \dot{W}_{\text{other}} = \frac{\partial}{\partial t} \int_V \left(u + \frac{v^2}{2} + gz \right) \rho dV$$

$$+ \int_{\text{area}} \left(u + \frac{v^2}{2} + gz + p\nu \right) \rho \vec{v} \cdot d\vec{A}$$

From the assumptions made, the governing equation can simplify to

$$0 = \frac{\partial}{\partial t} \int_V (u + gz) \rho dV + \int_{\text{area}} (u + gz + p\nu) \rho \vec{v} \cdot d\vec{A}$$

because there is no work/heat done by the system and that the velocity is slow, so that the square of the velocity is negligible. The velocity is included in the second area integral so we can calculate the mass flow rate. Also, we will make the assumption that the internal potential energy of the system, height, gravity, density, pressure are not functions of the volume of interest or the inflow area.

$$0 = \frac{\partial}{\partial t} \int_V (u + gz) \rho dV + (u + gz + p\nu)(-\rho v A) = \frac{\partial}{\partial t} \left[(u + gz) m \right] - (u + gz + p\nu)(\dot{m})$$

$$= \frac{\partial}{\partial t} (um + gzm) - (u + gz + p\nu)(\dot{m})$$

$$= \frac{\partial}{\partial t} (um) + \frac{\partial}{\partial t} (gzm) - u \frac{\partial m}{\partial t} - gz \frac{\partial m}{\partial t} - p\nu \frac{\partial m}{\partial t}$$

This is found by taking the area/volume integrals (making the assumption that none of the variables are a function of area/volume) and replacing the $\rho v A$ with $\dot{m}$. We also distribute the partial derivative across the summation and replace $\dot{m}$ with $\partial m / \partial t$. By making use of the product rule we can arrive at the following

$$0 = m \frac{\partial u}{\partial t} + u \frac{\partial m}{\partial t} + m \frac{\partial (gz)}{\partial t} + gz \frac{\partial m}{\partial t} - u \frac{\partial m}{\partial t} - gz \frac{\partial m}{\partial t} - p\nu \frac{\partial m}{\partial t} = m \frac{\partial u}{\partial t} - p\nu \frac{\partial m}{\partial t}$$

$$\dot{m} = \frac{\partial m}{\partial t} = \frac{m}{p\nu} \frac{\partial u}{\partial t}$$

if you make the assumption that the gravitational constant and height are not functions of time and rearrange the variables. Since we were told that the air behaves as an ideal gas, the following definitions apply

$$m = \rho V$$

$$p\nu = \frac{p}{\rho} = RT$$

$$\frac{\partial u}{\partial t} = C_v \frac{dT}{dt}$$

to obtain

$$\dot{m} = \frac{m}{pv}\frac{\partial u}{\partial t} = \frac{\rho V}{RT}C_v\frac{dT}{dt} = \frac{pV}{R^2T^2}C_v\frac{dT}{dt}$$

$$\dot{m} = \frac{0.98atm * 6L}{(287Nm/kg\ K)^2((37+273)K)^2}\left(717\frac{Nm}{kg\ K}\right)(0.0001\ K/s) = 3.24_E - 4\ g/\mathrm{min}$$

This is consistent with normal breathing.

The second law of thermodynamics is a statement about the disorder of a system. It states that the change of entropy of a system is greater than or equal to the amount of heat added to the system at a particular temperature:

$$dS \geq \frac{dQ}{T} \tag{3.49}$$

The time rate of change of entropy can therefore be defined as

$$\frac{dS}{dt} \geq \frac{\dot{Q}}{T} \tag{3.50}$$

for one specific volume of interest. When developing Eq. (3.22), we stated that for energy conservation the system property was S and the inherent property was s (entropy per unit mass). Substituting these values into Eq. (3.22), we have a statement for the conservation of energy related to the second law of thermodynamics:

$$\frac{dS}{dt} = \frac{\partial}{\partial t}\int_V s\rho dV + \int_{\text{area}} s\rho\vec{v} \cdot d\vec{A} \tag{3.51}$$

Substituting Eq. (3.50) into Eq. (3.51),

$$\frac{\partial}{\partial t}\int_V s\rho dV + \int_{\text{area}} s\rho\vec{v} \cdot d\vec{A} \geq \frac{\dot{Q}}{T} \tag{3.52}$$

Equation (3.52) is a statement of the second law of thermodynamics, which can be directly applied to fluid mechanics problems. In some instances, it may be useful to know that

$$\left(\frac{\dot{Q}}{T}\right)_V = \int_{\text{area}} \frac{\dot{Q}}{AT}dA \tag{3.53}$$

$\dot{Q}/A$ is the heat flux along one particular area. This is normally constant for one particular surface area of interest.

3.7 THE NAVIER–STOKES EQUATIONS

In the previous sections, we have applied various physical laws to a fluid volume of interest. However, to obtain an equation that describes the motion of the fluid at any time or location within the flow field, it may be easier to apply Newton's second law of motion to a particle. For a system such as this, Newton's second law becomes

$$d\vec{F} = dm\frac{d\vec{v}}{dt} \tag{3.54}$$

The derivation for the acceleration of a fluid particle has already been shown in Chapter 2. Using that relationship for particle acceleration, Newton's law becomes

$$d\vec{F} = dm\left(\frac{\partial\vec{v}}{\partial t} + u\frac{\partial\vec{v}}{\partial x} + v\frac{\partial\vec{v}}{\partial y} + w\frac{\partial\vec{v}}{\partial z}\right) \tag{3.55}$$

As discussed before, the forces on a fluid particle can be body forces or surface forces. To define these forces, let us look at the forces that act on a differential element with mass dm and volume $dV = dxdydz$. As done previously to describe the pressure acting on a differential element, we will assume that the stresses acting at the cubes center (denoted as p) are σ_{xx}, τ_{yx}, and τ_{zx} (Figure 3.19). Note that only the stresses that act in the x-direction are

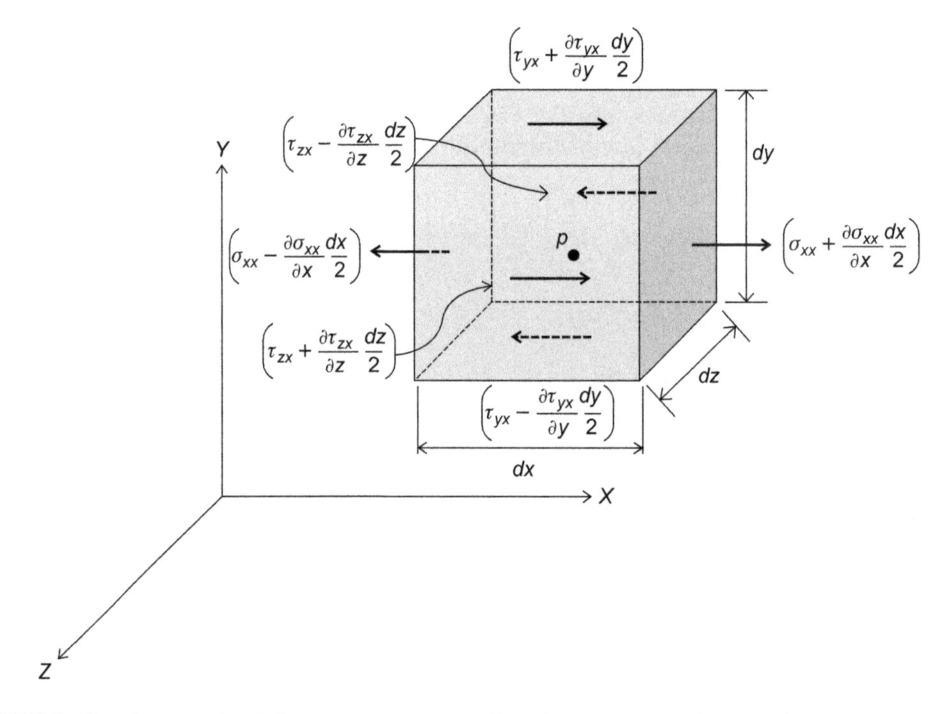

FIGURE 3.19 The normal and shear stresses acting in the x-direction on a differential fluid element. The stresses that act in the other Cartesian directions can be derived in a similar manner. Recall that only six of these stress values are independent for momentum conservation.

shown on this figure (and will be used in the derivation), but the same discussion applies to the other directions.

Summating the forces in the x-direction as a component of the total surface forces,

$$\begin{aligned} dF_{sx} &= \left(\sigma_{xx} + \frac{\partial \sigma_{xx}}{\partial x}\frac{dx}{2}\right)dydz - \left(\sigma_{xx} - \frac{\partial \sigma_{xx}}{\partial x}\frac{dx}{2}\right)dydz + \left(\tau_{yx} + \frac{\partial \tau_{yx}}{\partial y}\frac{dy}{2}\right)dxdz \\ &- \left(\tau_{yx} - \frac{\partial \tau_{yx}}{\partial y}\frac{dy}{2}\right)dxdz + \left(\tau_{zx} + \frac{\partial \tau_{zx}}{\partial z}\frac{dz}{2}\right)dxdy - \left(\tau_{zx} - \frac{\partial \tau_{zx}}{\partial z}\frac{dz}{2}\right)dxdy \\ &= \left(\frac{\partial \sigma_{xx}}{\partial x} + \frac{\partial \tau_{yx}}{\partial y} + \frac{\partial \tau_{zx}}{\partial z}\right)dxdydz \end{aligned} \tag{3.56}$$

Equation (3.56) is derived from the known stresses that act at point P, plus or minus the spatial gradient of the stress multiplied by the distance from P to the surface of the cube. To obtain a force, the stress that was quantified is then multiplied by the area over which it works. Using a similar analysis for each of the remaining two directions and assuming that the gravitational force is the only body force, the summation of the force in each direction becomes

$$\begin{aligned} dF_x = dF_{bx} + dF_{sx} &= \left(\rho g_x + \frac{\partial \sigma_{xx}}{\partial x} + \frac{\partial \tau_{yx}}{\partial y} + \frac{\partial \tau_{zx}}{\partial z}\right)dxdydz \\ dF_y = dF_{by} + dF_{sy} &= \left(\rho g_y + \frac{\partial \tau_{xy}}{\partial x} + \frac{\partial \sigma_{yy}}{\partial y} + \frac{\partial \tau_{zy}}{\partial z}\right)dxdydz \\ dF_z = dF_{bz} + dF_{sz} &= \left(\rho g_z + \frac{\partial \tau_{xz}}{\partial x} + \frac{\partial \tau_{yz}}{\partial y} + \frac{\partial \sigma_{zz}}{\partial z}\right)dxdydz \end{aligned} \tag{3.57}$$

For this formulation, we do not assume that gravity aligns with a particular Cartesian direction. Substituting Eq. (3.57) into Eq. (3.55) and writing the velocity in terms of its vector components gives us

$$\begin{aligned} \left(\rho g_x + \frac{\partial \sigma_{xx}}{\partial x} + \frac{\partial \tau_{yx}}{\partial y} + \frac{\partial \tau_{zx}}{\partial z}\right)dxdydz &= dm\left(\frac{\partial u}{\partial t} + u\frac{\partial u}{\partial x} + v\frac{\partial u}{\partial y} + w\frac{\partial u}{\partial z}\right) \\ \left(\rho g_y + \frac{\partial \tau_{xy}}{\partial x} + \frac{\partial \sigma_{yy}}{\partial y} + \frac{\partial \tau_{zy}}{\partial z}\right)dxdydz &= dm\left(\frac{\partial v}{\partial t} + u\frac{\partial v}{\partial x} + v\frac{\partial v}{\partial y} + w\frac{\partial v}{\partial z}\right) \\ \left(\rho g_z + \frac{\partial \tau_{xz}}{\partial x} + \frac{\partial \tau_{yz}}{\partial y} + \frac{\partial \sigma_{zz}}{\partial z}\right)dxdydz &= dm\left(\frac{\partial w}{\partial t} + u\frac{\partial w}{\partial x} + v\frac{\partial w}{\partial y} + w\frac{\partial w}{\partial z}\right) \end{aligned}$$

Using the relationship that $dm = \rho dV = \rho dxdydz$, the equations of motion become

$$\left(\rho g_x + \frac{\partial \sigma_{xx}}{\partial x} + \frac{\partial \tau_{yx}}{\partial y} + \frac{\partial \tau_{zx}}{\partial z}\right) = \rho\left(\frac{\partial u}{\partial t} + u\frac{\partial u}{\partial x} + v\frac{\partial u}{\partial y} + w\frac{\partial u}{\partial z}\right) = \rho\frac{Du}{Dt}$$

$$\left(\rho g_y + \frac{\partial \tau_{xy}}{\partial x} + \frac{\partial \sigma_{yy}}{\partial y} + \frac{\partial \tau_{zy}}{\partial z}\right) = \rho\left(\frac{\partial v}{\partial t} + u\frac{\partial v}{\partial x} + v\frac{\partial v}{\partial y} + w\frac{\partial v}{\partial z}\right) = \rho\frac{Dv}{Dt} \tag{3.58}$$

$$\left(\rho g_z + \frac{\partial \tau_{xz}}{\partial x} + \frac{\partial \tau_{yz}}{\partial y} + \frac{\partial \sigma_{zz}}{\partial z}\right) = \rho\left(\frac{\partial w}{\partial t} + u\frac{\partial w}{\partial x} + v\frac{\partial w}{\partial y} + w\frac{\partial w}{\partial z}\right) = \rho\frac{Dw}{Dt}$$

Equation (3.58) is the differential equations of motion, which are valid for any fluid that is a continuum and for any fluid that has the force of gravity as the only body force. In Chapter 2, we defined the normal stress as a function of hydrostatic pressure and the shear stresses as a function of viscosity and shear rate (which is a function of velocity). The following definitions apply for incompressible Newtonian fluids (complete derivations of these formula can be found in classical fluid mechanics textbooks):

$$\sigma_{xx} = -p - \frac{2}{3}\mu\nabla \cdot \vec{v} + 2\mu\frac{\partial u}{\partial x}$$

$$\sigma_{yy} = -p - \frac{2}{3}\mu\nabla \cdot \vec{v} + 2\mu\frac{\partial v}{\partial y}$$

$$\sigma_{zz} = -p - \frac{2}{3}\mu\nabla \cdot \vec{v} + 2\mu\frac{\partial w}{\partial z}$$

$$\tau_{xy} = \tau_{yx} = \mu\left(\frac{\partial u}{\partial y} + \frac{\partial v}{\partial x}\right) \tag{3.59}$$

$$\tau_{xz} = \tau_{zx} = \mu\left(\frac{\partial u}{\partial z} + \frac{\partial w}{\partial x}\right)$$

$$\tau_{yz} = \tau_{zy} = \mu\left(\frac{\partial v}{\partial z} + \frac{\partial w}{\partial y}\right)$$

where ∇ is the gradient operator and is defined as

$$\nabla f = \frac{\partial f}{\partial x}\vec{i} + \frac{\partial f}{\partial y}\vec{j} + \frac{\partial f}{\partial z}\vec{k}$$

for any function, f, in the Cartesian coordinate system (note that the gradient function can be calculated in any coordinate system for any function; we are just highlighting the Cartesian coordinate system for the derivation of the Navier–Stokes equations in a Cartesian system). Recall that the pressure is the mean of the principal stresses for a fluid with these criteria.

Substituting Eq. (3.59) into Eq. (3.58), the equations of motion become

$$\rho\frac{Du}{Dt} = \left(\rho g_x + \frac{\partial\left(-p - \frac{2}{3}\mu\nabla\cdot\vec{v} + 2\mu\frac{\partial u}{\partial x}\right)}{\partial x} + \frac{\partial\left(\mu\left(\frac{\partial u}{\partial y} + \frac{\partial v}{\partial x}\right)\right)}{\partial y} + \frac{\partial\left(\mu\left(\frac{\partial u}{\partial z} + \frac{\partial w}{\partial x}\right)\right)}{\partial z}\right)$$

$$= \rho g_x - \frac{\partial p}{\partial x} + \frac{\partial\left[\mu\left(2\frac{\partial u}{\partial x} - \frac{2}{3}\nabla\cdot\vec{v}\right)\right]}{\partial x} + \frac{\partial\left(\mu\left(\frac{\partial u}{\partial y} + \frac{\partial v}{\partial x}\right)\right)}{\partial y} + \frac{\partial\left(\mu\left(\frac{\partial u}{\partial z} + \frac{\partial w}{\partial x}\right)\right)}{\partial z}$$

$$\rho\frac{Dv}{Dt} = \left(\rho g_y + \frac{\partial\left(\mu\left(\frac{\partial u}{\partial y} + \frac{\partial v}{\partial x}\right)\right)}{\partial x} + \frac{\partial\left(-p - \frac{2}{3}\mu\nabla\cdot\vec{v} + 2\mu\frac{\partial v}{\partial y}\right)}{\partial y} + \frac{\partial\left(\mu\left(\frac{\partial v}{\partial z} + \frac{\partial w}{\partial y}\right)\right)}{\partial z}\right)$$

$$= \rho g_y - \frac{\partial p}{\partial y} + \frac{\partial\left[\mu\left(2\frac{\partial v}{\partial y} - \frac{2}{3}\nabla\cdot\vec{v}\right)\right]}{\partial y} + \frac{\partial\left(\mu\left(\frac{\partial u}{\partial y} + \frac{\partial v}{\partial x}\right)\right)}{\partial x} + \frac{\partial\left(\mu\left(\frac{\partial v}{\partial z} + \frac{\partial w}{\partial y}\right)\right)}{\partial z}$$

(3.60)

$$\rho\frac{Dw}{Dt} = \left(\rho g_z + \frac{\partial\left(\mu\left(\frac{\partial u}{\partial z} + \frac{\partial w}{\partial x}\right)\right)}{\partial x} + \frac{\partial\left(\mu\left(\frac{\partial v}{\partial z} + \frac{\partial w}{\partial y}\right)\right)}{\partial y} + \frac{\partial\left(-p - \frac{2}{3}\mu\nabla\cdot\vec{v} + 2\mu\frac{\partial w}{\partial z}\right)}{\partial z}\right)$$

$$= \rho g_z - \frac{\partial p}{\partial z} + \frac{\partial\left[\mu\left(2\frac{\partial w}{\partial z} - \frac{2}{3}\nabla\cdot\vec{v}\right)\right]}{\partial z} + \frac{\partial\left(\mu\left(\frac{\partial u}{\partial z} + \frac{\partial w}{\partial x}\right)\right)}{\partial x} + \frac{\partial\left(\mu\left(\frac{\partial v}{\partial z} + \frac{\partial w}{\partial y}\right)\right)}{\partial y}$$

Equation (3.60) is the full Navier–Stokes equations that are valid for any fluid. If we assume that the fluid is incompressible and the viscosity is uniform and constant, the equations simplify to

$$\rho\left(\frac{\partial u}{\partial t}+u\frac{\partial u}{\partial x}+v\frac{\partial u}{\partial y}+w\frac{\partial u}{\partial z}\right)=\rho g_x-\frac{\partial p}{\partial x}+\mu\left(\frac{\partial^2 u}{\partial x^2}+\frac{\partial^2 u}{\partial y^2}+\frac{\partial^2 u}{\partial z^2}\right)$$

$$\rho\left(\frac{\partial v}{\partial t}+u\frac{\partial v}{\partial x}+v\frac{\partial v}{\partial y}+w\frac{\partial v}{\partial z}\right)=\rho g_y-\frac{\partial p}{\partial y}+\mu\left(\frac{\partial^2 v}{\partial x^2}+\frac{\partial^2 v}{\partial y^2}+\frac{\partial^2 v}{\partial z^2}\right) \quad (3.61)$$

$$\rho\left(\frac{\partial w}{\partial t}+u\frac{\partial w}{\partial x}+v\frac{\partial w}{\partial y}+w\frac{\partial w}{\partial z}\right)=\rho g_z-\frac{\partial p}{\partial z}+\mu\left(\frac{\partial^2 w}{\partial x^2}+\frac{\partial^2 w}{\partial y^2}+\frac{\partial^2 w}{\partial z^2}\right)$$

Equation (3.61) is the form of the Navier–Stokes equations that will be used often in this textbook. In many biofluid mechanics examples, it is however more useful to solve the Navier–Stokes equations in a cylindrical coordinate system. The Navier–Stokes equations in cylindrical coordinates are as follows for incompressible fluids with a constant viscosity:

$$\rho\left(\frac{\partial v_r}{\partial t}+v_r\frac{\partial v_r}{\partial r}+\frac{v_\theta}{r}\frac{\partial v_r}{\partial \theta}-\frac{v_\theta^2}{r}+v_z\frac{\partial v_r}{\partial z}\right)=\rho g_r-\frac{\partial p}{\partial r}+\mu\left[\frac{\partial}{\partial r}\left(\frac{1}{r}\frac{\partial}{\partial r}(rv_r)\right)+\frac{1}{r^2}\frac{\partial^2 v_r}{\partial \theta^2}-\frac{2}{r^2}\frac{\partial v_\theta}{\partial \theta}+\frac{\partial^2 v_r}{\partial z^2}\right]$$

$$\rho\left(\frac{\partial v_z}{\partial t}+v_r\frac{\partial v_z}{\partial r}+\frac{v_\theta}{r}\frac{\partial v_z}{\partial \theta}+v_z\frac{\partial v_z}{\partial z}\right)=\rho g_z-\frac{\partial p}{\partial z}+\mu\left[\frac{1}{r}\frac{\partial}{\partial r}\left(r\frac{\partial v_z}{\partial r}\right)+\frac{1}{r^2}\frac{\partial^2 v_z}{\partial \theta^2}+\frac{\partial^2 v_z}{\partial z^2}\right]$$

$$\rho\left(\frac{\partial v_\theta}{\partial t}+v_r\frac{\partial v_\theta}{\partial r}+\frac{v_\theta}{r}\frac{\partial v_\theta}{\partial \theta}+\frac{v_r v_\theta}{r}+v_z\frac{\partial v_\theta}{\partial z}\right)=\rho g_\theta-\frac{1}{r}\frac{\partial p}{\partial \theta}+\mu\left[\frac{\partial}{\partial r}\left(\frac{1}{r}\frac{\partial}{\partial r}(rv_\theta)\right)+\frac{1}{r^2}\frac{\partial^2 v_\theta}{\partial \theta^2}+\frac{2}{r^2}\frac{\partial v_r}{\partial \theta}+\frac{\partial^2 v_\theta}{\partial z^2}\right]$$

(3.62)

The following two examples illustrate how to use either the Cartesian form or the cylindrical form of the Navier–Stokes equations. Remember that the usefulness of these equations is that the fully developed velocity profile of any flowing fluid can be determined. If the flow is one dimensional, then this can be easily solved by hand; with flows in multiple dimensions, solutions to coupled differential equations may be needed.

Example

Find an expression for the velocity profile and the shear stress (τ_{xy}) distribution for blood flowing in an arteriole with a diameter of 500 μm. Use the Navier–Stokes equations for Cartesian coordinates to solve this problem. The pressure driving this flow is given in Figure 3.20.

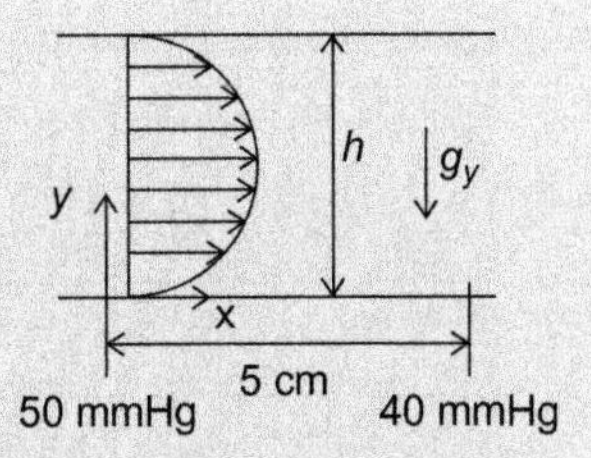

FIGURE 3.20 Pressure-driven flow in an arteriole for example problem.

Solution

To solve this problem, assume that $v_y = v_z = 0$ and v_x is a function of y only. Assume that the viscosity is constant and that the flow is incompressible and steady.

$$\rho\left(\frac{\partial u}{\partial t}+u\frac{\partial u}{\partial x}+v\frac{\partial u}{\partial y}+w\frac{\partial u}{\partial z}\right)=\rho g_x-\frac{\partial p}{\partial x}+\mu\left(\frac{\partial^2 u}{\partial x^2}+\frac{\partial^2 u}{\partial y^2}+\frac{\partial^2 u}{\partial z^2}\right)$$

$$0=-\frac{\partial p}{\partial x}+\mu\frac{\partial^2 u}{\partial y^2}=-\frac{\partial p}{\partial x}+\mu\frac{d^2 u}{dy^2}$$

$$\frac{d^2u}{dy^2}=\frac{1}{\mu}\frac{\partial p}{\partial x}\rightarrow d^2u=\frac{1}{\mu}\frac{\partial p}{\partial x}dy^2$$

$$\int d^2u=\int\frac{1}{\mu}\frac{\partial p}{\partial x}dy^2$$

$$du=\left(\frac{y}{\mu}\frac{\partial p}{\partial x}+c_1\right)dy$$

$$\int du=\int\left(\frac{y}{\mu}\frac{\partial p}{\partial x}+c_1\right)dy$$

$$u(y)=\frac{y^2}{2\mu}\frac{\partial p}{\partial x}+c_1y+c_2$$

To find the exact solution, we need boundary conditions for the particular flow scenario. Due to the no-slip boundary condition, we know that the velocity at both walls is zero and that the shear stress at the centerline is zero. Therefore, our boundary conditions are

$$u(0)=0$$

$$u(h)=0$$

$$\frac{du(h/2)}{dy}=0$$

Using two of these conditions to solve for the integration constants

$$u(0)=\frac{0^2}{2\mu}\frac{\partial p}{\partial x}+c_1*0+c_2=0$$

$$c_2=0$$

$$\frac{du(h/2)}{dy}=\frac{h/2}{\mu}\frac{\partial p}{\partial x}+c_1=0$$

$$c_1=-\frac{h}{2\mu}\frac{\partial p}{\partial x}$$

Substituting the values for these integration constants into the velocity equation

$$u(y) = \frac{y^2}{2\mu}\frac{\partial p}{\partial x} - \frac{h}{2\mu}\frac{\partial p}{\partial x}y = \frac{1}{2\mu}\frac{\partial p}{\partial x}(y^2 - hy)$$

For this particular flow scenario

$$\mu = 3.5\ cP$$

$$\frac{\partial p}{\partial x} = \frac{40\ mmHg - 50\ mmHg}{5\ cm - 0\ cm} = -2\ mmHg/cm$$

$$h = 500\ \mu m$$

$$u(y) = \frac{1}{2 * 3.5\ cP}(-2\ mmHg/cm)(y^2 - (500\ \mu m)y) = \frac{-3.81}{\mu ms}(y^2 - (500\ \mu m)y)$$

The shear stress profile for this particular flow is equal to

$$\tau_{xy} = \mu\left(\frac{\partial u}{\partial y} + \frac{\partial v}{\partial x}\right) = \mu\frac{\partial u}{\partial y}$$

$$\tau_{xy} = \mu\left(\frac{y}{\mu}\frac{\partial p}{\partial x} - \frac{h}{2\mu}\frac{\partial p}{\partial x}\right) = \frac{\partial p}{\partial x}\left(y - \frac{h}{2}\right)$$

Substituting the appropriate known values for this particular scenario

$$\tau_{xy} = \mu\left(\frac{y}{\mu}\frac{\partial p}{\partial x} - \frac{h}{2\mu}\frac{\partial p}{\partial x}\right) = \frac{\partial p}{\partial x}\left(y - \frac{h}{2}\right)$$

$$= -0.267\ dyne/cm^2\mu m(y - 250\ \mu m)$$

Example

Find an expression for the velocity profile and the shear stress distribution for blood flowing in an arteriole with a diameter of 500 μm. Use the Navier–Stokes equations for cylindrical coordinates to solve this problem. The pressure driving this flow is given in Figure 3.21.

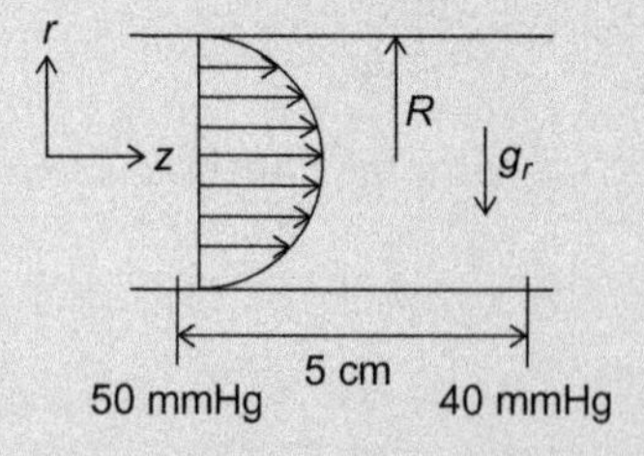

FIGURE 3.21 Pressure-driven flow in an arteriole with cylindrical coordinates for the in-text example. This is the same image as Figure 3.20, but choosing a different coordinate system to illustrate the usage of Cartesian coordinates versus cylindrical coordinates.

Solution

To solve this problem, assume that $v_r = v_\theta = 0$ and v_z is a function of r only. Assume that the viscosity is constant and that the flow is incompressible and steady.

$$\rho\left(\frac{\partial v_z}{\partial t} + v_r\frac{\partial v_z}{\partial r} + \frac{v_\theta}{r}\frac{\partial v_z}{\partial \theta} + v_z\frac{\partial v_z}{\partial z}\right) = \rho g_z - \frac{\partial p}{\partial z} + \mu\left[\frac{1}{r}\frac{\partial}{\partial r}\left(r\frac{\partial v_z}{\partial r}\right) + \frac{1}{r^2}\frac{\partial^2 v_z}{\partial \theta^2} + \frac{\partial^2 v_z}{\partial z^2}\right]$$

$$0 = -\frac{\partial p}{\partial z} + \frac{\mu}{r}\frac{\partial}{\partial r}\left(r\frac{\partial v_z}{\partial r}\right)$$

$$\frac{r}{\mu}\frac{\partial p}{\partial z} = \frac{d}{dr}\left(r\frac{dv_z}{dr}\right)$$

$$\int \frac{r}{\mu}\frac{\partial p}{\partial z}dr = \int d\left(r\frac{dv_z}{dr}\right)$$

$$\frac{r^2}{2\mu}\frac{\partial p}{\partial z} + c_1 = r\frac{dv_z}{dr} \rightarrow \frac{r}{2\mu}\frac{\partial p}{\partial z} + \frac{c_1}{r} = \frac{dv_z}{dr}$$

$$\int\left(\frac{r}{2\mu}\frac{\partial p}{\partial z} + \frac{c_1}{r}\right)dr = \int dv_z$$

$$v_z(r) = \frac{r^2}{4\mu}\frac{\partial p}{\partial z} + c_1\ln(r) + c_2$$

To find the exact solution, we need boundary conditions for the particular flow scenario. Due to the no-slip boundary condition, we know that the velocity at the wall is zero and that the shear stress at the centerline is zero. Therefore, our boundary conditions are

$$v_z(R) = 0$$

$$\frac{dv_z(0)}{dr} = 0$$

Note that we do not have a boundary condition of $v_z(-R) = 0$, because in cylindrical coordinates there is no negative radial direction. This location would be associated with 180° in the theta direction and $+R$ in the radial direction. Using these conditions to solve for the integration constants

$$\frac{dv_z(0)}{dr} = \frac{0}{2\mu}\frac{\partial p}{\partial z} + \frac{c_1}{0} = 0$$

Due to the discontinuity at $r = 0$, the only way for this equation to be valid is for $c_1 = 0$. Therefore, the discontinuity is removed. Using the second boundary condition

$$v_z(R) = \frac{R^2}{4\mu}\frac{\partial p}{\partial z} + c_2 = 0$$

$$c_2 = -\frac{R^2}{4\mu}\frac{\partial p}{\partial z}$$

Substituting the values for the integration constants into the velocity equation

$$v_z(r) = \frac{r^2}{4\mu}\frac{\partial p}{\partial z} - \frac{R^2}{4\mu}\frac{\partial p}{\partial z} = \frac{R^2}{4\mu}\frac{\partial p}{\partial z}\left(\left(\frac{r}{R}\right)^2 - 1\right)$$

For this particular flow scenario, using the same values as the previous example,

$$v_z(r) = \frac{R^2}{4\mu}\frac{\partial p}{\partial z}\left(\left(\frac{r}{R}\right)^2 - 1\right) = -11.9\ cm/s\left(\frac{r^2}{62,500\ \mu m^2} - 1\right)$$

The shear stress distribution is

$$\tau_{zr} = \mu\left(\frac{\partial v_r}{\partial z} + \frac{\partial v_z}{\partial r}\right) = \mu\frac{\partial v_z}{\partial r}$$

$$\tau_{zr} = \mu\left(\frac{r}{2\mu}\frac{\partial p}{\partial z}\right) = \frac{r}{2}\frac{\partial p}{\partial z} = -0.1332\ dyne/cm^2\mu m(r)$$

3.8 BERNOULLI EQUATION

The Bernoulli equation is a useful formula that relates the hydrostatic pressure, the fluid height, and the speed of a fluid element. However, there are a few important assumptions that are made to derive this formula, which makes this powerful equation not necessarily useful in many biofluid mechanics applications. Although as a back-of-the-envelope calculation, the Bernoulli equation can approximate real flow scenarios with reasonable accuracy. To derive this equation, the conservation of mass and conservation of momentum equations are simplified by making the assumptions that the flow is steady, incompressible, and inviscid (has no viscosity).

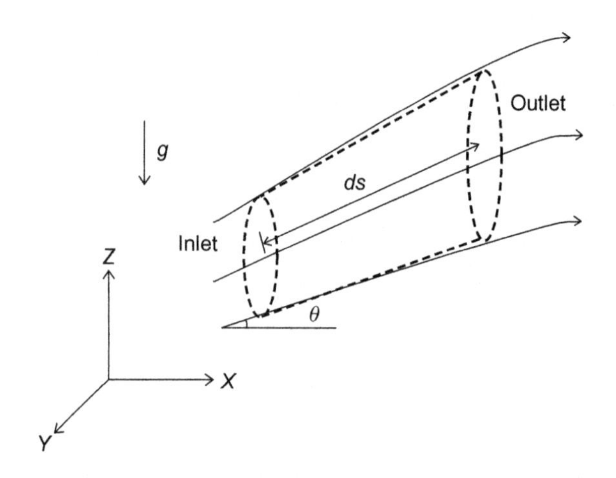

FIGURE 3.22 A differential volume of fluid following expanding streamlines (streamlines are the curved arrows in the figure). The expansion causes an increase in area at the outlet as compared to the inlet.

To derive the Bernoulli equation, let us follow a differential volume of fluid in an expanding streamline (Figure 3.22). The applicable fluid properties at the inlet will be denoted as p_i,v_i, A_i, and ρ. The applicable fluid properties at the outlet will be denoted as $p_i + dp_i$, $v_i + dv_i$, $A_i + dA_i$, and ρ. This same analysis can be conducted for a reducing streamline, where the solution would include negative differential changes, as necessary.

Applying the conservation of mass to this condition, we find that

$$0 = \frac{\partial}{\partial t}\int_V \rho dV + \int_{\text{area}} \rho\vec{v} \cdot d\vec{A}$$

$$0 = \int_{\text{area}} \rho\vec{v} \cdot d\vec{A} = -\rho v_i A_i + \rho(v_i + dv_i)(A_i + dA_i)$$

since we assumed that the flow is steady and incompressible, the time rate of change of mass within the control volume is equal to zero and density is not a function of the flow conditions. Again, the inflow conditions have a negative term and the outflow conditions are positive due to the direction of the velocity/area normal vector. Simplifying the previous equation (by multiplying), we can obtain

$$\rho v_i A_i = \rho(v_i A_i + v_i dA_i + A_i dv_i + dv_i dA_i) = \rho v_i A_i + \rho v_i dA_i + \rho A_i dv_i + \rho dv_i dA_i \tag{3.63}$$

Remember that the product of two differentials ($dv_i dA_i$) is going to be negligible compared to the remaining terms, which allows us to simplify Eq. (3.63) to

$$0 = v_i dA_i + A_i dv_i \tag{3.64}$$

Under the assumed conditions, the conservation of mass can simplify to Eq. (3.64). Now we will simplify the conservation of momentum equation in the streamline direction (s). The conservation of momentum states that

$$F_{bs} + F_{ss} = \frac{\partial}{\partial t}\int_V v_s \rho dV + \int_{\text{area}} v_s \rho\vec{v} \cdot d\vec{A}$$

for the streamline direction (subscript s). Because we are assuming steady flow, this formula simplifies to

$$F_{bs} + F_{ss} = \int_{\text{area}} v_s \rho\vec{v} \cdot d\vec{A} \tag{3.65}$$

Because the fluid is inviscid, the only forces that arise are from the pressure acting on the two surfaces and the surrounding fluid and the body force due to gravity. These forces are

$$\begin{aligned} F_{ss} &= p_i A_i - (p_i + dp_i)(A_i + dA_i) + \left(p_i + \frac{dp_i}{2}\right) dA_i = -A_i dp_i \\ F_{bs} &= g_s \rho dV = (-g\sin\theta)\rho\left(A + \frac{dA_i}{2}\right) ds = -\rho g\left(A_i + \frac{dA_i}{2}\right) dz \end{aligned} \tag{3.66}$$

because $\sin\theta\, ds = dz$. The differential volume term (dV in the F_{bs} equation) is an approximation of the volume using the midpoint area. The same analysis was used to approximate the pressure on the surrounding fluid; that is, use the midpoint pressure as the approximate pressure acting on the surfaces of the fluid. The flux term in Eq. (3.65) is equal to

$$\int_{\text{area}} v_s \rho \vec{v} \cdot d\vec{A} = v_i(-\rho v_i A_i) + (v_i + dv_i)(\rho(v_i + dv_i)(A_i + dA_i))$$

making use of the inflow/outflow conditions, as we have in the past. From the continuity equation (prior to simplifying the term)

$$v_i(-\rho v_i A_i) + (v_i + dv_i)(\rho(v_i + dv_i)(A_i + dA_i)) = v_i(-\rho v_i A_i) + (v_i + dv_i)(\rho v_i A_i) = \rho v_i A_i dv_i \quad (3.67)$$

Substituting Eqs. 3.66/3.67 into the momentum equation (Eq. (3.65))

$$-A_i dp - \rho g\left(A_i + \frac{dA_i}{2}\right)dz = -A_i dp_i - \rho g A_i dz = \rho v_i A_i dv_i$$

If we divide this equation by ρA_i, and simplify the velocity derivative (making use of a chain rule),

$$0 = v_i dv_i + \frac{dp_i}{\rho} + g dz = d\left(\frac{v_i^2}{2}\right) + \frac{dp_i}{\rho} + g dz$$

Integrating this equation and dropping the subscripts, we obtain the Bernoulli equation:

$$\frac{v^2}{2} + \frac{p}{\rho} + gz = \text{constant} \quad (3.68)$$

As we stated before, the Bernoulli equation is a powerful equation which relates the flow speed, the hydrostatic pressure, and the height to a constant. It can only be applied to a situation where the flow is steady, inviscid, and incompressible. In developing this relationship, we used a differential element, where these three criteria were valid. In most cases, it will not be easy to justify the use of the Bernoulli equation instead of the Navier–Stokes equations, the conservation of mass, and the conservation of momentum. However, as our example will show, we can use the Bernoulli equation as an approximation for various flow situations. In this simplified form, the Bernoulli equation is a statement of the conservation of energy for an inviscid fluid.

Example

Blood flow from the left ventricle into the aorta can be modeled as a reducing nozzle (Figure 3.23). Model both the left ventricle and the aorta as a tube with diameter of 3.1 and 2.7 cm, respectively. The pressure in the left ventricle is 130 mmHg and the pressure in the aorta is 123 mmHg. Blood is ejected from the left ventricle at a speed of 120 cm/s. Calculate the difference in height between these two locations.

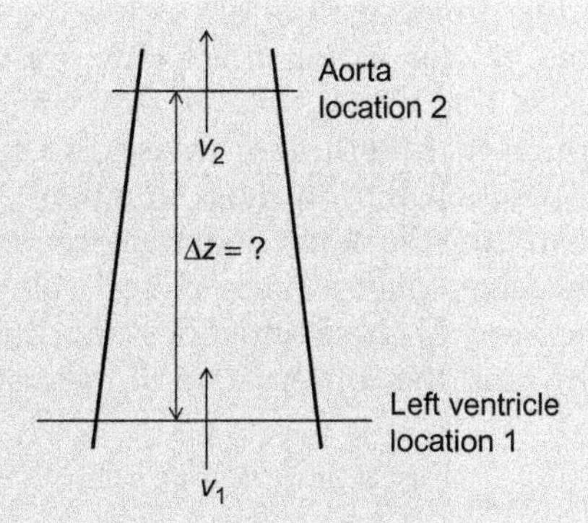

FIGURE 3.23 Schematic of the aorta downstream to the left ventricle. The aorta would experience a slight contraction within this area.

Solution

To solve this problem, we need to assume that the flow is steady, incompressible, and inviscid. Apply the conservation of mass to determine the blood velocity within the aorta.

$$0 = \int_{\text{area}} \rho \vec{v} \cdot d\vec{A} = -\rho v_1 A_1 + \rho v_2 A_2$$

$$v_2 = \frac{v_1 A_1}{A_2} = \frac{120\ cm \big/ s * \pi \left(\frac{3.1\ cm}{2}\right)^2}{\pi \left(\frac{2.7\ cm}{2}\right)^2} = 158\ cm/s$$

Note again, that we are seeing an acceleration of the fluid because of the decrease in area, but this does not match the physiology because blood vessels are not rigid. Using Bernoulli to solve for the difference in height

$$\frac{v_1^2}{2} + \frac{p_1}{\rho} + gz_1 = \frac{v_2^2}{2} + \frac{p_2}{\rho} + gz_2$$

$$g(z_2 - z_1) = \frac{v_1^2}{2} + \frac{p_1}{\rho} - \frac{v_2^2}{2} - \frac{p_2}{\rho}$$

$$z_2 - z_1 = \frac{\frac{v_1^2}{2} + \frac{p_1}{\rho} - \frac{v_2^2}{2} - \frac{p_2}{\rho}}{g}$$

$$= \frac{\frac{(120\ cm/s)^2}{2} + \frac{130\ mmHg}{1050\ kg/m^3} - \frac{(158\ cm/s)^2}{2} - \frac{123\ mmHg}{1050\ kg/m^3}}{9.81\ m/s^2} = 3.68\ cm$$

As you can imagine from the previous example, using the Bernoulli equation and the conservation of mass equations, we can estimate velocity, pressure, or height of a particular fluid. We want to emphasize that this is only an estimate in most real cases, because we need to make the assumption that the flow is inviscid to apply Bernoulli. Use caution when applying this powerful relationship. In most situations, if the blood vessel is large enough (e.g., aorta or vena cava), the Bernoulli equations can be applied, but as the vessel diameter reduces, the viscous forces play a more critical role in the flow. Therefore, the Bernoulli equations cannot be used in these situations and the Navier–Stokes equations, the conservation of momentum, and the conservation of mass should be applied.

END OF CHAPTER SUMMARY

3.1 The body forces acting on a differential fluid element are $d\vec{F}_b = \vec{g}dm = \vec{g}\rho dxdydz$. The surface forces acting on a differential fluid element are $d\vec{F}_s = -\nabla\vec{p}dxdydz$. For static fluids, where all acceleration terms are zero, the pressure gradient is equal to the gravitational acceleration multiplied by the fluid density. In Cartesian components, this is

$$-\frac{\partial p}{\partial x} + \rho g_x = 0$$

$$-\frac{\partial p}{\partial y} + \rho g_y = 0$$

$$-\frac{\partial p}{\partial z} + \rho g_z = 0$$

Most pressures that are recorded in biofluids are gauge pressures, which can be defined as

$$p_{\text{gauge}} = p_{\text{absolute}} - p_{\text{atmospheric}}$$

3.2 Buoyancy is the net vertical force that acts on a floating or an immersed object. The buoyancy forces can be defined as

$$F_z = \int_V dF_z = \int_V \rho g dV = \rho g V$$

3.3 A generalized formulation for the time rate of change of a system property can be represented as

$$\frac{dW}{dt} = \frac{\partial}{\partial t}\int_V w\rho dV + \int_{\text{area}} w\rho\vec{v}\cdot d\vec{A}$$

Applying this formulation to the conservation of mass, we would get

$$\left.\frac{dm}{dt}\right|_{\text{system}} = \frac{\partial}{\partial t}\int_V \rho dV + \int_{\text{area}} \rho\vec{v}\cdot d\vec{A} = 0$$

Depending on the particular flow conditions, the conservation of mass formula can be simplified in various ways.

3.4 The conservation of momentum can be represented as

$$\frac{dP}{dt} = \vec{F} = \vec{F}_b + \vec{F}_s = \frac{\partial}{\partial t}\int_V \vec{v}\rho dV + \int_{\text{area}} \vec{v}\rho\vec{v} \cdot d\vec{A}$$

Again, this can be simplified depending on the particular flow conditions.

3.5 The conservation of momentum could also include acceleration components. In Cartesian component form this would be

$$\vec{F}_{bx} + \vec{F}_{sx} - \int_V \vec{a}_{rx}\rho dV = \frac{\partial}{\partial t}\int_V u_{xyz}\rho dV + \int_{\text{area}} u_{xyz}\rho\vec{v}_{xyz} \cdot d\vec{A}$$

$$\vec{F}_{by} + \vec{F}_{sy} - \int_V \vec{a}_{ry}\rho dV = \frac{\partial}{\partial t}\int_V v_{xyz}\rho dV + \int_{\text{area}} v_{xyz}\rho\vec{v}_{xyz} \cdot d\vec{A}$$

$$\vec{F}_{bz} + \vec{F}_{sz} - \int_V \vec{a}_{rz}\rho dV = \frac{\partial}{\partial t}\int_V w_{xyz}\rho dV + \int_{\text{area}} w_{xyz}\rho\vec{v}_{xyz} \cdot d\vec{A}$$

3.6 It is common for heat exchange to occur within biological flows. To account for the conservation of energy within a fluid, the formulation would become

$$\dot{Q} - \dot{W} = \frac{\partial}{\partial t} e\rho dV + \int_{\text{area}} e\rho\vec{v} \cdot \left|d\vec{A}\right| = \frac{\partial}{\partial t}\int_V \left(u + \frac{v^2}{2} + gz\right)\rho dV + \int_{\text{area}} \left(u + \frac{v^2}{2} + gz\right)\rho\vec{v} \cdot d\vec{A}$$

The rate of work in this equation would need to account for all of the various work terms that may be applied to the fluid. In a simplified form, this would be

$$\dot{Q} + \int_{\text{area}} \vec{\tau} \cdot \vec{v} dA - \dot{W}_{\text{shaft}} - \dot{W}_{\text{other}} = \frac{\partial}{\partial t}\int_V \left(u + \frac{v^2}{2} + gz\right)\rho dV + \int_{\text{area}} \left(u + \frac{v^2}{2} + gz + p\nu\right)\rho\vec{v} \cdot d\vec{A}$$

The second law of thermodynamics can be applied to biofluids. It is represented as

$$\frac{\partial}{\partial t}\int_V s\rho dV + \int_{\text{area}} s\rho\vec{v} \cdot d\vec{A} \geq \frac{\dot{Q}}{T}$$

3.7 The Navier–Stokes equations are the solutions of Newton's second law of motion applied to fluid flow. For incompressible flows with a constant viscosity, these equations simplify to

$$\rho\left(\frac{\partial u}{\partial t} + u\frac{\partial u}{\partial x} + v\frac{\partial u}{\partial y} + w\frac{\partial u}{\partial z}\right) = \rho g_x - \frac{\partial p}{\partial x} + \mu\left(\frac{\partial^2 u}{\partial x^2} + \frac{\partial^2 u}{\partial y^2} + \frac{\partial^2 u}{\partial z^2}\right)$$

$$\rho\left(\frac{\partial v}{\partial t} + u\frac{\partial v}{\partial x} + v\frac{\partial v}{\partial y} + w\frac{\partial v}{\partial z}\right) = \rho g_y - \frac{\partial p}{\partial y} + \mu\left(\frac{\partial^2 v}{\partial x^2} + \frac{\partial^2 v}{\partial y^2} + \frac{\partial^2 v}{\partial z^2}\right)$$

$$\rho\left(\frac{\partial w}{\partial t} + u\frac{\partial w}{\partial x} + v\frac{\partial w}{\partial y} + w\frac{\partial w}{\partial z}\right) = \rho g_z - \frac{\partial p}{\partial z} + \mu\left(\frac{\partial^2 w}{\partial x^2} + \frac{\partial^2 w}{\partial y^2} + \frac{\partial^2 w}{\partial z^2}\right)$$

in Cartesian coordinates and

$$\rho\left(\frac{\partial v_r}{\partial t}+v_r\frac{\partial v_r}{\partial r}+\frac{v_\theta}{r}\frac{\partial v_r}{\partial \theta}-\frac{v_\theta^2}{r}+v_z\frac{\partial v_r}{\partial z}\right)=\rho g_r-\frac{\partial p}{\partial r}+\mu\left[\frac{\partial}{\partial r}\left(\frac{1}{r}\frac{\partial}{\partial r}(rv_r)\right)+\frac{1}{r^2}\frac{\partial^2 v_r}{\partial \theta^2}-\frac{2}{r^2}\frac{\partial v_\theta}{\partial \theta}+\frac{\partial^2 v_r}{\partial z^2}\right]$$

$$\rho\left(\frac{\partial v_z}{\partial t}+v_r\frac{\partial v_z}{\partial r}+\frac{v_\theta}{r}\frac{\partial v_z}{\partial \theta}+v_z\frac{\partial v_z}{\partial z}\right)=\rho g_z-\frac{\partial p}{\partial z}+\mu\left[\frac{1}{r}\frac{\partial}{\partial r}\left(r\frac{\partial v_z}{\partial r}\right)+\frac{1}{r^2}\frac{\partial^2 v_z}{\partial \theta^2}+\frac{\partial^2 v_z}{\partial z^2}\right]$$

$$\rho\left(\frac{\partial v_\theta}{\partial t}+v_r\frac{\partial v_\theta}{\partial r}+\frac{v_\theta}{r}\frac{\partial v_\theta}{\partial \theta}+\frac{v_r v_\theta}{r}+v_z\frac{\partial v_\theta}{\partial z}\right)=\rho g_\theta-\frac{1}{r}\frac{\partial p}{\partial \theta}+\mu\left[\frac{\partial}{\partial r}\left(\frac{1}{r}\frac{\partial}{\partial r}(rv_\theta)\right)+\frac{1}{r^2}\frac{\partial^2 v_\theta}{\partial \theta^2}+\frac{2}{r^2}\frac{\partial v_r}{\partial \theta}+\frac{\partial^2 v_\theta}{\partial z^2}\right]$$

in cylindrical coordinates. From either of these sets of equations, the fully developed fluid velocity profile can be directly solved for as a function of location within the flow field.

3.8 The Bernoulli equation is a useful formula that relates the pressure variation in a fluid to the height and the speed of the fluid element. However, this formulation is only valid for steady, incompressible, and inviscid flows. The Bernoulli equation states that

$$\frac{v^2}{2}+\frac{p}{\rho}+gz=\text{constant}$$

HOMEWORK PROBLEMS

3.1 A two-fluid manometer is used to measure the pressure difference for flowing blood in a laboratory experiment (Figure 3.24). Calculate the pressure difference between points *A* and *B* in the fluid. Assume that the density of blood is 1050 kg/m^3 and the density of water is 1000 kg/m^3.

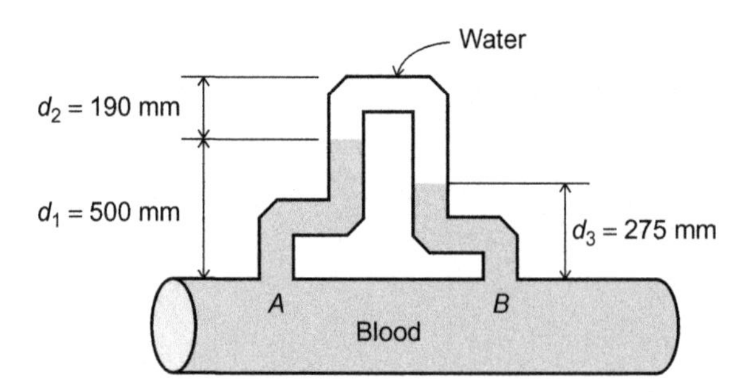

FIGURE 3.24 Figure for Homework Problem 3.1.

3.2 NASA is planning a mission to a newly found planet and will monitor the density of the new planet's atmosphere. Assume that NASA knows that atmosphere behaves as an ideal gas and that the planet's gravitational force is a function of altitude

$\left(g(z) = 18.7\ m/s^2\left(1 - \frac{z}{10{,}000\ m}\right)\right)$, where z is in m). The temperature of the atmosphere is constant at 250 K, and the gas constant is 340 Nm/kg K. Assume that the pressure at the planet's surface is 2 atm. Calculate the pressure and density at an altitude of 1, 5, and 9 km.

3.3 Calculate the hydrostatic pressure in the cranium and in the feet at the end of systole and the end of diastole for a hypertensive patient (end systolic pressure is equal to 185 mmHg and end diastolic pressure is equal to 145 mmHg). Assume that the blood density does not change significantly with height and that the cranium is 25 cm above the aortic valve and the feet are 140 cm below the aortic valve. Compare this with a normal patient (120 mmHg end systolic volume and 80 mmHg end diastolic volume).

3.4 A balloon catheter has been placed within a femoral artery of a patient to be passed to the coronary artery (use the same dimensions stated with Figure 3.6). Assume that the catheter consists of two components: (i) a chamber to hold the balloon, which is 2 mm in diameter and 1 cm in length (a perfect cylinder) and (ii) a tube 0.5 mm in diameter and the total length needed to transport the balloon to the opening locations. Calculate the buoyancy force on this catheter.

3.5 Consider the steady, incompressible blood flow through the vascular network as shown. Determine the magnitude and the direction of the volume flow rate through the daughter branch 2 (denoted as d_3 in Figure 3.25). The velocity at location 1 is inflow and the velocity at location 2 is outflow.

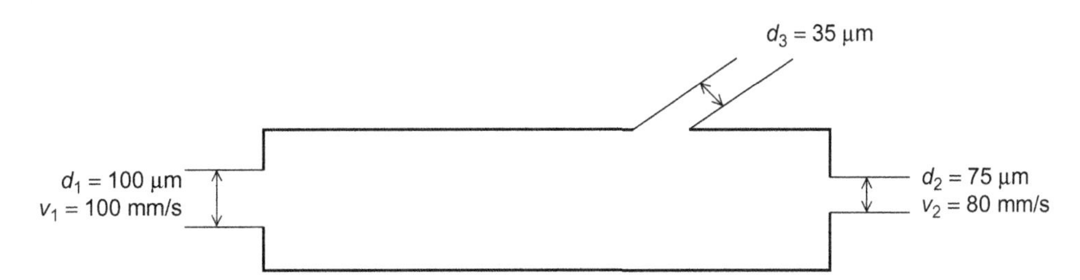

FIGURE 3.25 Figure for Homework Problem 3.5.

3.6 The variation of air viscosity as a function of temperature can be represented by the following equation:

$$\mu = \frac{\left(1.458 \times 10^{-6}\ kg/msK^{\frac{1}{2}}\right)T^{\frac{1}{2}}}{1 + \dfrac{110.4K}{T}}$$

Estimate changes in air viscosity as a function of altitude in Earth's atmosphere. Assume that the temperature is approximately −60°C, −40°C, and −80°C at an altitude of 10 km, 30 km and 75 km, respectively.

3.7 A biofluid with a density of 1080 kg/m^3 flows through the converging network as shown in Figure 3.26. Given that $d_1 = 15\ \mu m$, $d_2 = 9\ \mu m$, and $d_3 = 24\ \mu m$, with $v_1 = 5\ mm/s\ \vec{i}$ and $v_2 = 8\ mm/s\ \vec{j}$, determine the velocity v_3.

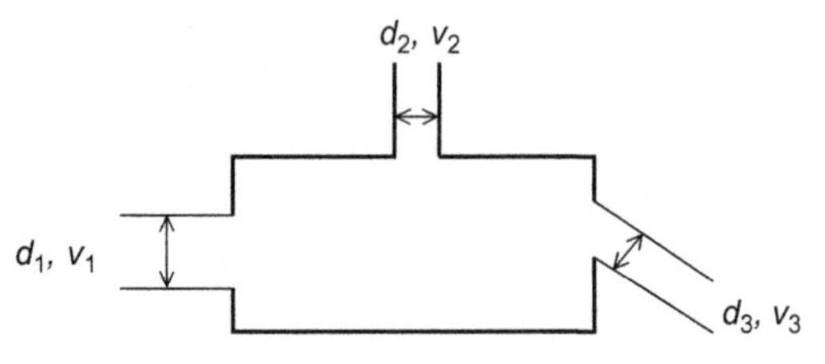

FIGURE 3.26 Figure for Homework Problem 3.7.

3.8 Using the same details for Problem 3.6, calculate the change in time rate of change of volume if v_3 is equal to 10 mm/s (as outflow). Assume that the density is not a function of time.

3.9 Air enters the lungs through a circular channel with a diameter of 3 cm and a velocity of 150 cm/s and a density of 1.25 kg/m^3. Air leaves the lungs through the same opening at a velocity of 120 cm/s and a density equal to that of the air within the lungs, which has decreased its net density, due to the lower density air entering the lungs. At the initial conditions the air within the lungs has a density of 1.4 kg/m^3, with a total volume of 6 L. Find the aggregate rate of change of the density of air in the lung assuming that your time step includes one inhale and one exhale (takes 15 s) and that there is no change in the lung volume over this time.

3.10 During peak systole, the heart delivers to the aorta blood that has a velocity of 100 cm/s at a pressure of 120 mmHg. The aortic root has a mean diameter of 25 mm. Determine the force acting on the aortic arch if the conditions at the outlet are a pressure of 110 mmHg and a diameter of 21 mm (Figure 3.27). The density of blood is 1050 kg/m^3. Assume that the aorta is rigid and that blood is incompressible. Ignore the weight of the blood vessel and the weight of blood within the blood vessel.

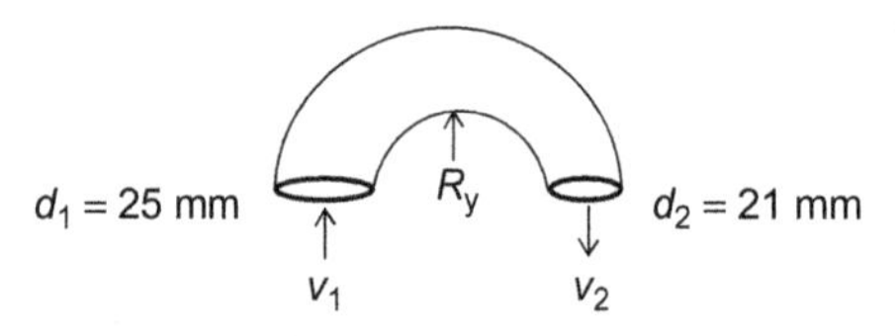

FIGURE 3.27 Figure for Homework Problem 3.10.

3.11 A reducing blood vessel has a 30° bend in it. Evaluate the components of force that must be provided by the adjacent tissue to keep the blood vessel in place. All necessary information is provided in Figure 3.28. Make sure to include the weight of blood within the blood vessel (the density of blood can be assumed to be 1050 kg/m^3).

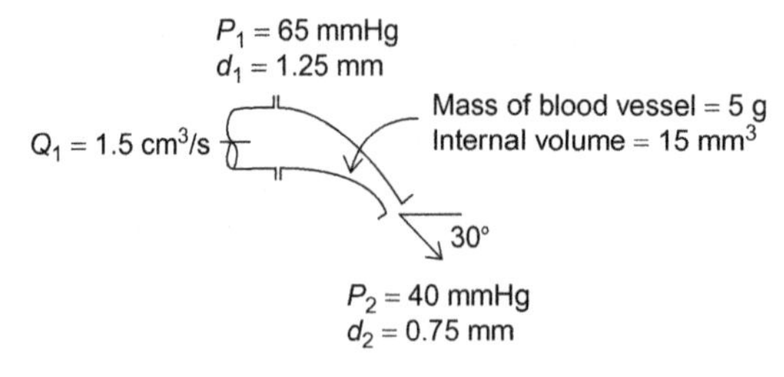

FIGURE 3.28 Figure for Homework Problem 3.11.

3.12 The following segment of the carotid artery (Figure 3.29) has an inlet velocity of 50 cm/s (diameter of 15 mm). The outlet has a diameter of 11 mm. The pressure at the inlet is 110 mmHg and at the outlet is 95 mmHg. Determine the reaction forces to keep this vessel in place. Assume that the blood density is 1050 kg/m^3 and ignore the mass of blood within the blood vessel and the mass of the blood vessel.

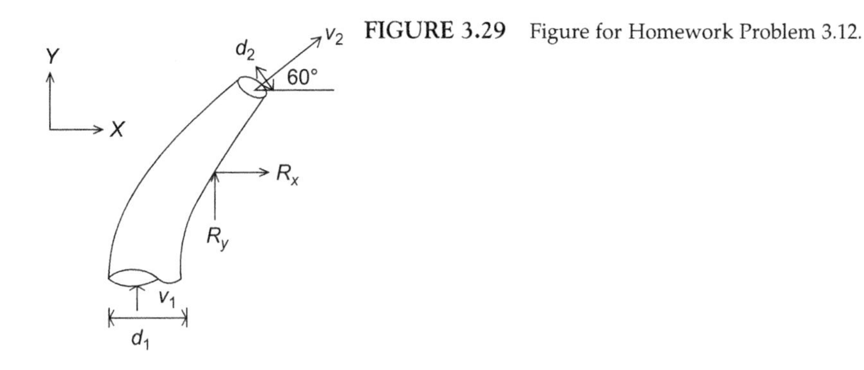

FIGURE 3.29 Figure for Homework Problem 3.12.

3.13 One of the first implantable mechanical heart valves was designed as a ball within a cage that acted as a check valve. Using the conservation of momentum (with acceleration), model the acceleration of the ball after it is hit by a jet of blood, being ejected from the heart (Figure 3.15). The ball has a turning angle of 45° and a mass of 20 g. Blood is ejected from the heart at a velocity of 110 cm/s, through an opening with a diameter of 25 mm. Determine the velocity of the ball at 0.5 s. Neglect any resistance to motion (except mass), assume that blood viscosity is 1050 kg/m^3 and there is no loss in area over this small distance.

3.14 During systole, blood is ejected from the left ventricle at a velocity of 125 cm/s. The diameter of the aortic valve is 24 mm, and there is no heat transfer or temperature change within the system. Assume that systole lasts for 0.25 s, that the height difference is 5 cm, and that there is no change in area within this distance. Determine the amount of work performed by the heart during systole and the power that the heart generates. The density of blood is 1050 kg/m^3 and we will assume that there is no internal energy changes associated with blood over this short time interval.

3.15 Air at standard atmosphere conditions (1 atm and 25°C) enters the lungs at 50 cm/s and leaves at a pressure of 1.1 atm, 37°C, and a velocity of 60 cm/s (with a constant mass flow rate of 1.2 g/s, e.g., density changes). The body removes heat from the lungs at a rate of 15 J/g. Calculate the power required by the lungs (this would be energy put gained by the system). Assume that there is no internal energy associated with this system.

3.16 The left common coronary artery has an axisymmetric constriction because of a plaque buildup (Figure 3.30). Given the upstream conditions of a velocity of 20 cm/s (systole) and 12 cm/s (diastole), calculate the velocity at the stenosis throat and the pressure difference between the stenosis throat and the inlet during systole and diastole. Assume that the Bernoulli principle can be used and that the density of blood is 1050 kg/m^3. Assume that there is no difference in height under these conditions.

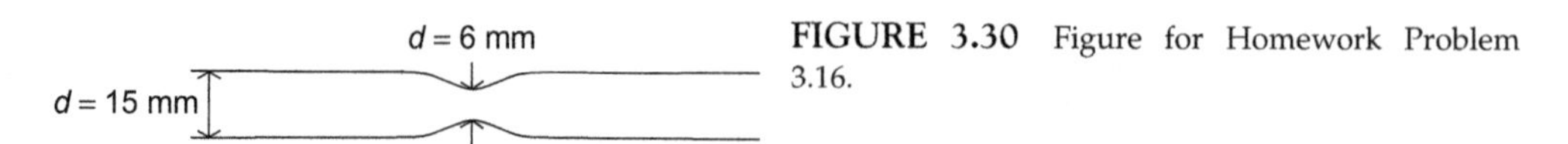

FIGURE 3.30 Figure for Homework Problem 3.16.

3.17 Blood flows through a 25% restricting (diameter reduces by 25%) blood vessel that experiences a 5 cm vertical drop (Figure 3.31). The blood pressure at the inlet is 65 mmHg, and the blood velocity is 50 cm/s. Calculate the blood velocity and the pressure at the outlet. Assume that the density of blood is 1050 kg/m^3.

FIGURE 3.31 Figure for Homework Problem 3.17.

3.18 The cross-sectional area of a diverging vein may be expressed as $A = A_1 e^{ax}$, where A_1 is the cross-sectional area of that inlet. Develop a relationship for the velocity profile within the vein (in terms of x, v_1). Also, develop a relationship for the pressure (if the inlet pressure is p_1) in terms of x. Assume that there is no variation in height and that the fluid properties remain constant within this channel.

3.19 Blood flows through a vertical tube with a kinematic viscosity of 3×10^{-4} m^2/s by gravity only (Figure 3.32). Solve the appropriate Navier–Stokes equations to find the velocity distribution $v_Z(r)$ and compute the average velocity.

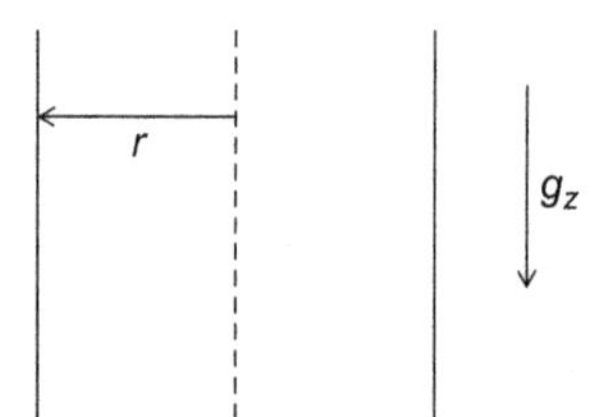

FIGURE 3.32 Figure for Homework Problem 3.19.

3.20 Solve Problem 3.19 assuming that the blood is flowing within a vertical parallel plate (i.e., calculate with the Cartesian Navier–Stokes equations), where the coordinate system is aligned with the wall and channel width is h (Figure 3.33).

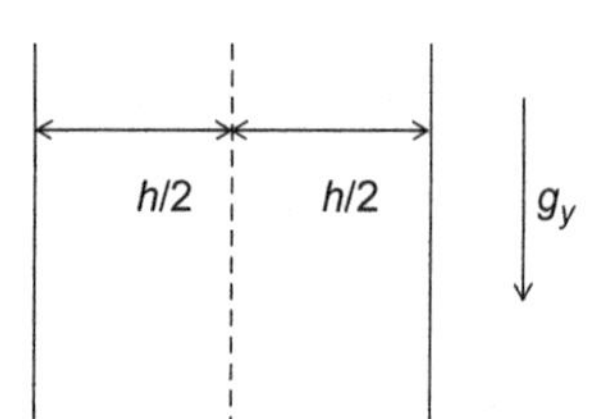

FIGURE 3.33 Figure for Homework Problem 3.20.

3.21 Upon inserting a catheter into the aorta, the blood flow must pass around the catheter. Assume that the catheter is placed directly in the centerline of the flow field (and is not moving at this instance in time; see Figure 3.34). Derive an expression for the velocity profile assuming that the flow is only pressure driven (dp/dz). The outer radius of the aorta is R and the outer radius of the catheter is kR.

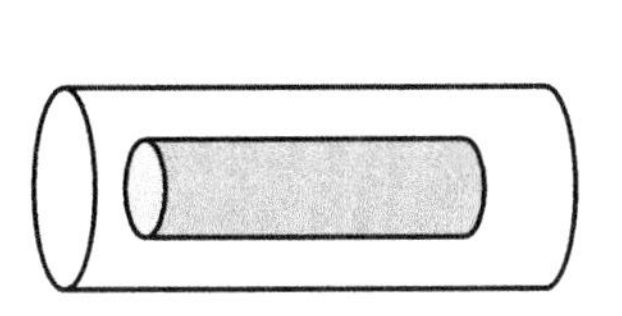

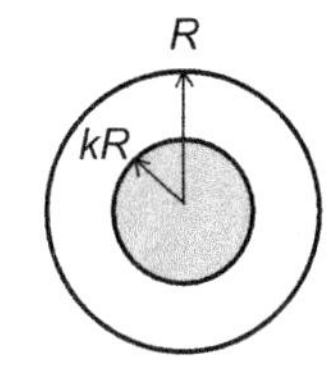

FIGURE 3.34 Figure for Homework Problem 3.21.

3.22 A 1.5-m person is standing underwater. First, calculate the amount of force acting on the person, assuming that the pressure at the surface of the water is 1 atm and that the person's head is just below the water surface. Second, repeat the calculation but assume that the person's head is 2 m below the surface. Approximate the volume of the person as a rectangular prism with width of 50 cm and depth of 20 cm. The density of water is 1000 kg/m^3.

3.23 Water leaves from a stationary hose and hits a moving vane with turning angle of $\theta = 40°$ (Figure 3.35). The vane moves away from the hose with a constant speed of $u = 20$ m/s and receives a jet that leaves the hose with a velocity of 125 m/s. The hose has an exit area of 0.025 m^2. Find the force that must be applied to maintain the speed of the vane. Assume that there is no loss/gain in area during the water motion, that the density of water is 1000 kg/m^3 and that the viscosity of water is 1 cP. Neglect the weight of water on the vane.

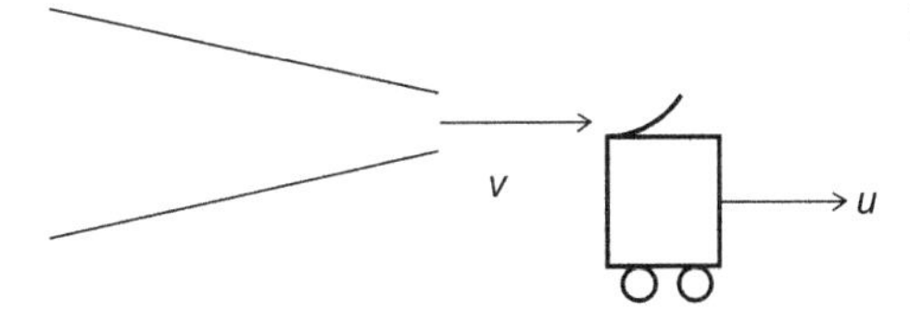

FIGURE 3.35 Figure for Homework Problem 3.23.

3.24 The heart is supplied with 5 L/min of blood from the vena cava (assume one blood vessel with a diameter of 20 cm). The aorta discharges the heart and has a diameter of 26 cm. Determine the pressure drop across the heart if it generates 4 μW during the process into the system. Assume that there is no internal energy in the fluid and that there is a negligible elevation difference between the outflow of the vena cava and the inflow of the aorta. The density of blood is 1050 kg/m^3.

3.25 The bladder acts as a pump which draws urine ($\rho = 1015$ kg/m^3) from the ureters and forces it to the urethra. The ureter has a diameter of 15 cm, and the urethra has a diameter of 5 cm. The height distance between the ureter and the urethra is 20 cm, and the pressure on the urethra is 1 atm. The average speed in the urethra is 3 mm/s. If the bladder has an efficiency of 80%, determine the power required to perform this process.

3.26 A branching blood vessel can be considered as a splitting vane that divides the flow into two streams. Find the mass flow rate ratio $\dot{m}_2/\dot{m}_3$ required to produce zero force in the y-direction (Figure 3.36). Determine the horizontal force under these conditions. Assume no energy loss due to branching and that velocity and diameters remain the same in each branch. There is no loss in area at the branch. The density of blood is 1050 kg/m^3.

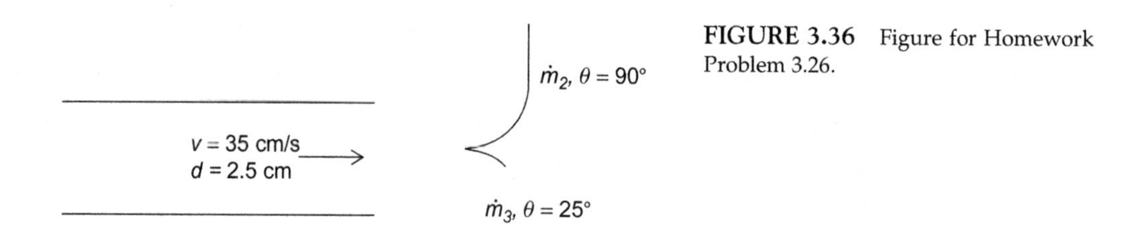

FIGURE 3.36 Figure for Homework Problem 3.26.

Reference

See references of Chapter 2.

PART II

MACROCIRCULATION

CHAPTER

4

The Heart

LEARNING OUTCOMES

1. Identify major structures that compose the heart
2. Explain how blood flows through the heart
3. Evaluate the action potential in the cardiac muscle as compared to nerve cells
4. Calculate the work that the heart undergoes to move blood through the cardiovascular system
5. Describe the cardiac conduction system
6. Relate the electrocardiogram waves to the contraction of heart muscle
7. Explain the pressure–volume relationship within the cardiac cycle
8. Model the motion of the heart in three dimensions
9. Calculate the stress distribution throughout heart wall tissue
10. Demonstrate the function of heart valves
11. Examine how blood flows through heart valves
12. Describe the tension–pressure relationship within heart valve leaflets
13. Compare salient disease conditions that are related to the heart and the heart valves

4.1 CARDIAC PHYSIOLOGY

The heart is a muscle that primarily acts to pump blood throughout the entire cardiovascular system (e.g., the systemic and pulmonary circuits). It is composed of two separate pumps that work in synergy to transport blood through the vascular system (Figure 4.1). One of these pumps (the "left pump") delivers oxygenated blood to all tissues that compose the body, while the other pump (the "right pump") delivers deoxygenated blood to the pulmonary circuit of the lungs to become oxygenated (this circuit does not provide oxygenated blood to the lung tissue; the systemic circulation via the bronchial arteries provide oxygenated blood to the lung tissue). Each of these pumps is composed of two pumping chambers: the atrium (plural, atria)

© 2015 Elsevier Inc. All rights reserved.

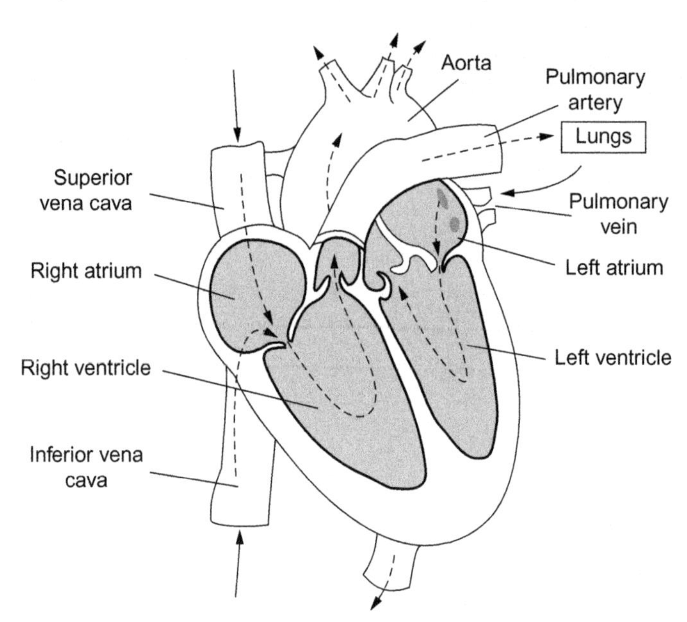

FIGURE 4.1 Anatomy of the heart depicting the direction of blood flow through the four heart chambers. The right side of the heart collects deoxygenated blood from the systemic circulation and delivers it to the pulmonary circulation. The left side of the heart collects oxygenated blood from the pulmonary circulation and delivers it to the systemic circulation.

and the ventricle. The atria are more of a passive pumping system that facilitates blood movement into the ventricles. The atria, therefore, act as primer pumps for the ventricles. The ventricles are more of an active pumping system that drives the blood flow throughout the body or pulmonary circuit of the lungs. In the next few sections, we will discuss the physiological controls on heart motion and how blood flows through the heart. Blood flow through the vascular system will be discussed in subsequent chapters.

To understand the physiology of the heart and how the heart functions, we first need to understand what cardiac muscle is and how it functions. Cardiac muscle cells (or cardiac myocytes) are interconnected in a very special way, allowing the cells to function as one. This is dissimilar to skeletal muscle, which readers may be more familiar with from previous physiology courses. Individual cardiac muscle fibers continually divide and merge, forming a highly interconnected mesh-like structure (Figure 4.2). The cell membranes of neighboring cardiac myocytes fuse at structures termed intercalated discs, which are typically present only along the longitudinal cell direction and not along the transverse direction (the intercalated discs are depicted as heavy black vertical lines penetrating the fibers in Figure 4.2). Many gap junctions are present in the intercalated discs allowing for a rapid ion communication between neighboring cardiac muscle cells. These gap junctions effectively reduce the intercellular resistance to ion transfer, by at least 2 orders of magnitude, as compared with ion movement across a cell membrane. Thus, there is a preference for communication molecules (e.g., ions) to move in the longitudinal direction to rapidly pass from one cardiac muscle cell to another cardiac muscle cell. Additionally, desmosomes are present within the intercalated discs of cardiac myocytes. Desmosomes are specialized cell–cell adhesion molecules that allow the cardiac myocyte contraction to mechanically pass between each cell and prevent the cells from "pulling" each other apart during a contraction.

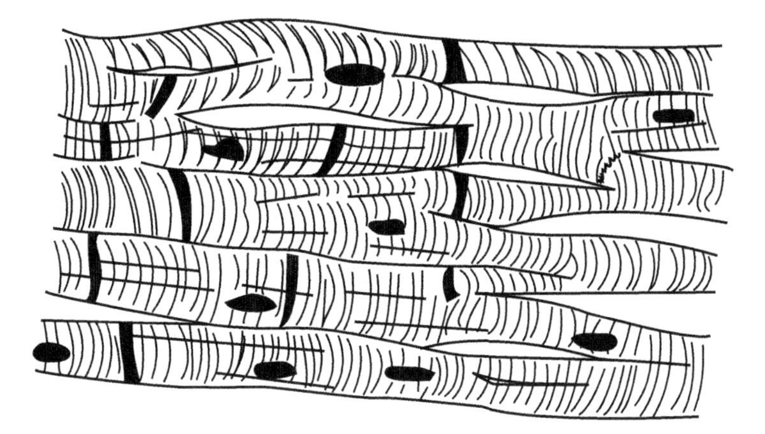

FIGURE 4.2 Schematic of cardiac muscle cells depicting the highly interconnected fibers that form a syncytium, and the intercalated discs that divide individual muscle cells. There are many gap junctions present within the intercalated discs to facilitate the movement of ions between cells, thus allowing the action potential to continuously flow from one cell to the next. *Source: Adapted from Guyton and Hall (2000).*

Although the physiology of cardiac muscle is quite different from skeletal muscle, cardiac muscle contracts in a very similar way to skeletal muscle. Again, readers may be familiar with this from a previous physiology course. Cardiac muscle fibers are striated and composed of similar sarcomere structures as skeletal muscles. Actin and myosin cross-bridges (in the form of myofibril units) produce the necessary connections to allow for muscle contraction. These proteins in the cardiac muscle unit are regulated with a very similar system to the troponin/tropomyosin system in skeletal muscles, which requires increased intracellular calcium to free actin binding sites. Again, similar to skeletal muscles, an action potential instigates contraction in cardiac muscle cells. Due to the presence of the intercalated discs (with many gap junctions) the action potential spreads from one cardiac myocyte to the neighboring myocytes with almost no impediment to the speed of the signal. This allows the formation of a muscle syncytium, which can work and produce contractions at the same time.

The coupling between excitation of the cardiac myocytes and the contraction of the muscle fibers is dictated by changes in the cell's cytosolic calcium concentration. Calcium is released from the sarcoplasmic reticulum and then binds to troponin, which allows the actin–myosin cross-bridge formation to begin. This occurs by a similar mechanism as skeletal muscle excitation–contraction coupling, however, there are some important differences. In cardiac muscle cells, the cell membrane (termed the sarcolemma) has many invaginations that surround the contractile elements (Figure 4.3). These invaginations are termed as transverse tubules or T-tubules, and they penetrate the muscle cell at the *z*-line of each sarcomere, therefore each sarcomere has effectively one T-tubule. The purpose of the T-tubule system is to bring the extracellular space close to the critical calcium containing intracellular space, termed the sarcoplasmic reticulum. The sarcoplasmic reticulum, which is analogous to the endoplasmic reticulum, is an organelle, which surrounds the entire myofibrils (and thus each sarcomere unit). The sarcoplasmic reticulum makes contact with the T-tubules at a specialized location termed the diad.

FIGURE 4.3 Spatial organization of the cardiomyocyte T-tubule system and sarcoplasmic reticulum. Invaginations of the sarcolemma for the T-tubule system, which associates with the sarcoplasmic reticulum at the dyad. *Source: Adapted from Feher (2012).*

At the diad unit, two very important calcium channels come into close proximity with each other. The dihydropyridine receptor (DHPR) can be found on the T-tubule membrane, whereas the ryanodine receptor (RyR) can be found on the sarcoplasmic reticulum membrane. Skeletal muscle and cardiac muscle contain DHPRs, but the protein subunits vary slightly, which is likely the cause of the different excitation–contraction coupling mechanisms between the two muscle types. In cardiac muscle cells, DHPR senses a depolarization of the T-tubule membrane and responds by activating the associated calcium channel. This small increase in intracellular calcium is insufficient to elicit a contraction response, however, the importance of this influx of calcium is that it is localized to the region of the RyR on the sarcoplasmic reticulum. The RyRs are responsive to increased cytoplasmic calcium and open allowing for a large increase in the intracellular calcium from the sarcoplasmic stores. This process is termed calcium-induced calcium release. A similar process can be observed in skeletal muscle contraction.

Cytoplasmic calcium regulates the formation of actin–myosin cross-bridges within the sarcomere unit. Briefly, under conditions when the cytoplasmic calcium concentration is low, tropomyosin blocks the myosin binding site on actin. When the intracellular calcium concentration is high, calcium binds to one of the troponin subunits (troponin C), causing a conformational change of troponin, in relation to actin and tropomyosin. This conformational change removes tropomyosin from the myosin binding site of actin, allowing for cross-bridge formation to occur, assuming there is sufficient ATP present to cause charging of the myosin head. Cross-bridge cycling is normally divided into four steps, where ATP binds to myosin, lowing its binding affinity to actin. ATP is hydrolyzed into ADP and an inorganic phosphate, changing the conformation of the myosin head. If the myosin binding site on actin is available (e.g., there is sufficient calcium present in the cytosol), then the charged myosin binds to actin. The conversion of ATP to ADP and inorganic phosphate induces a change in the myosin conformation, which induces a high affinity binding between myosin and actin. Finally, ADP and the inorganic phosphate are released, causing another conformational change of the myosin head where actin moves relative to myosin. This is termed the power stroke and is what elicits contraction. When another ATP molecule binds to the myosin head, myosin releases from the actin binding site. This process is schematized in Figure 4.4.

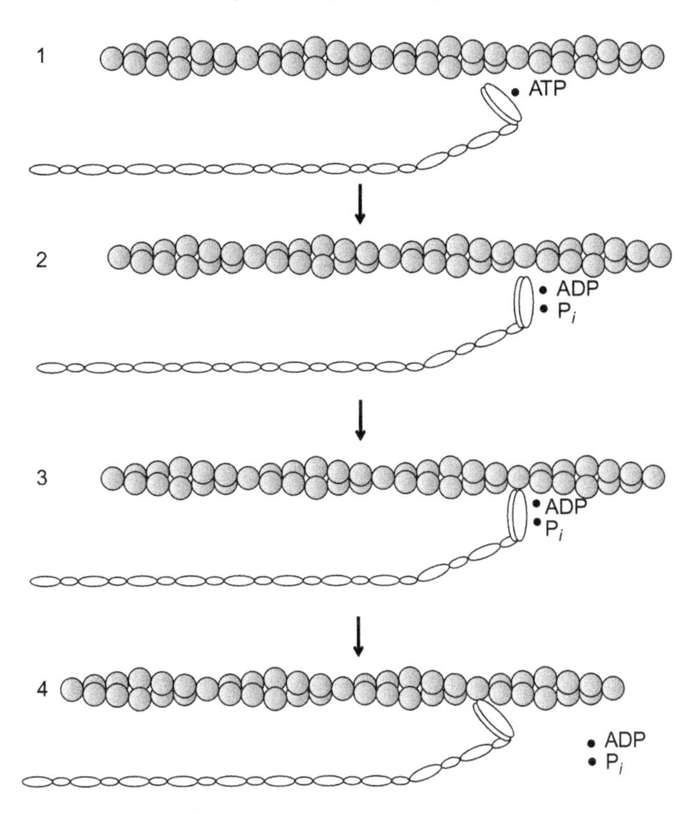

FIGURE 4.4 Schematic of the actin–myosin cross-bridge cycle. Note that troponin and tropomyosin are not depicted on this figure. (1) ATP binds to myosin, lowering the binding affinity between myosin and actin. (2) ATP is hydrolyzed into ADP and inorganic phosphate. Myosin head alters its conformation and stays in a "ready" state. In the absence of calcium the cycle will stop here. (3) In the presence of calcium, tropomyosin does not block the myosin binding site of actin and thus an actin–myosin cross-bridge is formed. (4) ADP and inorganic phosphate are released from the myosin head inducing a conformation change of myosin. This conformation change forces actin to move relative to myosin in a process termed the power stroke. In the absence of ATP, the cycle ends here (this is termed rigor).

This process of cross-bridge cycling will occur as long as the calcium concentration in the cytosol remains high. The RyR is typically open for a very short period, causing a burst of calcium release into the cytosol. This calcium is either pumped back into the sarcoplasmic reticulum (re-uptake of calcium) or out of the cell across the sarcoplasmic reticulum (extrusion of calcium). The sarcoplasmic reticulum contains smooth endoplasmic reticulum calcium ATPase pumps (SERCA pumps), which couple the movement of two calcium ions from the cytosol into the sarcoplasmic reticulum to the hydrolysis of ATP. This pump only functions if a regulatory protein, phospholamban, is colocalized with the SERCA pump and is phosphorylated. The sarcoplasmic reticulum can maintain a very high calcium concentration due to the presence of calsequestrin, which actively contains and "binds" to free calcium ions. Additionally, the sarcolemma contains a sodium–calcium exchanger and a plasma membrane calcium ATPase pump (PMCA),

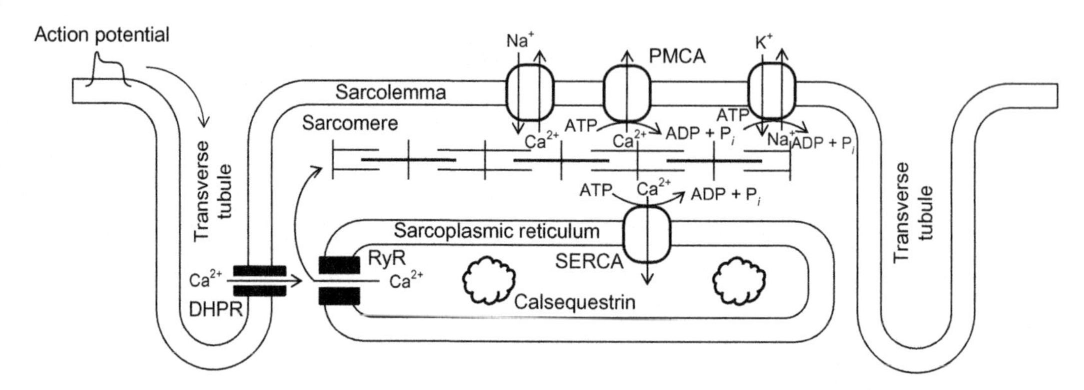

FIGURE 4.5 Overview of calcium-induced calcium release within cardiac muscle cells. First an action potential passes along the sarcolemma into the transverse tubule system. The action potential induces the opening of the DHPR, which allows a small amount of calcium into the muscle cell. This calcium induces the opening of the RyR found on the sarcoplasmic reticulum membrane, significantly increasing the cytosolic calcium concentration. This calcium is used to regulate cross-bridge formation (depicted in Figure 4.4). Calcium is then taken up by the SERCA to be stored in the sarcoplasmic reticulum and by the sodium calcium exchanger and PMCA on the sarcolemma.

along with the typical sodium–potassium ATPase. The exchanger transports three sodium ions down its concentration gradient (e.g., into the cell) in exchange for one calcium ion transported out of the cell. PMCA pumps one calcium ion out of the cell for every ATP that is hydrolyzed. Finally, the sodium–potassium ATPase returns the sodium to the external environment, in exchange for potassium influx and ATP hydrolysis. To a lesser extent cardiac myocyte mitochondria can uptake some calcium, through a sodium–calcium exchanger and a facilitated diffusion calcium channel. The entire pathway for calcium within a cardiac muscle cell is shown in Figure 4.5.

Now that we understand contractile mechanisms of the myocytes, we can begin to discuss how the heart regulates contraction at the organ level. The atrial and ventricular myocytes compose two separate syncytiums in the heart. When the atrial syncytium contracts as a whole, blood is ejected from the left and right atria into the left and right ventricles, respectively. When the ventricular syncytium contracts as a whole, blood is ejected from the two ventricles into the systemic or pulmonary circulatory system. These two syncytiums are separated by a fibrous tissue, which has a very high electrical resistance associated with it. The high-resistance tissue prevents the action potential signal from passing between the atria and the ventricles. Therefore, this prevents the ventricles from contracting at the same time as the atria and allows atrial contraction to fully fill the ventricles with blood. To couple atrial contraction with ventricular contractions, there is a specialized conduction pathway between these two chambers. This conduction pathway slows the action potential so that the atria can completely contract and eject sufficient blood to fill the ventricles before the ventricles begin to contract. Also, the conduction pathway brings the action potential to the apex of the heart, so that ventricle contraction proceeds from the bottom of the heart toward the top of the heart. This forces blood toward the aorta and the pulmonary arteries to exit the heart (see Section 4.2 for a detailed discussion on the cardiac conduction system).

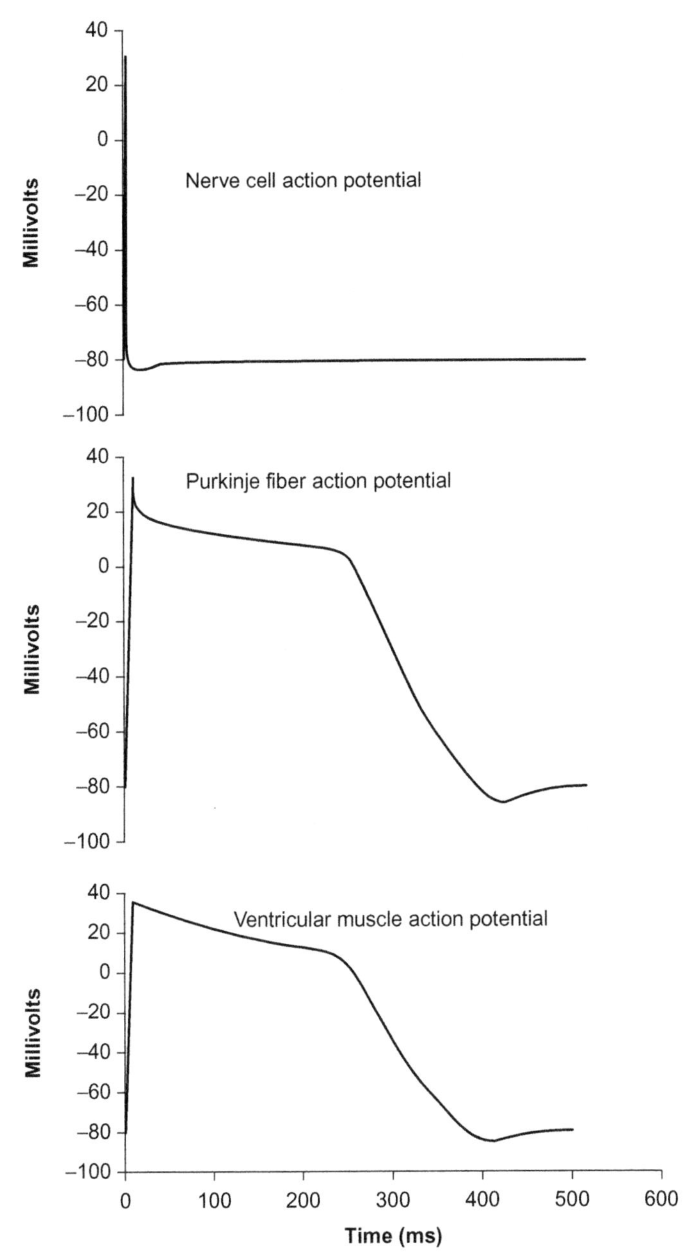

FIGURE 4.6 Comparison of the action potentials in nervous tissue and cardiac tissue. There is a significant elongation of the depolarization phase in cardiac tissues. Note that the cardiac action potential varies slightly based on what cardiac chamber the myocytes reside and that the myocytes of the SA node have a somewhat significantly different action potential waveform.

Another difference between cardiac muscle cells and skeletal muscle cells is that the cardiac action potential differs from skeletal muscle action potential, allowing for the cardiac myocytes to contract as a whole system, instead of as independent fibers (Figure 4.6). As with normal skeletal muscle action potentials, there is a rapid depolarization of the cells caused by an influx of sodium ions through rapid sodium channels. This depolarization increases the membrane potential from approximately negative 85 mV to approximately positive 15–20 mV. In a nerve cell, this is followed by a rapid repolarization (the duration of the action potential is less than 0.1 s) of the skeletal muscle. Cardiac action potentials, however, exhibit an elongated plateau phase in which the cells remain depolarized for approximately 0.25 s. This elongated plateau region is caused by a constant influx of calcium ions through slow calcium channels. During the plateau phase, the permeability of potassium ions decreases, significantly slowing the efflux of potassium ions out of the muscle. This altered potassium permeability is also not seen in skeletal muscle action potentials, where potassium efflux causes the rapid repolarization discussed above. After the slow calcium channels close, the cell membrane permeability toward potassium ions is increased, allowing for the rapid repolarization of the cell. After the cell has repolarized, there is a latent period of about 0.25 s, in which it is difficult to excite the cardiac myocytes again. After this latent period, the myocytes are able to contract again as normal. Except for the slow movement of calcium ions into the cardiac muscle cell, the ion movement during the cardiac and skeletal action potentials is similar.

This entire discussion has arrived at a point where it should be understood how the cardiac muscle contracts to effectively move blood throughout the cardiovascular system and the heart (Figure 4.1). To begin its path through the heart, deoxygenated blood enters from the superior vena cava or the inferior vena cava into the right atrium. The superior vena cava collects blood from the head, neck, and arms, while the inferior vena cava collects blood from the trunk and legs. Once the atrium contracts, blood passes through the right atrioventricular (AV) valve, also known as the tricuspid valve, into the right ventricle (as we will see later, blood actually passes through this valve during the majority of the cardiac cycle). This valve acts to prevent the backflow of blood into the right atrium, when the right ventricle contracts. Once the right ventricle contracts, blood is ejected through the right semilunar valve (also known as the pulmonary valve) into the respiratory circulation to become oxygenated. Blood is transported to the pulmonary circulation via the pulmonary arteries. Oxygenated blood returns into the left atrium via the pulmonary veins. As the atria contract, blood passes through the left AV valve, also known as the mitral valve, into the left ventricle. This valve also acts to prevent the backflow of blood during ventricular contraction and blood passes through it during a majority of the cardiac cycle. Upon left ventricle contraction, oxygenated blood is ejected into the systemic circulation, through the left semilunar valve (also known as the aortic valve). Note that all of the valves in the heart open and close passively, in response to pressure changes. During atrial contraction, the fluid pressure within each atrium becomes greater than the fluid pressure within the ventricles. This change in the pressure gradient direction helps to force remaining atrial blood to pass into the respective ventricles. As the atria relax and the ventricles contract, the pressure gradient reverses so that there is a higher pressure in the ventricles and a lower pressure in the atria. This exerts a force on the AV valves, causing them to close and thus prevent blood movement from the high pressure ventricles into the low pressure atria. As for the semilunar valves, when the ventricle pressure exceeds the aortic or pulmonary artery pressure, these valves open. When the ventricles relax, the pressure gradient reverses and the valves close to prevent the backflow of blood into the heart. As soon as the ventricular pressure reduces to a level below the atria pressure, the AV valves open allowing blood to pass directly from the venous circulation into the ventricles. Looking at the physiology of the heart valves, there are muscles attached to the AV valves (the papillary muscles). These muscles do not aid in valve opening, but they function to maintain the valves in a tightly closed position. During ventricle contraction, the papillary muscles contract to prevent the AV valves from opening in the reverse direction, into the atria. Without this preventative mechanism, blood would have a preference to flow back into the atria instead of into the aorta or pulmonary arteries, because the pressure in the atria is far lower than either of these vessels. For a complete discussion on the motion of the heart and the heart valve function, see Sections 4.4 and 4.5, respectively.

The work required for the heart to function is directly related to the pressure–volume relationship of the two ventricles. There is a very small component of energy needed to accelerate blood, but this is much less than the total work of the heart and is typically ignored when calculating heart work. In the pressure–volume relationship for the left ventricle, there are typically four phases that are discussed (Figure 4.7; this is shown for the left ventricle). The first phase (filling phase) is when the left ventricle is at rest and the left atrium is contracting, filling the left ventricle with blood. During this phase, the blood volume in the left ventricle increases from approximately 45 mL (end-systolic volume) to approximately 115 mL (end-diastolic volume). The pressure in the left ventricle increases from approximately 1 mmHg to approximately 5 mmHg during this time period. The second phase (isovolumic contraction) is the period during which the volume of the left ventricle does not change but the pressure increases rapidly. There is no change in left ventricular volume because both the mitral and aortic valves are closed. However, the left ventricular myocytes are contracting, which effectively increases the pressure to approximately 80 mmHg. This is the pressure in the aorta prior to valve opening (it is assumed that blood is incompressible and therefore there is no volume change). During the third phase (ejection phase), the pressure of the ventricle increases to approximately 120 mmHg, due to further contraction of the left ventricle myocytes and then reduces back to approximately 80 mmHg. The volume of the left ventricle also reduces because blood is being ejected into the aorta (the aortic valve is open). The last phase (isovolumic relaxation) is also classified by no change in left ventricular volume. During this phase, the mitral and

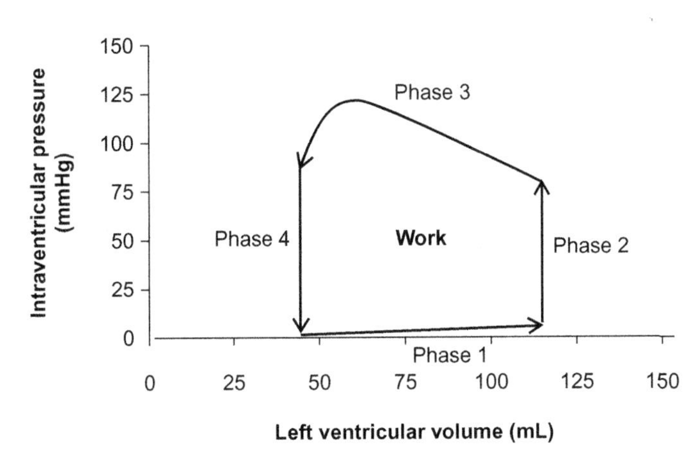

FIGURE 4.7 The pressure–volume relationship for the left ventricle, which describes the work that the left ventricle conducts. The work is the area within the pressure–volume curve. The work the heart conducts can drastically change under disease conditions. Note that work associated with the acceleration of blood is ignored on this figure and within the analysis of work conducted by the heart.

aortic valves are closed and the pressure reduces from approximately 80 mmHg to 1 mmHg. The reduction in pressure occurs because the left ventricle myocytes are relaxing. The pressure–volume loop forms a closed cycle, in which the work can be quantified by the area enclosed within the cycle. The right ventricular pressure–volume relationship is very similar to the left ventricle pressure–volume curve, except that the pressure is approximately five to six times lower than the left ventricle (the right ventricle curve is not shown). The reader is cautioned that the pressure–volume relationship changes drastically under different pathological conditions, where the systolic/diastolic pressure is not the "normal" 120 mmHg over 80 mmHg. Therefore, under a disease condition, the heart may work more than normal, potentially leading to other pathological conditions.

4.2 CARDIAC CONDUCTION SYSTEM/ELECTROCARDIOGRAM

The cardiac conduction system is composed of a specialized structure that generates impulses to excite the heart and a specialized conduction system to induce all ventricular myocytes contraction within a very short instant in time and only after atrial contraction completes (Figure 4.8). The sino-atrial (SA) node is the specialized structure that generates impulses within the heart. It is located within the right atrium wall immediately below the opening of the superior vena cava. The SA node is a group of specialized muscle cells that have little to no contractility associated with them, but they are directly connected to neighboring cardiac myocytes within the atrial syncytium.

The muscle cells that compose the SA node are surrounded by extracellular fluid with an abnormally high sodium concentration. Unlike other cardiac myocytes, the cells that compose the SA node have some cell membrane sodium channels that are always open in addition to the normal fast sodium channels, which open in response to an increasing membrane potential in the range of negative 40 to negative 50 mV. Due to the presence of continually open sodium channels (sometimes termed leak channels) and the continuous leak of sodium ions into the SA node cells, there is a slow increase in the SA node cells membrane potential from approximately negative 60 mV to approximately negative

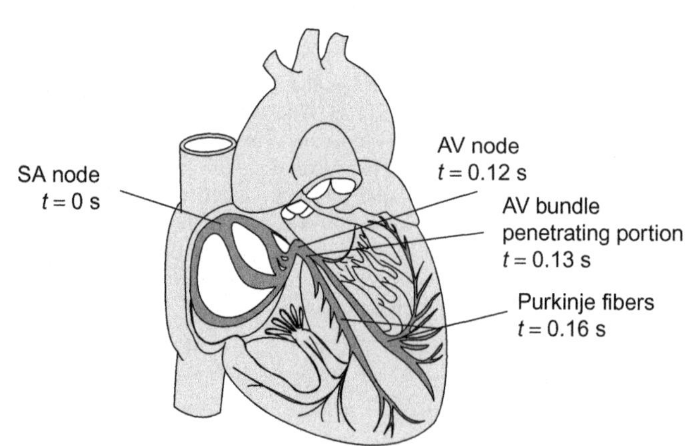

FIGURE 4.8 SA node and the Purkinje fiber system of the heart, showing the approximate times when the action potential signal reaches various locations within the conducting system. This conducting system allows for a rapid transmission of the cardiac action potential to each cardiac myocyte. For ease of viewing the interatrial band has been removed. This pathway conducts the action potential from the SA node to the left atrium rapidly.

40 mV. Once the membrane potential reaches approximately negative 40 mV (the SA node action potential threshold), the slow calcium channels and the fast sodium channels activate and open, allowing both calcium and sodium ions to pass rapidly into the myocytes that compose the SA node. Once this occurs, there is a rapid depolarization of the SA node cells, which leads to the action potential impulse throughout the atrial syncytium. Again, the cells that compose the SA node are directly connected with the neighboring atrial myocytes through extensive gap junction connections. Once the SA node has become depolarized, the action potential signal passes to neighboring myocytes through gap junctions, allowing for atria contraction. The conduction speed through the atria proceeds at approximately 0.3 m/s. The SA node cells also contain potassium channels, allowing for the efflux of potassium out of the cells, to bring the membrane potential back to approximately negative 60 mV. The constant sodium leak is counterbalanced by the presence of these potassium channels and the inactivity of the calcium and fast sodium channels (after the channels close), so that the membrane potential can approach the potassium equilibrium potential. The potassium and calcium channels act at the same time to prevent the SA node cells from continually experiencing a positive resting potential (near the sodium equilibrium potential), due to the continuous influx of sodium ions. The entire process of sodium leak into the SA node, depolarization of the SA node, and repolarization of the SA node occurs approximately once every second during normal resting conditions.

The SA node is also directly coupled to an interatrial band and three internodal pathways. The interatrial band connects the left and right atria, so that they can contract at approximately the same time. Remember that atrial contraction is more passive and the atria act as primer pumps for the ventricles. The interatrial band passes through the right atrium wall into the left atrium and is composed of specialized conducting cells that pass the depolarization signal to the left atrium at a rate of at least 1 m/s. The three internodal bands initiate at the SA node, follow along the right atrium wall (along the anterior, middle, and posterior sides), and terminate at the AV node. These fibers rapidly conduct the action potential signal (less than 0.04 s after they become activated by the action potential) to the AV node, which is responsible for instigating ventricle contraction.

The AV node is located in the wall of the right atrium and is primarily responsible for delaying the action potential signal from entering the ventricles. This allows both of the atria to fully prime the ventricles prior to ventricular contraction. The signal is delayed in the AV node for approximately 0.09 s before the action potential signal is allowed to pass through the penetrating portion of the AV node (total of approximately 0.13 s after SA node depolarization). The penetrating portion of the AV node passes directly through the highly resistant fibrous tissue that separates the atria and the ventricles. After passing through the fibrous tissue, there is another delay of approximately 0.03 s, in which time the action potential signal passes into the Purkinje fiber system to allow for rapid and synchronized depolarization of the ventricular muscle cells from the apex of the heart upwards. The cause of this slowed transmission in the AV node system is due to a decreased number of gap junctions between these cells. This effectively increases the resistance to action potential transmission as compared to the normal cardiac myocytes.

The Purkinje fibers act as a direct coupling between the AV node system and the ventricles. These fibers are very large and function to transmit the action potential signal very rapidly at speeds of up to 4 m/s. Recall that with an increase in fiber diameter there is a decrease in fiber resistance if other conditions remain the same. Immediately distal to the AV bundle, the Purkinje fibers split into the right and left branches of the Purkinje fibers. These branches follow the ventricular septum (the wall that separates the right and left ventricles) all the way to the apex of the heart. At the heart apex, these fibers branch into smaller fibers that penetrate the entire ventricular cardiac muscle mass. Therefore, this acts to pass the action potential signal directly to every muscle cell within the ventricular wall. The first cells that contract are those near the heart apex, and the wave of contraction passes upward and outward to the remaining cardiac muscle cells. As the reader may guess, the reason for the rapid conduction through the Purkinje fibers/ventricular mass is due to the presence of many gap junctions connecting the cells. This high conductivity allows all of the ventricular muscle cells to contract within 0.05 s of the cells at the apex. The same fibrous mass that blocks the atrial action potential signal from entering the ventricles blocks the ventricle action potential signal from entering the atria.

One of the most common ways to monitor the electrical system of the heart is through an electrocardiogram (ECG). To measure an ECG, electrical leads are placed at particular locations on the skin to measure the heart leak current that enters surrounding tissues. The normal ECG is composed of three waves termed the P wave, the QRS complex (sometimes this can be considered as three independent waves), and the T wave (Figure 4.9). Each of these waves represents a particular portion of the cardiac action potential. The P wave is associated with atrial depolarization and occurs immediately prior to atrial contraction. The P wave typically lasts for approximately 0.05 s under normal conditions. The QRS complex is associated with ventricular depolarization and occurs immediately prior to ventricular contraction. The QRS complex typically starts 0.15 s after the P wave starts and lasts for approximately 0.09 s. The T wave is associated with ventricular repolarization, and it occurs once the ventricular muscle cells recover from depolarization. This wave lasts for approximately 0.05 s, occurring approximately 0.25 s after the QRS complex ends. The atrial repolarization wave does not show up on the normal ECG because it occurs approximately 0.15 s after atrial depolarization. The atrial depolarization wave is a

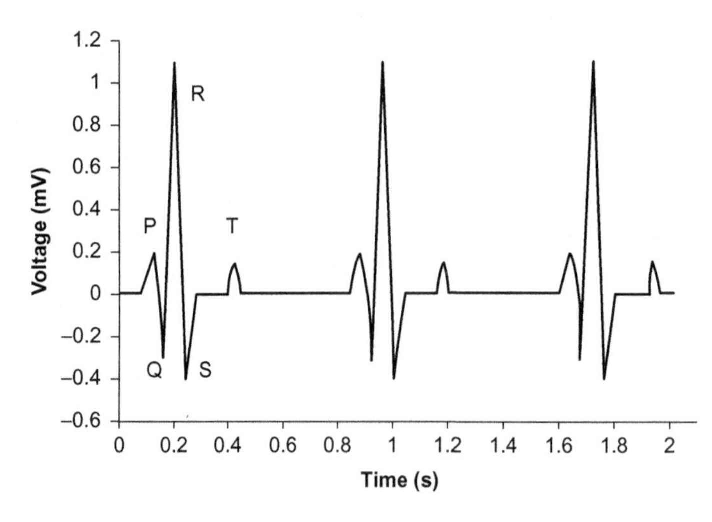

FIGURE 4.9 Schematic of a normal ECG trace, depicting the various waveforms used to determine if the heart is functioning properly. The P wave is associated with atria contraction, the QRS complex is associated with ventricular contraction, and the T wave is associated with ventricular repolarization. Atrial repolarization is masked by the much larger QRS complex.

small wave, and it is hidden by the much larger QRS complex that is occurring at the same time.

The normal ECG is characterized by the length of time of each wave, the time between each wave (typically called the interval, e.g., the P–Q interval or the Q–T interval), and the voltage associated with each wave. Many pathological heart conditions can be diagnosed by changes in the ECG signal. For instance, an increased voltage associated with the QRS complex can be attributed to an increased muscle mass. Decreased QRS voltage is typically associated with a decrease in muscle mass potentially because of infarctions present within the muscle mass. If the conduction system is blocked or damaged, then it is typical to see a change in the onset time or the length associated with the QRS complex. Under these pathological conditions, the QRS complex can elongate to approximately 0.15 s or more. It is also typical to see an inverted T wave when the QRS complex is significantly elongated.

4.3 THE CARDIAC CYCLE

The cardiac cycle describes all of the events that occur during one heartbeat and during the latent time until the next heartbeat. It makes the most sense to describe these events starting from the initiation of an action potential within the SA node (see Section 4.2). The cardiac cycle consist of two phases: diastole and systole. Cardiac myocytes do not contract during diastole, and this is when the heart chambers fill with blood. During systole, the myocytes contract and eject blood from the heart. We use the term systole and diastole interchangeably with ventricular systole and ventricular diastole. The contraction cycle of the atria can be described by their own systolic/diastolic pattern, however, the ventricular cycle is much more significant for the functionality of the heart.

To depict the cardiac cycle, the aortic pressure, the left ventricular pressure, the left atrium pressure, the left ventricular volume, and the ECG are overlaid onto one figure that is plotted against time (Figure 4.10, a similar figure can be drawn for the right side of the heart, with pressure values that are approximately 1/6th the value of the left side). Recall that the ECG P wave is associated with atrial depolarization. During this time, the mitral valve is open and the left atrium forces the remaining blood into the left ventricle, effectively priming the left ventricles for contraction. This priming action occurs, because during much of left ventricular diastole, the left atrial pressure is higher than the left ventricular pressure, which suggests that the mitral valve is open. Any blood that enters the left atria from the venous pulmonary circulation passes directly into the left ventricles, hence, the steady rise in ventricular volume during the diastolic portion of the cardiac cycle. During atrial systole, 10–20 mL of blood is forced into the ventricles, which acts to expand the ventricular muscle mass. This expansion causes an elastic recoil response, in addition to the ventricle muscle contraction, to aid in moving blood through the cardiovascular system.

Upon the onset of the QRS complex, there is a rapid increase in ventricular pressure due to ventricle contraction. This is associated with the closing of the mitral valve and isovolumic contraction of the ventricle. The left ventricle contracts for a short amount of time without losing volume because both the mitral valve and the aortic valve are closed.

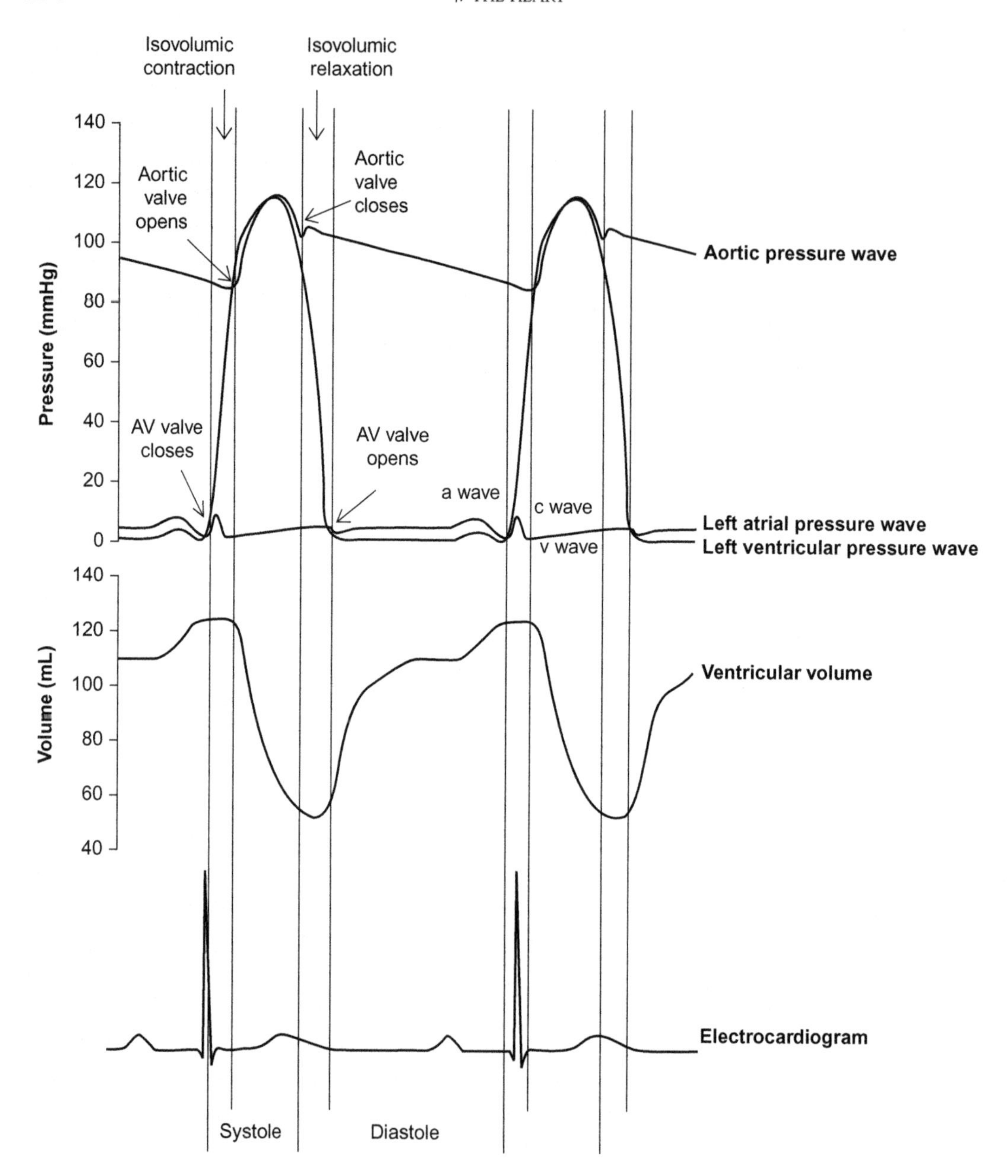

FIGURE 4.10 Pressure and volume waves associated with the left side of the heart. This figure depicts the relationship between the ECG and the contraction and filling of the cardiac tissue. Various important points are noted such as valve opening and ventricular systole versus ventricular diastole. The named waves that are observed during the atrial waveform are also shown in this figure.

As the QRS complex ends, the aortic valve opens (due to the left ventricle pressure surpassing the aortic pressure) and thus the left ventricle ejects blood into the systemic circulation. The duration of ventricle contraction is termed systole. The volume of blood in the ventricles reduces from approximately 120 mL to approximately 45 mL, which is termed the residual ventricular volume. During systole, the T wave is recorded, and this is when the ventricles begin to relax. At this time, the vascular pressure (e.g., aortic) is still lower than the ventricular pressure, so that blood continues to be ejected out of the heart for a few milliseconds. Toward the end of the T wave, the aortic valve closes (because the left ventricular pressure drops below the aortic pressure) and the ventricle enters the isovolumic relaxation phase, which marks the beginning of diastole. After a few milliseconds, the pressure in the ventricles returns to approximately 1 mmHg and the mitral valve opens again. At this point, diastole continues until the ventricles begin to contract and the mitral valve closes once again. As mentioned above, during the entire period of diastole, even though the left atrium is not contracting, the left ventricle is filling with blood. In fact, approximately 75% of the blood that enters the atrium passes directly into the ventricle without the aid of atrial contraction. During atrial contraction, the remaining blood volume enters the ventricles. Note that the discussion was for the left side of the heart, which is shown in Figure 4.10. A similar discussion for the right side of the heart could be conducted, but the pressure would be reduced as compared with the left side of the heart.

The pressure in the atria remains fairly constant (and low) during the entire cardiac cycle. However, there are three major changes that occur within the atrial pressure waveform, and they are denoted as the a (atrial contraction), c (ventricular contraction), and v (venous filling) waves. The a wave is associated with atrial contraction and occurs immediately after the P wave of the ECG. During the a wave, both atria experience an increase in pressure of about 6–7 mmHg, with the left atrium experiencing a slightly higher pressure increase than the right atrium. The c wave corresponds to the beginning of ventricular contraction and occurs immediately after the QRS complex of the ECG. This is caused primarily by the ventricular pressure acting on the AV valves. Also, at the beginning of systole, there is a small amount of blood backflow into the atria because the valves have not yet closed. Combined, these two changes induce an increase in atrial pressure. The v wave is a steady increase in atrial pressure that occurs during ventricular contraction, which is caused by venous blood entering the atria. When the AV valves reopen, this increased pressure aids in blood movement directly into the ventricles without atrial contraction.

Figure 4.10 depicts the cardiac cycle for the left ventricle, the left atrium, and the aorta. The aortic pressure curve is what is estimated when a patient has his or her blood pressure taken, or more accurately, the maximum aortic pressure and the pressure at which the aortic valve opens. At the point that the aortic valve opens, the pressure in the aorta increases due to blood being forced into the vascular system from the left ventricle. At peak systole, the blood pressure in the aorta reaches approximately 120 mmHg under normal conditions. At the point that the aortic valve closes, the pressure in the aorta is approximately 100 mmHg. At the time of valve closure, the pressure increases by approximately 5–10 mmHg, due to aortic elastic recoil and blood passing from the apex of the aortic arch back towards to the aortic valve. The backflow of blood occurs because the left ventricular pressure has dropped below the aortic pressure and now the pressure gradient

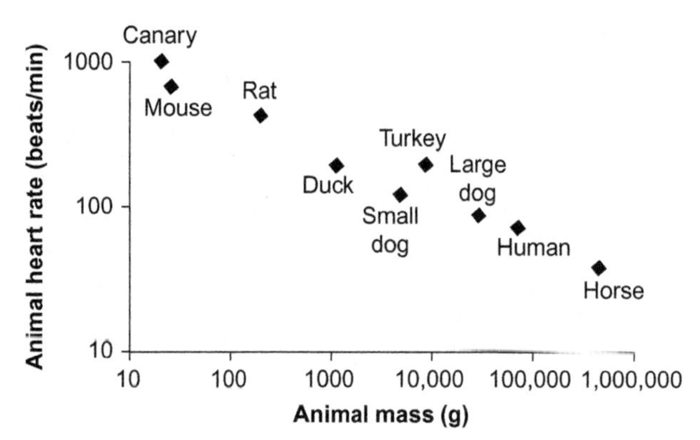

FIGURE 4.11 The relationship between animal mass and heart rate. As the mass of the animal increases, there is a general decrease in heart rate. The relationship between these two measurements can be correlated to many different properties of the animal, as described in the text.

favors blood moving down from the aortic arch towards both the abdominal aorta and the aortic arch. The slight raise in the aortic pressure is termed the dicrotic notch. The following rise in aortic pressure is referred to as the dicrotic wave. Then due to the viscoelastic recoil of the aorta, there is a slow, but continual, decrease in aortic pressure during diastole. At the end of diastole and the isovolumic contraction phase, the aortic pressure is approximately 80 mmHg. Once the left ventricular pressure surpasses this, the aortic valve opens and the pressure increases again to approximately 120 mmHg.

An additional interesting point is the sound that the heart makes during the cardiac cycle. A physician can listen to these sounds via a stethoscope. These sounds are caused by first the closure of the AV valves and then the closure of the semilunar valves. The sounds are generated by the valve vibrations during the closing process. As mentioned previously, the AV valves can bulge into the atrium and the semilunar valves can bulge into the ventricles immediately after closure. This is caused by an increase and a reversal in the pressure gradient across the leaflets. With respect to the AV valves, the papillary muscles and the tendons that attach to the valves (the chordate tendineae) experience recoil, inducing valve leaflet vibration. The semilunar valves themselves are highly elastic and experience recoil due to the pressure difference. This elastic recoil generates an audible sound.

As we know, the heart beats approximately 72 beats per minute for an average human. Compared with other animals, one can see that heart rate is inversely proportional to body mass (Figure 4.11). There are many models that can be used to describe this relationship, some of which account for if the animal is warm-blooded or cold-blooded, what the daytime activity of the animal is, and how the animal developed from an evolutionary standpoint. For this textbook, it is important to keep in mind that, in general, animals with a lower mass will have a higher heart rate.

4.4 HEART MOTION

During cardiac myocyte contraction, the heart as a whole moves in three-dimensional space. This movement affects the blood flow within the heart and also the blood flow

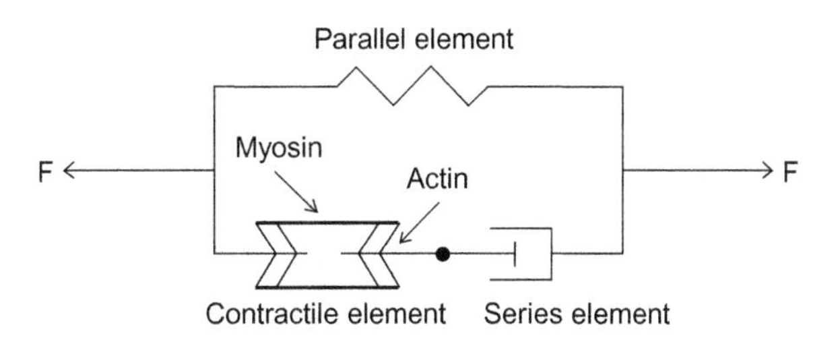

FIGURE 4.12 A.V. Hill's model for heart muscle contraction, which couples the active contraction component (actin–myosin cross-bridge formation), a viscoelastic relaxation, and a parallel elastic resistance to motion. This is perhaps the most often used model of cardiac muscle contraction and has many forms of quantification.

within the coronary blood vessels. Before discussing the motion of the heart, one must first understand how to model muscle contraction. In 1939, A.V. Hill modeled the active contraction of heart muscle using a three-component model (Figure 4.12). The three components of this model are a parallel element, a contractile element, and a viscoelastic series element. The parallel element models the connective tissue of the heart. This component is typically modeled as a purely elastic material, and the empirical values that describe its responses to a load can be calculated from standard length–tension curves of explanted muscle. These constants will be variable based on the activity of the muscle, the age of the person, the type of muscle fibers, etc., and thus it is difficult to provide one value for this constant. The contractile element models the actin–myosin mechanism that actively causes muscle contraction. This component of the system is modeled as a temporally active element that can generate tension and/or displacements. Known values for cross-bridge formation, including force generation and displacements, can depict this element. Again, it is somewhat difficult to provide empirical values for this value since it is dependent on many different factors within the system. The series element models the viscoelasticity of the actin–myosin cross-bridges and the viscoelasticity of the system as a whole. The empirical values that describe this component of the system can also be quantified by stimulating muscle to contract and monitoring the relaxation of the system (Figure 4.13).

Since 1939, many experiments have been conducted in order to determine the empirical constants that quantify this model. A full description of these experiments and the mathematical derivation of these formulae can be found in Y.C. Fung's biomechanics texts (see Further Readings). To summarize the salient results, the total force acting within the muscle (F) can be formulated as a summation of the parallel element force (P) and the contractile/series element force (S):

$$F = P + S \tag{4.1}$$

P is a function of the length of the muscle fiber. It is proportional to the number of sarcomeres present and can be formulated as the length of one myosin fiber plus the length of two actin molecules minus the overlap of the myosin and actin proteins. Assuming that all sarcomeres have the same physical structure, the parallel element force (P) can then be multiplied by the fiber length (or number of sarcomeres). If the sarcomere is under a stressed state, there is an added constant to account for the viscoelastic extension (η). S has

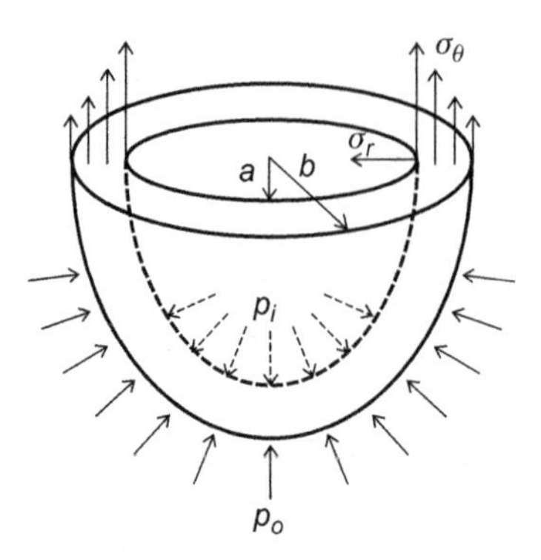

FIGURE 4.13 Schematic of the radial and circumferential stress in the left ventricle. This assumes that the cardiac tissue is homogeneous in size (and a perfect cylindrical shell) and homogeneous in mechanical properties, and that the pressure loading conditions are uniform. These formulations also neglect the blood flow through the ventricles. Note that we have tried to depict that the circumferential stresses act into and out of the paper, whereas the radial stresses act towards (or away from) the center of the sphere.

been shown to be an exponential function that is dependent on the initial tension in the fiber and the extension of the fibers. Hill's model neglects the kinetics of the biochemical reactions that need to take place in order for actin and myosin to form cross-bridges between each other. These (and other) necessary components of the heart model have been incorporated into theoretical models by others, which represent the tension and length curves reasonably well. However, Hill's model provides a fairly good prediction of how cardiac tissue will respond to forces.

A second form of the Hill equation relates the tension generated within a muscle under maximum tetanic contraction to the velocity of contraction and thermodynamic properties of the tissue. This equation takes the form of

$$F = \frac{\beta(F_0 + q)}{v + \beta} - q \tag{4.2}$$

where F is the force (as tension) that is generated within the muscle, F_0 is the maximum tetanic load generated within the muscle, v is the velocity of the contraction, q is the thermodynamic coefficient that relates to the heat lost during contraction, β is defined as $\frac{qv_0}{F_0}$, where v_0 is the maximum velocity of contraction which occurs when F is equal to zero. One can observe that this equation would predict a rapid contraction velocity when the muscle is under low loads, whereas the velocity of contraction would be slower under higher load. This loss in velocity appears to be due to both biochemical interactions during the contraction period (e.g., reduced tension generation when the sarcomere is shorter and an increase in the amount of intracellular fluid that must be moved in order to generate these motions).

Combining all of the observations from the Hill model, we can state that the force generated in the parallel element plus the force generated in the contractile element is equal to the total force generated in the muscle (e.g., Eq. (4.1)). Additionally, the force generated in the contractile element is equal to the force generated in the series element. As for the length of the muscle, the total muscle length is equal to the length of the parallel element, which is also equal to the length of the contractile element plus the series element. Clearly, mathematical representations of each of these components are needed to accurately model this, but as stated above, the modeling of these elements is dependent on many factors that are not always constant with time and between people.

To understand heart motion, we will briefly discuss the solid mechanics associated with the heart. As learned in previous solid biomechanics courses, biological tissue is nonhomogenous, is viscoelastic, is subjected to a very complex loading condition, and has a complex three-dimensional geometry. This makes its analysis relatively difficult. However, to first understand the loading conditions and how the elasticity can affect blood flow, let us first consider the simplest case. If the left ventricle can be modeled as a homogenous isotropic spherical shell, then the radial stress can be defined as

$$\sigma_r(r) = \frac{p_o b^3(r^3 - a^3)}{r^3(a^3 - b^3)} + \frac{p_i a^3(b^3 - r^3)}{r^3(a^3 - b^3)} \tag{4.3}$$

Recall that the radial stress acts "through" the muscle wall from some center point in the heart outward. The circumferential stress, which is greatest along the inner surface of the sphere, can be defined as

$$\sigma_\theta(r) = \frac{p_o b^3(2r^3 + a^3)}{2r^3(a^3 - b^3)} - \frac{p_i a^3(2r^3 + b^3)}{2r^3(a^3 - b^3)} \tag{4.4}$$

Recall that the circumferential stress acts along the circumference of the heart muscle wall at every location through the wall. The variables in Eqs. (4.3) and (4.4) are the inner radius of the sphere (a), the outer radius of the sphere (b), internal pressure (p_i), and external pressure (p_o) (Figure 4.10, circumferential stress is depicted to act into or out of the textbook page). As the reader can see, Eq. (4.4) is maximized when r is equal to a (remember that b must be greater than a). Looking at these two formulas, one can see that these wall stresses would induce a stress on the fluid next to the ventricle wall. In fact, the internal pressure, acting on the inner heart wall, can be equal to the left ventricular hydrostatic pressure (which is a function of time, see Section 4.3), but this is not necessarily accurate due to residual stresses within the cardiac muscle fibers. However, it is not as simple as only equating the hydrostatic and normal pressures. In fact, the systolic pressure is determined by the circumferential stress in the ventricular wall, which is caused by muscle contraction. As the sarcomere length increases, the maximal circumferential stress also increases, because with a longer sarcomere length, the sarcomere can perform more work (at least to a certain extent, which is limited by the size of the sarcomere). A higher circumferential stress must be balanced by the internal fluid pressure, so an increase in sarcomere length also increases the left ventricular pressure waveform (the most critical component is the peak systolic pressure, i.e., 120 mmHg). Some have shown that the number of sarcomeres in the heart remain fairly constant with time. Therefore, with an increase in the internal radius of the heart, the muscle tension and the blood pressure increase. This is seen in pathological conditions when the heart enlarges but there is little increase in the number of cardiac myocytes. As a reference note, the external pressure is equal to the pleura pressure, which is slightly lower than atmospheric pressure (negative 4 to negative 2 mmHg).

Example

Calculate the circumferential stress distribution in the left ventricle during peak systole (assuming that the ventricle is a spherical shell). The inner radius and outer radius of the sphere are 3.5 and 4 cm, respectively. The external pressure is −3 mmHg.

Solution

$$\sigma_\theta = \frac{p_o b^3(2r^3 + a^3)}{2r^3(a^3 - b^3)} - \frac{p_i a^3(2r^3 + b^3)}{2r^3(a^3 - b^3)}$$

$$\sigma_\theta = \frac{(-3\ mmHg)(4\ cm)^3(2r^3 + (3.5\ cm)^3)}{2r^3((3.5\ cm)^3 - (4\ cm)^3)} - \frac{(120\ mmHg)(3.5\ cm)^3(2r^3 + (4\ cm)^3)}{2r^3((3.5\ cm)^3 - (4\ cm)^3)}$$

$$\sigma_\theta = 252.64\ mmHg + \frac{1.065J}{r^3}$$

When $r = a$

$$\sigma_\theta = 252.64\ mmHg + \frac{1.065\ J}{(3.5\ cm)^3} = 438.9\ mmHg$$

When $r = b$

$$\sigma_\theta = 252.64\ mmHg + \frac{1.065\ J}{(4\ cm)^3} = 377.5\ mmHg$$

The stress distribution within the ventricular wall is described by

$$\sigma_\theta = 252.64\ mmHg + \frac{1.065\ J}{r^3}$$

with 3.5 cm $\leq r \leq$ 4 cm.

As the example illustrates, this formulation predicts a differential workload carried out by the muscle cells that line the inner wall versus those that line the outer wall of the heart. In fact, it indicates that the cells along the endocardium (inner wall) would produce more work than the cells along the epicardium (outer wall). To meet this criterion, there should be an increased concentration of ATP, calcium, and blood supply to the inner wall, which has never been measured in a laboratory setting, although many have investigated this problem. In order to account for this, it has been suggested that there is a residual stress within cardiac muscles, even under an unloaded condition. These residual stresses and strains balance the uneven workload requirement that has been previously summarized and therefore the need for increased ATP, calcium, and blood. As this discussion and the example show, these formulas are fairly limited in their accuracy due to the simplifying assumptions made about the homogeneous isotropic mechanical properties and the geometric properties of the left ventricle.

To couple the solid mechanics of heart motion and the fluid mechanics of blood flow within the heart, the use of computation fluid dynamics would be required (see

Chapter 14). Fluid structure interaction and multiscale modeling should be incorporated within this model to make the prediction as accurate as possible. There have been attempts to compute an analytic solution to this problem, but they are generally overly simplified. For instance, most will use an idealized geometry for the heart (similar to the spherical shell described previously) combined with homogenous mechanical properties and incompressible invisicid fluid flow. Without going into much detail about these models here, the pressure and the fluid velocity in the ventricle can be solved by using Bernoulli's equation. From Section 3.8, we learned that the Bernoulli equation can only be applied to very specialized flow situations and that the actual physiological scenario does not fit the assumptions necessary to apply the Bernoulli equation. Lastly, these solutions typically show that the fluid is nonturbulent, which may not be the most accurate depiction for blood flowing into/out of the left ventricle, where short duration turbulent-like properties may develop. From fluid dynamics courses, recall that many accelerating viscous flows through a constriction will become turbulent at particular locations within the flow regime, but this developed turbulence is typically not self-sustaining. These locations are typically characterized by flow separation and recirculation zones. A more thorough analysis of these types of turbulent flows can be found in other classical fluid mechanics textbooks.

The goal of this discussion is to illustrate how solid mechanics and fluid mechanics can be coupled to determine the pressure forces that are acting on blood within the heart and that can even drive fluid flow through the heart and the cardiac vasculature. For a course at this level, to calculate a solution, many assumptions need to be made concerning the geometry, the mechanical properties, and the fluid properties. With these assumptions, a meaningful solution can still be reached (see Bernoulli Equation, Section 3.8), but it is more accurate to minimize the assumptions made. As stated before, solutions that use this complex approach will be discussed in Chapter 14.

4.5 HEART VALVE FUNCTION

As discussed in previous sections, the primary function of the heart valves is to prevent the backflow of blood during muscle mass contraction. The mechanism by which the valves accomplish this is fairly interesting, and this section will briefly discuss valve function. We will discuss the mitral valve and the aortic valve, because this is where most of the research has been conducted. The function of the tricuspid valve mimics the mitral valve (except that it has a trileaflet structure versus the bileaflet structure of the mitral valve) and the function of the pulmonary valve mimics the aortic valve. Also, recall that a major difference between the left side and the right side of the heart is that the pressure gradients across the valves are much greater on the left side of the heart. The anatomy of the mitral valve and the aortic valve is shown in Figure 4.14 and is discussed in the following section.

Recall that the mitral valve regulates the blood flow between the left atrium and the left ventricle. The mitral valve consists of two thin membranes, which are connected by tendons (the chordate tendineae) to the papillary muscles. The papillary muscles in turn are connected to the left ventricular muscle mass. The mitral valve remains open, allowing blood to flow into the left ventricle, for approximately 75% of the cardiac cycle (this is

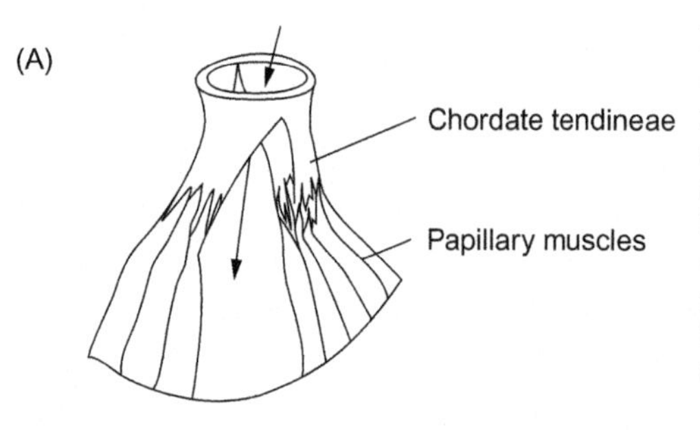

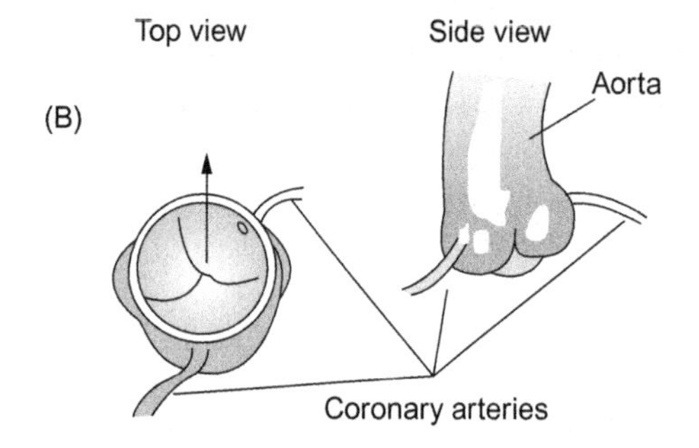

FIGURE 4.14 The mitral valve (A) and the aortic valve (B) determine the blood flow characteristics and blood flow direction through the left side of the heart. The chordae tendineae and the papillary muscles do not open or close the mitral valve, but instead prevent the leaflets from bulging into the atria during ventricular contraction. The coronary arteries come off the aortic valve sinuses and feed the cardiac tissue with blood. As described in Section 4.6, the flow through the coronary circulation opposes normal flow.

during ventricular diastole, Figure 4.15). For approximately two-thirds of this open time, blood flows passively through the mitral valve with a fairly steady and nonturbulent flow profile. This steady flow is due to a pressure gradient of 2–4 mmHg from the venous circulation through the left atrium into the left ventricle. Notice on Figure 4.10 that the majority of blood volume enters the left ventricle during this period. Once the atrium begins to contract (atrial systole), the flow through the valve becomes jet like and recirculation zones form behind the mitral valve leaflets within the left ventricle. As the valves begin to close, the jet flow increases (there is a decrease in the valve opening cross-sectional area) and the turbulence behind the valve leaflets deteriorates because of the increased area behind the valve. The valves begin to close, prior to ventricular contraction due to an increased blood volume in the heart, which physically pushes on the leaflets. This is predominately caused by the priming action of atrial contraction, an increase in size of the left ventricle as well as a decrease in the driving force for fluid flow into the ventricle. Also during valve closure, a majority of the fluid flow becomes stagnant in the upper portion of the valve (atrial side), although there still is a small jet of blood entering the ventricle. Upon valve closure (approximately 50 ms prior to true left ventricular contraction), there is no flow between the atrium and the ventricle.

The pressure gradient across the valve is the driving force for mitral valve opening and closing. During ventricular diastole, the intraventricular pressure reduces to a value below the intraatrial pressure, which causes blood within the atrium to push against the valve inducing valve opening. Additionally, blood entering from the venous circulation aids in increasing the local pressure of the atria. Once blood begins to enter the ventricle, the intraventricular pressure slowly rises, eventually surpassing the pressure within the atrium. This adverse pressure gradient slows the blood flow into the ventricles and helps

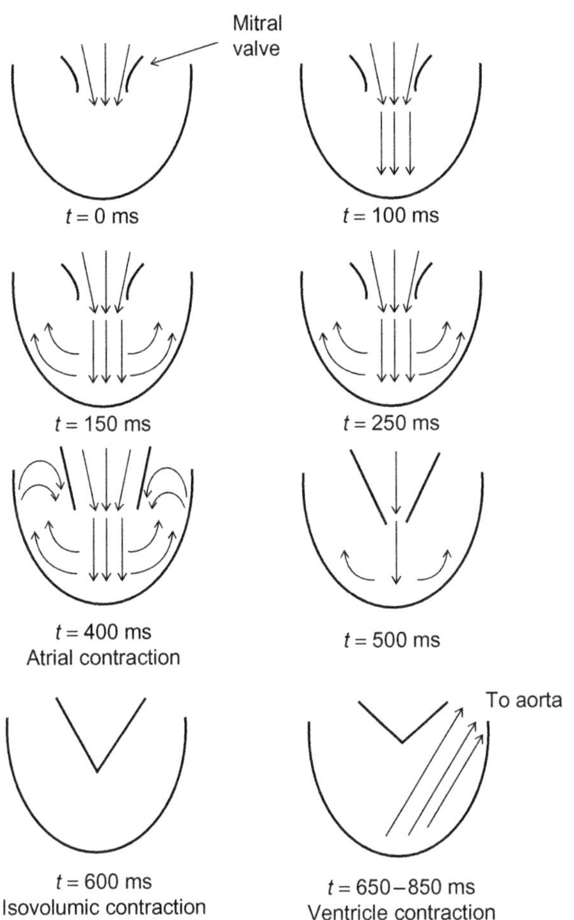

FIGURE 4.15 Schematic of blood flow through the mitral valve during the cardiac cycle (the papillary muscles/chordae tendineae are not shown). The times shown are approximate, but depict the relative change in blood flow characteristics with time. Note that the blood flow through the mitral valve is largely laminar, even though there is some mixing that may occur behind the valve leaflets.

to close the valve leaflets. An interesting occurrence during mitral valve function is that when the ventricles are in a relaxed state, the valve orifice is larger than that when the ventricles are actively contracting. This aids in blood flow through the valve and aids the valve achieving a secure closed position.

The papillary muscles and the chordae tendinae do not function to open or close the valve; instead they prevent the reversing of the valve direction during ventricular systole (i.e., into the atrium). These muscles are also a structural extension of the valve leaflets and can assume some of the stresses and strains that the leaflets experience during the cardiac cycle. When both the mitral valve and the aortic valve are closed (isovolumic contraction and isovolumic relaxation), which is coupled to ventricular contraction or relaxation, the papillary muscles have a small but regulatory role on intraventricular pressure. Recall that during this time, the majority of the pressure is generated from the contraction of the cardiac muscles or is relieved by the relaxation of the muscle cells. The tension generated by the papillary muscles can change the radius of curvature of the mitral valve leaflets, which may change the surface area of the ventricular wall. This is directly related to the pressures that are exerted on the internal and external membrane of the leaflet. Therefore, by changing the tension in the papillary muscles, the pressure within the ventricle and the atrium must be changed to balance the forces. The relationship that governs the pressure–tension relationship of the mitral valve leaflets is

$$p_i - p_e = \frac{T_i}{r_i^2} + \frac{T_e}{r_e^2} \tag{4.5}$$

where p is the pressure, T is the tension, and r is the radius of curvature for the internal surface (i) or the external surface (e) of the valve leaflet, assuming uniform mechanical and geometric properties. The papillary muscles contract at the same time as the ventricular muscle mass, since the action potential signal passes to them through the same Purkinje fiber system that induces ventricular muscle mass contraction.

Example

Calculate the tension needed on the interior side of the mitral valve leaflet to maintain the valve in the closed state immediately after valve closure. Assume that at valve closure there is a pressure difference of 1.5 mmHg between the ventricle and the atrium. The atrial pressure is 3 mmHg, and the radii of curvature for the inner and outer membrane of the leaflet are 1 and 1.2 cm, respectively. Assume that the mitral valve leaflet is one-third of a spherical shell with a uniform width of 2 mm.

Solution

To calculate the tension on the outer leaflet we will first need to calculate the exposed area of the leaflet. With the assumption that the leaflet is a spherical shell, the area of interest is

$$L = 2\pi * 1.2\ cm * \left(\frac{120^\circ}{360^\circ}\right) = 2.51\ cm$$

$$A = 2.51\ cm * 2\ mm = 50.2\ mm^2$$

where L is the arc length of the exterior portion of the leaflet. From previous discussion, the pressure in the atrium will be equal on all surfaces, including the leaflet. Therefore, the tension that the leaflet must have to maintain its shape is

$$T_e = 3\ mmHg * 50.2\ mm^2 = 20\ mN$$

The tension on the inner leaflet surface is

$$p_i - p_e = \frac{T_i}{r_i^2} + \frac{T_e}{r_e^2}$$

$$T_i = r_i^2\left(p_i - p_e - \frac{T_e}{r_e^2}\right) = (1\ cm)^2\left(1.5\ mmHg - \frac{20\ mN}{(1.2\ cm)^2}\right) = 6\ mN$$

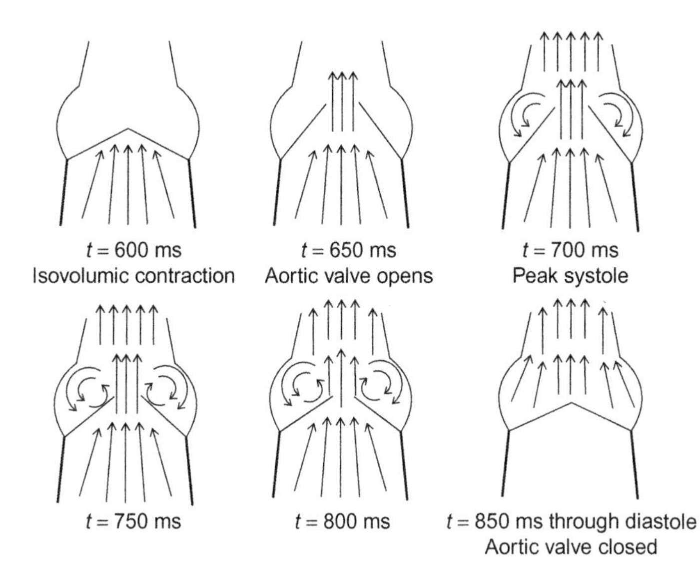

FIGURE 4.16 Schematic of blood flow through the aortic valve during the cardiac cycle (the times overlap and continue from Figure 4.15). Again, the times are relative, but give a general idea of blood flow through the aortic valve. Again, the flow may approach turbulence through the aortic valve, but this turbulent flow is not self-sustaining and would dissipate instantaneously.

The aortic valve consists of three thin leaflets; each in a crescent shape. Immediately distal to the three leaflets there is an enlargement of the aorta (which is normally around 27 mm in diameter) termed the sinuses of Valsalva. The aortic sinus plays an important role in valve closure, which will be discussed later. The aortic valve remains closed until the pressure within the left ventricle exceeds the pressure in the aorta (under normal conditions this is approximately 80 mmHg). Upon valve opening, the blood flow is split into two separate streams (Figure 4.16). The first stream comprises the majority of the blood and enters the ascending aorta to enter the systemic circulation. The second portion of the blood flow is directed into the valve sinus. This blood flows in a slow vortex behind the valve leaflets and eventually rejoins the blood in the ascending aorta, once the valve closes. Immediately after peak systole, the ventricular muscles begin to relax and the intraventricular pressure begins to reduce from approximately 115 mmHg. However, due to the viscoelastic mechanical properties of the aorta, the pressure in the aorta does not fall rapidly but continues to increase for a few milliseconds (reaching approximately 120 mmHg under normal conditions). This causes a backward flow pressure gradient; the pressure is higher at the tip of the aortic valve leaflet and is lower at the valve orifice. Due to this backward pressure gradient, the blood that was already in the sinus prevents the tip of the aortic valve from opening more fully in response to the reverse pressure gradient. The reverse pressure gradient also causes more blood to flow into the aortic valve sinus, which aids in forcing the leaflet into the secure closed position. Blood passes into the sinus at a higher rate than the ventricle, because the effective open area of the sinus is greater than the orifice. Furthermore, as more blood enters the sinus, the open area of the valve orifice decreases until the valve is fully closed. Once the valve is completely closed, the blood in the sinus enters the aortic mainstream flow and is delivered to the body. The aortic valve is open for approximately 25% of the cardiac cycle.

The Gorlin equation is an approximation to the aortic valve opening area, as a function of the cardiac output, the heart rate, the hydrostatic pressure, and the ejection period. The cardiac output, the heart rate, and the ejection period will approximate the flow across the valve leaflet during ejection. This formulation takes the form of

$$Valve\ Area = \frac{Cardiac\ Output}{44.3(Heart\ Rate)(Systolic\ Ejection\ Period)(\sqrt{Mean\ Pressure\ Gradient})} \tag{4.6}$$

If the cardiac output is measured in mL/min, the heart rate is measured in beats per minute, the ejection period is measured in seconds, and the mean gradient is measured in mmHg, then the valve area will be in units of cm^2. Typically, the ejection period is 0.33 s, and the mean pressure gradient is 50 mmHg.

4.6 DISEASE CONDITIONS

In most of the following chapters, the authors have chosen to include a section for disease conditions. The authors want to preface this particular disease condition section by stating that most of the diseases of the cardiovascular system are related to each other and in some instance new links between particular diseases are still emerging. With that said, the authors had to make a choice about which chapter to include the discussion within. This was not an easy decision because of the overlap between many of the cardiovascular conditions, which have wide ranging effects at multiple locations throughout the cardiovascular system. As an example, one of the major causes of coronary artery disease is atherosclerosis, but atherosclerosis is a disease that can affect the entire vascular system. One of the major downstream effects of coronary artery disease is myocardial infarction. Due to these overlapping effects, it was not easy to choose where to include the particular discussion for these three diseases. In this instance our choice was to include coronary artery disease and myocardial infarction in this disease condition section and we leave the extensive discussion of the more general atherosclerosis for Chapter 5.

4.6.1 Coronary Artery Disease

The coronary arteries supply the cardiac muscle with blood (Figure 4.17). The left main coronary artery and the right main coronary artery branch off from the ascending aorta approximately 0.5–1 cm above the aortic valve. The left coronary artery is responsible for providing the left ventricular muscle mass with blood (along with sections of the right ventricle and the left atrium). The left coronary artery delivers blood to approximately 75% of the cardiac tissue. The remaining portions of the cardiac muscle are supplied with blood from the right coronary artery. The coronary blood vessels branch into smaller vessels that penetrate the muscle mass to supply blood to the entire heart. Because the heart never rests, it requires about 5% of the total cardiac output, which amounts to approximately 250 mL/min, to maintain its function. The majority of the venous coronary blood flow is returned directly to the right side of the heart via the coronary sinus. Also, a small portion of deoxygenated blood is returned to all chambers of the heart via thebesian veins.

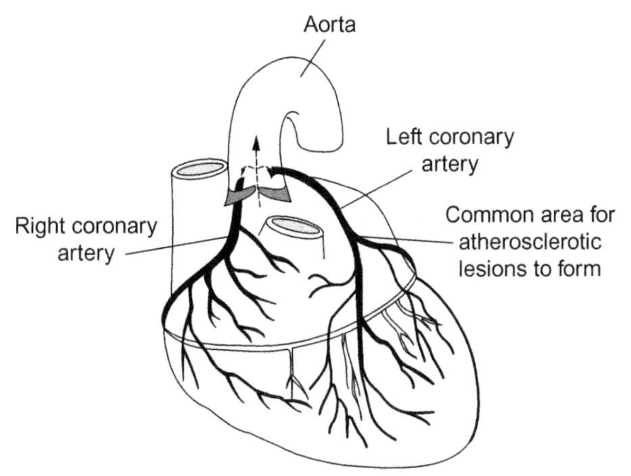

FIGURE 4.17 The coronary arteries that supply the entire cardiac muscle with blood. These vessels are principal locations for atherosclerotic lesions and other cardiac diseases. Depending on the severity of the damage to these vessels, blood flow to the cardiac muscle cells can be severely impaired.

The thebesian veins are similar to open-ended blood vessels, which penetrate the cardiac mass and open to all four heart chambers.

The flow within the coronary blood vessels is quite interesting, because the majority of the blood flow occurs during ventricular diastole and the flow almost ceases during systole. This is the reverse of what happens in most other vascular beds. The reason for this reversed flow is that during ventricular systole the cardiac muscle contracts, which causes a compression of the coronary blood vessels. This compression is so severe that the resistance to blood flow increases significantly, and the blood flow pressure gradient is not large enough to overcome the resistance to flow. During diastole, the cardiac muscle relaxes and relieves the constriction of the coronary blood vessels. Therefore, the resistance to flow reduces and blood flow through the coronary circulation initiates. This is especially prevalent in the blood vessels that supply the left ventricular muscle mass as compared to the right ventricular muscle mass, because the muscle contraction is much stronger there.

Coronary artery disease is still the leading cause of death in the Western world. Coronary artery disease lumps together pathologies that lead to a reduction in coronary blood flow, which can eventually lead to myocardial infarction (see Section 4.6.2). In the United States, the most common cause of coronary artery disease is atherosclerosis. Atherosclerosis can be initiated by genetic factors or by lifestyle choices (such as eating foods high in cholesterol) and these risk factors tend to be a mixture of both modifiable and nonmodifiable elements. Either way, atherosclerosis initiates through the deposition of large quantities of cholesterol within the subendothelial space. Cholesterol deposits can be found throughout the body, but many vascular beds can adapt to this reduction in blood flow. Due to the fact that the coronary arteries are not as adaptable as other vascular beds and that the heart requires a large quantity of blood at all times, any cholesterol deposition within the coronary blood vessels is potentially a major problem for the patient. This becomes even more prevalent under conditions of physical (exercise, extreme cold temperature) or emotional stress.

After cholesterol deposition occurs, there are a few processes that proceed in parallel. The first process that can happen is that cholesterol molecules can coalesce to form larger deposits. As this is occurring, macrophages can invade the subendothelial space to ingest the cholesterol deposits. Macrophages that ingest a large quantity of cholesterol transform into foam cells that are basically cells full of fat/cholesterol. Generally, foam cells are large in size and they cannot migrate out of the subendothelial space. Along with these two processes, fibrous tissue and smooth muscle cells (that both naturally surround the blood vessels) begin to proliferate and sometimes can become calcified. This occurs in response to the increased macrophage population in the subendothelial space. The net result of these three processes (cholesterol coalesce, foam cell production, and cell proliferation) is that a plaque forms that extends into the blood vessel lumen. A blockage in the vessel lumen decreases blood flow, because the resistance to flow increases. If the blood vessel becomes calcified, it becomes less pliant, hence more brittle, and if it is severe enough, the vessel potentially cannot withstand the normal pressure forces exerted by the fluid. This can lead to the vessel bursting.

One of the first clinical manifestations of coronary artery disease is through angina pectoris, which is a term that refers to chest pain resulting from oxygen deficiency in a region(s) of the heart. Oxygen deficiencies occur when oxygen demand surpasses oxygen supply for an extended period of time or when there is an overall decrease in the oxygen delivery. Angina is most commonly associated with coronary artery diseases/atherosclerosis of the coronary arteries, but others pathologies of the heart have been shown to induce angina. Cardiomyocyte oxygen demand can increase if the myocardial tension increases (this is a function of blood pressure), the heart rate increases or the contractility of the heart increases.

One of the common ways to detect the presence of coronary atherosclerosis is through the use of arteriography. Arteriography is conducted by injecting a radio-opaque contrasting agent into the blood. This is followed by imaging techniques, typically x-ray fluoroscopy, to allow for the visualization of the inside diameter of the contrasted blood vessels. Using this technique, physicians can classify the extent of plaque formation and make a determination of whether or not an intervention surgery is needed. For the most part, physicians will not recommend an intervention if the luminal diameter is constricted to 70% or less of the normal diameter. Reports have shown that a constriction of 50%, which is defined as an important luminal diameter constriction, can alter the coronary blood flow under exertion but the resting coronary blood flow is not altered. A luminal diameter reduction of 70% or more is defined as a significant luminal diameter constriction, can alter both the coronary blood flow under exertion and the resting coronary blood flow. Most physicians will recommend intervention surgery under these conditions (e.g., angioplasty or grafting).

Atherosclerosis development is very common in the first few centimeters of the coronary arteries. This is most likely caused by the three-dimensional structure of the coronary arteries (Figure 4.17). Immediately after branching from the aorta, there is a high degree of tortuosity (twists/bends) and tapering within the coronary vessels. Because the blood flow in these vessels is very fast, these rapid geometric changes induce localized regions of turbulence with the blood vessel. Regions of turbulence can be characterized by flow separation, stagnation points, and recirculation zones, in which the flow rate is reduced, allowing cholesterol enough time to migrate into the subendothelial space. Recall from our earlier discussions that this is not true turbulence, even though some localized mixing may be observed.

Reduced blood flow throughout the coronary vessels is devastating to the muscle mass of the heart (see Section 4.6.2). However, if the blockage is not severe enough or if it progresses slowly enough, the coronary arterioles can protect the heart tissue for some time. At the time of a blockage, the dilation of the downstream coronary arterioles can counterbalance the flow restriction for approximately 24–48 h. For a long-term solution, the coronary collateral circulation will need to adapt and supply the undernourished tissue with blood. The collateral circulation of the coronary vessels allows for a large redundancy of flow to one particular region of cardiac muscle meaning that at a minimum two large arterioles supply the same capillary bed with blood. If one of these arterioles is blocked, the second arteriole (and potentially others in the vicinity) will remodel so that the blood flow to that region returns to approximately 100% of the normal value. In most cases, atherosclerotic lesions progress slowly enough to allow for a dilation of downstream arterioles and a remodeling of collateral vessels to occur. These two processes potentially allow a patient with a coronary lesion to survive for years after the onset of the disease.

Some of the interesting phenomena regarding the coronary collateral circulation is the recruitment of these vessels during a time of need. In the most simplistic relationship, the flow through a vessel is a function of the pressure gradient and the resistance within that vessel. Under most cases, the resistance to flow is so large in the collateral circulation that the vast majority of flow proceeds through the normal coronary circulation. However, as atherosclerotic lesions begin to develop, the resistance to flow skews from the normal circulation to the collateral circulation. Emerging evidence suggests that new shear forces that are applied to the coronary collateral vessels are responsible for the remodeling that occurs under pathological conditions.

4.6.2 Myocardial Infarction

One of the immediate effects of a blocked coronary artery is decreased blood flow into the region downstream of the blockage. If the muscle that is being supplied by the blocked vessel can no longer maintain normal cardiac contractions, this region will become infarcted. Clinically, this is typically defined as irreversible cellular injury that leads to necrosis (cell death) as a consequence of prolonged ischemia (reduced oxygen). Without a blood supply for 1–2 h, cardiac muscle cells begin to die. As stated above, collateral blood flow can return to the infarcted area. However, as significant tissue damage may have been done, this blood has little purpose there because the damaged cells cannot contract as strongly as before, if they can contract at all. Blood that enters an infarcted region stagnates because the major vessels have been closed off from circulation through a separate remodeling process. The presence of this stagnated blood in the infarcted area alters the contractile/mechanical properties of the heart. In extreme cases, the endothelial cells become very permeable and leak blood into the heart wall.

There are two major mechanisms by which myocardial infarction occurs; this first is a prolonged increase in myocardial oxygen demand with a significant decrease in coronary artery flow (which is common during coronary artery atherosclerotic lesion formation and was discussed above) and the second is a decrease in the oxygen delivery to the myocardium. The second cause of myocardial infarction is typically found in patients with hypotension (low blood pressure that cannot overcome the resistance to flow) or spasms of the coronary artery (which can alter the fluid dynamics through the coronary circulation). These issues can be identified in patients via an ECG, where it is common to observe a significant Q wave (greater than 0.04 s in duration), a reduced amplitude of the R wave (e.g., by 30% or more) or a more prominently peaked T wave. A second, slightly more specific manner in which the occurrence of myocardial infarction can be diagnosed is through the presence of myocardial specific creatine phosphokinase (MB-CPK). The increase in MB-CPK peaks at approximately 10 h after an acute myocardial infarction event can be observed within approximately 1 h after the myocardial event, but returns back to normal after 24 h of the event.

With a minor block in the coronary vessels, it is unlikely that cardiac tissue will become infarcted. Cardiac tissue is supplied with approximately four times as much blood as it needs to survive. So even with a blockage that reduces blood flow by half, it is unlikely that muscle tissue will die immediately, if collateral circulation does not take up the slack of the blocked vessel. However, if the blockage is severe enough or the collateral circulation does not supply the newly infarcted region with enough blood, the muscle tissue will begin to die.

Once a large region of the heart becomes infarcted, there are four likely outcomes that can each lead to death. The first two outcomes are consequences of the heart not pumping enough blood to the systemic circulation. As stated above, an infarcted region of the heart does not contract as strongly as is necessary. If the region of the heart that is infarcted is large, it is possible that the total cardiac output reduces. A lower cardiac output can and does affect many locations throughout the body. A decreased blood supply to any organ will change its function and potentially lead to death. Also, when the ventricular muscle mass is not moving enough blood, the blood must pool somewhere. Because the total blood volume does not change and the arterial system is not highly compliant, the excess blood is typically held within the veins and the pulmonary circulation and may even back-up into the heart. When the blood volume in the venous and pulmonary circulation systems increases significantly, the overall pressure of the cardiovascular system increases (see Section 5.11.1). This especially becomes detrimental when the capillary pressure increases and blood cannot be delivered to the body. If the heart begins to "store" some blood, the cardiac output severely reduces.

The remaining two outcomes are problems associated with the heart due to the lack of nutrients being supplied to the infarcted region. A common cause of death associated with myocardial infarction is ventricular fibrillation. With zero or negligible quantities of blood being supplied to the cardiac tissue, there is a rapid reduction of potassium ions from the muscle cells. Without potassium ions, a cell cannot become repolarized after an action potential has passed. This leads to a basic breakdown in the membrane potential because the ionic concentration gradients are incorrect, causing the cells to not function properly. All of these processes lead to fibrillation of the tissue. The last major possible occurrence is rupture of the cardiac tissue. An infarcted region by definition is composed of dead muscle cells. A few days after the infarction occurs, the tissue will begin to degenerate and thin. If the majority of the heart is functioning at its optimal capacity, then it is likely that the thinned region will become unable to withstand the pressure generated by cardiac contraction. At the onset of this process, the infarcted region begins to bulge outward, and eventually, if not corrected, the infarcted region can burst. Blood would then pool in the pericardial space and begin to compress other regions of the heart. In a short time, the weaker atria will collapse and little to no blood will enter into the ventricles or be pumped out by the ventricles. The patient would die from a reduction in cardiac output.

4.6.3 Heart Valve Diseases

As discussed before, the heart valves regulate the flow of blood through the heart and prevent the backflow of blood into the various heart chambers. A disease of the heart valve obstructs these functions. There are two common ways for a disease of the heart valve to develop. The first is associated with rheumatic fever, and the second is associated with an increased calcium concentration in the body. Rheumatic fever is a disease in which a bacterium enters the body and produces a variety of proteins. These proteins are similar to some of the innate proteins within the body, and therefore, antibodies produced by the host's immune system attack the bacterial proteins and some host proteins. The heart valve leaflets are likely to be attacked by these antibodies, causing severe damage to their structure. Normally, the mitral valve and the aortic valve are the first to be attacked because they are exposed to more flow/mechanical disturbances during normal functioning (due to the higher pressure forces acting on these valves). If the disease progresses far enough, it is common to see lesions within the tricuspid and the pulmonary valves.

There are two stages of valve leaflet destruction that can occur during rheumatic fever. During the early stages of the disease, the leaflets of adjacent valves can become stuck together due to the protein destruction and remodeling. In this case, the leaflets can typically close fully, but they cannot open fully, causing a stenosed valve. Blood is not ejected as efficiently through a stenosed valve as a normal valve. Clearly, if the valve is stenosed enough, not enough blood will enter the ventricles and/or the systemic or pulmonary circulation. During later stages of rheumatic fever, scar tissue can form along valve leaflets due to the extensive destruction of their protein structure. As the reader knows, scar tissue is mechanically stiffer than the very pliant valve leaflets. This prevents the valve leaflets from opening and closing fully. A valve that does not close properly causes backflow of blood into the atria or in the ventricles, and this type of valve is said to experience regurgitation. Regurgitation occurs during the cardiac cycle when the valves should be closed but they are not fully closed. The end product again is a reduced cardiac output. In most cases of rheumatic fever, there is some degree of valve stenosis along with regurgitated flow.

Another common heart valve disease occurs when the valve leaflets become calcified. It is currently not clear what the mechanism behind this phenomenon is. However, what occurs is the formation of a bone-like matrix (i.e., hydroxyapatite) within the valve leaflets. Similar to the rheumatic fever disease scenario, calcified valve leaflets are much stiffer than normal leaflets. This will induce some combination of valve stenosis and regurgitation flow through the valve and effectively reduce the cardiac output.

Other consequences of heart valve disease must be taken in consideration as to which valve has become diseased and which chamber of the heart must compensate by undergoing hypertrophy or dilation. When a semilunar valve becomes stenotic or regurgitant or when an AV valve is regurgitant the primary burden is placed on the appropriate

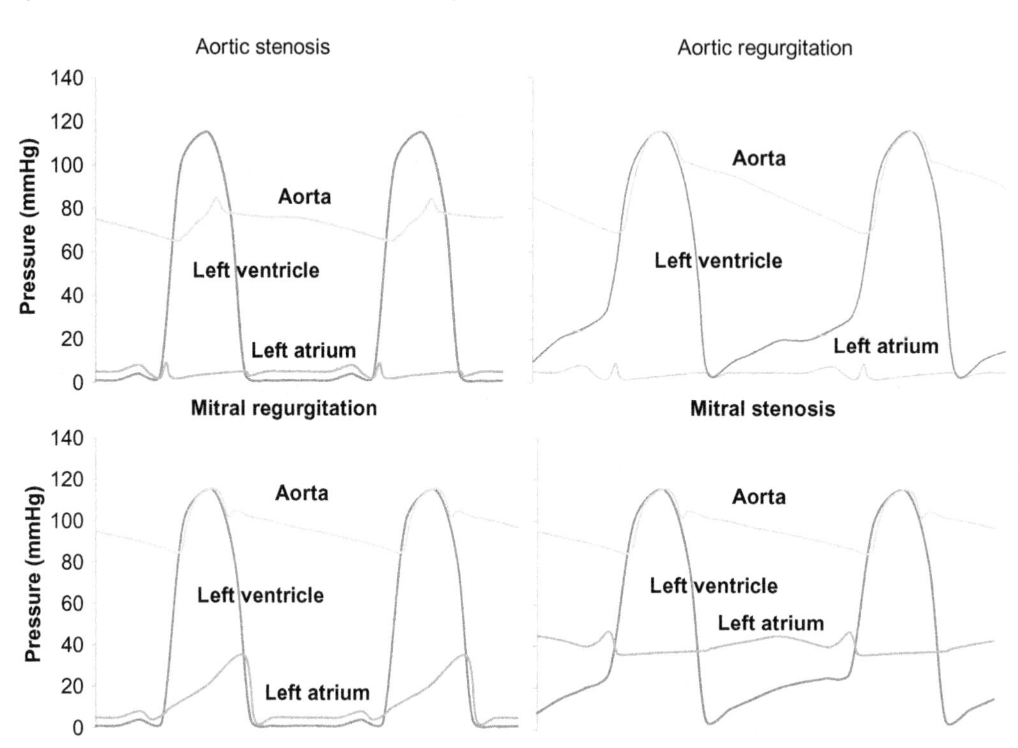

FIGURE 4.18 Pressure relationship in the left side of the heart and in the aorta during valve diseases of the left side of the heart. The normal relationship for the pressure waveforms can be found in Figure 4.10. The pathologies associated with aortic stenosis (aorta pressure reduction) and mitral regurgitation (left atrium pressure increase) are predominately observed during ventricular systole, whereas the pathologies associated with aortic regurgitation (left ventricle pressure increase/aortic pressure decrease) and mitral stenosis (left atrium pressure increase/left ventricle pressure increase) are predominately observed during ventricular diastole.

ventricle (depending on which side of the heart is experiencing the disease). With aortic stenosis, there are major changes to the pressure waveform in the systemic circulation, which alters the delivery of blood throughout the entire systemic circulation. If the mitral valve is regurgitant, there are major secondary effects observed in the venous pulmonary circulation, and there is a reduced oxygenated blood delivery to the systemic circulation. Stenotic AV valves place a larger primary burden on the atrium and there are secondary effects observed within the corresponding venous circulation. For instance, when the tricuspid valve becomes stenotic, there is a systemic increase in the venous circulation pressure.

Changes observed to the pressure waveforms on the left side of the heart, under various heart valve diseases are shown in Figure 4.18. Under aortic stenosis conditions, the left ventricular pressure waveform appears normal, but the aortic pressure does not mimic the left ventricle waveform during ventricular systole. This primarily leads to left ventricular hypertrophy, altered coronary blood flow, and altered myocardial oxygen demand. Under aortic valve regurgitation conditions, both the aortic and left ventricular pressure waveforms are altered from the normal conditions. The backflow of blood into the left ventricle significantly increases the left ventricular pressure, while reducing the aortic pressure. Therefore, the ejection period of blood during ventricular contraction is increased and an overload of blood within the ventricle is observed. Mitral valve regurgitation puts an excess burden on the left ventricle, because there is a leak between the ventricle and the atrium during ventricular systole. The heart has to increase its stroke volume (and cardiac output potentially through heart rate changes) to make up for the loss of blood volume to the systemic circulation. There is also a large abnormal increase in the left atrium pressure waveform during ventricular systole. Mitral stenosis progressively elevates the atrium pressure and somewhat reduces the ability for the ventricle to maintain a normal stroke volume. The atrium contraction cannot overcome the incompressibility of blood and thus the atria loses its ability to act as a primer pump. As one can imagine, the downstream effects of these changes on the corresponding vascular beds are numerous, including systemic vasoconstriction, with a decrease in the ability to maintain temperature and cyanotic peripheral tissue. The changes on the right side of the heart will be analogous to those discussed on the left side of the heart, except remember that the pressure are approximately sixfold less within the pulmonary circulation than in the systemic circulation.

END OF CHAPTER SUMMARY

4.1 The human heart consists of two separate pumps that work in synergy to transport deoxygenated blood to the pulmonary circulation and oxygenated blood to the systemic circulation. Each pump is composed of a primer pump, the atrium, and an active pump, the ventricle. The cardiac muscle cells are highly interconnected to allow for the transfer of electrical signals between the muscle cells and to prevent the heart muscle from tearing itself apart. Troponin and tropomyosin are present within cardiac muscle to regulate actin–myosin cross-bridge formation, in a similar manner to skeletal muscle. The major difference between cardiac muscle tissue and skeletal muscle is that cardiac muscle contains voltage-gated calcium channels which help to elongate the depolarization phase of the cardiac action potential. This elongation allows the atria to contract as one and then the ventricles to contract as one, instead of having a wave of contractions through the cardiac tissue. The work that the heart undergoes to contract can be depicted by the pressure–volume relationship of each heart chamber. The area within this closed-loop circuit describes the work of heart contraction.

4.2 The cardiac action potential is generated in the SA node, which is a highly specialized structure that is surrounded by extracellular fluid with an abnormally high sodium concentration. This sodium continually enters the cells that comprise the SA node and approximately every 60 s generates an action potential. The cells of the SA node are directly connected with neighboring cardiac myocytes to allow for the coupling of the atria with the SA node. There are three pathways along the right atrium wall, which help to pass the action potential signal from the SA node to the AV node. The AV node is responsible for initiating ventricular contraction, but the AV node slows down the conduction speed of the cardiac action potential so that the atria can finish contracting prior to ventricle contraction. The action potential passes from the AV node to the Purkinje fiber system, which rapidly transmits the action potential to the apex of the heart and then penetrates into the ventricular muscle mass. This causes the bottom portion of the ventricle to contract first, pushing the blood toward the valves located superior to the ventricle. The ECG is a common way to measure the electrical potential within the cardiac tissue. The ECG signal is composed of three waves: the P wave, which is associated with atrial contraction; the QRS complex, which is associated with ventricular contraction; and the T wave, which is associated with ventricular repolarization. Changes in the intervals between these waves, the lengths of the waves, and the electrical potential of these waves can help physicians diagnose various cardiac conditions.

4.3 The cardiac cycle describes all of the events that occur during one heartbeat and is divided into two phases, diastole and systole (typically related to ventricular relaxation and contraction, respectively, but it can be discussed regarding the atria as well). Diastole is characterized by cardiac myocytes that are not actively contracting, whereas systole is characterized by active cardiac myocyte contraction. During ventricular diastole, there is a rapid reduction in the intracardiac pressure and there is a rise in the blood volume for the chambers within the heart. During ventricular systole, there is a rapid increase in cardiac pressure accompanied by a rapid decrease in the blood volume.

4.4 During active contraction, the entire heart moves in three-dimensional space, which can be modeled by changes in the forces/pressures within the heart. A.V. Hill modeled the heart contraction as a three-component model that is comprised of the active contraction mechanism (actin–myosin), a viscoelastic relaxation, and a parallel resistive element. Hill's model can be used to calculate the force generated during contraction. The solid mechanics of the heart are also very important to understand when modeling heart motion. Assuming that the heart is a perfect spherical shell with homogeneous material properties, the radial stress and the circumferential stress generated within the cardiac tissue can be described by

$$\sigma_r(r) = \frac{p_o b^3(r^3 - a^3)}{r^3(a^3 - b^3)} + \frac{p_i a^3(b^3 - r^3)}{r^3(a^3 - b^3)}$$

$$\sigma_\theta(r) = \frac{p_o b^3(2r^3 + a^3)}{2r^3(a^3 - b^3)} - \frac{p_i a^3(2r^3 + b^3)}{2r^3(a^3 - b^3)}$$

respectively.

4.5 There are four valves present within the heart: the mitral valve (between the left atrium and left ventricle), the tricuspid valve (between the right atrium and right ventricle), the aortic valve (between the left ventricle and the aorta), and the pulmonary valves (between the right ventricle and the pulmonary artery). The function of the mitral and tricuspid valves is very similar, and the same is true for the aortic and pulmonary valves. The mitral and tricuspid valves remain open for approximately 75% of the cardiac cycle, so that blood flows freely from the atria to the ventricles. As the ventricles begin to contract, these valves will close and the aortic and pulmonary valves will open. The mitral and tricuspid valves are connected to the chordae tendineae and the papillary muscles which prevent the valves from bulging (and potentially opening) into the atria. The most important function of the valve leaflets is to determine the direction of blood flow through the heart. The pressure–tension relationship for the valve leaflets can be represented as

$$p_i - p_e = \frac{T_i}{r_i^2} + \frac{T_e}{r_e^2}$$

The Gorlin equation can approximate the aortic valve area as

$$Valve\ Area = \frac{Cardiac\ Output}{44.3(Heart\ Rate)(Systolic\ Ejection\ Period)(\sqrt{Mean\ Pressure\ Gradient})}$$

4.6 Coronary artery disease is characterized by a reduction in the coronary artery blood flow, which can eventually lead to cardiac muscle tissue death. The coronary arteries are responsible for supplying the cardiac tissue with the oxygen/nutrients it needs to contract during the entire cardiac cycle. Myocardial infarction is characterized by a section of cardiac tissue that begins to die because it does not receive enough blood. This region will not contract as strongly as possible, and this can have a drastic effect on the blood delivery to the vascular system (depending on the size of the infarcted region). Heart valve disease is characterized by a destruction of the heart valves, which can potentially affect blood flow through the heart. If the valves do not open fully, there will likely be a reduction in blood flow to the cardiovascular system, and if the valves do not close fully, then there is the likelihood for backflow of blood into the heart. This can be observed by changes to the pressure waveforms within the heart chambers and/or the respective vessels.

HOMEWORK PROBLEMS

4.1 Blood returning to the human heart in one of the pulmonary veins will drain first into the _______, after which it passes into the ______ and then enters the _____ circulation.

4.2 The occurrence of cardiovascular diseases is dependent on the country of origin. For instance, Greece has a much lower cardiovascular disease rate than the United States. List a few possible reasons for this dependence. Would it be possible to test the reasons that you list, and how would you test them?

4.3 During diastole, heart chambers experience

a. contractions that push blood into the downstream chamber
b. a decrease in pressure, associated with muscle relaxation
c. an opening of the left and right semilunar valves
d. an increase in intracellular calcium

4.4 Discuss the differences between the cardiac muscle and skeletal muscle. Why are the action potentials different in these two types of muscle?

4.5 Do the chordae tendinae play an active role in heart valve motion? What is the purpose of these structures?

4.6 You are sitting outside on a warm day and are losing a great deal of fluid volume through sweat. You decide to have your blood pressure taken and find that it is higher than normal. Why is this?

4.7 If your alarm clock wakes you up and you rise rapidly, why may you feel light-headed? Under other conditions, such as voluntary waking without an alarm clock, this does not occur, why?

4.8 A patient with congestive heart failure (severe myocardial infarction) has swollen ankles and feet. What is the relationship between heart failure and fluid loss from the vascular system?

4.9 A patient has an ECG and the physician finds that there are two P waves, followed by the QRS complex and then the T wave. What is a possible reason behind this?

4.10 What are the principle valves in the heart and what are their functions?

4.11 Describe the salient aspects of the cardiac cycle.

4.12 Under normal physiological conditions, the pressure–volume relationship for the heart is described in Figure 4.7. Approximate the work associated with the left ventricle under these conditions. Also, what is the cardiac work associated with a patient that is experiencing high blood pressure, whose phase 1 starts at 2 mmHg and ends at 8 mmHg. Phase 2 ends when the left ventricular pressure reaches 110 mmHg, and at this point the aortic valve opens. The maximum pressure associated with phase 3 reaches 175 mmHg and ends at 125 mmHg, and follows a similar path as described in Figure 4.4. Phase 4 closes the work loop. Assume that the volumes under the disease condition remain the same.

4.13 Discuss the time rate of change experienced in the left ventricular volume filling curve during the cardiac cycle. Would this temporally varying quantity differ during a disease condition?

4.14 A patient has a murmur in his ventricle that produces a loud gushing sound at the beginning of systole. Which valve is most likely causing this sound and what is a possible remedy for this murmur?

4.15 Calculate and plot the radial and circumferential stress distribution in the left ventricle at the end of systole ($p = 80$ mmHg; assume that the ventricle is a spherical shell). The inner radius of the heart is 3.2 cm and the outer radius of the heart is 3.8 cm. The external pressure surrounding the heart is -1 mmHg. Under a disease condition where the heart muscle thickens, calculate the radial and the circumferential stress distribution in the left ventricle at the end of systole. Under these conditions the pressure at the end of systole remains the same, but the inner wall radius is 3 cm and the outer wall radius is 4.2 cm. Compare this to normal conditions and comment.

4.16 Recalculate the stress distribution through the cardiac wall with an inner radius of 3.5 cm and an outer radius of 4 cm; however, the internal pressure during peak systole is 150 mmHg. All other conditions are as listed in Problem 4.15.

4.17 Use the Cartesian Navier–Stokes equations to approximate the flow through the left ventricle during peak systole. Assume that the gravitational effects on the flow are negligible and that the opening orifice for blood to flow through is 25 mm (aorta). The width of the left ventricle can be approximated as 2 cm and the total length from the apex of the heart to the aortic valve of 4 cm. Determine the maximum velocity at both the aorta and within the ventricle. Blood viscosity is 3.5 cP assume that the pressure gradient across this flow is -1 mmHg/cm.

4.18 During heart valve degeneration, the thickness of the valves can decrease. Calculate the tension on the inner leaflet to maintain the valve in a closed position. Assume that during valve closure, there is a pressure difference of 3 mmHg, with an atrial pressure of 5 mmHg. The radii of curvature for the inner leaflet is 1 cm and that for the outer leaflet is 1.05 cm. There is a uniform width of 3 mm across the leaflets. Assume that the leaflet is approximately 1/3 of sphere.

4.19 Under a condition where the heart valves stiffen, the tension that they can withstand reduces to 2 mN on the exterior surface and 0.5 mN on the interior surface. What is the radius of curvature needed for the leaflets to maintain this pressure? Assume that all pressure conditions and geometric considerations are the same as in Problem 4.18.

4.20 *Modeling*: Design a two-dimensional aortic valve geometry (including the aorta) using a computational fluid dynamics program. Compute the velocity field behind the valve at peak systole (assume that the pressure gradient is 15 mmHg across the section of the valve shown in Figure 4.12).

4.21 What is the cardiac output of a person with an aortic root area of 0.5 cm^2, heart rate of 72 beats per minute, ejection period of 0.3 s and a pressure gradient of 45 mmHg?

4.22 The answer to the previous question somewhat underestimates the cardiac output of a person: (1) illustrate the dependence of cardiac output approximation on both the ejection period and the pressure gradient, using the aortic root area and heart rate as provided in the Problem 4.21, (2) discuss your findings for (1), and (3) discuss the meaning of the constant of proportionality in the Gorlin equation (e.g., 44.3).

4.23 The cardiac conduction system is designed with some very specific considerations as to the speed of conduction from the SA node to the apex of the heart. Knowing the conduction delay between the various segments within the cardiac conduction system, estimate the conduction speed through each portion of the cardiac conduction system. Estimate the length of each segment.

References

[1] B. Bellhouse, F. Bellhouse, Fluid mechanics of model normal and stenosed aortic valves, Circ. Res. 25 (1969) 693.

[2] B.J. Bellhouse, Fluid mechanics of a model mitral valve and left ventricle, Cardiovasc. Res. 6 (1972) 199.

[3] A.J. Brady, Length dependence of passive stiffness in single cardiac myocytes, Am. J. Physiol. 260 (1991) H1062.

[4] A.J. Brady, Mechanical properties of isolated cardiac myocytes, Physiol. Rev. 71 (1991) 413.

[5] K.D. Costa, P.J. Hunter, J.M. Rogers, J.M. Guccione, L.K. Waldman, A.D. McCulloch, A three-dimensional finite element method for large elastic deformations of ventricular myocardium: I–Cylindrical and spherical polar coordinates, J. Biomech. Eng. 118 (1996) 452.

[6] J. Feher, Quantitative Human Physiology: An Introduction, Elsevier, Waltham, MA, USA, 2012.

[7] Y.C. Fung, Mathematical representation of the mechanical properties of the heart muscle, J. Biomech. 3 (1970) 381.

[8] Y.C. Fung, Biorheology of soft tissues, Biorheology 10 (1973) 139.

[9] J.M. Guccione, A.D. McCulloch, L.K. Waldman, Passive material properties of intact ventricular myocardium determined from a cylindrical model, J. Biomech. Eng. 113 (1991) 42.

[10] J.M. Guccione, L.K. Waldman, A.D. McCulloch, Mechanics of active contraction in cardiac muscle: part II–Cylindrical models of the systolic left ventricle, J. Biomech. Eng. 115 (1993) 82.

[11] J.M. Guccione, A.D. McCulloch, Mechanics of active contraction in cardiac muscle: part I–Constitutive relations for fiber stress that describe deactivation, J. Biomech. Eng. 115 (1993) 72.

[12] J.M. Guccione, K.D. Costa, A.D. McCulloch, Finite element stress analysis of left ventricular mechanics in the beating dog heart, J. Biomech. 28 (1995) 1167.

[13] A.C. Guyton, Determination of cardiac output by equating venous return curves with cardiac response curves, Physiol. Rev. 35 (1955) 123.

[14] A.V. Hill, The heat of shortening and the dynamic constants of muscle, Proc. R Soc. London (Biol.) 126 (1939) 136.

[15] J.D. Humphrey, F.C. Yin, A new constitutive formulation for characterizing the mechanical behavior of soft tissues, Biophys. J. 52 (1987) 563.
[16] J.D. Humphrey, F.C. Yin, Biaxial mechanical behavior of excised epicardium, J. Biomech. Eng. 110 (1988) 349.
[17] J.D. Humphrey, F.C. Yin, Biomechanical experiments on excised myocardium: theoretical considerations, J. Biomech. 22 (1989) 377.
[18] J.D. Humphrey, R.K. Strumpf, F.C. Yin, Determination of a constitutive relation for passive myocardium: II. Parameter estimation, J. Biomech. Eng. 112 (1990) 340.
[19] J.D. Humphrey, R.K. Strumpf, F.C. Yin, Determination of a constitutive relation for passive myocardium: I. A new functional form, J. Biomech. Eng. 112 (1990) 333.
[20] J.D. Humphrey, R.K. Strumpf, F.C. Yin, Biaxial mechanical behavior of excised ventricular epicardium, Am. J. Physiol. 259 (1990) H101.
[21] W.C. Hunter, J.S. Janicki, K.T. Weber, A. Noordergraaf, Systolic mechanical properties of the left ventricle. Effects of volume and contractile state, Circ. Res. 52 (1983) 319.
[22] C.S.F. Lee, L. Talbot, A fluid mechanical study on the closure of heart valves, J. Fluid Mech. 91 (1979) 41.
[23] R.B. Panerai, A model of cardiac muscle mechanics and energetics, J. Biomech. 13 (1980) 929.
[24] C.S. Peskin, Numerical analysis of blood flow in the heart, J. Comp. Phys. 25 (1977) 220.
[25] C.S. Peskin, A.W. Wolfe, The aortic sinus vortex, Fed. Proc. 37 (1978) 2784.
[26] J.G. Pinto, Y.C. Fung, Mechanical properties of the heart muscle in the passive state, J. Biomech. 6 (1973) 597.
[27] H. Suga, K. Sagawa, Mathematical interrelationship between instantaneous ventricular pressure–volume ratio and myocardial force–velocity relation, Ann. Biomed. Eng. 1 (1972) 160.
[28] H. Suga, K. Sagawa, A.A. Shoukas, Load independence of the instantaneous pressure–volume ratio of the canine left ventricle and effects of epinephrine and heart rate on the ratio, Circ. Res. 32 (1973) 314.
[29] H. Suga, K. Sagawa, Instantaneous pressure–volume relationships and their ratio in the excised, supported canine left ventricle, Circ. Res. 35 (1974) 117.
[30] H. Suga, A. Kitabatake, K. Sagawa, End-systolic pressure determines stroke volume from fixed end-diastolic volume in the isolated canine left ventricle under a constant contractile state, Circ. Res. 44 (1979) 238.
[31] S. Winegrad, Calcium release from cardiac sarcoplasmic reticulum, Annu. Rev. Physiol. 44 (1982) 451.
[32] S. Winegrad, Mechanism of contraction in cardiac muscle, Int. Rev. Physiol. 26 (1982) 87.
[33] S. Winegrad, Endothelial cell regulation of contractility of the heart, Annu. Rev. Physiol. 59 (1997) 505.
[34] A.Y. Wong, P.M. Rautaharju, Stress distribution within the left ventricular wall approximated as a thick ellipsoidal shell, Am. Heart J. 75 (1968) 649.

CHAPTER

5

Blood Flow in Arteries and Veins

LEARNING OUTCOMES

1. Differentiate the anatomy/physiology of the arterial system and venous system
2. Explain the distinct layers that compose the blood vessel wall
3. State the general path of blood through the systemic circulation
4. Describe the function of the venous valve system
5. Discuss how blood moves through the venous system
6. Compare the characteristics and functions of cells that compose the cellular component of blood
7. Explain the different constituents of the plasma component and describe their function
8. Compare the different types of white blood cells
9. Interpret various characteristics of blood and how they play a role in rheology
10. Define the relationship between pressure, flow, and resistance in the arterial system
11. Calculate the volumetric flow rate using the Hagen–Poiseuille formulation
12. Develop a relationship between the pressure, the flow, and the resistance in the venous system
13. Formulate relationships for the volumetric flow rate of blood within veins
14. Apply the Windkessel model to blood flows
15. Discuss wave propagation in the arterial circulation
16. Compare the wave speed within the arterial vessel wall with different assumptions
17. Interpret the cause of pressure changes at bifurcations
18. Calculate the optimal design for bifurcations
19. Illustrate flow separation and how flows skew at bifurcations
20. Use the pressure gradient to determine the likelihood for flow separation

 © 2015 Elsevier Inc. All rights reserved.

21. Model flow through blood vessels using a taper or steps
22. Identify and express the critical components within the Navier–Stokes equations for a tapered blood vessel
23. Analyze pulsatile blood flow conditions
24. Explain the mathematical principle behind turbulent flows
25. Categorize common disease conditions that affect the vascular system

5.1 ARTERIAL SYSTEM PHYSIOLOGY

The entire cardiovascular system is composed of the heart (see Chapter 4) and blood vessels that have the primary function of transporting blood throughout the entire body. In this textbook, we will divide the blood vessels into three main categories: the arteries, the veins (for vein physiology, see Section 5.2), and the capillaries (see Chapters 6 and 7). What is interesting to note is the distribution of blood within these systems. Under normal conditions, approximately 7% of the blood volume is held within the heart, approximately 10% of the blood volume is in the arteries (including the aorta), approximately 10% of the blood volume is in the arterioles/capillary beds, approximately 10% of the blood volume is in the pulmonary circulation, and the remaining approximately 60% of the blood volume is held within the venous system. Therefore, the vast majority of the blood in the entire body is in the venous system, which therefore acts as a blood reservoir. More about the venous system will be discussed in Section 5.2.

An artery is defined as any vessel that transports blood away from the heart (note that this does not mean that the blood within an artery has to be oxygenated; i.e., the pulmonary arteries carry deoxygenated blood away from the heart).The term *artery* comes from the Greek words αηρτηρειν, which literally translates to "to contain air." Prior to the knowledge of blood, it was believed that the arteries transported air throughout the body. Within both the systemic and pulmonary circulatory systems, arterial transport is carried out under high pressure, as compared with their respective venous systems. As discussed in the previous chapter, this high pressure is primarily generated from the contraction of the heart muscle during the cardiac cycle and the lower compliance of arteries as compared with veins. The pressure pulse generated during heart contraction applies a large amount of force throughout the cardiovascular system that acts on the arteries themselves, along with the typical pressure that acts as a driving force for flow. In order to be able to withstand these high pressure forces generated by the heart, arteries are composed of thick and relatively muscular walls.

In fact, the arterial wall is composed of three distinct layers: the tunica intima, the tunica media, and the tunica adventitia (Figure 5.1). The tunica intima is the interior layer of the blood vessel wall and therefore is the only layer in contact with blood (under normal conditions). This layer is composed of endothelial cells and a small connective tissue layer. The connective tissue within the tunica intima is mostly composed of elastic fibers and helps secure the endothelial cells in place. In most arteries, this layer is not smooth, due to the constant contraction and dilation (pressure pulse) and the elastic recoil of the blood vessels; arteries must expand in response to heart contraction. The tunica media is

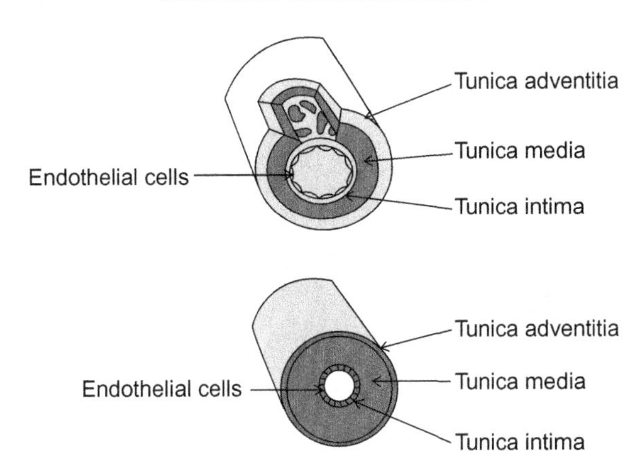

FIGURE 5.1 Anatomical structure of large elastic arteries and muscular arteries, respectively. In both of these types of blood vessels, the tunica media, which is composed of smooth muscle cells, is thick to withstand the high pressures within the arterial system. The tunica adventitia is composed of elastic fibers and helps to anchor the blood vessel to the surrounding tissue. The internal layer, the tunica intima, is composed of endothelial cells that are in contact with blood.

composed of smooth muscle cells organized in concentric rings. These muscle cells are responsible for altering the blood vessel diameter in response to neuronal input and local humoral control as well as maintaining the integrity of the blood vessels, while exposed to the high pressures. The tunica media is typically the thickest layer within the arterial wall (especially for large and medium-sized arteries), allowing the arteries to have sufficient mechanical properties to withstand the large pressure forces generated by the heart. The tunica adventitia is composed of connective tissue, mainly collagen and elastin. In tissues, this arterial wall layer is responsible for anchoring the blood vessel to adjacent tissue. The tunica adventitia is normally nearly as thick as the tunica media layer. Figure 5.1 illustrates a large artery (top panel) and a muscular artery (bottom panel).

The systemic circulation begins at the ascending aorta, immediately distal to the aortic valve (Figure 5.2). Within a few centimeters, the aorta makes a 180° turn (this is called the aortic arch) that leads to the descending aorta. The aortic arch contains branches for blood flow to the head (carotid arteries) and the arms (subclavian arteries). As the aorta descends toward the pelvis, it continually tapers and directly branches to deliver blood to many of the major abdominal organs (i.e., renal arteries, hepatic arteries, bronchial arteries, among others). At the pelvic region, the aorta branches into the left and right iliac arteries, which supply the legs with blood, as well as the sacral artery. The aorta can be described as twisting (moves through different planes), tapering, and branching throughout its entire length and therefore, the blood flow through the aorta is very complex. Furthermore, blood velocity and pressure is normally the highest in this vessel and the compliance of the aorta is relatively low. However, nearly all of the secondary arteries to the smallest arterioles and every artery in between are fairly straight (at least in two dimensions, radial and theta; it can vary in the z-dimension) and do not taper significantly until there is a branching point. At every branching point, the diameter of the branching vessel(s) (termed daughter vessel[s]) and the diameter of the main branch are reduced. The reduction in

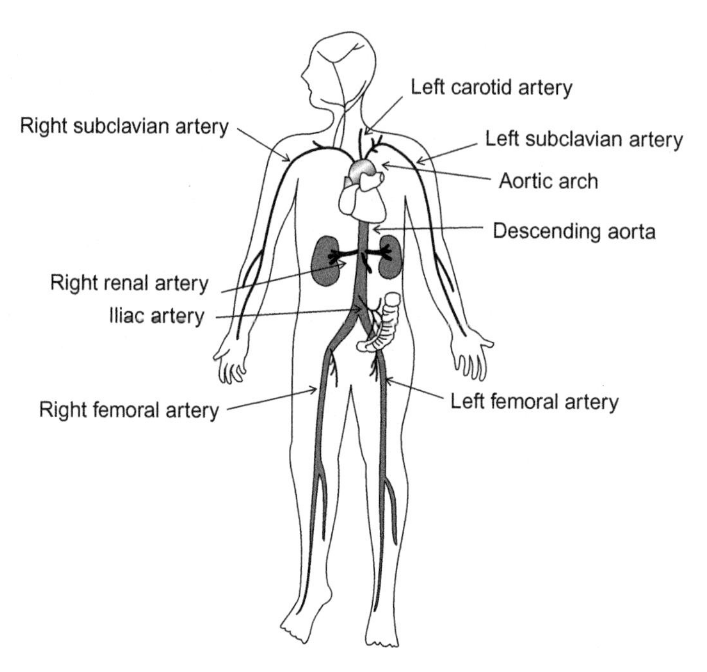

FIGURE 5.2 A schematic of the major arteries within the systemic circulation. After passing from the heart, into the aorta, oxygenated blood can enter the carotid arteries (feed the head), the subclavian arteries (feed the arms), the renal arteries (feed the kidneys), the iliac arteries (feed the pelvic region), or the femoral arteries (feed the lower limbs). There are other vessels that branch off the aorta that are not included within this figure.

blood vessel diameter is not constant (e.g., the vessel diameter is not always halved at a branch point) at each branch point and in many instances the daughter branches do not have the same diameter. For a comparison, the diameter of the ascending aorta is approximately 30 mm in humans. The diameter of the descending aorta, immediately distal to the aortic arch, is approximately 27 mm. By the time the aorta has reached the diaphragm, the diameter has reduced to approximately 20 mm. As a comparison, the iliac and carotid arteries have a diameter of approximately 10 mm and these vessels maintain a diameter similar to this until the first branching point. Therefore, within a relatively short length (approximately 30 cm), the aorta experiences a reduction in diameter of approximately one-third and therefore an approximate five-ninths reduction in the cross-sectional area. This is significant for the flow conditions throughout the vessel. Again, the secondary arteries maintain a fairly constant diameter until another vessel branches off from the original vessel.

After blood has passed through the arteries, it enters the smaller arterioles, which are the last branches of the arterial system prior to blood entering the microcirculation. The arterioles are also under high pressure (approximately 40–80 mmHg), to aid in the rapid blood movement throughout the arterial system, and therefore, they are also composed of a thick muscular layer. Not only do the muscles in the arterioles act to maintain the integrity of the blood vessel under high pressure; they also regulate blood flow into the capillary beds. As a side note, the muscles in the larger arteries do not regulate blood flow (e.g., they do not actively contract/dilate to dictate microvascular perfusion); they only act to maintain the blood vessel integrity. When the arteriolar muscles contract, blood flow can be halted in that vessel and shunted to another region that has a larger blood demand. Physically, blood is shunted from the region that has undergone a contraction, due to an increase in the vessel resistance. The region that the blood has been shunted to typically has a lower resistance and thus flow would proceed to that region. Under a dilatory relaxation, the arteriolar diameter can be increased by three to four times, which significantly increases the blood flow to its downstream capillary bed (see Section 5.5 for details on flow resistance). To some extent, these vessels are the controllers/regulators of blood flow throughout the entire body.

Blood velocity through the arterial circulation is directly proportional to the aggregate cross-sectional area of the particular vascular bed. This is a direct application of the Conservation of Mass formulation. Notice that we stated that it is proportional to the aggregate cross-sectional area of the vascular bed and not a cross-sectional area of a representative blood vessel. Under Conservation of Mass, if the cross-sectional area decreases, then the flow velocity increases, and vice versa. However, we know that the blood flow is slower in smaller vessels as compared with the feeding larger vessels. This is because one must account for the total aggregate cross-sectional area of all similar blood vessels. For instance, there is one aorta with an average cross-sectional area of approximately 3 cm^2. Combining all of the small arteries in the body (diameter range from 0.5 to 1.5 cm), there is an average aggregate cross-sectional area of approximately 25 cm^2. Combining all of the arterioles in the body (diameter range from 20 to 500 μm), there is an average aggregate cross-sectional area of approximately 50 cm^2. For comparison, capillaries account for a combined cross-sectional area of greater than 2000 cm^2, with the total combined venules accounting for an aggregate cross-sectional area of approximately 250 cm^2 and the large veins (vena cava) accounting for a cross-sectional area of approximately 10 cm^2. For quantity comparisons, one aorta branches into approximately 200 arteries (large and small), which eventually branch into approximately 160,000 arterioles, which branch into 35 billion capillaries, which converge into approximately 10,000,000 venules, which converge into 50 veins, which lead to 1 inferior vena cava, and 1 superior vena cava.

5.2 VENOUS SYSTEM PHYSIOLOGY

A vein is defined as any blood vessel that transports blood towards the heart (note that this does not mean that the blood within a vein is deoxygenated; i.e., the pulmonary veins carry oxygenated blood towards the heart).Veins also perform the critical task of acting as a reservoir for blood within the circulatory system (remember that approximately 60% of the total blood volume is housed within the venous system). In contrast to the arterial circulation, venous transport is carried out under much lower pressures (typically less than 5 mmHg). Due to the low pressure in the venous system, the walls of blood vessels in the venous circulation are thinner than the arterial walls, but they are still relatively muscular. The diameter of veins is smaller but comparable to that of arteries at the same location within the vascular tree. Contraction and dilation of muscles within the venous wall alter the amount of blood that can be held in the venous reservoir, but this does not regulate blood velocity or vascular perfusion as significantly as it does in the arterial circulation.

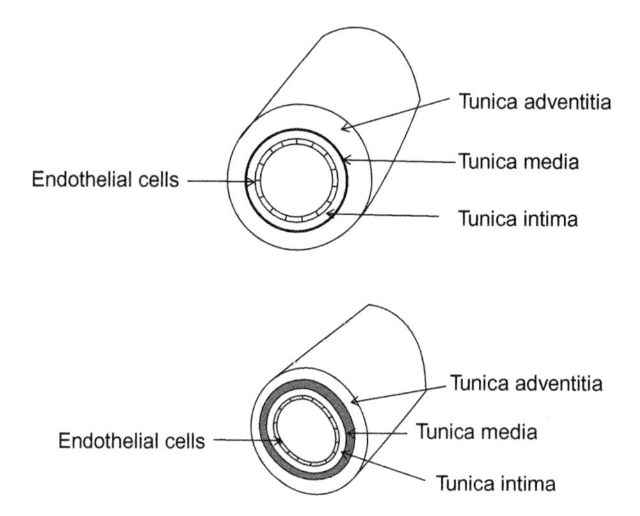

FIGURE 5.3 Anatomical structure of large elastic veins and medium-sized veins, respectively. The tunica adventitia is composed of elastic fibers and helps to anchor the blood vessel to the surrounding tissue. This is the thickest layer in veins. The tunica media is composed of smooth muscle cells and is smaller in veins as compared to arteries because they are under a lower pressure. The internal layer, the tunica intima, is composed of endothelial cells that are in contact with blood.

In the venous circulation, blood flows much slower and in response to skeletal muscle contraction (as we will see later).

Similar to the arterial wall, the venous wall is composed of three layers, the tunica intima, the tunica media, and the tunica adventitia (Figure 5.3). These layers have the same general anatomy as the arterial layers described previously, with the following differences. The endothelial layer within the tunica intima is typically smooth because veins do not experience significant dilation or contraction as often as arteries. The tunica media is relatively thin, as compared with the arterial wall and it is composed of smooth muscle cells and collagen. This layer is still responsible for controlling blood vessel diameter, however, the internal blood pressures are much lower in the veins than the arteries. The tunica adventitia is the thickest layer and is composed of collagen, elastin, and some smooth muscle cells. This layer is responsible for vessel anchoring to nearby tissues.

Although there is a common structure between arteries and veins, there are some important differences to discuss. Generally, the walls of arteries are thicker than those of veins. This difference is primarily manifested in the tunica media, because arteries are composed of significantly more smooth muscle cells and more elastic fibers as compared with veins. Again, however, the veins have a similar size as the paired arteries. Arteries are normally cylindrical in shape, whereas veins do not have a typical structure because of the low hydrostatic pressure throughout the system. Also, when under high pressure, it is more beneficial for blood vessels to have a uniform (or regular) circular geometry, so that the pressure forces are equally distributed at all locations and therefore one location does not experience an extremely high load. Otherwise, this would induce a weak spot within the vessel wall that would be more likely to fail. Veins do not have as high a pressure force to distribute across the vessel wall, as compared with arteries. Lastly, veins contain valves to prevent the backflow of blood toward the capillary beds. Again, since the venous system is under low pressure and a significant portion of venous flow is from the lower extremities upwards towards the heart, blood would tend to pool within the lower limbs because the pressure gradient cannot overcome the hydrostatic pressure gradient. Thus blood would preferentially pool within the low regions of the limbs.

Venous circulation begins as blood passes out from the capillary beds and enters the venules. These blood vessels act to collect and store blood, as well as to pass the blood along to the large veins. Therefore, venous flow is convergent flow, meaning that two (or more) blood vessels combine to form a large vessel. This continually occurs within the venous system until blood enters the heart via the superior (blood from the head and upper torso) and inferior (blood from the lower torso) vena cava.

Large veins have very little resistance to flow when they are fully open. However, under normal conditions, the veins in the chest cavity are compressed to some extent. Compression of veins occurs due to the low pressures within the venous system, the higher pressure from other surrounding organs or the interstitial space acting on the veins and veins bending over bones or other stiffer biological materials (e.g., the subclavian vein bends over the first rib bone). This compression partially regulates venous return into the heart by increasing flow resistance, however, contraction of local muscles can overcome this resistance to allow for venous return to occur. As we discussed in Chapter 3, the venous flow is also a function of the hydrostatic pressure within the fluid, because the pressure head (driving force ∇P) for flow is so low, the hydrostatic pressure gradient becomes very important. An average standing adult experiences an approximately 100 mmHg pressure difference between the heart and the toes. The forcing dynamic pressure head (∇P) in the venous system cannot overcome this pressure difference itself, and that is why if you stand still for an extended period of time, your lower limbs can cramp, blood pools in the lower limbs, and the "pins and needles" feeling occurs upon motion due to reperfusion of the tissue. Conversely, when standing, the pressure in your head is approximately 10 mmHg lower than the pressure of the right atrium. This aids in constant blood flow from the head back to the heart, and therefore, there is little pooling of the blood in the cranium and neck and a constant circulation of nutrients and wastes to/from the head. As you should know, increased pressure on the brain, including blood pressure, can have dramatic effects that can lead to death and/or improper brain function.

To help blood flow from the lower limbs back toward the heart, the venous system can act as a pump. The venous pump uses the valves that were discussed briefly to prevent the backflow of blood toward the capillary beds. In fact, without the valve system, it would be nearly impossible for the blood to ever flow back into the heart when a person is in a standing or a sitting position. The opening and closing of the venous valves help to regulate and induce blood flow through the majority of the venous system. This valve movement is facilitated via skeletal muscle contraction, and the increase in local fluid pressure induced by skeletal muscle contraction. Every time a skeletal muscle contracts, it compresses the veins that are in close proximity to the muscle, forcing the blood to flow through the downstream valves. Similar to the heart valves, the venous valves are designed in such a way that blood flows toward the heart and not toward the capillaries (Figure 5.4); the valves are unidirectional. As the blood evacuates one of the isolated venous chambers (a section of vessel

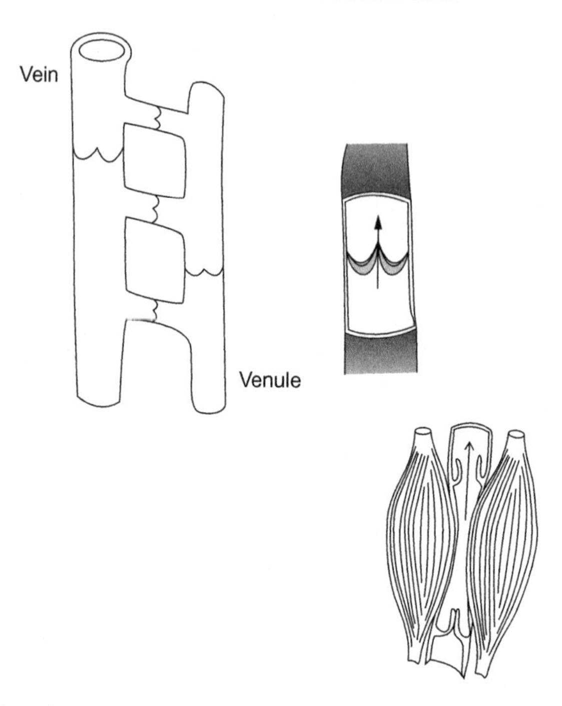

FIGURE 5.4 Venous valves determine the direction of blood within the venous system, preventing blood flow back toward the capillaries. Skeletal muscle contraction helps to transiently increase the pressure within veins to propel blood toward the heart. Arrows depict blood flow direction. The venous valve system is useful in the lower extremities when blood pressure cannot overcome the hydrostatic pressure.

between two valves), the overall pressure of that portion of the venous blood vessel is lowered, which facilitates blood flow into that section during the next muscle contraction. This pumping action is efficient under normal conditions, so that enough blood returns back to the heart and so there is not an excess of blood stored in the venous system. As a side note, when the valves within the venous system do not function properly, it is common for a person to develop varicose veins. Varicose veins are characterized by an overall increase in the diameter of the veins which causes the valves to not close properly. Therefore, under these conditions, blood can flow toward the capillaries and then the hydrostatic pressure continually increases within the damaged venous section. Skeletal muscle contraction cannot overcome this increased pressure under severe cases, and therefore, blood begins to pool within the damaged venous section. Pooled blood causes a distention of the vein beneath the skin, and this is what is seen as varicose veins.

5.3 BLOOD CELLS AND PLASMA

Blood is composed of two major components: the cellular component and the plasma component (Table 5.1). In an average adult, the blood volume is approximately 5 L, of which approximately 55–60% is plasma and the remaining portion is cellular. More than

TABLE 5.1 Blood Composition

Cellular Component (~40%)	Cell Type	Cell Concentration	Characteristic Shape/ Dimensions
	Red blood cell (erythrocyte—~99.7%)	~5,000,000/μL	Biconcave discs/8 μm diameter 2.5 μm thickness
	White blood cell (leukocyte—~0.2%)	~7500/μL	Spherical 20–100 μm diameter
	Platelet (thrombocyte—~0.1%)	~250,000/μL	Ellipsoid 4 μm long axis 1.5 μm short axis
Plasma Component (~60%)	**Composition**	**Major Contributors**	**Function**
	Water (~92%)	H_2O	Reduce viscosity
	Plasma proteins (~7%)	Albumin (~60%)	Osmotic pressure
		Globulins (~35%)	Immune function
		Fibrinogen (~3%)	Clotting
		Others (~2%)	Enzymes/hormones
	Other solutes (~1%)	Electrolytes	Homeostasis
		Nutrients	Cellular energy
		Wastes	Excretion

99% of the cellular component is composed of red blood cells. The most common way to quantify the percent of blood that is cellular is by quantifying the packed red blood cell volume, which is termed the hematocrit. Red blood cells can be packed via centrifugation, which forces the heavier cells to one location in a tube and the lighter fluid to a different location. Since the white blood cells and platelets make up a small percentage of the cellular volume and are not readily visible after centrifugation, the packed red blood cell volume is typically quantified. An adult male hematocrit is typically greater than that of an adult female, approximately 43% for males and approximately 40% for females. Some research groups have speculated that the reason that heart disease is more common in males is due to the increased hematocrit as compared to females (these groups suggest that a higher iron concentration is correlated with and can induce various cardiovascular disease. A second thought is that the altered balance between testosterone and estrogen may lead to altered cardiovascular disease responses.). Anemia (or reduced red blood cell volume) can be easily diagnosed through a hematocrit measurement. Patients with anemia can have a hematocrit that ranges anywhere from 10% to 35%. Anemia, although it is characterized by a low red blood cell count, is also associated with a decrease transport of oxygen and difficulties in maintaining core body temperature. Our discussion will continue with a discussion on the salient aspects of blood, starting with the cellular component and then the plasma component.

Red blood cells (or erythrocytes) have multiple functions in the body, but one of the most critical functions these cells perform is the delivery of oxygen to and the removal of carbon dioxide from all cells of the body. This function is performed through the use of hemoglobin (which is a metallo-organic protein) and carbonic anhydrase, which are both associated with red blood cells. Hemoglobin is the primary transporter of oxygen within the red blood cells. Red blood cells are basically filled with hemoglobin molecules. Carbonic anhydrase, an enzyme that resides within red blood cells, catalyzes the conversion reaction of carbon dioxide to water. This allows the blood to carry carbon dioxide in the form of bicarbonate ions (HCO_3^-). As you should be aware, oxygen/carbon dioxide exchange occurs within the capillaries of the systemic and pulmonary circulatory systems and will be discussed in detail in Chapters 7 and 9. Details of oxygen transport and function of carbonic anhydrase are provided within those discussions. Another important function that the red blood cells perform is the maintenance of the blood pH, through the use of hemoglobin as a buffer. The buffering capacity of hemoglobin is relatively large and it is able to maintain the blood pH between 7.35 and 7.45, by associating with any free hydrogen ions and/or releasing those ions under basic conditions.

Red blood cells are biconcave discs with a mean diameter in the range of 8 μm and a thickness of approximately 2.5 μm at the thickest point and approximately 0.8 μm in the center (Figure 5.5). Although we have not discussed capillaries yet, the average diameter of capillaries is approximately 6–8 μm. Therefore, the shape of red blood cells can and must change in order for these cells to squeeze through capillaries (a complete discussion on capillaries and red blood cell perfusion of capillaries can be found in Part 3 of this textbook). Due to the normal biconcave shape of red blood cells, there is an excess amount of

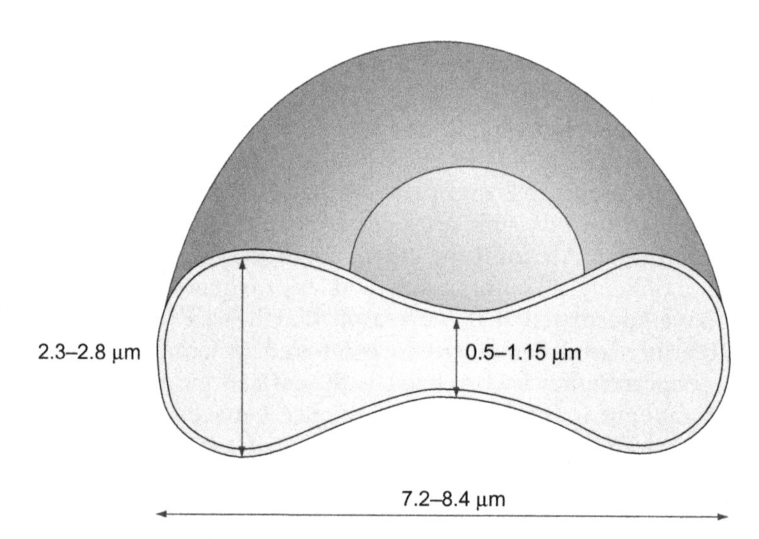

FIGURE 5.5 Cross-sectional view of a mature red blood cell. We also include the average range for various dimensions of a mature red blood cell. Each red blood cell contains hemoglobin molecules that facilitate the transport of oxygen and carbon dioxide throughout the cardiovascular system. There are approximately 5,000,000 red blood cells/μL of blood, and each red blood cell remains in circulation for approximately 120 days.

red blood cell membrane allowing the cell to deform into many other shapes. Under normal conditions, there are approximately 5,000,000 red blood cells/μL of blood (males tend to have a concentration slightly higher than this quantity and females tend to have a concentration slightly lower than this quantity). Under normal conditions, the quantity of hemoglobin per red blood cell is maintained at its maximum (i.e., saturated), to efficiently and effectively transport oxygen throughout the cardiovascular system. The quantity of hemoglobin within the red blood cell is only limited by the production of hemoglobin (in the liver) and not the transport of hemoglobin to form red blood cells. At its maximum production rate, the concentration of hemoglobin in each red blood cell is approximately 30 g/dL. The average red blood cell can stay in circulation for approximately 120 days before the cell (especially the hemoglobin and other proteins) is recycled.

The majority of red blood cells are produced in the marrow of bones during adult life (during fetal development, the liver/spleen/yolk sac produces the majority of red blood cells). In contrast to what many believe, most of the red blood cells are produced in the vertebrae and sternum during later stages of life (greater than 20–25 years old); only during the early stages of life do the marrow of long bones produce significant quantities of red blood cells. All blood cells are formed from hematopoietic stem cells, which differentiate and specialize into specific committed stem cells, which individually develop into different blood cell types. The major regulator of red blood cell production in the body is erythropoietin, which is a 34-kDa cytokine. Under hypoxic conditions (low tissue oxygenation), the formation of erythropoietin is increased within the kidneys and liver. Erythropoietin circulates to the red blood cell producing marrow and induces the differentiation of hematopoietic stem cells into proerythroblasts, which differentiate into erythrocytes within 1–2 days. For the complete differentiation of proerythroblasts into erythrocytes, vitamin B_{12} and folic acid are required as cofactors of the conversion reaction. Both of these vitamins are needed for DNA synthesis, which prevents nuclear maturation in the proerythroblast stage (red blood cells do not contain nuclei). Without these two vitamins, erythrocyte-like cells can be produced, but they can only stay within circulation for 30–40 days, inducing a form of anemia caused by a production of "inefficient" and high turnover red blood cells. Causes of low tissue oxygenation are anemia, destruction of hemoglobin, and reduced blood flow, among others, all of which have been shown to lead to increases in erythropoietin production. When there is a high tissue oxygenation level, the production of erythropoietin is slowed, so that there is not an excess of red blood cells/hemoglobin circulating within the cardiovascular system.

White blood cells (or leukocytes) are the primary cells that protect the body from foreign particles. There are two ways that leukocytes can perform this task: (i) leukocytes can directly destroy foreign particles or (ii) leukocytes can produce antibodies that aid in the immune response. There are six types of white blood cells in the body, each with its own specialized function. They are neutrophils, lymphocytes, monocytes, eosinophils, basophils, and plasma cells (some include platelets as a specialized type of white blood cell, however, for clarity we will separate the discussion on platelets). There are approximately 7000–8000 white blood cells/μL of blood. Of those cells, neutrophils account for more than 60% of the total, lymphocytes account for approximately 30%, monocytes account for approximately 5%, eosinophils account for approximately 2.5%, basophils account for 0.5%, and plasma cells approximately 0.1%. All white blood cells are formed in the bone marrow, except for lymphocytes and plasma cells, which are formed in lymph glands/spleen (see Chapter 8). Compared to red blood cells, white blood cells have a relatively short life span, which is typically only 5 days. The exceptions to this are monocytes and lymphocytes; monocytes can migrate into tissues within 1 day of being in the blood, and lymphocytes continually migrate between the blood and the tissue for approximately 1 month. Monocytes that enter the tissue can swell and become macrophages and can stay in the extracellular space for months. Some plasma cells are an exception as well.

We will now briefly describe the immune responses of these cells, but a more complete discussion can be found in other textbooks focused on inflammatory processes. Neutrophils and monocytes engulf invading particles directly through phagocytosis. These cells contain many lysosomes to digest and destroy the foreign particles. Lymphocytes play a role in acquired immunity, which is the inflammatory reaction against specific invading particles within the body. Some specialized lymphocytes have a memory for foreign particles that have previously entered into the body. Eosinophils exhibit a phagocytotic response toward parasites and release hydrolytic enzymes to neutralize the parasite invasion. Basophils play an important role in allergic reactions, and as such they can release histamine, serotonin, and heparin directly into the bloodstream. Plasma cells are antibody-producing cells and act in response to specific antigens. Each plasma cell can form one antibody and acts as a memory for the immune system. Depending on the type of plasma cell, they can reside in the circulation for months or within the bone marrow for years (termed long lived plasma cells). Loss or low counts of white blood cells do not typically affect the inflammatory response because the function of many of these cell types overlaps. Therefore, if one cell type is not functioning properly, a second type of white blood cell can take up the slack. Leukemia is a cancer characterized by the overproduction of white blood cells within the body. Unfortunately, these cells do not function properly and therefore multiple inflammatory processes are lost. The extent of leukemia dictates the amount of inflammatory reactions that are lost.

Platelets (or thrombocytes) are the primary cells for hemostasis. They are cellular fragments of megakaryocytes, which are derived from the same hematopoietic stem cells that lead to the formation of red blood cells and white blood cells. Platelets are typically ellipsoid in shape with a long axis of approximately 4 μm and a short axis of approximately 1.5 μm. The normal blood platelet concentration is approximately 250,000 platelets/μL, but can vary somewhat from this value. Platelets do not contain nuclei or many of the other common cellular organelles, but they do contain a large amount of calcium, ADP, prostaglandins, and coagulation cofactors. The platelet cell membrane may be the most important structure of the entire platelets. It contains many glycoproteins which adhere to other platelets, endothelial cells, extracellular proteins, and proteins during clot formation. Also, the charge of the cell membrane can accelerate or decelerate coagulation kinetics. The average life span of platelets is 12 days within circulation, at which time they are removed by the spleen.

Hemostasis is a complex process that culminates in the formation of insoluble fibrin. At this point, a fibrin mesh forms that is similar in structure to gauze, which acts to seal off the wound, preventing blood loss. The coagulation cascade is a series of enzymatic reactions, in which an active enzyme cleaves an inactive zymogen into its active form. The newly activated enzyme cleaves another zymogen into its active form and so on. This process continues until thrombin cleaves soluble fibrinogen into the insoluble fibrin, forming the mesh to prevent blood loss. Platelets are integral in these coagulation reactions because they provide multiple necessary cofactors for these reactions to occur. For instance, platelets provide a negative phospholipid surface (activated cell membranes, which express negative phospholipids which typically reside on the inner leaflet of the cell membrane) for coagulation complexes to form on (coagulation complexes include an activated enzyme and its specific inactive zymogen and any possible cofactors). Platelets also provide particular cofactors to accelerate the coagulation reactions (especially Factor Va and calcium). This helps in localizing the coagulation factors needed for clot formation and helps to increase the kinetics of the reactions within a localized region. Without these two critical roles that platelets play, fibrinogen will not be cleaved to fibrin rapid enough to prevent a major loss of blood. Lastly, platelets can aggregate with each other and various plasma proteins, forming a temporary mesh until the fibrin mesh has stabilized through the actions of Factor XIII. Negative feedbacks on hemostasis include the removal of platelets that are participating in these processes and antithrombotic proteins that cleave activated zymogens. If the platelet count is low, coagulation does not proceed as rapid as normal and minor injuries can cause severe blood loss. This is similar to hemophilia (loss or low concentration of Factor VIII), which is characterized by prolonged bleeding. It is important to note that shear stress can alter the platelet physiology significantly and that under disturbed blood flow conditions (e.g., high shear stresses, recirculation zones, oscillating stresses), platelets may accelerate cardiovascular disease progression.

The second major component of blood is plasma (Table 5.1). This includes everything other than cells that is within the blood. The major component of plasma is water. The remaining components are mostly electrolytes, sugars, urea, phospholipids, cholesterol, and proteins. The plasma concentration of electrolytes is very similar to that of the interstitial fluid because these compartments are only separated by the very permeable capillary wall (discussed in Section 3). Therefore, these compartments reach or approach equilibrium relatively quickly along the blood vessel length. Sodium ions are the most abundant positively charged ions in the plasma (although potassium, calcium, and magnesium are not negligible). Chloride is the most abundant negatively charged ions (but bicarbonate, phosphate, and sulfate are also present). The total plasma osmolarity is approximately 300 mOsmolar/L H_2O. Approximately 2% of plasma osmolarity is accounted for by sugars within the plasma. Cholesterol contributes 4–5% of the plasma osmolarity (in a normal diet) and phospholipids contribute approximately 6–7% to the plasma osmolarity. Proteins account for approximately 1 mOsmolar/L H_2O of plasma, but account for 7% of the total composition of plasma. The most crucial and most abundant protein within the blood is albumin, which accounts for over 60% of the total plasma protein concentration. Albumin has the primary function of maintaining the osmotic pressure of blood. It acts to balance the mass transfer across the capillary wall (see Chapter 7). The next major protein contributors to plasma are the globulins, which account for approximately 35% of the plasma protein concentration. Antibodies (immunoglobulins) and transport globulins are the major constituents of this portion of blood. Antibodies play a role in immunology and transport globulins can bind to hormones, metallic ions, and steroids, among others, to transport these molecules throughout the body. The remaining plasma protein composition is made up of fibrinogen (approximately 3%) and all of the other proteins. Fibrinogen is the precursor to fibrin and forms a mesh during clot formation, as discussed previously. The other plasma proteins have a wide range of functions but include insulin, thyroid-stimulating hormone, and all of the coagulation proteins.

The liver is the primary organ that synthesizes plasma proteins, including albumin, fibrinogen, and some of the hormones. Antibodies are produced by white blood cells as discussed earlier. The other components of plasma are typically synthesized in cells, transported between cells, or ingested through the digestive system. The relative concentration of these components is regulated tightly, but can change relatively fast depending on many factors. For instance, after eating a candy bar, your blood glucose levels can increase significantly. However, under normal conditions, the body regulates this concentration closely, and the blood glucose level should return back to its normal level rather quickly.

Plasma does not have a function per se because it is an amalgam of many different biologically active components. In general, the water portion of plasma is used to reduce the viscosity of the cellular component of blood. This in effect reduces the resistance to flow and even can be considered to allow for blood flow to occur (i.e., cellular matter itself would not "flow"). A major function of the remaining plasma components is to maintain equilibrium with the interstitial space, which aids in homeostasis. Sugars are used as the nutrient source for cells. Cholesterol can be used within the cell membrane to locally increase its rigidity so the cell can withstand forces better. Also, it is thought that highly specialized signaling occurs within this cholesterol rich area of the cell membrane. Proteins have specific functions and each protein may have a different task. Therefore, plasma in effect has a very critical function for the human body. Serum is a special component of plasma, in which the protein fibrinogen has been removed. This is most likely due to clot formation, but can be made in the lab to study various properties of plasma.

5.4 BLOOD RHEOLOGY

Rheology is the study of flowing materials combined with the study of the physical properties of the particular flowing material. In this section, we will discuss a more accurate way to model the viscosity of blood, which accounts for some of the rheological properties of blood. As we have seen in previous sections, blood is a special non-Newtonian fluid that is also composed of at least two phases. Recall that, blood is a combination of cells, proteins, and electrolytes, among other compounds, all suspended in a fluid medium. Plasma (the portion of blood not including cells) behaves as a Newtonian fluid (e.g., constant viscosity under shear forces), under physiological conditions, and has a viscosity of approximately 1.2 cP. Whole blood, however, does not have a constant viscosity, exhibits a yield stress, and is composed of many different components. The viscosity of whole blood varies with shear rate, hematocrit, temperature, and disease conditions (Figure 5.6), and this is predominantly due to the presence of cells, the interactions of cells with each other and the fluid medium, and the presence of other charged compounds within the fluid. Figure 5.6 illustrates (i) that plasma (Hct = 0) has a constant viscosity within the shear rate tested, (ii) viscosity increases with increasing hematocrit, and (iii) viscosity decreases with increasing shear rate when cells are

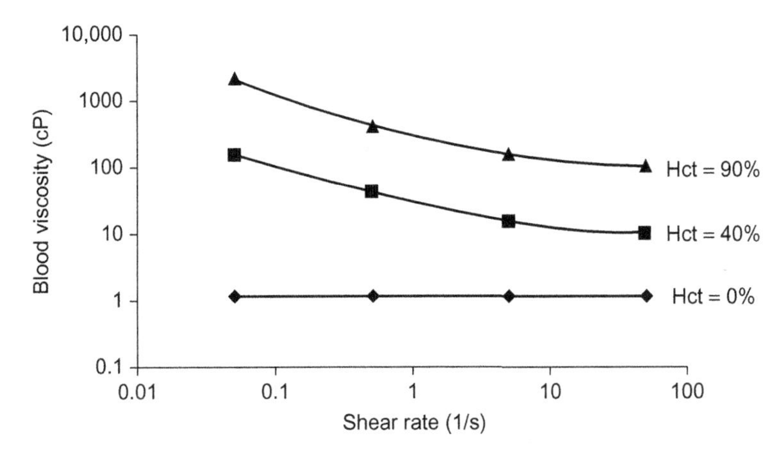

FIGURE 5.6 The variation in blood viscosity as a function of shear rate and blood hematocrit (Hct). This figure illustrates that without the cellular component the blood viscosity remains constant, at approximately 1.2 cP, over the range of shear rates depicted on the figure. When cells are present in the blood, the viscosity is no longer constant and increases with decreasing shear rate. This figure is a summary of data collected by Chien et al.

present. There have been many investigations into the rheology of blood which have supported the idea that the presence of cells and charged molecules causes blood to behave as a non-Newtonian fluid. Importantly, many groups have tried to determine the yield stress for blood, but have seen that it is difficult to quantify blood motion as the shear rate approaches zero for many reasons including settling of cells, those settled cells interact with each other in different ways and the controls of the experimental systems tend to be quite difficult. However, using a Casson model for blood flow combined with measured values for shear stress at given shear rates, the yield stress for blood can be extrapolated. The Casson model was introduced in 1959 for a classical fluid mechanics approach and was observed to fit blood flow well, because it accounts for the interaction of solid and fluid particles as a two-phase fluid. This model incorporates both a yield stress for blood and the shear-thinning viscous properties of blood. The Casson model for blood states that

$$\sqrt{\tau} = \sqrt{\tau_y} + \sqrt{\eta\dot{\gamma}} \tag{5.1}$$

is a relationship that can describe the dependence of shear stress on shear rate ($\dot{\gamma}$), a yield stress (τ_y) (which varies based on hematocrit), and apparent viscosity (η), which is an experimentally fit constant to approximate the fluid's viscosity. Under most physiological conditions, the yield stress of blood can be approximated as 0.05 dyne/cm^2. Therefore, when a shear stress less than 0.05 dyne/cm^2 is applied to blood, blood will not flow or will only flow as a rigid body.

We have already learned that under normal laminar, steady, fully developed flow conditions in a cylindrical tube, the shear stress at the centerline is zero. Therefore, blood at the centerline would need to move as a rigid body, because this force has not exceeded the yield stress for blood (in fact this would theoretically be true for all fluids). Extending

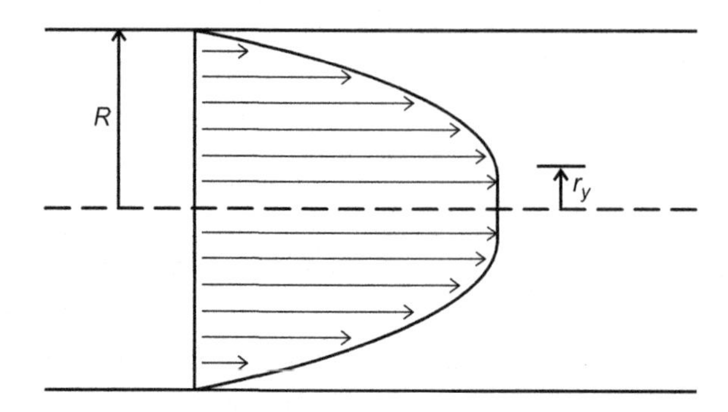

FIGURE 5.7 The velocity profile of blood flowing through a cylindrical vessel using the Casson model. The profile is blunter than a purely Newtonian fluid's velocity profile because the fluid that is between the centerline and r_y flows as a solid. At r_y, the shear stress surpasses the yield stress of the fluid (τ_y) and the viscous forces take effect.

this thought further, the shear stress experienced by blood will surpass 0.05 dyne/cm^2 (the assumed yield stress of blood) at a finite distance from the vessel centerline. For modeling purposes, we state that at a distance of r_y, the shear stress will exceed τ_y. Therefore, blood that is beyond this location (closer to the vessel walls) will flow as a "normal" viscous fluid, leading to a pseudo-parabolic velocity profile that has a blunted centerline velocity, and fluid that is closer to the centerline will only flow as a rigid body (Figure 5.7). This is caused by the principle that the blood viscosity will be zero (or blood can be modeled as an inviscid fluid) when the yield stress within the flow field has not been surpassed and thus blood moves as a rigid body. This becomes very apparent within the microcirculation, where red blood cells (which are semi-rigid objects) stream towards the centerline of the blood vessel.

Mathematically, this phenomenon can be represented using the Casson model. We already know that

$$\tau = -\frac{r}{2}\frac{dp}{dx}$$

and therefore, the Casson model can be rewritten as

$$\sqrt{-\frac{r}{2}\frac{dp}{dx}} = \sqrt{\tau_y} + \sqrt{\eta}\sqrt{\dot{\gamma}} \tag{5.2}$$

Solving Eq. (5.2) for shear rate we can obtain

$$-\frac{du}{dr} = \dot{\gamma} = \frac{1}{\eta}\left(\sqrt{-\frac{r}{2}\frac{dp}{dx}} - \sqrt{\tau_y}\right)^2 \tag{5.3}$$

If we integrate this function from r_y to R (tube radius), over the appropriate reduced cross-sectional area (e.g., the cross-sectional area is not πR^2, but it is $\pi(R^2 - r_y^{\,2})$), we get the following function for the velocity profile, keeping in mind that the fluid velocity is constant between the centerline ($r = 0$) and r_y:

$$u(r) = \begin{cases} -\frac{1}{4\eta}\frac{dp}{dx}\left(R^2 - r^2 - \frac{8}{3}r_y^{0.5}(R^{1.5} - r^{1.5}) + 2r_y(R - r)\right) & r_y \leq r \leq R \\ -\frac{1}{4\eta}\frac{dp}{dx}\left(\sqrt{R} - \sqrt{r_y}\right)^3\left(\sqrt{R} + \frac{1}{3}\sqrt{r_y}\right) & r \leq r_y \end{cases} \tag{5.4}$$

To obtain this formulation, we make use of many of the same assumptions that were discussed in previous sections for blood flow through vessels (e.g., constant cross-sectional area, constant apparent viscosity, rigid containing vessel, the flow is steady). Using these assumptions, r_y is a particular constant value and thus the velocity profile of a fluid that is modeled with the Casson approximation can be represented by Figure 5.7.

The Casson model is a very useful representation of blood flow through the arterial system, but it tends to complicate the computations by hand. The following sections discuss a method to simplify the calculations, but these methods may not accurately represent the blood flow conditions.

A second approximation for blood flow characteristics that has become popular is the power law model for blood flow. This model also takes into account the non-Newtonian properties of the fluid which include approximations for shear thinning and yield stress behavior. The power law for blood flow can be represented by

$$\tau = a\dot{\gamma}^b \tag{5.5}$$

where a and b are empirically defined power law constants for blood flow. The consistency index for fluids is represented within a; as a increases, the viscosity of the fluid also increases. Non-Newtonian characteristics are represented in b; as b diverges from a value of 1, the fluid exhibits more non-Newtonian characteristics. Equation (5.5) is similar in form to Eq. (2.39), where $b > 1$ would be a shear-thickening fluid and $b < 1$ would be a shin-thinning fluid. The power law for blood flow does not represent the yield stress of blood flow. The Herschel–Bulkley model extends the power law relationship to include a yield stress and can be represented as

$$\begin{aligned} \tau &= a\dot{\gamma}^b + \tau_y & \tau \geq \tau_y \\ \dot{\gamma} &= 0 & \tau \leq \tau_y \end{aligned} \tag{5.6}$$

where the variables in Eq. (5.6) represent the same quantities as discussed for Eqs. (5.1) and (5.5). This model is very similar to Casson model for blood flow and has been used in many biological fluid flows.

Because plasma behaves as a Newtonian fluid (as shown in Figure 5.6, when the hematocrit is set to 0), the non-Newtonian fluid properties are more than likely attributed to the cellular component of blood. Recall that by definition, plasma contains all of the dissolved solutes that are found in blood (e.g., proteins, ions, sugars, fats, etc.). The first contributing factor to non-Newtonian behavior is the ability of red blood cells, the majority of the cellular component, in the fluid phase to form aggregates termed rouleaux (Figure 5.8). Interestingly, the presence of rouleaux in the blood is dependent on the shear rate at which the fluid is flowing, hematocrit and temperature. At lower shear rates, rouleaux are more

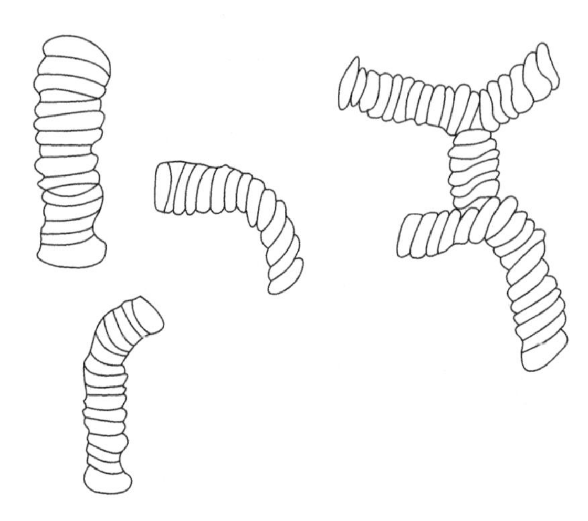

FIGURE 5.8 Red blood cell aggregates termed rouleaux, which can form linear or branched structures. The number of red blood cells in a rouleaux can vary from 2 to approximately 50. Red blood cells in the microcirculation typically traverse through capillaries within rouleaux structures.

prevalent, likely due to the greater time that cells can come into contact with each other and/or proteins in the blood. It has been speculated that as the shear rate approaches zero, blood will act as a viscoelastic solid, because all of the red blood cells will form into one large aggregate. As you may recall from solid mechanics courses, viscoelastic solids have a definite yield stress, which may account for the yield stress of blood. Also, at low shear rates, the formation of rouleaux consumes energy when they tumble through the fluid. This can be recorded as an effective increase in the viscosity of blood, whereas at higher shear rates, when the rouleaux break apart, the apparent viscosity of blood decreases.

Two second contributing factor to the non-Newtonian behavior of fluids is due to the deformability of the red blood cells. The deformability of red blood cells arises due to the composition of the cell membrane and the overall rigidity of the cytoskeleton. Although, many model red blood cells as solid particles within the fluid phase, red blood cells exhibit many viscoelastic properties. A report by Goldsmith investigated the effects of particle rigidity and particle concentration (or hematocrit) on the fluids apparent viscosity. This report showed that at a shear rate sufficient to prevent aggregation ($>100\,s^{-1}$), a suspension consisting of rigid latex particles in water, rigid rubber discs in glycerol, red blood cells in blood, or sickled red blood cells in blood (all particles had a similar size as red blood cells) exhibited an increase in the apparent viscosity as the percent of particles to fluid increased. However, for high hematocrit ratios, all suspensions, excluding those of normal red blood cells, approached an apparent viscosity of infinity, whereas normal red blood cells at a hematocrit of 0.9 approached 20 cP. This suggests that red blood cells can deform to allow for fluid motions to occur. The data from this report is redrawn in Figure 5.9. It has been observed and shown that deformability is a function of the red blood cell membrane. Unlike many droplets and suspended particles, the pressure that acts on the internal cell membrane and the external cell membrane must be equal because of the biconcave shape of the cell, if we assume that the bending rigidity of the cell

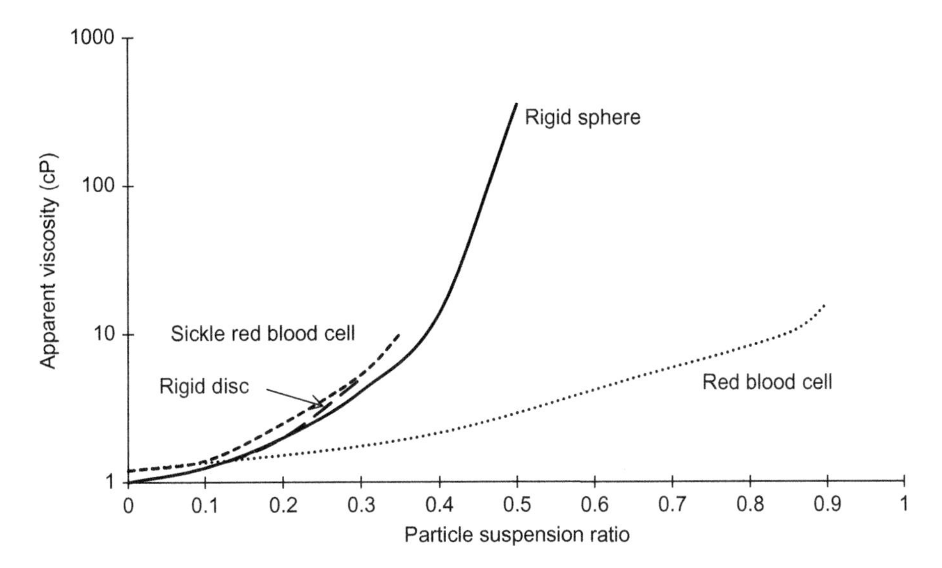

FIGURE 5.9 Investigations to determine the apparent viscosity of a suspension at a given shear rate. It has been shown that as the particle suspension ratio (e.g., the hematocrit for blood) increases, the apparent viscosity of the solution also increases. Investigations into the mechanical properties of the red cell membrane have shown that the fluid properties of the membrane prevent a massive increase in apparent viscosity, with marginal increase in particle suspension ratio. Rigid spheres or rigid discs, with a similar size/area as red blood cells show this large increase in apparent viscosity with small changes in suspension ratio. Additionally, sickle red cells lose their fluid-like properties.

membrane is negligible. Under this scenario, the cell membrane in the normal biconcave shape is under negligible (or zero) stress. The shape of the red blood cell also suggests that large deformations can be accomplished without stretching the cell membrane and without the input of significant energy. The packing density of red blood cells, due to the deformability and the unique shape, can be higher with less energy input than suspended particles.

5.5 PRESSURE, FLOW, AND RESISTANCE: ARTERIAL SYSTEM

This and the following section will reiterate some of the important formulas that govern the flow within the arterial system and the venous system (Section 5.6) from the previous chapters. Our goal for these two sections is to discuss some of the common simplifying assumptions that are made when modeling flows through the arterial and the venous circulation. In the simplest case, the flow through a blood vessel (Q) is determined by the pressure difference across the vessel (ΔP) and the resistance to flow throughout the vessel (R). This relationship follows an Ohm's law formulation, where a driving force (ΔP, in our case) is proportional to the flow through the system (Q, in our case).

$$Q = \frac{\Delta P}{R} = \frac{P_1 - P_2}{R} \tag{5.7}$$

Under normal resting conditions, the blood flow rate is approximately 5 L/min. This means that all of the blood within the body traverses the circulatory system once a minute and that the flow through a given section of the cardiovascular tree must equate to 5 L/min. Pressure difference is something that we have discussed numerous times before. Pressure is relatively easy to measure and is typically in units of mmHg. Recall that this is the driving force for flow and not the static pressure difference across tissue. Resistance to flow, however, is something that cannot be measured; it can however be directly calculated from Eq. (5.7). Be cautioned that there are many properties that will affect Q and R, and Eq. (5.7) is overly simplified to some extent. Some of the assumptions that are built into Eq. (5.7) are that the fluid flow and fluid properties have no temporal variation and that there is a constant uniform circular cross-section within a rigid blood vessel.

Example

Calculate the resistance to blood flow within the descending aorta and the inferior vena cava. Assume that the pressure difference between the distal portion of the aortic arch and the iliac arteries is 20 mmHg. The pressure difference within the inferior vena cava is 3 mmHg. Assume that the flow rate through both vessels is 4.5 L/min.

Solution

Descending Aorta	Inferior Vena Cava
$R = \frac{\Delta P}{Q} = \frac{20\ mmHg}{4.5\ L/min}$	$R = \frac{\Delta P}{Q} = \frac{3\ mmHg}{4.5\ L/min}$
$R = 355.5\ dyne * s/cm^5$	$R = 53.3\ dyne * s/cm^5$

This problem illustrates that to maintain the same flow rate at a higher pressure head, the resistance to flow must be larger. If the resistance to flow was the same in this example, then the flow would have to be much greater with a larger pressure gradient. This is clearly an oversimplification of the biology, but can provide some insight into the flow through the cardiovascular system.

A more useful approximation for volumetric flow rate (Q) is the Hagen–Poiseuille's law. The general Poiseuille solution is obtained by solving for the volumetric flow rate for laminar fluid flow within a cylindrical tube. Recall from Chapter 3 that the general velocity profile for pressure-driven cylindrical flow, with constant fluid properties and a rigid containing vessel, is given by

$$v_z(r) = \frac{R^2}{4\mu}\frac{\partial p}{\partial z}\left(\left(\frac{r}{R}\right)^2 - 1\right)$$

The volumetric flow can be calculated by integrating the velocity profile over the cross-sectional area of the vessel, as in

$$Q = \int_0^R 2\pi r v_z(r) dr = \int_0^R 2\pi r \left[\frac{R^2}{4\mu}\frac{\partial P}{\partial z}\left(\left(\frac{r}{R}\right)^2 - 1\right)\right] dr = 2\pi \frac{R^2}{4\mu}\frac{\partial P}{\partial z}\int_0^R r\left[\left(\frac{r}{R}\right)^2 - 1\right] dr$$

$$= \frac{\pi R^2}{2\mu}\frac{\partial P}{\partial z}\int_0^R \left(\frac{r^3}{R^2} - r\right) dr = \frac{\pi R^2}{2\mu}\frac{\partial P}{\partial z}\left[\frac{r^4}{4R^2} - \frac{r^2}{2}\right]_0^R = \frac{\pi R^2}{2\mu}\frac{\partial P}{\partial z}\left[\frac{R^4}{4R^2} - \frac{R^2}{2} - 0 + 0\right] \tag{5.8}$$

$$= -\frac{\pi R^4}{8\mu}\frac{\partial P}{\partial z}$$

As the partial derivative in space approaches the tube length (which is what can be measured), the spatial partial derivative of pressure becomes

As

$$\partial z \rightarrow L : \frac{\partial P}{\partial z} \rightarrow -\frac{\Delta P}{L}$$

where

$$\Delta P = P_1 - P_2$$

and

$$\partial P = P_2 - P_1$$

Therefore, the volumetric flow rate can be approximated as

$$Q = \frac{\pi \Delta P R^4}{8\mu L} \tag{5.9}$$

where $\Delta P / L$ is the pressure gradient across a tube of length L and radius R. To clearly state the assumptions that were used to derive this formulation, the fluid within the tube must have a constant viscosity of μ and a uniform circular cross-sectional area over the entire length, L. Additionally, none of the fluid properties can vary with time; however, this formula can be used for one discrete time period of a temporally varying flow. As we have seen, the Hagen–Poiseuille solution is derived from integrating the velocity of the fluid (solved from the Navier–Stokes equations) with respect to blood vessel area. In other words, this is a weighted average for each section of fluid that has the same velocity. The lamina of fluid against the wall would be characterized by the circumference of the tube (or total arc length of an irregular shaped container) with a velocity of zero. The lamina in the center of the tube would have a differential circular area with the maximum velocity of the fluid. Using the generalized integral relationship, one can derive a formula for volumetric flow rate for any shaped container under pressure-driven flow. From Eq. (5.9), one can define resistance, R, from Eq. (5.7) as

$$R = \frac{8\mu L}{\pi R^4} \tag{5.10}$$

What is interesting to note about Eqs. (5.9) and (5.10) is that small changes in the vessel radius cause very large changes in the flow rate by altering the resistance significantly. A doubling of the blood vessel radius would be associated with a 16-fold increase in the volumetric flow rate throughout the tube, assuming all of the other parameters remain the same. This is not a trivial discussion, because radius doubling is quite possible when a blood vessel passes from a maximally constricted state to a maximally dilated state. It is common for the resistance blood vessels (arterioles) to experience changes in vessel diameter on the order of fourfold (again this would be maximally constricted to maximally dilated), which theoretically would increase the volumetric flow rate 256 times. Also what one can infer from this is that the blood vessel radius does not have to change significantly to increase (or decrease) flow to a region that has non-normal (low or high) tissue oxygenation. This is important for energy usage (or consumption) and storage for the smooth muscle cells (you may recall from other classes that the cell cannot effectively store energy in the form of ATP, it is almost a use-it or lose-it scenario). If flow rate did not change rapidly with small changes in vessel diameter, then blood vessels would need to dilate or constrict significantly to counterbalance the tissue oxygenation level changes. This would be associated with large energy demands that would effectively deplete the smooth muscle cells of energy, and then they would be unable to respond to changes in the tissue needs in the near future. Instead, the biological system is designed so that minimal energy is required for vast blood flow changes.

Example

Calculate the volumetric flow rate within an arteriole with a length of 100 μm and a radius of 35 μm. The pressure difference across the arteriole is 10 mmHg. Also calculate the change in radius needed to reduce the volumetric flow rate by 5% and to increase the volumetric flow rate by 10%. In this vessel, the effective blood viscosity is 2.8 cP. Assume that when the radius changes there are no other changes in the other fluid parameters.

Solution

Using the Hagen–Poiseuille's formulation to calculate the volumetric flow rate, we get

$$Q = \frac{\pi \Delta P R^4}{8\,\mu L} = \frac{\pi(10\,mmHg)(35\,\mu m)^4}{8(2.8\,cP)(100\,\mu m)} = 0.168\,mL/\min$$

5% of this flow rate is 0.00842 mL/min; therefore,

5% Reduction	**10% Increase**
$Q = 0.160\,mL/\min$	$Q = 0.185\,mL/\min$
$R = \sqrt[4]{\frac{8Q\mu L}{\pi \Delta P}}$	$R = \sqrt[4]{\frac{8Q\mu L}{\pi \Delta P}}$
$= \sqrt[4]{\frac{8(0.160\,mL/\min)(2.8\,cP)(100\,\mu m)}{\pi(10\,mmHg)}}$	$= \sqrt[4]{\frac{8(0.185\,mL/\min)(2.8\,cP)(100\,\mu m)}{\pi(10\,mmHg)}}$
$= 34.5\,\mu m$	$= 35.8\,\mu m$
$\Delta R = 0.5\,\mu m$ (1.5% change)	$\Delta R = 0.8\,\mu m$ (2.3% change)

The previous problem illustrates that relatively small changes in an arteriole radius (e.g., a 1.5% change) can significantly affect the flow rate (5% change). The take home message is that very small changes in vessel diameter cause large changes in flow.

We would like to revisit the Casson model of blood flow to show that the volumetric flow rate associated with this more realistic approximation of blood rheology obtains a solution that is analogous to the Hagen–Poiseuille solution. If we compute the volumetric flow rate of the Casson velocity profile (Eq. (5.4)), as given by

$$Q = 2\pi \int_0^R u(r) r dr$$

we obtain

$$Q = \frac{\pi R^4}{8\eta} \frac{dp}{dx} F(\Psi) \tag{5.11}$$

The function of Ψ is defined as

$$F(\Psi) = -\left(1 - \frac{16}{7}\Psi^{0.5} + \frac{4}{3}\Psi - \frac{1}{21}\Psi^4\right) \tag{5.12}$$

and

$$\Psi = \left(\frac{2\tau_y}{R}\right)\left(-\frac{dp}{dx}\right)^{-1} \tag{5.13}$$

Comparing Eqs. (5.9) and (5.11), we can see that by using the Casson model of blood flow, the volumetric flow rate becomes a function of the fluid yield shear stress. This would be anticipated because if the shear stress applied to the fluid was below this threshold, the fluid would not flow or would move as a rigid body. However, there are distinct similarities between Eqs. (5.9) and (5.11). If we can assume that our fluid has a negligible a yield stress, the Hagen–Poiseuille solution may be applicable (if the other assumptions are valid). Also notice that if the fluid has no yield stress ($\tau_y = 0$), then we would obtain the exact Hagen–Poiseuille solution as shown in Eq. (5.8) (if $\tau_y = 0$ then $\Psi = 0$ and $F(\Psi) = -1$).

We have already learned that the pressure variation in a normal human aorta is 80–120 mmHg. A comparison between blood pressure and mass shows that as the mass of the animal increases, the mean systolic blood pressure also increases (Figure 5.10). As we discussed with changes in heart rate based on animal mass, correlations for this variation can be made based on the activities and evolutionary history of the animals.

5.6 PRESSURE, FLOW, AND RESISTANCE: VENOUS SYSTEM

The mathematical formulation that was derived in the previous section for volumetric flow rate through the arterial system can be used to approximate the volumetric flow rate in the venous system as well. However, it is critical here to determine whether the assumptions that have been made in deriving Hagen–Poiseuille's law are still valid. In most instances, veins/venules are not circular, and therefore, this formulation will only

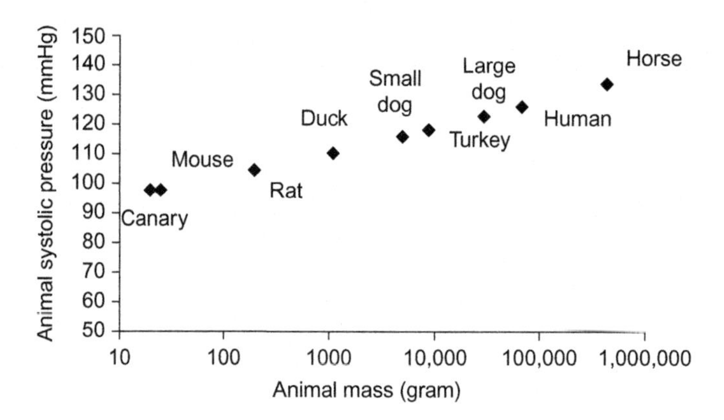

FIGURE 5.10 The relationship between animal mass and mean systolic pressure. As the mass of the animal increases, there is a general increase in systolic pressure. The relationship between these two measurements can be correlated to many different properties of the animal, as described in the text.

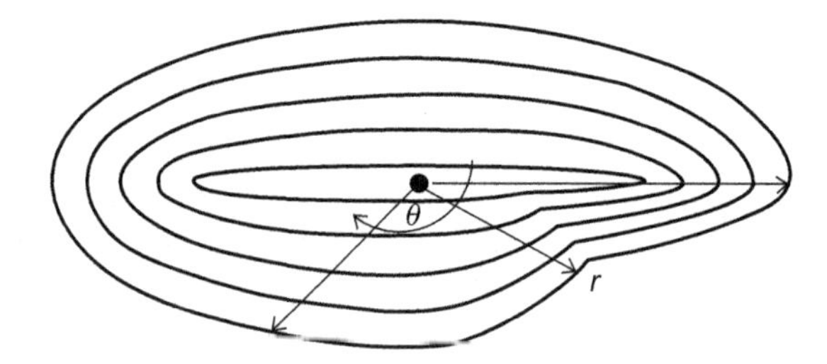

FIGURE 5.11 Schematic of a vein with a nonstandard geometry, shown by varying radial's length at various angles, θ. The method of choosing laminas of known average velocities still apply to these vessels; however, one would need to take account for the different areas of each lamina. Additionally, one would need to consider whether or not the fluid properties change as a function of the different radial directions.

provide an approximation to the volumetric flow rate. However, like arteries, the approximate radius along the vessel length is relatively constant between converging branches, but veins are not rigid. Therefore, let us revisit the derivation of the volume flow rate, using cylindrical coordinates the velocity profile within a vein would become

$$v_z(r) = \frac{r^2}{4\mu}\frac{\partial P}{\partial z} + c_1 \ln(r) + c_2 \tag{5.14}$$

by solving the cylindrical Navier–Stokes equations (if $v_z(r)$ is a function of radius only) and assuming that flow is only pressure driven. The integration constants can only be found if the boundary conditions of the particular flow conditions are known (we typically assumed that the shear rate along the centerline was zero, $\frac{dv_z}{dr}(r=0)=0$, and that the velocity along the wall was zero $v_z(r=R)=0$). With a randomly shaped object, the second boundary condition may become difficult to define because the cross-sectional area is not uniform (if the cross-sectional area is not constant over the tube length than the centerline moves as well, making the first boundary condition difficult to determine). For instance, in Figure 5.11, which depicts an irregular shaped vein, the blood velocity along the entire wall is zero but the radial coordinate describes the wall is not constant with changes in the angular, θ, direction. Therefore, the solution for the volumetric flow rate would become

$$\int_A \left(\frac{r^2}{4\mu} \frac{\partial p}{\partial z} + c_1 \ln(r) + c_2 \right) dA \tag{5.15}$$

which may be approximated for laminas of a particular thickness, if the velocity can be assumed to be uniform within that lamina and the velocity changes with theta are negligible. This would use a weighted average formulation. The key here is that we must be able to neglect changes in velocity in the theta direction to apply this type of approximation and that we must be able to approximate sections of the vessel area with a "uniform" velocity.

Example

Calculate the average volumetric flow rate for the following blood vessel that is represented as an ellipse. The blood vessel has known average velocity values and geometric values for laminae of fluid (r_L is the radius for the long axis and r_S is the radius for the short axis, Figure 5.12).

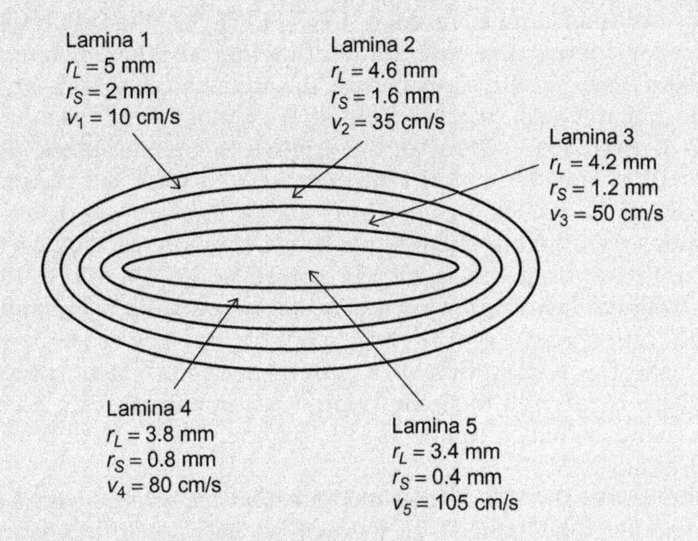

FIGURE 5.12 Figure associated with example problem.

Solution

First, we will calculate the area associated with each fluid lamina:

$$A_1 = \pi(5\ mm)(2\ mm) - \pi(4.6\ mm)(1.6\ mm) = 8.29\ mm^2$$
$$A_2 = \pi(4.6\ mm)(1.6\ mm) - \pi(4.2\ mm)(1.2\ mm) = 7.29\ mm^2$$
$$A_3 = \pi(4.2\ mm)(1.2\ mm) - \pi(3.8\ mm)(0.8\ mm) = 6.28\ mm^2$$
$$A_4 = \pi(3.8\ mm)(0.8\ mm) - \pi(3.4\ mm)(0.4\ mm) = 5.28\ mm^2$$
$$A_5 = \pi(3.4\ mm)(0.4\ mm) = 4.27\ mm^2$$

Using the average velocity for each lamina, the volumetric flow rate can be found:

$$Q = Q_1 + Q_2 + Q_3 + Q_4 + Q_5 = v_1A_1 + v_2A_2 + v_3A_3 + v_4A_4 + v_5A_5$$

$$Q = 8.29\ mm^2(10\ cm/s) + 7.29\ mm^2(35\ cm/s) + 3.28\ mm^2(50\ cm/s) + 5.28\ mm^2(80\ cm/s) + 4.27\ mm^2(105\ cm/s) = 913.7\ mL/\text{min}$$

Note that this type of approach likely oversimplifies the flow conditions, but would be necessary if the assumptions made when deriving the Hagen–Poiseuille formulation cannot be met.

5.7 WINDKESSEL MODEL FOR BLOOD FLOW

The previous two sections discussed one of the simplest ways to model the flow of blood through the circulation; with only relating the pressure, flow, and resistance within particular sections of the vascular tree. However, this analysis is somewhat overly simplified considering the magnitude of biological phenomenon that can occur that can potentially alter the flow properties through blood vessels and the physical properties of the blood vessel and the cells that cannot be modeled very accurately with these types of approaches. Many of the original theories of blood flow focused on the arterial circulation and the majority of them started with the assumption that the arterial circulation can be modeled as perfectly elastic pressure chambers. The rationale for this assumptions was the knowledge that blood flow out of the heart was pulsatile but blood flow into the tissues was largely nonpulsatile. Therefore, the vessels that transport blood from the heart to the peripheral tissue must have the ability to dampen out the pulsatility introduced from the heart contraction. The most simplistic of these models would make the assumption of a purely perfectly elastic pressure chamber. This type of analysis is referred to as the Windkessel model after the German word for air compression chamber, which can be used to dampen out pulsatility in classical fluid mechanics applications.

The first assumption of the Windkessel model is that the arterial tree is composed of a series of interconnected tubes, each with its own capacity to store blood. The inflow of each tube is provided by a time varying inflow flow rate and the steady outflow of each tube is governed by the inflow, the overall blood vessel resistance, the capacity for the blood vessel to accommodate more blood, the properties of the blood vessel, and the pressure gradient across the vessel. An overall schematic of arterial flow based on the Windkessel model assumptions is shown in Figure 5.13. To begin an analysis of blood flow using the Windkessel model, we start with a mass balance across the volume of interest. In this case, the mass balance can be summarized as

$$\text{Inflow} - \text{Outflow} = \text{Time Rate of Change of Blood Accumulation}$$

$$Q_i(t) - Q_o = \frac{da}{dt} \tag{5.16}$$

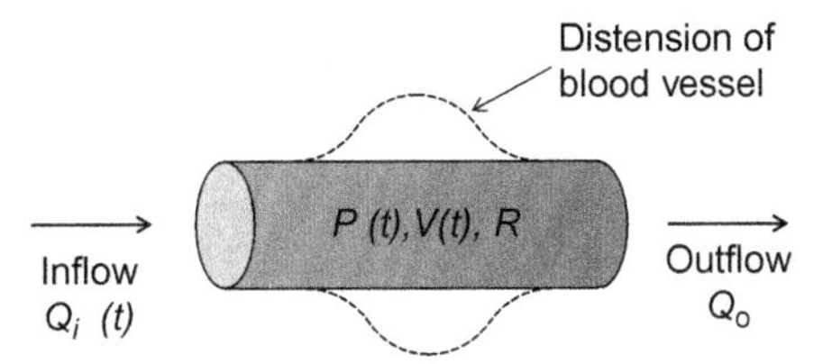

FIGURE 5.13 Schematic of a distensible arterial tree which can accommodate increases in blood flow via an increased volume. If we assume that the outflow flow rate has little variation in time, then we can analyze this type of vessel using a Windkessel approach as outlined in the text.

The inflow would need to be a given and would be represented as a function of time. If we assume that the outflow is into the peripheral tissue, where the pulsatility is dampened out, then the Hagen–Poiseuille formulation can be used to quantify the outflow flow rate as

$$Q_o = \frac{\Delta P}{R}$$

The time rate of change of blood accumulation is a little more complex to analyze. First, we would have to be able to relate the pressure and volume changes to the mechanical properties of the blood vessel. This is typically quantified in terms of a distensibility value, which is defined as

$$D = \frac{dV}{d\Delta P} \tag{5.17}$$

The time rate of change of blood accumulation (a) can be defined from the time rate of change of the blood vessel volume. Using some mathematical manipulations and substitutions the rate of accumulation can equate to

$$\frac{da}{dt} = \frac{dV}{dt} = \frac{dV}{d\Delta P}\frac{d\Delta P}{dt} = D\frac{d\Delta P}{dt} \tag{5.18}$$

Therefore the overall Windkessel formulation, with the assumption that the outflow has no pulsatility, can be mathematically represented as

$$Q_i(t) - \frac{\Delta P}{R} = D\frac{d\Delta P}{dt} \tag{5.19}$$

This equation is a differential equation in terms of time and pressure. As such, to solve this problem, the inflow function would need to be given to obtain a solution for how the pressure varies as a function of time through the section of vessel that is being analyzed.

Example

Calculate the pressure variation in the arterial circulation using a Windkessel model. Assume that the inflow pressure waveform can be represented by the following piecewise continuous function:

$$Q_i = \begin{cases} \sin\left(\frac{\pi t}{0.3}\right) & 0 \le t \le 0.3 \\ 0 & 0.3 < t \le 0.9 \end{cases}$$

Assume that the pressure gradient is equal to Ps at the beginning of the systolic phase of the flow (first phase of the piecewise continuous function) and P_D at the beginning of the diastolic phase of the flow (second phase of the piecewise continuous function). The distensibility and resistance are constant values over the time of this problem.

Solution

First, we will setup the governing equation for this system and write the known initial conditions for the *systolic phase* of the waveform. We have assumed that the outflow has no pulsatility because this section of the blood vessel will dampen any incoming pulsatility out prior to the outflow.

$$Q_i(t) - \frac{\Delta P}{R} = D\frac{d\Delta P}{dt}$$

$$\sin\left(\frac{\pi t}{0.3}\right) - \frac{\Delta P}{R} = D\frac{d\Delta P}{dt}$$

$$\Delta P(0) = P_S$$

This can be rewritten as the following:

$$\frac{d\Delta P}{dt} + \frac{\Delta P}{DR} - \frac{1}{D}\sin\left(\frac{\pi t}{0.3}\right) = 0$$

This is a first-order linear differential equation, which can be solved with various methods. We have chosen to use the Laplace transform and the convolution principle to arrive at a solution. We will not go through all of the calculations here but the major steps are shown. After taking the Laplace transform and rearranging the terms, we get

$$\overline{\Delta P} = \frac{1}{s + (1/DR)}\left(\frac{\pi/0.3}{s^2 + (\pi/0.3)^2}\right) + \frac{P_S}{s + (1/DR)}$$

where $\overline{\Delta P}$ represents the time varying pressure waveform in the Laplace domain. The second-term on the right-hand side of the equation has a relatively common inverse Laplace transform, but the first term on the right-hand side requires the use of the convolution principle to find its inverse Laplace transform. In this case, the convolution of the first term on the right-hand side can be represented as

$$\Delta P(t)_{\text{term1}} = \frac{1}{D}\left[e^{-t/DR} * \sin\left(\frac{\pi t}{0.3}\right)\right] = \frac{1}{D}\int_0^t e^{-(t-\tau)/DR}\sin\left(\frac{\pi\tau}{0.3}\right)d\tau$$

If one works through this convolution, which would require two "integration by parts stages," one finds that this equation "simplifies" to

$$\frac{1}{1+(0.3/DR\pi)^2}\left[\left(\frac{-0.3}{\pi}\right)e^{t/DR}\cos\left(\frac{\pi t}{0.3}\right)+\frac{1}{DR}\left(\frac{0.3}{\pi}\right)^2 e^{t/DR}\sin\left(\frac{\pi t}{0.3}\right)+\frac{0.3}{\pi}\right]$$

Incorporating this term as the inverse Laplace transform of the first term on the right-hand side of the equation in the Laplace domain, we can get a function for the change in pressure with respect to time, during the systolic phase of the cardiac cycle. This includes the inverse Laplace transform of the second term on the right-hand side of the equation as well. This equation is

$$\Delta P(t)_{systolic}=\frac{1}{1+(0.3/DR\pi)^2}\left[\left(\frac{-0.3}{\pi}\right)e^{t/DR}\cos\left(\frac{\pi t}{0.3}\right)+\frac{1}{DR}\left(\frac{0.3}{\pi}\right)^2 e^{t/DR}\sin\left(\frac{\pi t}{0.3}\right)+\frac{0.3}{\pi}\right]+P_S e^{-t/DR}$$

Using a similar procedures for the diastolic phase, we start by writing the governing equation

$$Q_i(t)-\frac{\Delta P}{R}=D\frac{d\Delta P}{dt}$$

$$0-\frac{\Delta P}{R}=D\frac{d\Delta P}{dt}$$

$$\Delta P(0.3)=P_D$$

This differential equation can be solved by using a separation of variable method as follows:

$$-\int_{0.3}^{t}\frac{dt}{DR}=\int_{0.3}^{P}\frac{d\Delta P}{\Delta P}$$

$$-\left(\frac{t-0.3}{DR}\right)=\ln\left(\frac{P}{P_D}\right)$$

$$P(t)_{\text{diastolic}}=P_D e^{-(t-0.3)/DR}$$

Therefore, the pressure waveform over one cardiac cycle can be represented as

$$P(t)=\begin{cases}\dfrac{1}{1+(0.3/DR\pi)^2}\left[\left(\dfrac{-0.3}{\pi}\right)e^{t/DR}\cos\left(\dfrac{\pi t}{0.3}\right)+\dfrac{1}{DR}\left(\dfrac{0.3}{\pi}\right)^2 e^{t/DR}\sin\left(\dfrac{\pi t}{0.3}\right)+\dfrac{0.3}{\pi}\right]+P_s e^{-t/DR} & 0\le t\le 0.3\\ P_D e^{-(t-0.3)/DR} & 0.3<t\le 0.9\end{cases}$$

The function that was obtained in the example problem can be plotted against time for a cardiac cycle to see how the pressure waveform changes as a function of the variables (Figure 5.14). We plotted various cases, but what can be observed is that (i) there is essentially no transition between the systolic and diastolic pressure waveforms under normotensive conditions, (ii) under hypertensive conditions, the curves are slightly changed (and shifted upward) and there is a small transition at systolic and diastolic values (this

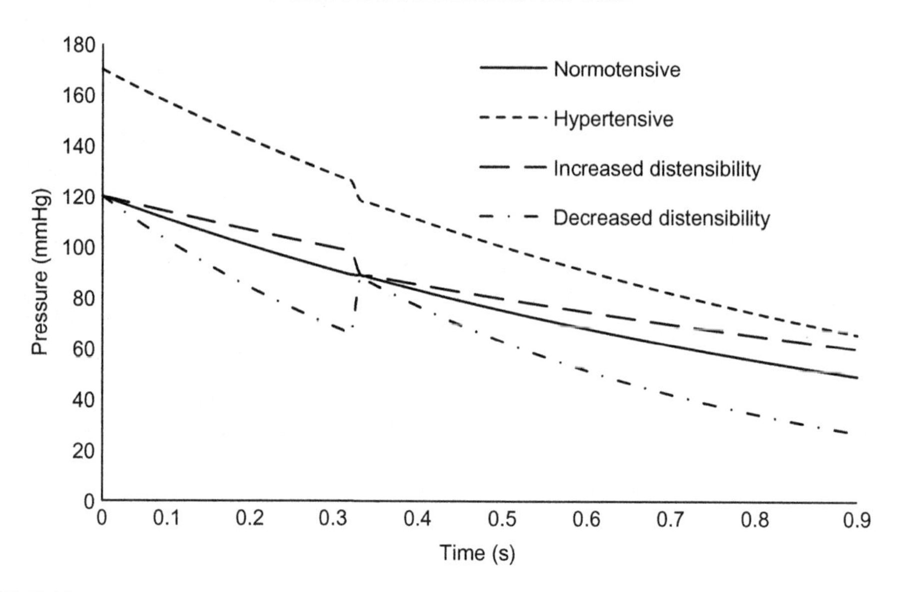

FIGURE 5.14 Figure associated with the example problem. Note that the pressure values were set to $P_S = 120$ mmHg and $P_D = 80$ mmHg for all cases, except the hypertensive case ($P_S = 170$ mmHg and $P_D = 120$ mmHg). The distensibility was either increased or decreased by 50% to arrive at the other curves.

can be removed by altering the peak systolic and diastolic values, which we did not attempt to do to smooth the curve), (iii) with an increased distensibility (increased by 50%), the pressure does not dissipate as quickly because the vessel is less likely to accommodate changes in the pressure forces, and (iv) with a decreased distensibility (decreased by 50%), the pressure dissipates quickly because the vessel is more likely to accommodate changes in the pressure forces. Note that the jumps at the beginning of diastole are caused because we left the diastolic value the same as the normotensive case, however, if we made the pressure waveform curves piecewise continuous (by finding a peak diastolic value that matches the pressure value at the end of systole), then the changes in the pressure waveform curves would be more pronounced.

Although the Windkessel model can provide us with analysis techniques that can be used for flow through the arterial tree, there are a number of limitations to this modeling approach. First, the Windkessel model assumes that changes to the pressure waveforms are transmitted instantaneously (e.g., there is no wave propagation) and that reflectance of the waveform does not occur (we did not include a bifurcation in this model, but could). Additionally, all elastic properties of the blood vessel are summarized in the distensibility and need to be known. The blood vessels cannot exhibit viscoelastic properties. We also assumed that the resistance and the distensibility were not time varying (e.g., include the viscoelastic properties). The Windkessel model can be extended to these more complex scenarios but the computations associated with the model would likely need computational algorithms to solve. However, as with some of the earlier models, the Windkessel model provides a somewhat accurate approximation of the pressure variations in the arterial tree.

5.8 WAVE PROPAGATION IN ARTERIAL CIRCULATION

Wave propagation in arteries is primarily concerned with the displacement of the arterial wall in response to the progression of pressure/velocity waves through the blood vessel. This displacement is primarily due to the elasticity of the vessel wall combined with the large pressure variances that are observed in the arterial tree. To fully represent this condition, a fairly complex set of mathematical equations must be solved simultaneously. This solution would describe an instance of fluid structure interactions, where the fluid forces impact the wall and the wall responds to these changes (solutions that make few assumptions are discussed in Chapter 14). Additionally, the recoil and other mechanical properties of the vessel will impact the flow conditions. To begin to understand the concepts of wave propagation, let us first consider a blood vessel that is cylindrical and straight with mechanical properties that are homogenous and elastic. Also, we will first make the assumption that the fluid is incompressible, inviscid, and relatively slow (in the main direction of flow). A pressure pulse at one end of the blood vessel will cause a geometric change in the vessel, which propagates along the tube length at a particular speed. In most instances, the wavelength is much larger than the wave amplitude, which allows us to assume that the flow velocity is still one-dimensional. If the amplitude of the wave was large, with respect to the wavelength, fluid would need to enter/leave the space created/removed when the wave passed through the blood vessel wall, and this velocity typically would not be in the same direction of the majority of the fluid flow. Therefore, a coupled Navier–Stokes solution would be required to determine the velocity profile (e.g., instead of solving the Navier–Stokes in one direction, you would need to solve the problem in multiple directions).

Combining all of these assumptions and using the continuity equation and the Navier–Stokes equation, one can determine the speed of wave propagation in an artery. First we should consider the forces acting on the blood vessel during the pressure pulse through the vessel (Figure 5.15). We will assume that the forces balance in the y-direction and therefore, they are not considered on this figure. In the x-direction, the summation of the pressure forces is

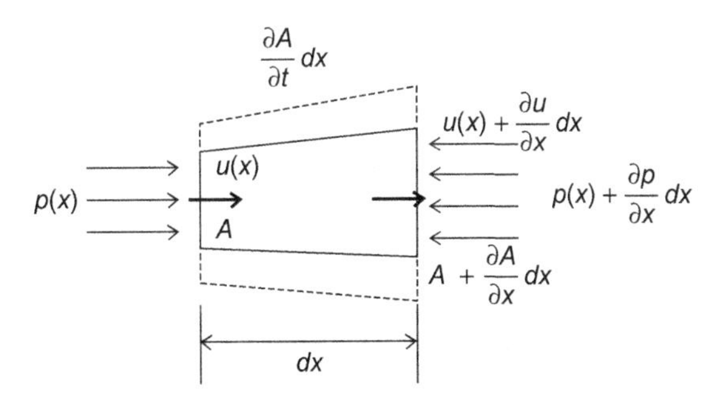

FIGURE 5.15 Wave propagation within a deformable homogenous artery. With a deformable boundary, one would need to take into account the change in area, with time, of the fluid element. The material properties of the arterial wall need to be included within this formulation as well.

$$\sum F_x = p_{inlet}A_{inlet} - p_{outlet}A_{outlet} + p_{walls}A_{walls}$$

$$p(x)A - \left[p(x) + \frac{\partial p}{\partial x}dx\right]\left[A + \frac{\partial A}{\partial x}dx\right] + p(x)\frac{\partial A}{\partial x}dx = -A\frac{\partial p}{\partial x}dx \tag{5.20}$$

if the second-order differential terms are ignored. The pressure acting on the walls may be a component of the pressure forces that act perpendicular to the flow. Using Newton's second law of motion, the net force acting on a differential element is equal to the mass multiplied by the acceleration of the object. Writing this in equation form, we get (using density/volume)

$$-A\frac{\partial p}{\partial x}dx = (\rho A dx)\left(\frac{\partial u}{\partial t} + u\frac{\partial u}{\partial x}\right)$$

$$0 = \frac{1}{\rho}\frac{\partial p}{\partial x} + \frac{\partial u}{\partial t} + u\frac{\partial u}{\partial x} \tag{5.21}$$

which is a simplified form of the Navier–Stokes equation (weight acts in the y-direction on this figure). The Conservation of Mass on this element must take into account the increase in area of the element during time. Assuming the density remains constant within the entire element (e.g., our assumption that the fluid is incompressible), Conservation of Mass states (again neglecting second-order terms), where u is the x-velocity

$$uA = \left(u + \frac{\partial u}{\partial x}dx\right)\left(A + \frac{\partial A}{\partial x}dx\right) + \frac{\partial A}{\partial t}dx \tag{5.22}$$

$$uA = uA + u\frac{\partial A}{\partial x}dx + A\frac{\partial u}{\partial x}dx + \frac{\partial A}{\partial t}dx$$

$$u\frac{\partial A}{\partial x} + A\frac{\partial u}{\partial x} + \frac{\partial A}{\partial t} = \frac{\partial A}{\partial t} + \frac{\partial}{\partial x}(uA) = 0 \tag{5.23}$$

The first term in Eq. (5.22) is the inflow, the second term is the outflow, and the last term accounts for the increase in area as a function of time. To determine the wave propagation throughout the blood vessel, the material properties of the vessel wall must be known and related to the fluid properties. In some instances, we can make the assumption that the changes in the blood vessel radius (r_i) is linearly proportional to the blood pressure acting on the vessel wall (p_i), using the following relationship:

$$r_i = r_{i,0} + \alpha p \tag{5.24}$$

where $r_{i,0}$ is the radius of the blood vessel at a zero external pressure and α is a constant that describes the material properties of the blood vessel. Equations (5.21), (5.23), and (5.24) govern the wave propagation speed throughout a blood vessel (with the assumptions that we have made, note that Eq. (5.24) is overly simplifying the biological mechanical properties/responses to forces, but this can give us a starting point for the analysis of the wave propagation through the vessel). Substituting the area into Eq. (5.23) and remembering that the fluid velocity is relatively slow, Eqs. (5.21) and (5.23) simplify to

$$0 = \frac{1}{\rho}\frac{\partial p}{\partial x} + \frac{\partial u}{\partial t} \tag{5.25}$$

$$0 = \frac{\partial(\pi r_i^2)}{\partial t} + \frac{\partial}{\partial x}(u\pi r_i^2) = 2\pi r_i\frac{\partial r_i}{\partial t} + u\pi\frac{\partial r_i^2}{\partial x} + \pi r_i^2\frac{\partial u}{\partial x} = 2\pi r_i\frac{\partial r_i}{\partial t} + \pi r_i^2\frac{\partial u}{\partial x} = \frac{2}{r_i}\frac{\partial r_i}{\partial t} + \frac{\partial u}{\partial x} \tag{5.26}$$

For clarity, the two appropriate assumptions are that the blood velocity is close to zero ($u = 0$) and that the wave amplitude is much smaller than the wavelength. Combining Eq. (5.24) with Eq. (5.26), we get

$$0 = \frac{2}{r_i}\frac{\partial r_i}{\partial t} + \frac{\partial u}{\partial x} = \frac{2}{r_i}\frac{\partial}{\partial t}(r_{i,0} + \alpha p) + \frac{\partial u}{\partial x} = \frac{2\alpha}{r_i}\frac{\partial p}{\partial t} + \frac{\partial u}{\partial x} \tag{5.27}$$

To solve this in the form of the wave equation, we must take the temporal derivative of Eq. (5.27) and subtract this from the spatial derivative of Eq. (5.25). With some algebraic modifications of the proportionality constants, we can get the wave speed (c):

$$\frac{\partial}{\partial x}\left[0 = \frac{1}{\rho}\frac{\partial p}{\partial x} + \frac{\partial u}{\partial t}\right] \rightarrow 0 = \frac{1}{\rho}\frac{\partial^2 p}{\partial x^2} + \frac{\partial^2 u}{\partial x \partial t}$$

$$\frac{\partial}{\partial t}\left[0 = \frac{2\alpha}{r_i}\frac{\partial p}{\partial t} + \frac{\partial u}{\partial x}\right] \rightarrow 0 = \frac{2\alpha}{r_i}\frac{\partial^2 p}{\partial t^2} + \frac{\partial^2 u}{\partial t \partial x}$$

$$0 = \frac{1}{\rho}\frac{\partial^2 p}{\partial x^2} + \frac{\partial^2 u}{\partial t \partial x} - \frac{2\alpha}{r_i}\frac{\partial^2 p}{\partial t^2} - \frac{\partial^2 u}{\partial t \partial x} = \frac{1}{\rho}\frac{\partial^2 p}{\partial x^2} - \frac{2\alpha}{r_i}\frac{\partial^2 p}{\partial t^2} = \frac{\partial^2 p}{\partial x^2} - \frac{2\alpha\rho}{r_i}\frac{\partial^2 p}{\partial t^2} \tag{5.28}$$

Putting Eq. (5.28) into the form of the wave equation,

$$0 = \frac{\partial^2 p}{\partial x^2} - \frac{1}{c^2}\frac{\partial^2 p}{\partial t^2}$$

$$c = \sqrt{\frac{r_i}{2\alpha p}} \tag{5.29}$$

Using the same analysis, we can derive the equation for wave speed in various types of blood vessels. For instance, if the blood vessel wall is purely elastic and relatively thin compared to the diameter and perfectly circular, then the wave speed is

$$c = \sqrt{\frac{Eh}{4pr_i}} \tag{5.30}$$

where E is the elastic modulus of the vessel wall and h is the wall thickness. If more of the assumptions are relaxed, we can develop more realistic approximations for the wave propagation speed within an arterial wall, but that is beyond the scope of this discussion.

Example

Calculate the wave speed using Eqs. (5.29) and (5.30) to determine if the assumptions that were made to obtain these formulas are acceptable. Assume that the pressure within a blood vessel with a 12-mm radius and a thickness of 100 μm is 100 mmHg. The compliance of this blood vessel is 6.7 μm/Pa, and the Young's Modulus is 15 kPa.

Solution

$$c = \sqrt{\frac{r_i}{2\alpha p}} = \sqrt{\frac{12\ mm}{2(6.7\ \mu m/Pa)(100\ mmHg)}} = 25.9\ cm/s$$

$$c = \sqrt{\frac{Eh}{4\ pr_i}} = \sqrt{\frac{15\ kPa(100\ \mu m)}{4(100\ mmHg)(12\ mm)}} = 4.8\ cm/s$$

Considering that we are only concerned with the flow at one instant in time and we are neglecting the effects of slow velocity, the second approximation is more reasonable than the first solution. Also, we made many assumptions regarding the mechanical and geometric properties of the blood vessel. In formulating Eq. (5.29), too many assumptions were made, and this calculation for wave speed within the blood vessel wall is not close to the physiological scenario. *In vivo*, wave speeds are on the order of 1–10 mm/s so the second approximation is much more reasonable.

We will include a brief discussion on the dissipation of the pulsatility of the pressure and velocity waveforms in arteries due to the reflection and transmission of wave energy at bifurcations (and around turns). In large arteries, the discussion up to this point is relatively valid because of the thickness of the vessel wall and the relatively small wave amplitude as compared to wavelength. Consider a standard bifurcating blood vessel (Figure 5.16). The parent branch not only contains the incident wave but also contains a portion of the wave that is reflected backward into the parent branch. Consider this the portion of the velocity

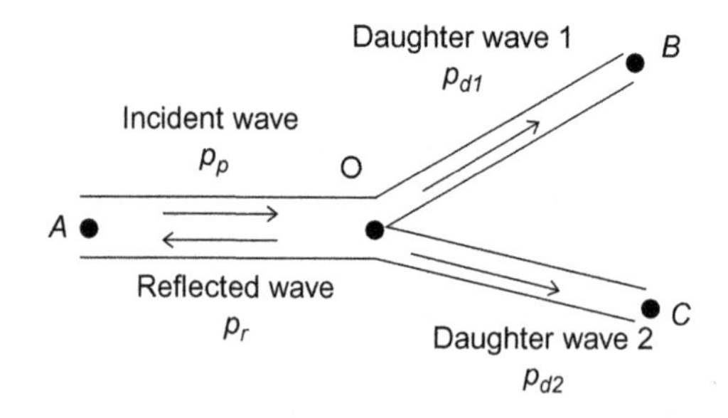

FIGURE 5.16 Pressure wave reflectance and transmittance at a bifurcation. The pressure pulse is reflected back along the mother branch, but this is normally out of phase with the next incident wave and therefore the entire pressure would be reduced. The energy of the transmitted waves do not summate to the energy of the incident wave.

or pressure waveform that travels perfectly straight and hits the perfectly sharp point where the two daughter branches meet for this simplified geometry (in a "real" blood vessel there is a more significant area that the incoming pressure waveforms can impact on). Each of the two daughter branches will contain a portion of the wave that is transmitted down the vascular tree. There will be some energy loss because of the energy used when the wave hits the vessel wall, rebounds off the wall, and causes a physical distortion to the wall. Mathematically, the pressure (p) must have a single value at point O in Figure 5.16, and the volumetric flow rate (Q) at this same point must be continuous. Therefore,

$$p_p - p_r = p_{d1} + p_{d2} \tag{5.31}$$

and

$$Q_p - Q_r = Q_{d1} + Q_{d2} \tag{5.32}$$

Equations (5.31) and (5.32) are basic conservation laws. The volumetric flow rate Q can be quantified as

$$Q = Au = \frac{A}{\rho c} p \tag{5.33}$$

under these conditions, where A is the cross-sectional area, u is the flow velocity, ρ is the blood density, c is the wave speed, and p is the pressure. The characteristic impedance (Z) of the blood vessel is defined as

$$Z = \frac{\rho c}{A} \tag{5.34}$$

Interestingly, if we use this new relationship for pressure-driven flow, we can calculate the amplitudes of the pressure waveforms immediately after a simple two-dimensional bifurcation.

$$ZQ = p$$

$$\frac{p_p - p_r}{Z_p} = \frac{p_{d1}}{Z_{d1}} + \frac{p_{d2}}{Z_{d2}} \tag{5.35}$$

The impedance can be different within each vessel, primarily due to the area changes or the velocity within the channel. Therefore,

$$\frac{p_r}{p_p} = \frac{Z_p^{-1} - (Z_{d1}^{-1} + Z_{d2}^{-1})}{Z_p^{-1} + (Z_{d1}^{-1} + Z_{d2}^{-1})} \tag{5.36}$$

$$\frac{p_{d1}}{p_p} = \frac{p_{d2}}{p_p} = \frac{2Z_p^{-1}}{Z_p^{-1} + (Z_{d1}^{-1} + Z_{d2}^{-1})} \tag{5.37}$$

Equations (5.35) through (5.37) show that there will be a reduction in the pressure pulse through the arterial system, because of energy that is reflected back into the parent branch and is reflected on through the daughter branches. This analysis can be extended for more complex bifurcating systems and more complex mechanical/physical properties (e.g., Eq. (5.34)).

5.9 FLOW SEPARATION AT BIFURCATIONS AND AT WALLS

Before we move into the discussion of flow separation, it is important to first discuss some of the important flow considerations at bifurcations. We have already learned that a very efficient way to control blood flow through the vascular system is through small changes in the blood vessel radius (Hagen–Poiseuille equation). Let us consider the simplest case, where one vessel branches into two daughter vessels (Figure 5.17). The flow in the main branch has to be divided into the two daughter branches. The question remains, how is the bifurcation designed to optimize the amount of work that the blood vessel completes in order to supply the tissue with sufficient blood compared to the rate of energy used by the tissue? Meaning how far can tissue be from a capillary to efficiently deliver nutrients and remove wastes.

The first and easiest way to optimize the delivery of blood through this bifurcation is for points *A*, *B*, and *C* (refer to Figure 5.17) to all lie within one plane. If these points are in one plane, the blood vessels connecting each of these points to point *O* could potentially be aligned with that plane, thus minimizing the length of the vessel (however, realistically points *A*, *B*, *C*, and *O* would not be in the same plane). What is also known is that the lengths of each vessel are dependent on the location of the center of the bifurcation point (e.g., point *O*). For instance, by moving the branch point (*O*) to the right (but keeping it in the same plane, which is represented by the textbook's sheet of paper), branch 1 elongates and branches 2 and 3 reduce in length. At the same time, the branch angles, θ and ϕ, get larger. If the branch point is moved along the $\overline{OC}$ line or the $\overline{OB}$ line, various combinations of changes in the vessel lengths/angles can be obtained and optimized. In a study conducted in 1926, C.D. Murray proposed a mathematical representation to optimize the work at a bifurcation. These equations became known as Murray's law for the minimum work of a bifurcating network, and subsequently, they have been related to the vascular system. They are represented as

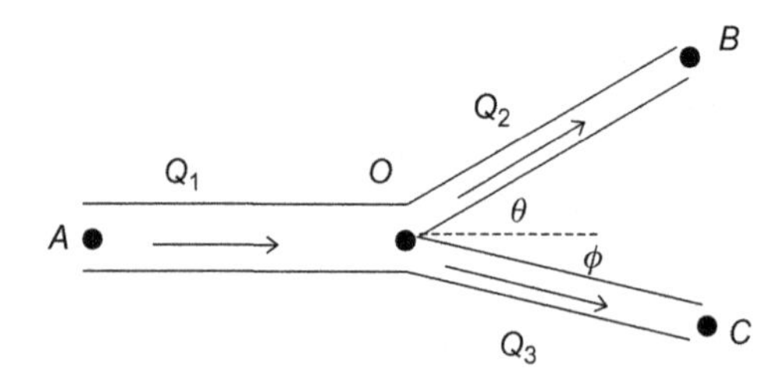

FIGURE 5.17 Flow rate at a bifurcation would be divided into two daughter branches. The geometry (angles and radii) of the bifurcation determines the division of the flow within the two daughter branches. By assuming that the work of supplying tissue with oxygen for each vessel should be minimized, one can derive a relationship between the vessels.

$$r_1^3 = r_2^3 + r_3^3$$

$$\cos\theta = \frac{r_1^4 + r_2^4 - r_3^4}{2r_1^2 r_2^2}$$

$$\cos\phi = \frac{r_1^4 - r_2^4 + r_3^4}{2r_1^2 r_3^2} \quad (5.38)$$

$$\cos(\theta + \phi) = \frac{r_1^4 - r_2^4 - r_3^4}{2r_2^2 r_3^2}$$

These formulas have been validated with experimental physiological studies, and the fit is exceptionally well. Now the important point to remember is, although we can optimize the work needed throughout a bifurcation, there are still many fluid dynamics principles that may not be optimized.

Example

Calculate the radius of one daughter branch knowing that the radius of the parent vessel is 175 μm and the radius of the other daughter branch is 125 μm. Draw the bifurcation, to scale, after calculating the bifurcation angles. Also, validate the last equation from Murray's law, which relates both bifurcation angles to the radii of the three blood vessels (Figure 5.18).

Solution

$$r_1^3 = r_2^3 + r_3^3$$

$$(175\ \mu m)^3 = (125\ \mu m)^3 + r_3^3$$

$$r_3 = \sqrt[3]{(175\ \mu m)^3 - (125\ \mu m)^3} = 150.5\ \mu m$$

$$\theta = \cos^{-1}\left(\frac{r_1^4 + r_2^4 - r_3^4}{2r_1^2 r_2^2}\right) = \cos^{-1}\left(\frac{(175\ \mu m)^4 + (125\ \mu m)^4 - (150.5\ \mu m)^4}{2(175\ \mu m)^2(125\ \mu m)^2}\right) = 45.6^\circ$$

$$\phi = \cos^{-1}\left(\frac{r_1^4 - r_2^4 + r_3^4}{2r_1^2 r_3^2}\right) = \cos^{-1}\left(\frac{(175\ \mu m)^4 - (125\ \mu m)^4 + (150.5\ \mu m)^4}{2(175\ \mu m)^2(150.5\ \mu m)^2}\right) = 29.5^\circ$$

$$\cos(\theta + \phi) = \cos(45.6^\circ + 29.5^\circ) = \cos(75.1^\circ) = 0.256$$

$$\frac{r_1^4 - r_2^4 - r_3^4}{2r_2^2 r_3^2} = \frac{(175\ \mu m)^4 - (125\ \mu m)^4 - (150.5\ \mu m)^4}{2(125\ \mu m)^2(150.5\ \mu m)^2} = 0.256$$

Therefore,

$$\cos(\theta + \phi) = \frac{r_1^4 - r_2^4 - r_3^4}{2r_2^2 r_3^2}$$

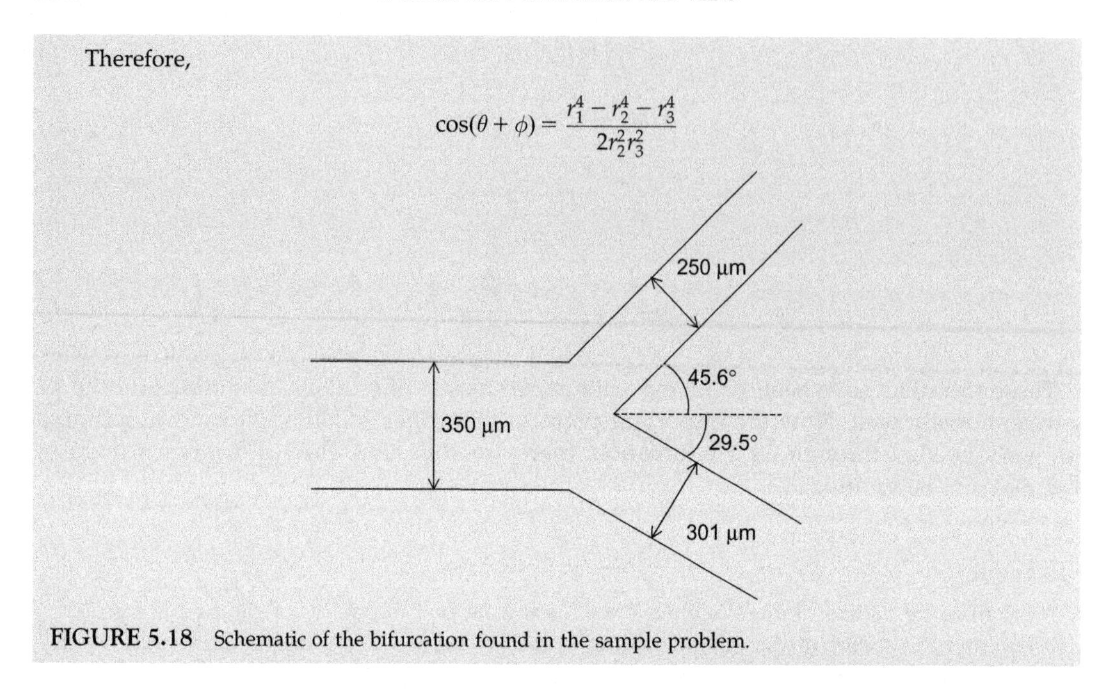

FIGURE 5.18 Schematic of the bifurcation found in the sample problem.

Flow separation is a common occurrence at bifurcations (especially within the arterial circulation where the pressures/velocities are high) and near the wall when there are rapid geometric changes that expand the cross-sectional area of the blood vessel. Flow separation is a phenomenon that occurs when the fluid pulls away from the containing wall forming a region that has very slow flow associated with it (Figure 5.19). The fluid in this region typically forms a recirculation pattern in which the fluid is trapped within that area. Flow separation occurs when the fluid velocity is relatively fast and cannot expand with the geometry rapidly. The formation of recirculation zones is a property of all viscous fluids and cannot be ignored in physiological settings. The fluid that is trapped within the recirculation zone stays within the recirculation zone because it cannot overcome the flowing fluid pressure to reenter the mainstream flow. Although, diffusion of dissolved solutes may occur along the boundaries. In general, flow separation occurs when the angle between the main fluid flow direction and the wall is relatively large.

Also, a common way to generate flow separation in biofluids mechanics problems is when there is an adverse pressure gradient causing convection throughout the fluid laminae. In a pressure-driven viscous flow, the momentum equation at the wall is (see Chapter 3 for the derivation of this equation)

$$\left.\frac{\partial^2 u}{\partial y^2}\right|_{y=0} = \frac{1}{\mu}\frac{\partial p}{\partial x} \tag{5.39}$$

To maintain the flow regime without recirculation zones, the wall curvature must have the same sign as the pressure gradient. Under normal conditions, the pressure gradient $\frac{\partial p}{\partial x}$ is negative, and therefore, the normal parabolic flow profile at the wall matches this sign, satisfying the momentum equation. However, as the pressure gradient approaches zero,

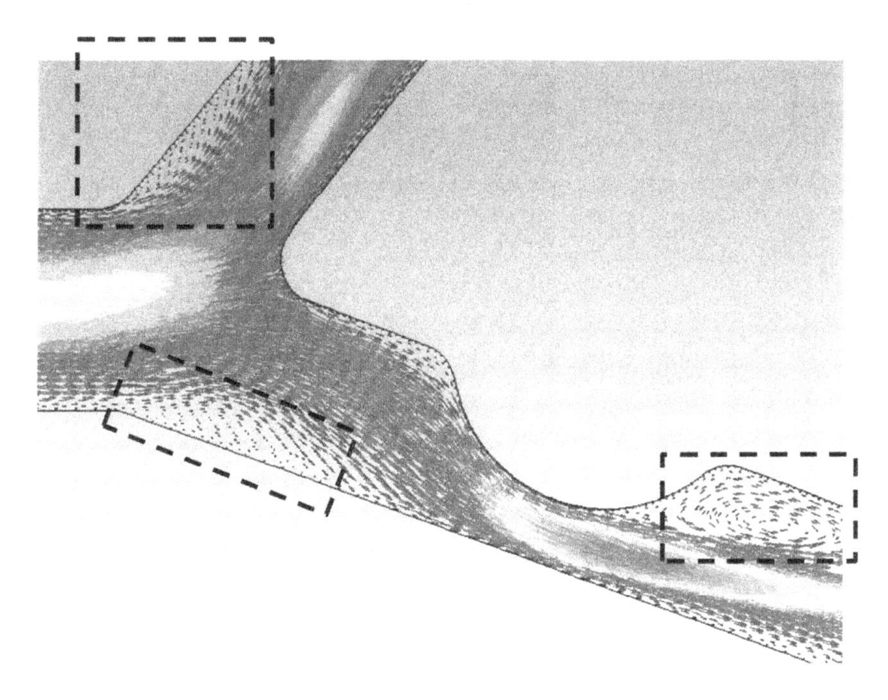

FIGURE 5.19 Flow separation (region enclosed by dashed lines) downstream of a bifurcation and a stenosis (rapid area expansions). Flow separates from the wall partially due to the skewing of the velocity profile toward one of the walls of the blood vessel. By quantifying the pressure gradient along the wall, the likelihood for flow separation can be predicted.

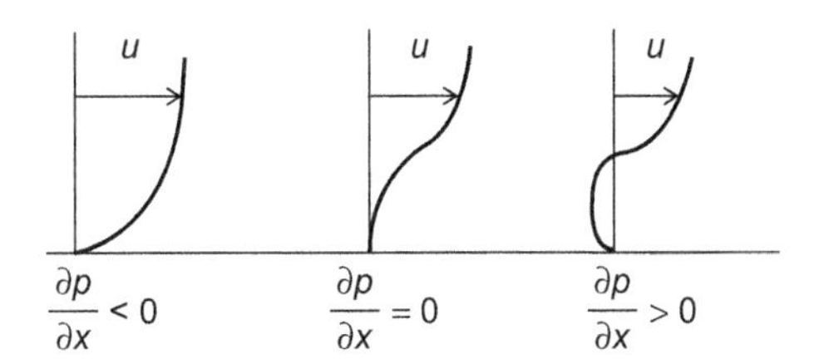

FIGURE 5.20 Effects of a pressure gradient on the flow profile near the wall. With an adverse (or positive) pressure gradient along the wall, the flow would tend to separate. With a negative or zero wall pressure gradient, the flow would not likely separate.

the flow begins to separate from the wall because the signs no longer are the same. As the gradient becomes positive, the flow will completely separate from the wall, forming a recirculation zone (Figure 5.20). From this image we can see that flow separation tends to occur at a region where the near wall fluid flow rate decreases.

Another interesting phenomenon that occurs at bifurcations is that the flow skews toward one of the walls, if the bifurcation angles are large enough. Due to the velocity of the fluid entering the bifurcation, more flow typically moves toward the inner walls of the bifurcation and therefore the fluid moves away from the outer walls (Figure 5.21). Also, the flow at the outer wall typically slows down, because the core flow is pushed towards the inner wall of

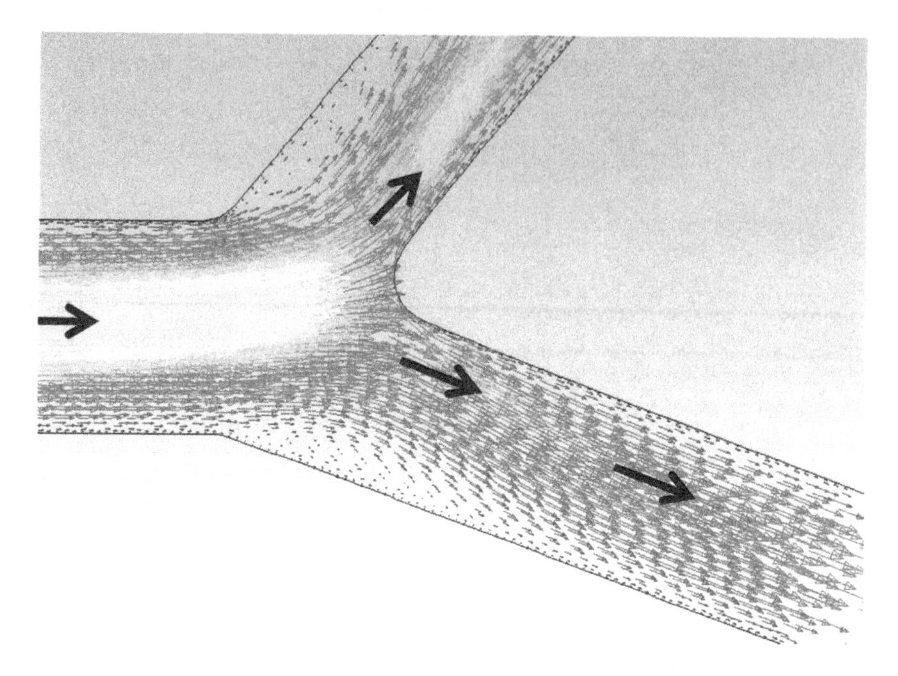

FIGURE 5.21 Flow skew at a bifurcation, where the highest velocity blood flow is no longer at the centerline (highest velocity is marked by black arrows). The extent of flow skewing is determined by the geometry of the bifurcation.

this skewed velocity profile. Additionally, as can be observed in Figure 5.21, there is a rapid spatial increase in the cross-sectional area of the blood vessel at a bifurcation (flow would decelerate), which then, just as rapidly, decreases again into the two cross-sectional areas of the daughter branch. The flow in these skewed regions is not fully developed; it does not have the typically parabolic flow profile. The highest velocity is no longer at the centerline but it moves closer to the inner bifurcation wall. The entrance length is a fluid property that quantifies the distance that is required for the fluid to become fully developed again after it has become skewed. As an approximation, if the vessel cross-sectional area remains constant after the bifurcation, within two–five tube diameters, the flow would be fully developed again. The significance of flow skewing at a bifurcation is that the wall shear stress is not what one would anticipate based on a fully developed flow. Along the inner wall, the shear stress increases and along the outer wall the shear stress decreases. This will be discussed in detail later, but cardiovascular diseases tend to initiate in locations where the shear stress diverges from the normal conditions (either heightened or reduced shear stress).

5.10 FLOW THROUGH TAPERING AND CURVED CHANNELS

In most of the examples that we have discussed in this chapter, we made the assumption that the blood vessel is circular with a constant diameter and that the vessel remains straight (in Chapter 3, we illustrated some examples without these assumptions, but the

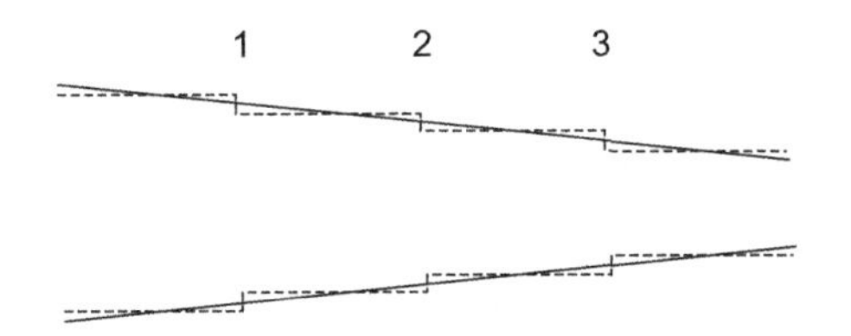

FIGURE 5.22 The modeling of a continuous taper with evenly spaced steps. To calculate the velocity profile through a continuous taper, one would need to determine the exact taper function, which may not be as simple as the examples shown in the text. However, if the taper is complicated, it can be approximated with discrete steps.

rate of change of the taper was not considered in those problems). However, this assumption does not match the real scenarios; most blood vessels experience some changes in diameter between bifurcations and do not remain in the same plane in three-dimensional space (although many blood vessels are relatively straight and not very twisted). In most blood vessels, the taper along the vessel length is relatively small until the branch point occurs; however, it is fairly easy to accommodate the effects of taper within our calculations if the taper can be approximated mathematically. Also, the curves within vessels can easily be accounted for as we have previously shown in Chapter 3.

The first assumption that can be made about a tapered blood vessel is that it tapers in discrete steps (Figure 5.22). To model a taper with discrete instantaneous step changes in diameter, the assumption would be made that the velocity, pressure, shear stress, among other fluid parameters are all constant throughout each section at one instant in time. The analysis of changes within these flow variables would need to be conducted only at the discrete steps, marked as 1, 2, and 3 on Figure 5.22. These calculations can also be carried out independent of each other (only the output flow variables of each step would be used to calculate the input flow variables within the next step). Common reasons that this type of discrete step analysis technique would be conducted instead of considering the exact blood vessel taper is because (i) the change in area cannot be represented accurately or (ii) the change in area is too complicated to compute by hand and the use of numerical methods is needed.

To be able to model a taper and compute the fluid parameters of interest, one must use the same formulas for Conservation of Mass and Conservation of Momentum throughout the entire blood vessel, which were developed in previous sections. For a simple example, a continuous taper that reduces the blood vessel radius by half over the entire vessel length, can be modeled as

$$r(x) = r_i\left(1 - \frac{x}{2L}\right) \tag{5.40}$$

where L is the tube length and r_i is the blood vessel radius at the inlet. By using Eq. (5.40) within the Conservation of Mass, Momentum, or Energy formulation, quantities of interest can easily be calculated. In an analogous situation, the velocity of a fluid element at a particular location along the blood vessel length can be given as a function of position (instead of radius as a function of position) and the other fluid parameters of interest can be calculated from this given value.

Example

The velocity of a fluid element along the tapered channels centerline (Figure 5.23) is given by $\vec{v} = v_i\left(1 + \frac{2x}{L}\right)\hat{i}$, where v_i is the centerline inlet velocity. Calculate the acceleration of any particle along the centerline (i.e., as a function of x) and the position of a particle (as a function of time) that is located at $x = 0$ at time zero.

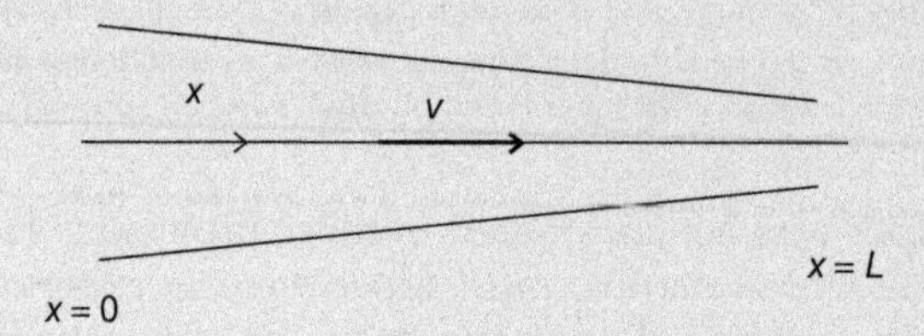

FIGURE 5.23 Model of a continuous taper for the example problem.

Solution

To find the acceleration of a particle, one must use the total derivative of velocity, developed in Chapter 2 (for the x-direction):

$$\frac{D\vec{v}}{Dt} = \frac{\partial u}{\partial t} + u\frac{\partial u}{\partial x} + v\frac{\partial u}{\partial y} + w\frac{\partial u}{\partial z}$$

For this situation, u is only a function of x, and v and w are equal to zero.

$$\frac{D\vec{v}}{Dt} = u\frac{\partial u}{\partial x} = v_i\left(1 + \frac{2x}{L}\right)\left[\frac{\partial}{\partial x}\left(v_i\left(1 + \frac{2x}{L}\right)\right)\right] = v_i\left(1 + \frac{2x}{L}\right)\left[\frac{\partial}{\partial x}\left(\frac{2v_ix}{L}\right)\right]$$

$$= v_i\left(1 + \frac{2x}{L}\right)\left(\frac{2v_i}{L}\right) = \frac{v_i^2}{L}\left(2 + \frac{4x}{L}\right)$$

The acceleration of any particle along the centerline is

$$a_x(x) = \frac{v_i^2}{L}\left(2 + \frac{4x}{L}\right)$$

To determine the location of a particle, we first know that at time 0, the particle is located at position 0 (or at the inlet) and has a velocity of v_i. At some later time, the particle will be located at L and will have a velocity of $3v_i$. From Chapter 2, we learned that the velocity of a particle is the time derivative of its position. Using this relationship, we can develop a formula for the location of a fluid element as a function of time.

$$u_x = \frac{dx}{dt} = v_i\left(1 + \frac{2x}{L}\right)$$

To solve this equation, first we separate the variables:

$$v_i dt = \frac{dx}{\left(1 + \frac{2x}{L}\right)}$$

Integrate this equation with the bounds that at time zero the particle is located at zero and at some time t the particle is located at some point x:

$$\int_0^t v_i dt = \int_0^x \frac{dx}{\left(1+\frac{2x}{L}\right)}$$

$$v_i t\Big|_0^t = \frac{L}{2}\ln\left(1+\frac{2x}{L}\right)\Big|_0^x$$

$$v_i t = \frac{L}{2}\ln\left(1+\frac{2x}{L}\right)$$

$$\ln\left(1+\frac{2x}{L}\right) = \frac{2v_i t}{L}$$

$$1+\frac{2x}{L} = e^{\frac{2v_i t}{L}}$$

$$x(t) = \frac{L}{2}\left(e^{\frac{2v_i t}{L}}-1\right)$$

This example illustrates the use of an exact solution for a tapering blood vessel.

As stated above, most blood vessels have a small taper and/or curvature in the body. Curving blood vessels do not add much complexity to the problems that we have discussed above. Similar to the previous examples and the examples in Chapter 3, continuity, momentum conservation, and energy conservation can be used to calculate the velocity profile, the forces needed to hold the particular vessel in place, the energy needed to maintain the flow, the pressure within the fluid, acceleration, shear stress, and others fluid properties. The Navier–Stokes equations can be used to determine the velocity profile as a function of geometric changes. The following example illustrates the use of the Navier–Stokes equations to solve for the velocity profile in a curved tapering vessel.

Example

Determine the velocity profile in the three sections of the blood vessel shown in Figure 5.24. Assume that the only portion that is tapering is within the bend. Also assume that gravity only acts in the y-direction and that this is pressure-driven flow, throughout the entire vessel.

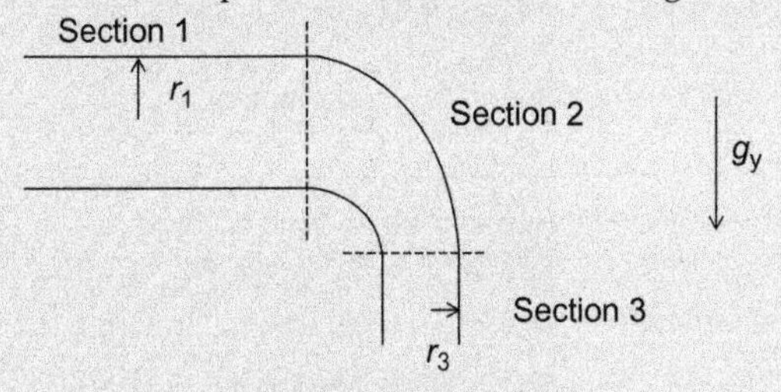

FIGURE 5.24 Cross-section of a tapered curving blood vessel. This problem shows the difficulty of adding a tapering blood vessel with changes in orientation.

Solution

Sections 1 and 3 should be relatively easy to solve because there is no taper and no bend in these sections. Also, the sections can be solved independently because we know that the outflow of one section will be the inflow of the next section.

For Section 1, if we choose the Cartesian Navier–Stokes equation

$$\rho\left(\frac{\partial u}{\partial t}+u\frac{\partial u}{\partial x}+v\frac{\partial u}{\partial y}+w\frac{\partial u}{\partial z}\right)=\rho g_x-\frac{\partial p}{\partial x}+\mu\left(\frac{\partial^2 u}{\partial x^2}+\frac{\partial^2 u}{\partial y^2}+\frac{\partial^2 u}{\partial z^2}\right)$$

$$0=-\frac{\partial p}{\partial x}+\mu\left(\frac{\partial^2 u}{\partial y^2}\right)$$

$$\frac{\partial^2 u}{\partial y^2}=\frac{1}{\mu}\frac{\partial p}{\partial x}$$

$$\frac{\partial u}{\partial y}=\frac{y}{\mu}\frac{\partial p}{\partial x}+c_1$$

$$u(y)=\frac{y^2}{\mu}\frac{\partial p}{\partial x}+c_1 y+c_2$$

Using the boundary conditions of

$$u(0)=0 \text{ and } u(2r_1)=0$$

the solution of the velocity profile becomes

$$u(y)=\frac{y^2}{\mu}\frac{\partial p}{\partial x}-\frac{r_1}{\mu}\frac{\partial p}{\partial x}y$$

Solving the Navier–Stokes equations for Section 3, again using the Navier–Stokes equation for the y-direction.

$$\rho\left(\frac{\partial v}{\partial t}+u\frac{\partial v}{\partial x}+v\frac{\partial v}{\partial y}+w\frac{\partial v}{\partial z}\right)=\rho g_y-\frac{\partial p}{\partial y}+\mu\left(\frac{\partial^2 v}{\partial x^2}+\frac{\partial^2 v}{\partial y^2}+\frac{\partial^2 v}{\partial z^2}\right)$$

$$0=\rho g_y-\frac{\partial p}{\partial y}+\mu\left(\frac{\partial^2 v}{\partial x^2}\right)$$

$$\frac{\partial^2 v}{\partial x^2}=\frac{1}{\mu}\frac{\partial p}{\partial y}-\frac{\rho}{\mu}g_y$$

$$\frac{\partial v}{\partial x}=\left(\frac{1}{\mu}\frac{\partial p}{\partial y}-\frac{\rho}{\mu}g_y\right)x+c_1$$

$$v(x)=\left(\frac{1}{\mu}\frac{\partial p}{\partial y}-\frac{\rho}{\mu}g_y\right)\frac{x^2}{2}+c_1 x+c_2$$

Using the boundary conditions of

$$v(0)=0 \text{ and } v(2r_3)=0$$

the solution of the velocity profile becomes

$$v(x) = \left(\frac{1}{\mu}\frac{\partial p}{\partial y} - \frac{\rho}{\mu}g_y\right)\frac{x^2}{2} + r_3\left(\frac{\rho}{\mu}g_y - \frac{1}{\mu}\frac{\partial p}{\partial y}\right)x$$

The solution for Section 2 becomes more difficult because you would need to solve two coupled partial differential equations (u is a function of x/y and v is a function of x/y).

$$\rho\left(\frac{\partial u}{\partial t} + u\frac{\partial u}{\partial x} + v\frac{\partial u}{\partial y} + w\frac{\partial u}{\partial z}\right) = \rho g_x - \frac{\partial p}{\partial x} + \mu\left(\frac{\partial^2 u}{\partial x^2} + \frac{\partial^2 u}{\partial y^2} + \frac{\partial^2 u}{\partial z^2}\right)$$

$$\rho\left(\frac{\partial v}{\partial t} + u\frac{\partial v}{\partial x} + v\frac{\partial v}{\partial y} + w\frac{\partial v}{\partial z}\right) = \rho g_y - \frac{\partial p}{\partial y} + \mu\left(\frac{\partial^2 v}{\partial x^2} + \frac{\partial^2 v}{\partial y^2} + \frac{\partial^2 v}{\partial z^2}\right)$$

$$\rho\left(u\frac{\partial u}{\partial x} + v\frac{\partial u}{\partial y}\right) = -\frac{\partial p}{\partial x} + \mu\left(\frac{\partial^2 u}{\partial x^2} + \frac{\partial^2 u}{\partial y^2}\right)$$

$$\rho\left(u\frac{\partial v}{\partial x} + v\frac{\partial v}{\partial y}\right) = \rho g_y - \frac{\partial p}{\partial y} + \mu\left(\frac{\partial^2 v}{\partial x^2} + \frac{\partial^2 v}{\partial y^2}\right)$$

The first two equations that are listed are the Navier–Stokes equations for the two applicable directions. The second two equations are the reduced form of the Navier–Stokes equations and as the reader can see, these equations are coupled. The easiest solution to the second two equations is through the use of numerical methods, which will not be addressed here. Because we did not provide the exact details about the flow properties, we will leave the discussion at this point. However, if you can make the assumption that the x-direction velocity component does not vary much with x (i.e., $\frac{\partial u}{\partial x} \approx 0$) and that the y-direction velocity component does not vary much with y (i.e., $\frac{\partial v}{\partial y} \approx 0$), then this problem will simplify significantly. However, around a bend and considering that there is a taper in this vessel, these assumptions may not be accurate.

5.11 PULSATILE FLOW AND TURBULENCE

In most of the examples that we have discussed so far, we have always made the assumption that blood flow is steady. Due to the cardiac cycle (Section 4.3), the blood flow rate that is being ejected from the heart into the vasculature is not constant. In fact, the systemic blood flow mimics the left ventricular pressure waveform in time within the arterial circulation (Figures 5.25 and 5.26). Figure 5.25 illustrates that at one location within the arterial vascular tree, the velocity waveform will mimic the cardiac output and will show a somewhat consistent waveform. Figure 5.26 illustrates the variation in the pressure waveform throughout the vascular tree. Notice that within the arterial circulation there is a pulsatility that is dampened out before the capillary circulation. In the large arteries, a significant variation in the blood pressure can be observed that has a range of approximately 40 mmHg under normal physiological conditions. The variation between the peak systolic pressure (typically around 120 mmHg) and the minimum diastolic pressure (typically around 80 mmHg) is termed the pulse pressure. The mean arterial pressure is the

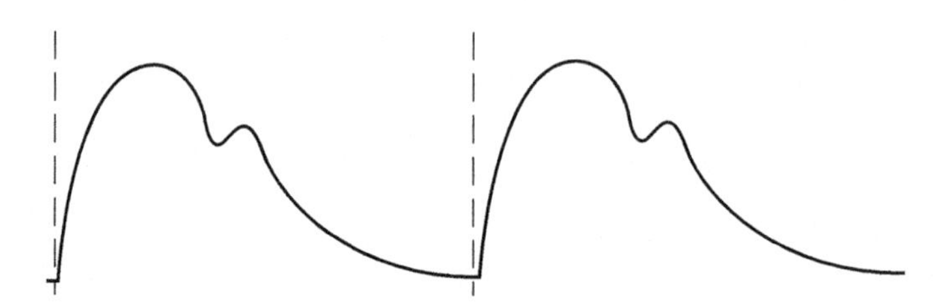

FIGURE 5.25 A schematic of the velocity waveform in a large artery, such as the aorta, showing that the velocity changes temporally and is pulsatile. As we can see, this mimics the pressure pulse of the left ventricle, but is reasonably constant with time.

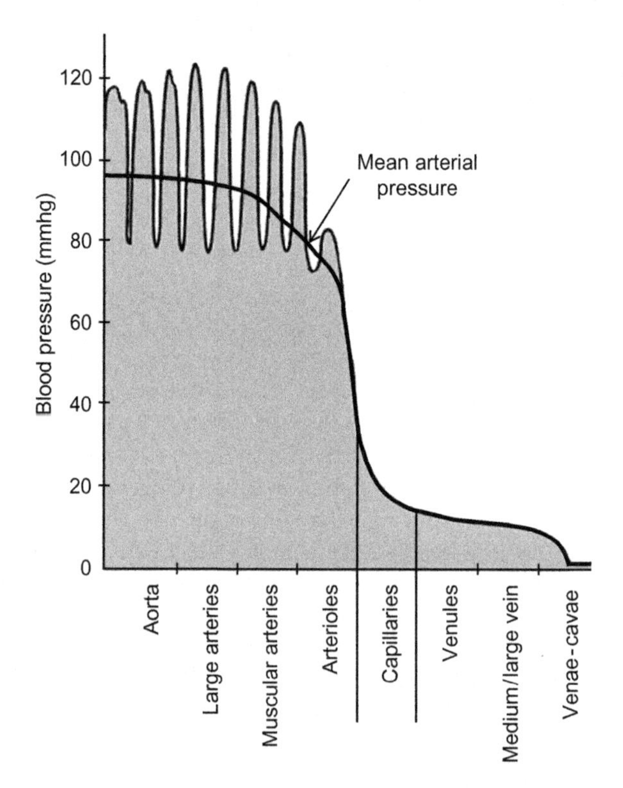

FIGURE 5.26 Pressure variation within the systemic circulation, illustrating that (i) there is a variation in pressure spatially along the systemic circulation and (ii) that within a specific level within the systemic circulation, the pressure varies with the systolic and diastolic contractions of the heart. The mean arterial pressure is the effective pressure in the systemic circulation.

effective pressure in the vasculature and can be calculated as a time weighted average of the peak systolic and minimum diastolic pressures, as follows:

$$\text{Mean Arterial Pressure} = \text{Diastolic Pressure} + \frac{\text{Pulse Pressure}}{3} \tag{5.41}$$

In the arterioles, a temporally varying pressure waveform can be observed, however, the waveform looks less like the left ventricular pressure waveform and the pulsatility begins to significantly dampen out. In the capillaries and the venous circulation, the pressure pulse is negligible and the flow is largely nonpulsatile. Thus, the question remains, does the pulsatility in large vessels affect the calculations that we have been carrying out in previous sections?

One can argue that the steady flow assumption does or does not affect the calculations that we have done so far. Since the flow is pulsatile, then the change in velocity with regard to time cannot be zero (see any Navier–Stokes example where we ignored the temporal variation in velocity; $\frac{\partial v}{\partial t}$). In one sense, the calculations of the fluid properties that we have conducted never describe the actual condition within the body. However, by restricting ourselves to one instant in time, the calculations can be carried out by ignoring the temporal derivatives. If we were interested in a reasonably small time scale with a reasonable time step, we can carry out the computations a few times, using the output of one time step as the input to the subsequent time step. The effect of this is that one loses the complete transmittance of the fluid throughout the vessel (e.g., if one was interested in how a blood cell moves with time, the information may be lost), but the solution of the problems can be carried out (i.e., you lose some information but gain some other information). By restricting the calculation to one instant in time, the assumption is made that the speed of blood everywhere throughout the blood vessel length is the same at that time. However, we have already discussed that due to the fluid flow and the elastic properties of the blood vessel, there is wave propagation within the walls of the blood vessel and within the fluid itself. Additionally, the conservation laws tell us that the fluid velocity cannot be the same along the entire blood vessel length, if there is a change in cross-sectional area of the blood vessel. Additionally, we typically assume that the fluid has a viscosity, such that the velocity is not uniform within one cross-section. Therefore, when looking at one section of a blood vessel, the velocity profile will not be constant in magnitude and direction (at least in larger vessels) both within one cross-section and throughout the vessel. By considering the pulsatile nature of blood velocity, it more than likely requires the use of numerical methods to calculate flow properties (see Chapter 14 for more details) unless the inflow velocity waveform is overly simplified. The way to work around this is to define the inlet velocity as a function of time at discrete time points. Then solve for the flow properties at those discrete time points, using the output of one solution as the input for the subsequent iteration over the length of interest.

Another factor that we have largely ignored is that blood flow can become turbulent, especially during the deceleration phase immediately after peak systole, when the blood velocity is the greatest and is rapidly reducing in velocity. Note that while the generation of turbulence is possible during this time, it is likely to only occur in the ascending aorta and/or the aortic arch, due to the combination of the rapidly decelerating flow and the complex geometry. Additionally, the turbulence that may generate at this time is not true turbulence (as defined in Section 2.12) because the turbulence would not be self-sustaining and would almost instantaneously dissipate back towards laminar flow. Even with these two special conditions in the ascending aorta, it is unlikely for mixing to occur and thus we typically consider biological flows within the vascular system to be laminar over the entire duration. When flow becomes turbulent, there are random fluctuations within the fluid velocity and pressure. By nature of being random, they cannot be calculated or predicted, but they can be observed. For turbulent flows, we would discuss the velocity as a mean velocity plus some random velocity fluctuation term at each spatial point. It is unrealistic to be able to define the entire flow field as a function of time and space under true turbulent conditions. Computational approximations of the turbulent flow field within computer simulations are needed to "accurately" represent turbulent flows. For hand calculations of turbulent flows to be carried out, one must be willing to make significant assumptions about the flow properties. Normally, there are two methods that are employed in order to approximate turbulent flows. The first would be the use of theoretical statistical methods, which aims to correlate the observed fluctuations in particular ways. Typically, one would look at frequency correlations or space–time correlations of fluctuating variables. The second approach used to approximate turbulent flows is the semiempirical modeling of the turbulent mean quantities. Empirical modeling relies solely on observations and/or experiments to reach a conclusion about the observations, whereas semiempirical modeling must rely on some observation/experiments and some theory. With this approach, many of the fluid dynamics principles that were developed in the earlier sections are extended to include various turbulent properties. Using this approach, the mean velocity in the x-direction and the fluctuation in the x-component of velocity, at one spatial location, would be defined as

$$\overline{u} = \frac{1}{T}\int_{t_0}^{t_o+T} u dt \qquad u = \overline{u} + u' \tag{5.42}$$

where any fluid quantity with a prime (′) superscript denotes that we are talking about the fluctuating (random) component of that variable (as in Section 2.12, we are using velocity as the example, but this analysis would hold for any fluid variable). For completeness, we will include some of the salient fluid mechanics equations that can be used in turbulent flow modeling. As discussed previously, the continuity equation for incompressible fluid flows can be defined as

$$\left.\frac{dm}{dt}\right|_{system} = \int_{area} \vec{v} \cdot d\vec{A} = 0 \tag{5.43}$$

if we can assume that the volume of interest does not change with time. Making these assumptions, we can substitute the turbulent velocity components (from Eq. (5.42)) into Eq. (5.43) to get

$$\frac{\partial u}{\partial x} + \frac{\partial v}{\partial y} + \frac{\partial w}{\partial z} = \frac{\partial(\overline{u} + u')}{\partial x} + \frac{\partial(\overline{v} + v')}{\partial y} + \frac{\partial(\overline{w} + w')}{\partial z} = 0 \tag{5.44}$$

or

$$\frac{\partial \overline{u}}{\partial x} + \frac{\partial \overline{v}}{\partial y} + \frac{\partial \overline{w}}{\partial z} = 0 \tag{5.45}$$

if Eq. (5.44) is time-averaged. If Eq. (5.45) is subtracted from Eq. (5.44) but we do not time average the result, one obtains

$$\frac{\partial u'}{\partial x} + \frac{\partial v'}{\partial y} + \frac{\partial w'}{\partial z} = 0 \tag{5.46}$$

The importance of Eqs. (5.45) and (5.46) is that both the mean velocity component and the fluctuating velocity component must each separately satisfy continuity equations. If the fluid was compressible, these equations would be coupled via the mean density fluctuation term, which would be a function of both the mean velocity components and the fluctuating velocity components. Extending this analysis to the time-average turbulent Navier–Stokes equation, we obtain

$$\rho \frac{D\vec{v}}{Dt} + \rho \frac{\partial}{\partial x_j}\left(\overline{u_i' u_j'}\right) = \rho \vec{g} - \nabla \overrightarrow{\overline{p}} + \mu \nabla^2 \overrightarrow{\overline{v}} \tag{5.47}$$

which is used extensively in computational fluid dynamics. The term $\overline{u_i' u_j'}$ is the turbulent inertial tensor, exists in all turbulent flows, and is one of the major reasons why calculations of turbulent flows becomes so difficult. Since this component is a tensor component, it iterates through nine variables in the standard Cartesian coordinate system (one must choose i and j for each x, y and x-direction). Unfortunately, these nine components can only be defined by unattainable knowledge regarding the turbulent flow properties and knowledge about how these properties relate to local conditions (e.g., geometric and surface roughness). If the turbulent flow can be simplified to a two-dimensional problem, then luckily, the turbulent inertial tensor is composed of only one complicated value; $\overline{u'v'}$ (assuming an x/y coordinate system). Even with this simplification, there are no physical or empirical observations that can help us to solve for this value. Thus, most of the simulations and work associated with turbulent flows are semiempirical at best.

Equation (5.47) can be rearranged so that the turbulent inertial tensor is included within the stress terms on the right-hand side of the equation (unfortunately, the turbulent inertial tensor components are not stresses but some insights into the formula can be obtained from this manipulation). Rearranging Eq. (5.47), we get

$$\rho \frac{D\vec{v}}{Dt} = \rho \vec{g} - \nabla \overrightarrow{\overline{p}} + \nabla \cdot \tau_{ij} \tag{5.48}$$

where

$$\tau_{ij} = \underbrace{\mu \left(\frac{\partial u_i}{\partial x_j} + \frac{\partial u_j}{\partial x_i}\right)}_{\text{Laminar}} - \underbrace{\rho \overline{u_i' u_j'}}_{\text{Turbulent}} \tag{5.49}$$

This formulation suggests that the turbulent inertial tensor is composed of a laminar Newtonian viscous stress plus an apparent turbulent stress. Unfortunately, the turbulent stress term is still unknown and some observations are needed to analyze these equations. These fluctuating quantities can only be resolved after the flow has occurred and the fluctuations can be measured with respect to the mean velocity.

To continue our brief discussion of turbulence, we would like to define the energy equation for incompressible turbulent flows with constant properties. Using similar mathematical manipulations, as discussed above, we can time average the energy equation to obtain the following

$$\rho c_p \frac{D\overline{T}}{dt} = -\frac{\partial}{\partial x_i}(q_i) + \overline{\Phi} \tag{5.50}$$

where

$$\overline{\Phi} = \overline{\frac{\mu}{2}\left(\frac{\partial \overline{u}_i}{\partial x_j} + \frac{\partial u_i'}{\partial x_j} + \frac{\partial \overline{u}_j}{\partial x_i} + \frac{\partial u_j'}{\partial x_i}\right)^2} \tag{5.51}$$

and

$$q_i = \underbrace{-k\frac{\partial \overline{T}}{\partial x_i}}_{\text{Laminar}} + \underbrace{\rho c_p \overline{u_i' T'}}_{\text{Turbulent}} \tag{5.52}$$

Similarly, to the above discussion with the effective turbulent stress terms, this rearrangement generated a number of new terms to the energy equation. The new dissipation term ($\overline{\Phi}$) is quite complex, even if we can simplify the flow to a two-dimensional case, as shown. Additionally, the conduction and turbulent convection terms were combined into a pseudo-heat flux vector (q_i). This new term contains both the molecular flux due to laminar flow plus the new turbulent flux. Again, at the onset of the flow we have no knowledge about these fluctuating terms, they can only be observed or semiempirically formulated.

There are two more elegant formula that have been derived in the case of turbulent flows. These values are quite important when designing devices that many introduce turbulence into the flow fields (for instance some cardiovascular implantable devices may need to consider the following quantities in their designs). The turbulent kinetic energy can be defined as

$$\begin{aligned}\frac{D(KE)}{Dt} = &-\frac{\partial}{\partial x_i}\left[u_i'\left(\frac{1}{2}u_i'u_j' + \frac{p'}{p}\right)\right] - \overline{u_i'u_j'}\frac{\partial \overline{u}_j}{\partial x_i} \\ &+ \frac{\partial}{\partial x_i}\left[\nu u_j'\left(\frac{\partial u_i'}{\partial x_j} + \frac{\partial u_j'}{\partial x_i}\right)\right] - \nu\frac{\partial u_j'}{\partial x_i}\left(\frac{\partial u_i'}{\partial x_j} + \frac{\partial u_j'}{\partial x_i}\right)\end{aligned} \tag{5.53}$$

where the term on the left-hand side of the equation is the time rate of change of the turbulent kinetic energy, the first term on the right-hand side is the kinetic energy convective diffusion, the second term on the right-hand side is the kinetic energy production, the third term on the right-hand side is the work done by turbulent viscous stresses, and finally the fourth term on the right-hand side is the turbulent viscous dissipation. Similarly, the Reynolds stresses can be represented as

$$\begin{aligned}\frac{D\overline{u_i'u_j'}}{Dt} = &-\left(\overline{u_j'u_k'}\frac{\partial \overline{u}_i}{\partial x_k} + \overline{u_i'u_k'}\frac{\partial \overline{u}_j}{\partial x_k}\right) - 2\nu\frac{\partial u_i'}{\partial x_k}\frac{\partial u_j'}{\partial x_k} + \frac{p'}{p}\left(\frac{\partial u_i'}{\partial x_j} + \frac{\partial u_j'}{\partial x_i}\right) \\ &- \frac{\partial}{\partial x_k}\left[\overline{u_i'u_j'u_k'} - v\frac{\partial \overline{u_i'u_j'}}{\partial x_k} + \frac{p'}{p}\left(\delta_{jk}u_i' + \delta_{ik}u_j'\right)\right]\end{aligned} \tag{5.54}$$

where the left-hand term is the time rate of change of the Reynolds stress, the first term on the right-hand side is the generation of stresses, the second term on the right-hand side is the dissipation of stresses, the third term on the right-hand side is the effects of pressure and strain on stresses, and the fourth term on the right-hand side is the diffusion of the Reynolds stresses. Again, unfortunately for analyzing flows with Eqs. (5.53) and (5.54), the majority of these terms are unknown prior to the flow. Some empirical correlations have been made for these equations but for the most part, they must be observed to be able to characterize the changes in the turbulent kinetic energy and the changes in the Reynolds stresses. The numerical approximations of turbulent biofluid mechanics properties will be discussed in Chapter 14.

Since one cannot discretely define the properties of turbulent flows, most biofluid mechanics engineers discuss the common characteristics of turbulence, which were described in Section 2.12, if turbulence appears in the particular case that is being studied. Briefly recall that turbulent flows are characterized by being random, mixing flows, where mass and heat transfer occurs between fluid laminae. Turbulent flows also are characterized by very large inertial forces as compared to the internal viscous forces (e.g., the Reynolds number is normally in the range of 10,000). Turbulent flows also require a large amount of energy to maintain the turbulence, since a true fully developed turbulent flow is self-sustaining. The energy is needed to overcome the increased frictional forces that are found due to lamina mixing.

5.12 DISEASE CONDITIONS

5.12.1 Arteriosclerosis/Stroke/High Blood Pressure

In Section 4.6, we began a discussion of coronary artery disease and myocardial infarction as instigated through a plaque formation. Here we will more fully discuss the disease of arteriosclerosis and how it is manifested within the arterial system. In general, arteriosclerosis can be characterized by a thickening and a stiffening of the arterial wall. As we have discussed in this chapter, a decrease in the cross-sectional area of blood vessels can have multiple effects in the body. Flow can accelerate around the constriction, causing enhanced shear stresses or pressure gradients within the fluid. Flow can also be stagnated around the constriction if the pressure gradient is not large enough to overcome the enhanced resistance to flow. Flow can separate from the wall, which also decreases the mean fluid velocity and may enhance recirculation zone formation. When a mean flow velocity decreases, there is the potential for a reduction in the tissue oxygenation along with a decrease in the glucose delivered to cells. Also, stiffening of the arterial wall decreases the compliance of the vessel, and it may not be able to withstand the high pressure forces that act on the vessel. Arteriosclerosis is manifested in two main forms: calcification of the arterial wall or atherosclerosis. Calcification is associated with the deposition of calcium salts between the tunica intima and the tunica media and a breakdown of smooth muscle cells. As discussed in Chapter 4, atherosclerosis is the deposition of lipids within the arterial wall. Monocytes that enter the vessel wall engulf the lipids, transform into foam cells, and attach themselves to the vessel wall. Foam cells can release cytokines and growth factors which aid in plaque formation.

Risk factors for arteriosclerosis include a diet high in cholesterol, high blood pressure, cigarette smoking, diabetes, obesity, being elderly, and being male (possibly from increased iron levels or decreased estrogen levels). Most of these risk factors can be addressed by having the patient make a lifestyle change. However, if the disease progresses too far, it is common for medical doctors to perform a balloon angioplasty to flatten the plaque and to use stenting to prevent restenosis, or bypass surgery to redirect flow. A major focus of biofluids research efforts is on the initiation and development of cardiovascular diseases for a number of reasons. First, cardiovascular diseases account for approximately half of all of the deaths in the United States. Many of the disease interventions are palliative, stressing lifestyle changes without addressing the root cause of the disease. Interventions focused on lifestyle changes can be largely ineffective since it is a requirement for the patient to adhere to the lifestyle change guidelines. Additionally, deficiencies of the cardiovascular system can translate to many other organ systems and thus, have wide ranging systemic effects.

Arteriosclerosis is an inflammatory disease, which arises due to altered fluid dynamic forces, which cause an injury to the vascular wall, and/or oxidation of lipoproteins (which function to transport lipids through water-based solutions, such as blood). When one of these functions occurs, it is common to see lipid deposition in the subendothelial space and proliferation of vascular smooth muscles cells, which combine to increase the thickness of the intima. Once lipoproteins become oxidized, it is common to observe enhanced expression of endothelial cell receptors that are associated with white blood cell and/or platelet adhesion. This helps to localize inflammatory cells to the inflamed region, but in the case of arteriosclerosis, ends up damaging the tissue further.

Localization of white blood cells, in particular monocytes, is one of the primary determinants of whether or not arteriosclerosis will progress. In the early stages of lesion formation, monocytes extravasate from the blood and differentiate into tissue macrophages. These macrophages engulf and retain the oxidized LDL cholesterol, eventually forming foam cells. These foam cells can release paracrines, which attract addition macrophages to the region of plaque formation and cause the proliferation of vascular smooth muscle cells. If these functions occur for a significant amount of time without intervention, the intima begins to thicken and becoming hyperplastic. Intimal hyperplasia can have sever effects on flow around the lesion and may eventually occlude the flow, downstream of the forming lesion. These occlusions have varied effects, depending on their location and thus, what vessels are downstream of the occlusion. For reference, there are typically four common sites for atherosclerotic lesion formation to occur within the systemic cardiovascular system. These sites include, the coronary arteries (which provide blood to the heart), the branch points of the carotid arteries from the aortic arch (which provide blood to the brain), the branch points for the renal arteries from the aorta (which provide blood to the kidney), and the branch points for the common iliac arteries where the aorta terminates (this provides blood to the lower limbs). Note that these locations are characterized by complex geometries.

When a blood vessel that supplies the brain with blood becomes atherosclerotic, there is a large possibility for stroke to develop. A stroke occurs when the blood supply to the brain is interrupted. This should be differentiated from brain hypoxia, which occurs when there is a reduction of oxygen delivery to the brain (this may occur when the atmospheric pressure reduces or during chronic pulmonary disease) but the blood flow through the brain vasculature remains relatively stable. Cerebral ischemia, which can lead to stroke, occurs when there is an interruption in the blood supply to the brain. If this occurs over a relatively short period of time (5–10 min), after which the normal blood flow returns to the brain, it is atypical to see tissue death. However, if blood flow to the brain is interrupted for a long period of time, then it is common to see tissue death and stroke development.

To prevent an interruption in brain blood supply, the brain vascular system has a unique structure called the Circle of Willis. The Circle of Willis receives blood supply from the carotid arteries and the vertebral arteries, and then it feeds all of the major brain blood vessels with blood. Therefore, a major blockage prior to or within the Circle of Willis will more than likely not affect the overall blood supply to the brain because of this redundant system. However, a severe block distal to the Circle of Willis can cause a decrease in blood flow to particular sections of the brain. A brain that does not receive any blood for 5–10 s will instigate unconsciousness. It is also possible to generate a stoke, due to emboli within the vascular system. Emboli can form in a number of ways (some of these will be discussed later) but can be considered as large particulate matter that flows within the vasculature. If this particulate matter becomes lodged within a significant feeding vessel to the brain, then it is common to observe stroke formation.

High blood pressure (or hypertension) is a risk factor for many cardiovascular diseases. Patients with hypertension tend to have a mean arterial pressure that is greater than 110 mmHg (the normal mean arterial pressure during one cardiac cycle is approximately 90 mmHg). Typically, this relates to a diastolic pressure greater than 90 mmHg and a systolic pressure greater than 140 mmHg. In severe cases, the systolic pressure can approach 200 mmHg, with the diastolic pressure approaching 150 mmHg. Interestingly, mean systolic pressure has shown to increase with age (from approximately 120 mmHg at age 20, to approximately 160 mmHg by age 80). This is found in both males and females, however, females tend to have a mean systolic pressure a few mmHg below the age-matched males. Additionally, the spread of mean blood pressure found around the mean blood pressure increases with age; at 20 years old, 90% of the population is found to have a mean blood pressure between ~100 and 150 mmHg, whereas at 80 years old, 90% of the population is found to have a mean blood pressure between ~125 and 220 mmHg. These ranges may be considered normal depending on body mass, size, activity level, etc., so the definitions of when hypertension develops are somewhat difficult to make. Issues associated with hypertension are not necessarily related to the condition itself, but instead is related to conditions that it may invoke on other regions of the vascular system.

Any increases in blood pressure can lead to a higher chance of death predominantly from (i) an increased work conducted by the heart, (ii) hemorrhaging, or (iii) vessel rupture. An increased heart work can easily lead to heart failure because the cardiac muscle will use an excessive amount of energy to maintain the work and require a larger percentage of the blood supply. At some time, when the cardiac tissue exhausts all of the energy, the heart will not be able to contract anymore. Hemorrhaging associated with high blood pressure is common in the kidneys. This will lead to kidney failure because of the blood that is being pooled within the kidney. Kidney failure will always lead to death. Blood vessel rupture is also a serious problem if it occurs in the brain or in the cardiac tissue. As we have previously discussed, this can lead to stroke or cardiac ischemia.

A common cause for high blood pressure is excessive salt in the diet. With excess salt in the body, the osmotic balance of the blood and the extracellular fluid is shifted. To counter this, a person will typically increase the volume of fluids that they drink and produce more antidiuretic hormone (see Chapter 12), which both act to increase the volume of fluid within the extracellular space. With an increase in extracellular fluid, the external pressure acting on the blood vessels increases; that is, there is a preference for the blood vessels to collapse. Therefore, to maintain the same flow throughout the body, the mean arterial pressure must be increased.

Interestingly, atherosclerosis has been found in mummified remains, suggesting that the presence of atherosclerotic regions is not a new disease, solely based on changes in the lifestyles/diets within the modern world. The term atherosclerosis was first used in the early 1900s and was contrasted from arteriosclerosis by focusing more on the lipid rich diseases, whereas arteriosclerosis is a more general disease which only requires that blood vessels thicken and stiffen. A second way to differentiate between atherosclerosis and arteriosclerosis is that arteriosclerosis tends to occur in concentric and diffuse patterns, whereas atherosclerosis tends to occur asymmetrical at particular points.

5.12.2 Platelet Activation/Thromboembolism

Another condition that typically occurs in parallel with cardiovascular diseases is enhanced platelet activation. Platelets can become activated after exposure to a short duration high-intensity shear stress or other altered shear stresses (such as low magnitude oscillatory shear stress). Activated platelets express negative phospholipids on the exterior leaflet of their cell membrane, release vasoactive compounds, release cytokines, and release growth factors. The products that platelets release upon activation have the effect of activating more platelets, inducing a chain reaction that activates many platelets. Due to all of these processes, it is possible for platelets to begin to aggregate onto the endothelial cell wall and to other platelets, if the activated platelets are not removed from the circulation.

Platelet aggregates are basically small blood clots that are termed a thrombus. A thrombus can act like a plaque because it blocks the blood flow and reduces the cross-sectional area of the blood vessel. At this point, it is possible for fibrinogen to be cleaved into fibrin and adhere to activated platelets within the aggregate. This increases the intercellular strength of the thrombus, making it harder to remove it from the circulation. As fibrin adheres to the thrombus, red blood cells and platelets can be caught within the growing thrombus. When the thrombus becomes larger, the shear forces that act on it also increase (due to the acceleration of blood around the thrombus). If the shear forces exceed the strength of the intercellular/intermolecular bonds, a portion of the thrombus can break off and enter the circulation. A portion of the thrombus that breaks off of the main thrombus is termed a thromboembolus or embolus (pl. emboli; in general, emboli can be any abnormal solid mass flowing within the cardiovascular system, see the discussion above).

Emboli that enter the circulation can have many devastating effects on the patient. Emboli have size and mass associated with them. As they pass through the circulatory system, it is possible that emboli with different sizes become lodged in different sections of the cardiovascular system. This will cease the blood flow to that particular region and depending on the region, it can cause death (a discussion similar to the downstream effects of lesion growth applies here). If the emboli become stuck in the cerebral circulation, a stroke can occur. If they become stuck in the cardiac circulation, a heart attack or myocardial infarction can occur. In other instances, free-flowing emboli can recruit more platelets to them and potentially adhere back to a damaged endothelial cell wall. Whatever the outcome of the embolus is, it can have devastating effects within the body.

5.12.3 Aneurysm

An aneurysm (Figure 5.27) is a bulge in the wall of an artery that is typically caused by a weakening of the muscles within the vascular wall. Because arteries carry blood under high pressure, they are more susceptible to aneurysm formation. This typically occurs when a patient has high blood pressure, which exceeds the viscoelastic mechanical properties of the blood vessel wall. The bulge in the blood vessel looks like a bubble within the vascular system. If the vessel wall has weakened to such a great extent that it cannot maintain the blood pressure any longer, the aneurysm ruptures. This typically occurs in all aneurysms which have developed for a long period of time, because as the wall begins to weaken, it cannot withstand the pressure forces and weakens further. As the blood vessel wall continues to weaken, the bulge in the blood vessel becomes larger and the likelihood for rupture to occur increases. A blood vessel that ruptures will cause a severe hemorrhage, which may eventually lead to death.

Depending on the blood vessel that has experienced an aneurysm, the effect varies. Unfortunately, the arteries within the brain and the abdominal aorta are very susceptible

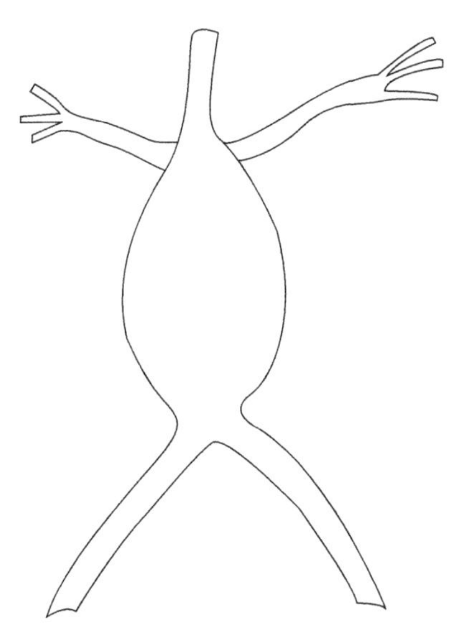

FIGURE 5.27 Schematic of a saccular aortic aneurysm, which is a bulging of the blood vessel wall. Normally, the wall bulges because there is a deterioration of the muscular mass in the wall and due to the constant pressure loading the wall begins to bulge. In severe cases, the wall can deteriorate to such an extent that the vessel breaks and blood begins to pool in the extravascular space.

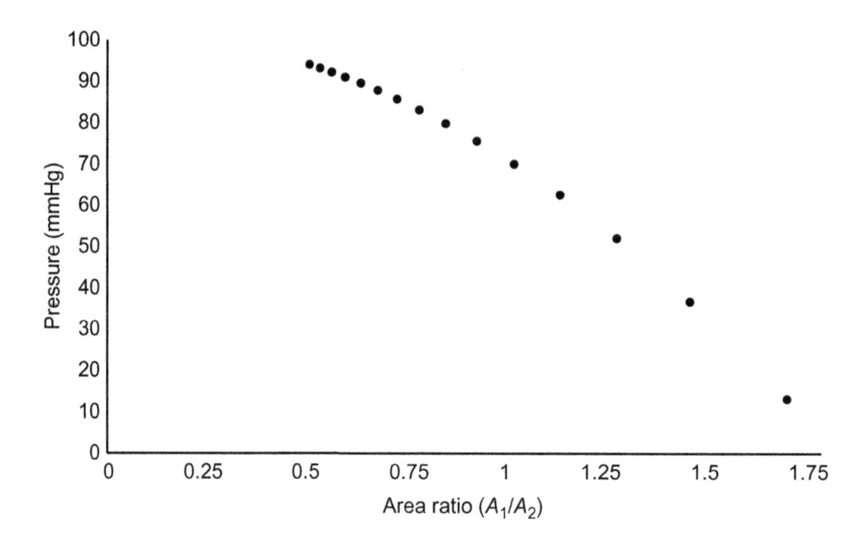

FIGURE 5.28 Relationship between the ratio of the inlet area (A_1) to the outlet area (A_2) and the pressure recorded at the outlet area. This assumes that there is no variation within height, viscous dissipation can be ignored and that there are no temporally varying components to this flow. As the ratio increases (as would be seen during a stenosis), the pressure decreases and as the ratio decreases (as would be seen during an aneurysm), the pressure increases.

to aneurysm formation due to their constant high pressure loading (and unloading). A ruptured aneurysm in the brain can cause stroke or increased pressure on the brain, while those in the aorta will cause excessive bleeding into the chest cavity. Blood in the chest increases the pressure on all of the organs present within the chest cavity and prevents them from functioning properly. In either case, because the blood vessel is either supplying the brain or some of the chest cavity organs with nutrients (depending on the location of the aneurysm, although approximately 95% of abdominal aneurysms occur distal to the renal arteries), the effect will be devastating.

A true aneurysm is one in which the intima of the blood vessel has bulged outside past the media and the adventitia. A false aneurysm is caused by blood leaking and clotting in a small space next to the blood vessel. The morphology of aneurysms varies, and the most common have been termed saccular (resembling a sphere in three-dimensional space) or fusiform (resembling a cylinder in three-dimensional space). Typical risk factors for aneurysm include high blood pressure, diabetes, tobacco smoke, and alcohol.

It is interesting to apply some of the fundamental biofluid mechanics equations to the study of both arteriosclerosis lesion formation and the development of an aneurysm. In both cases, a simplified conservation of mass and Bernoulli principle can be used. Under these conditions, these rules can simplify to

$$A_1 v_1 = A_2 v_2$$

$$\frac{p_1}{\rho} + \frac{v_1^2}{2} = \frac{p_2}{\rho} + \frac{v_2^2}{2}$$

if we assume that the density is constant, gravitational and viscous effects are negligible and that the flow is steady. Rearranging these equations, we can obtain the following:

$$P_2 = P_1 + \left(\frac{\rho}{2}\right)\left[v_1^2 - \left(\frac{A_1}{A_2}v_1\right)^2\right] \tag{5.55}$$

which shows that as the area ratio $\left(\frac{A_1}{A_2}\right)$ increases (e.g., A_2 decreases, as during a stenosis) the local pressure within the vessel decreases. This helps to accelerate the fluid through the constriction. Additionally, as the area ratio decreases (e.g., A_2 increases, as during an aneurysm) the local pressure within the vessel increases. Therefore, during an aneurysm, the pressure within the aneurysm increases, which can help to rupture the already weakened arterial wall. A curve of this relationship is shown in Figure 5.28.

END OF CHAPTER SUMMARY

5.1 Arteries are blood vessels that transport blood away from the heart. The arterial wall is composed of three distinct layers: the tunica intima, the tunica media, and the tunica adventitia. The tunica intima is the inner layer of the vessel wall and is composed of endothelial cells. The tunica media is composed of smooth muscle cells and is typically the thickest layer within an artery. The tunica adventitia is composed mostly of connective tissue. The arterial side of the vascular system is under high pressure. The arteries have a mean pressure in the range of 90–100 mmHg, whereas the arterioles have a mean pressure that ranges from 40 to 80 mmHg. Arterioles are the major resistance vessels in the circulatory system.

5.2 Veins are blood vessels that transport blood towards the heart and the venous wall is also composed of three layers. The major difference between the arterial wall and the venous wall is the thickness of the tunica media; the arterial wall is typically thicker than the venous wall. Another major difference between the arterial anatomy and the venous anatomy is that veins contain valves to prevent the backflow of blood toward the capillary beds. Flow through the veins is carried out under low pressure and is facilitated through the contraction of skeletal muscles neighboring the veins termed the venous pump.

5.3 Blood is composed of two major components, the cellular component and the plasma component. Under normal conditions, the cellular component accounts for approximately 40% of blood. The cellular component is divided into the red blood cells (greater than 99% of all cells), the white blood cells (approximately 0.2%), and the platelets (approximately 0.1%). Red blood cells are primarily responsible for the delivery of oxygen and carbon dioxide. White blood cells are primarily responsible for protection against foreign particles (e.g., inflammatory responses). Platelets are responsible for hemostasis. The plasma component consists of water, proteins, and other solutes such as sugars and ions, among others.

5.4 Rheology is the study of a flowing material combined with the study of the physical properties of that material. Plasma behaves as a Newtonian fluid and has a viscosity of approximately 1.2 cP. Whole blood, however, does not have a constant viscosity. The viscosity of whole blood varies with shear rate, hematocrit, temperature, and disease conditions, and this is predominantly due to the presence of cells and other compounds within the fluid. The Casson model for blood states

$$\sqrt{\tau} = \sqrt{\tau_y} + \sqrt{\eta\dot{\gamma}}$$

as a relationship between shear stress and shear rate, where τy is a constant yield stress (which varies based on hematocrit), and η is an experimentally fit constant, which approximates the fluids viscosity. Under most physiological conditions, the yield stress of blood can be approximated as 0.05 dyne/cm^2. The velocity profile of a fluid modeled with the Casson flow can be described by

$$u(r) = \begin{cases} -\dfrac{1}{4\eta}\dfrac{dp}{dx}\left(R^2 - r^2 - \dfrac{8}{3}r_y^{0.5}(R^{1.5} - r^{1.5}) + 2r_y(R-r)\right) & r_y \le r \le R \\ -\dfrac{1}{4\eta}\dfrac{dp}{dx}\left(\sqrt{R} - \sqrt{r_y}\right)^3\left(\sqrt{R} + \dfrac{1}{3}\sqrt{r_y}\right) & r \le r_y \end{cases}$$

Red blood cells can form aggregates termed rouleaux. Interestingly, the presence of rouleaux in the blood is dependent on the shear rate, at which the fluid is flowing. At lower shear rates, rouleaux are more prevalent, likely due to the greater time that cells can come into contact with each other and/or proteins in the blood.

5.5 The pressure, flow, and resistance in the arterial system are related by

$$Q = \frac{\Delta P}{R}$$

The volumetric flow rate can also be approximated using the Hagen–Poiseuille solution:

$$Q = \frac{\pi \Delta P r^4}{8\mu L}$$

which is valid for laminar flow within a cylindrical tube.

5.6 The formulas derived for the relationship between pressure, flow, and resistance in the arterial system can also be used in the venous system. However, it is unlikely for the veins to be perfectly cylindrical. In this case, the Navier–Stokes solution would need to be solved for the randomly shaped vein, which can find a relationship between pressure, flow, and resistance. The general Navier–Stokes for cylindrical components under these conditions would be

$$v_z(r) = \frac{r^2}{4\mu}\frac{\partial p}{\partial z} + c_1 \ln(r) + c_2$$

where the integration constants would need to be obtained for the particular conditions. It is possible to do a weighted average of laminae of fluid with known average velocities.

5.7 The Windkessel model is a powerful representation of flow throughout the arterial circulation. It makes the assumption that the inflow properties can be defined with a time-varying component, the outflow properties are steady and the difference in these two terms relates to the dissipation/accumulation of blood and energy within the arterial circulation.

5.8 Wave propagation within the arterial circulation can occur in two ways. The first is the pressure pulse that displaces the arterial wall, where a differential area would need to be considered within the properties. The mechanical properties of vessel wall would also need to be considered in these formulations. Making assumptions about the homogeneity of the arterial wall mechanical properties, the wave speed can be computed from

$$c = \sqrt{\frac{Eh}{4pr_i}}$$

The second wave that can be propagated throughout the arterial system is the fluid pressure (or velocity) at bifurcations. The geometry of the bifurcation would determine the relative division of the pressure between the branches. However, one must consider that some of the energy is reflected back along the main feed branch. In this case, the relative pressure within the daughter branches and the main branch can be solved by

$$\frac{p_r}{p_p} = \frac{Z_p^{-1} - (Z_{d1}^{-1} + Z_{d2}^{-1})}{Z_p^{-1} + (Z_{d1}^{-1} + Z_{d2}^{-1})}$$

$$\frac{p_{d1}}{p_p} = \frac{p_{d2}}{p_p} = \frac{2Z_p^{-1}}{Z_p^{-1} + (Z_{d1}^{-1} + Z_{d2}^{-1})}$$

5.9 At bifurcations, the flow will also be separated based on the geometry of bifurcation. By minimizing the work that the branches conduct, a set of formulas have been developed to describe the most ideal branch angles. These formulas are

$$r_1^3 = r_2^3 + r_3^3$$

$$\cos\theta = \frac{r_1^4 + r_2^4 - r_3^4}{2r_1^2 r_2^2}$$

$$\cos\phi = \frac{r_1^4 - r_2^4 + r_3^4}{2r_1^2 r_3^2}$$

$$\cos(\theta + \phi) = \frac{r_1^4 - r_2^4 - r_3^4}{2r_2^2 r_3^2}$$

Flow separation occurs at bifurcations because the wall pressure gradient becomes positive, and therefore, the flow would tend to move upstream along the wall. Another occurrence at bifurcations is that the flow tends to skew toward the inner walls of the bifurcation. This means that the maximal velocity is no longer along the centerline of the blood vessel.

5.10 Most blood vessels experience a significant taper along their length. This can be modeled with the exact formulation for the wall geometry or can be simplified with steps at specified distances along the vessel length. Regardless of the choice, it would need to be considered within the formulations.

5.11 Within the majority of the arterial circulation, the velocity and pressure waveforms vary with time. This is directly related to the contraction (and hence the pressure pulse) of the left ventricle as well as the elastic properties of the blood vessels themselves. Due to the pulsatile nature of the blood velocity, temporal changes in the fluid parameters must be considered in biofluid mechanics problems. However, there are ways to work around the temporal changes in blood velocity and blood pressure if one considers the discrete steps in time and numerically couples the time steps. It is also possible for turbulence to develop within the arterial system at very specific locations and very specific times. However, true turbulence rarely forms since any turbulent flows would dissipate nearly instantaneously and are thus not self-sustaining. In special cases one may want to analyze the flow conditions using turbulent properties. In these instances, turbulent flows are typically modeled with statistical methods, where the time averaged fluctuations in fluid properties must be considered, or with semiempirical modeling, which looks at a combinations of observations and theory. For a general fluid property, the formulation would be

$$\overline{u} = \frac{1}{T}\int_{t_o}^{t_o+T} u dt$$

$$u' = u - \overline{u}$$

Some of the formulations for turbulent were described within the text for completeness.

5.12 Arteriosclerosis, stroke, and high blood pressure are common diseases that affect the vascular tree. Arteriosclerosis is characterized by a decrease in the cross-sectional area of arterial blood vessels. If a blood vessel that supplies the brain becomes arteriosclerotic, it is possible for stroke to develop. High blood pressure is linked to arteriosclerosis because the heart would need to develop a high pressure head to force blood through the vascular tree. Platelets can activate in the presence of high shear stress, which can be generated when the cross-sectional area of the blood vessels decreases. If platelets begin to aggregate, there is the potential for emboli to form, which can occlude other regions of the vascular system. Aneurysm occurs when there is a thinning of the vascular wall that eventually leads to a distention of the blood vessel because it cannot maintain the high pressure within the arterial circulation.

HOMEWORK PROBLEMS

5.1 Blood is primarily transported through the venous system via

a. increasing the intravascular blood pressure
b. muscle contractions
c. venous valve system
d. b and c

5.2 Which of the following changes would have the most significant effect on increasing the vascular resistance?

a. doubling the blood vessel diameter
b. halving the blood vessel length
c. doubling the blood viscosity
d. halving the blood vessel diameter
e. doubling the blood vessel length

5.3 Discuss the salient anatomical differences between arteries and veins, and detail how this relates to the function of each blood vessel.

5.4 Why is the blood flow to some organs (such as the stomach, intestines) discontinuous and the blood flow to other organs (such as the brain, heart) continuous?

5.5 The plasma component of blood contributes approximately ____ percent to the total blood volume, of which water accounts for _____ percent of the total plasma volume.

5.6 Discuss the important proteins that are found in plasma. Why are the concentrations of these proteins so tightly regulated?

5.7 Discuss coagulation and the role of platelets in the coagulation process.

5.8 When you are dehydrated, it would cause an (increase or decrease) in the hematocrit. What effect would this have on blood flow, viscosity, blood pressure, etc.?

5.9 The three cellular components of blood have different anatomical structures that relate to their functions. Briefly discuss the important structures for each cellular component.

5.10 Calculate the difference in blood flow rate for a 5-cm section of a blood vessel with a pressure difference of 25 mmHg that experiences an increase in blood vessel diameter from 100 to 200 μm. Also calculate the blood flow rate for a constriction that reduces the diameter to 50 μm. For each of these three cases, what is the vascular resistance? Assume that blood viscosity is 3.5 cP.

5.11 When a patient experiences high blood pressure, discuss what happens to the vascular flow rate. What changes must occur to allow the vascular flow rate return to normal conditions? Assume that the peak systolic blood pressure for the hypertensive patient is 190 mmHg and that the diastolic pressure is 130 mmHg. Compare this flow with the normotensive case (120/80 mmHg).

5.12 Calculate the wave propagation speed for a 100-μm-thick blood vessel with a radius of 10 mm. This blood vessel has a pressure of 85 mmHg and a Young's modulus of 12.5 kPa. Under disease conditions, the vascular wall can stiffen so that the Young's modulus approaches 30 kPa. Calculate the change in the wave speed under these conditions. What blood vessel property is likely to change in response to the increase in stiffness and why?

5.13 Model the pressure wave propagation at a branching junction where one parent vessel branches into two daughter vessels (such as Figure 5.16). Assume that the diameter of the parent branch is 100 μm, the diameter of one daughter branch is 60 μm, and the diameter of the other daughter branch is 45 μm. Assume that the thickness of each vessel is the same and is 10 μm, that the incident pressure within each branch is 50 mmHg (for wave propagation), and that the Young's modulus of each vessel is 15 kPa. Calculate the reflected pressure and the pressure within each daughter branch.

5.14 Calculate the radius of the parent branch knowing that the radii of the two daughter branches are 400 and 435 μm. Determine the bifurcation angles and draw an accurate figure of this branch.

5.15 Calculate the radius of the daughter branches and the angles of bifurcation if the parent branch has a radius of 500 μm.

5.16 Can Murray's law be extended to a more complicated branching pattern (i.e., two daughter branches that branch in three dimensions or three daughter branches)? If so, derive the applicable relationships.

5.17 The velocity through a tapering channel can be estimated as $\vec{v} = v_i\left(1 - e^{\frac{2x}{L}}\right)\hat{i}$. Calculate the acceleration of any particle along the centerline and the position of a particle that is located at $x = 0$ at time 0, where L is the vessel length and v_i is the inflow centerline velocity.

5.18 The radius of a continuously tapering blood vessel can be modeled as $r(x) = r_i\left(e^{-\frac{x}{2L}}\right)$. Determine the velocity and the acceleration of the fluid within this blood vessel.

5.19 Under atherosclerotic conditions, the aorta can reduce in cross-sectional area. Is it more likely for the blood flow to remain laminar or become turbulent? What would happen to the blood flow if there is an aneurysm present within the aorta?

5.20 Show that the volumetric flow rate for a Casson model of blood flow is

$$Q = \frac{\pi R^4}{8\eta}\left[-\frac{dp}{dx} - \frac{16}{7}\left(\frac{2\tau_y}{R}\right)^{0.5}\left(-\frac{dp}{dx}\right)^{0.5} + \frac{4}{3}\left(\frac{2\tau_y}{R}\right)^{4}\left(-\frac{dp}{dx}\right)^{-3}\right]$$

5.21 If the elastic deformation of the blood vessel wall is relatively small, what happens to the pressure flow relationship? Show the approximate function and describe the physiology behind this.

5.22 If the diameter of a person's aorta is abnormally large, would the flow through the aorta more likely be laminar or turbulent? What if the heart rate is increased, but flow rate is maintained the same?

5.23 In a Casson model of flow, the yield shear stress varies with hematocrit as $\tau_y = (0.325{*}Hct)^3$, where *Hct* is the hematocrit (e.g., 0.4). Plot and describe this relationship for various cardiovascular diseases.

References

[1] J.W. Adamson, Regulation of red blood cell production, Am. J. Med. 101 (1996) 4S.

[2] R.K. Andrews, J.A. Lopez, M.C. Berndt, Molecular mechanisms of platelet adhesion and activation, Int. J. Biochem. Cell Biol. 29 (1997) 91.

[3] M. Anliker, M.B. Histand, E. Ogden, Dispersion and attenuation of small artificial pressure waves in the canine aorta, Circ. Res. 23 (1968) 539.

[4] T. Aoki, D.N. Ku, Collapse of diseased arteries with eccentric cross section, J. Biomech. 26 (1993) 133.

[5] E.O. Attinger, Wall properties of veins, IEEE Trans. Biomed. Eng. 16 (1969) 253.

[6] J.C. Buckey Jr., F.A. Gaffney, L.D. Lane, B.D. Levine, D.E. Watenpaugh, S.J. Wright, et al., Central venous pressure in space, J. Appl. Physiol. 81 (1996) 19.

[7] C.G. Caro, J.M. Fitz-Gerald, R.C. Schroter, Arterial wall shear and distribution of early atheroma in man, Nature 223 (1969) 1159.

[8] C.G. Caro, J.M. Fitz-Gerald, R.C. Schroter, Atheroma and arterial wall shear. Observation, correlation and proposal of a shear dependent mass transfer mechanism for atherogenesis, Proc. R Soc. Lond. B Biol. Sci. 177 (1971) 109.

[9] C.G. Caro, J.M. Fitz-Gerald, R.C. Schroter, Proposal of a shear dependent mass transfer mechanism for atherogenesis, Clin. Sci. 40 (1971) 5P.

[10] S. Chien, S. Usami, H.M. Taylor, J.L. Lundberg, M.I. Gregersen, Effects of hematocrit and plasma proteins on human blood rheology at low shear rates, J. Appl. Physiol. 21 (1966) 81.

[11] K. Dai, H. Xue, R. Dou, Y.C. Fung, On the detection of messages carried in arterial pulse waves, J. Biomech. Eng. 107 (1985) 268.

[12] M.D. Deshpande, D.P. Giddens, R.F. Mabon, Steady laminar flow through modelled vascular stenoses, J. Biomech. 9 (1976) 165.

[13] M.H. Friedman, O.J. Deters, Correlation among shear rate measures in vascular flows, J. Biomech. Eng. 109 (1987) 25.

[14] D.L. Fry, D.M. Griggs Jr., J.C. Greenfield Jr., Myocardial mechanics: tension–velocity–length relationships of heart muscle, Circ. Res. 14 (1964) 73.

[15] Y.C. Fung, S.Q. Liu, Elementary mechanics of the endothelium of blood vessels, J. Biomech. Eng. 115 (1993) 1.

[16] D.N. Granger, G.W. Schmid-Schoenbein, Physiology and Pathophysiology of Leukocyte Adhesion, American Physiological Society, Oxford University Press, New York, 1995.

[17] A.C. Guyton, C.E. Jones, Central venous pressure: physiological significance and clinical implications, Am. Heart J. 86 (1973) 431.
[18] W.F. Hamilton, P. Dow, An experimental study of the standing waves in the pulse propagated through the aorta, Am. J. Physiol. 125 (1939) 48.
[19] E. Jones, M. Anliker, I.D. Chang, Effects of viscosity and constraints on the dispersion and dissipation of waves in large blood vessels. II. Comparison of analysis with experiments, Biophys. J. 11 (1971) 1121.
[20] E. Jones, M. Anliker, I.D. Chang, Effects of viscosty and constraints on the dispersion and dissipation of waves in large blood vessels. I. Theoretical analysis, Biophys. J. 11 (1971) 1085.
[21] A. Kamiya, T. Togawa, Adaptive regulation of wall shear stress to flow change in the canine carotid artery, Am. J. Physiol. 239 (1980) H14.
[22] A. Kamiya, R. Bukhari, T. Togawa, Adaptive regulation of wall shear stress optimizing vascular tree function, Bull. Math. Biol. 46 (1984) 127.
[23] T. Karino, M. Motomiya, Flow through a venous valve and its implication for thrombus formation, Thromb. Res. 36 (1984) 245.
[24] G.S. Kassab, K. Imoto, F.C. White, C.A. Rider, Y.C. Fung, C.M. Bloor, Coronary arterial tree remodeling in right ventricular hypertrophy, Am. J. Physiol. 265 (1993) H366.
[25] G.S. Kassab, Y.C. Fung, Topology and dimensions of pig coronary capillary network, Am. J. Physiol. 267 (1994) H319.
[26] G.S. Kassab, Y.C. Fung, The pattern of coronary arteriolar bifurcations and the uniform shear hypothesis, Ann. Biomed. Eng. 23 (1995) 13.
[27] A.L. King, Elasticity of the aortic wall, Science 105 (1947) 127.
[28] D.N. Ku, D.P. Giddens, C.K. Zarins, S. Glagov, Pulsatile flow and atherosclerosis in the human carotid bifurcation. Positive correlation between plaque location and low oscillating shear stress, Arteriosclerosis 5 (1985) 293.
[29] M. Landowne, Characteristics of impact and pulse wave propagation in brachial and radial arteries, J. Appl. Physiol. 12 (1958) 91.
[30] J.A. Maxwell, M. Anliker, The dissipation and dispersion of small waves in arteries and veins with viscoelastic wall properties, Biophys. J. 8 (1968) 920.
[31] E.W. Merrill, E.R. Gilliland, G. Cokelet, H. Shin, A. Britten, R.E. Wells Jr., Rheology of human blood, near and at zero flow. Effects of temperature and hematocrit level, Biophys. J. 3 (1963) 199.
[32] G. Minetti, P.S. Low, Erythrocyte signal transduction pathways and their possible functions, Curr. Opin. Hematol. 4 (1997) 116.
[33] E. Monos, V. Berczi, G. Nadasy, Local control of veins: biomechanical, metabolic, and humoral aspects, Physiol. Rev. 75 (1995) 611.
[34] C.D. Murray, The physiological principle of minimum work applied to the angle of branching of arteries, J. Gen. Physiol. 9 (1926) 835.
[35] C.D. Murray, The physiological principle of minimum work. II. Oxygen exchange in capillaries, Proc. Natl. Acad. Sci. USA 12 (1926) 299.
[36] C.D. Murray, The physiological principle of minimum work. I. The vascular system and the cost of blood volume, Proc. Natl. Acad. Sci. USA 12 (1926) 207.
[37] C.F. Notarius, S. Magder, Central venous pressure during exercise: role of muscle pump, Can. J. Physiol. Pharmacol. 74 (1996) 647.
[38] M.J. Thubrikar, S.K. Roskelley, R.T. Eppink, Study of stress concentration in the walls of the bovine coronary arterial branch, J. Biomech. 23 (1990) 15.
[39] J.R. Womersley, Oscillatory flow in arteries: effect of radial variation in viscosity on rate of flow, J. Physiol. 127 (1955) 38.
[40] K.K. Wu, Platelet activation mechanisms and markers in arterial thrombosis, J. Intern. Med. 239 (1996) 17.
[41] M. Zamir, Shear forces and blood vessel radii in the cardiovascular system, J. Gen. Physiol. 69 (1977) 449.

PART III

MICROCIRCULATION

CHAPTER

6

Microvascular Beds

LEARNING OUTCOMES

1. Contrast capillary physiology versus arterial and venous physiology
2. Describe the important structures within a capillary bed
3. Examine mass transfer across the capillary wall
4. Discuss endothelial cell physiology
5. Discuss smooth muscle cell physiology
6. Identify the mechanisms for information transfer between endothelial cells and smooth muscle cells
7. Assess the theories for local control of blood flow
8. Model the pressure distribution throughout microvascular beds
9. Model the velocity distribution throughout microvascular beds
10. Develop the Stokes equation to describe flow through microvascular beds
11. Analyze the drag force on particles immersed within a fluid
12. Discuss the relevant physiology of the interstitial space
13. Formulate a relationship between flow through the capillary wall and the pressure within the capillary and the interstitial space
14. Compare the hematocrit in capillaries to larger blood vessels
15. Describe the Fahraeus–Lindquist effect
16. Illustrate the plug flow principle in capillaries
17. Model the accumulation of species within the microcirculation
18. Analyze blood flow within capillaries as a two-phase fluid
19. Describe the interactions of blood cells with the endothelial cell wall
20. Explain disease conditions that affect microvascular beds

© 2015 Elsevier Inc. All rights reserved.

6.1 MICROCIRCULATION PHYSIOLOGY

Throughout the first two sections of this textbook, we have restricted our discussion of fluid mechanics to flows that remain within their particular container (e.g., large blood vessel). This means that mass continuity only includes inflows, outflows, changes in the container volume, or changes in the fluid density. However, in the microcirculation, there is transport of fluid and molecules out of the blood vessel into the interstitial space and vice versa across the wall of the container. One may consider that we should just add a lumped sum inflow and outflow parameter to include the flow across the vessel wall. However, to model this phenomenon, we typically do not add a discrete inflow or outflow along the vessel wall (although this is possible); permeability concepts will need to be included in the formulation. The benefit of this type of approach is that permeability as a function of time and space (along the vascular wall) can now be easily included into the analysis, instead of a single permeability value. We will see in later sections of this chapter that the permeability through the vascular wall is not constant. This transport across the vascular wall is arguably the most crucial function that the entire cardiovascular system performs. Without this function, few cells within the human body would be able to obtain oxygen or glucose, and they would not be able to remove cellular wastes from their local environment.

Microvascular beds are composed of capillaries, which are small thin blood vessels. In fact, the vascular wall of capillaries is composed of a single layer of endothelial cells without any substantial connective tissue or smooth muscle cell layer. These cells are highly permeable and partially regulate the transport of molecules across the vascular wall. It has been estimated that there are on the order of 35 billion capillaries within the systemic circulation, with a total surface area of approximately 2000 m^2, all of which are available for molecular transport/exchange. Nearly all cells are within 50 μm of a capillary, which is within the free diffusion distance for all essential nutrient transportation (this will be discussed more in Chapter 7). Cells outside of this range would require some form of facilitated (and possibly active) transport to obtain nutrients and remove wastes. Clearly, the efficiency of transport would decrease significantly if this was the case and a significant percentage of metabolic energy would go to these transport functions.

At the point where the arterial circulation has branched into arterioles that are on the order of 10–15 μm in diameter, the regulation of the blood flow through these vessels becomes critically important to the capillary beds. Vessels within this diameter range are typically called metarterioles (or terminal arterioles). These are the last blood vessels that have a smooth muscle cell layer prior to blood entering the capillary bed (most capillaries have an internal diameter ranging between 5 and 8 μm). At the point where a metarteriole diverges into a capillary, there is one last smooth muscle cell that encircles the blood vessel. This smooth muscle cell is termed the precapillary sphincter. Like all sphincters, it closes the lumen of the particular vessel (or channel) when it constricts and opens the lumen when it dilates. Therefore, in the case of precapillary sphincter constriction, the resistance to flow through the particular downstream capillary network increases significantly and the pressure gradient typically cannot overcome this resistance to flow. Therefore, in the microcirculation when a precapillary sphincter constricts, blood is

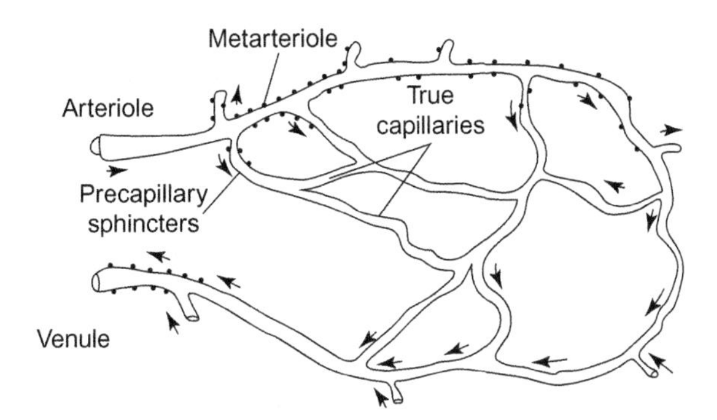

FIGURE 6.1 Typical structure of a capillary bed, which is fed by an arteriole. Metarterioles are the last vessels on the arterial side that have a smooth muscle layer. These vessels highly regulate the flow through the microvascular beds. The last endothelial cell associated with a smooth muscle cell is termed the precapillary sphincter, and it is this cell that determines blood flow through capillaries. True capillaries are defined with divergent flow at their inlet and convergent flow at their outlet. Capillaries feed into venules. *Adapted from Guyton and Hall (2000).*

shunted from the capillary bed downstream of that sphincter to a different microvascular bed because the resistance to flow through the constricted sphincter is too large to permit blood flow. Once blood has passed through a capillary, blood vessels begin to converge into venules, making their way back to the heart via the inferior and superior vena cava. This leads to an important definition. A true capillary is defined by divergent flow at its inlet and convergent flow at its outlet (Figure 6.1).

It is important to note here (although we will discuss this more fully in Section 6.3) that the metarterioles (and hence the precapillary sphincters) are in a close proximity to the capillaries themselves. This allows for the local changes in nutrient concentrations (oxygen, glucose, among others) or waste concentrations (carbon dioxide, metabolites) within the downstream capillary bed to partially regulate the dilation and constriction of the precapillary sphincter. Due to the closeness of the capillaries and the precapillary sphincters, the "needs" of the tissue can rapidly, within a few seconds, be addressed. There is no large demand for an overarching "command center" for every capillary bed within the cardiovascular system and furthermore, the response time of such a center would potentially be too large to accommodate tissue needs.

There are four common types of capillary bed organizations in the cardiovascular system and they are classified by their ability to compensate for arterial or venous blockages. The supply to the tissue can be defined as either independent or overlapping (Figure 6.2). Independent arrangements of capillary beds suggest that one and only one capillary bed supplies one tissue region with blood, whereas an overlapping arrangement would have multiple capillary beds supplying one tissue region with blood. In either of these arrangements, the terminal arterioles and/or postcapillary venules may be disconnected or connected with each other. Disconnected systems do not allow mixing or shunting of blood between the microcirculation regions whereas connected systems, also termed redundant systems, allow for mixing of blood between microcirculation regions. In the redundant system, the vessels are said to be anastomotic and this special redundant circulation is termed the collateral circulation. Each of these systems are viable options to perfuse the

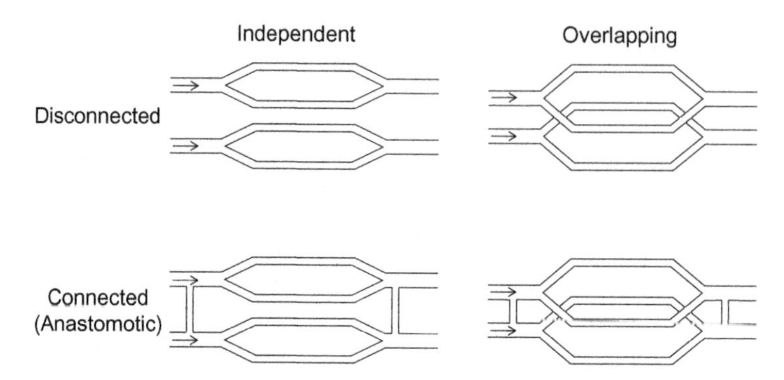

FIGURE 6.2 Schematic of different arrangements of capillary networks. Capillary networks can be independent of neighboring networks or they can overlap neighboring networks. Additionally, neighboring networks can be disconnected or connected (termed anastomotic). Each of these organizations have advantages and disadvantages, which include redundancy (for anastomotic and overlapping networks, but for different reasons) or ease of downstream control for independent and disconnected networks. Most tissues have a combination of these different types of networks. Note that the collateral circulation typically has little to no flow and is only used under pathological conditions.

tissue but there are some very important implications regarding the different designs of these systems. Comparing the extreme cases, we can see that a disconnected independent network is very susceptible to any damage upstream of the capillary network. For instance, an occlusion upstream would prevent any blood from reaching the downstream tissue. However, the control of flow to a downstream tissue region is very easy with these types of arrangements. Comparing to the other extreme case, it would be relatively difficult to control the tissue perfusion within an anastomotic overlapping system. However, these systems are not susceptible to upstream blockages. Each of these types of systems, which are depicted in Figure 6.2, can be found throughout the body.

Flow through capillary beds is typically not continuous. Depending on the tissue that the capillary bed is feeding, the flow can alternate between on and off every few seconds or every minute, in a process termed vasomotion. As stated above, the changes in flow are controlled by the dilation and constriction of the precapillary sphincters. Tissue oxygenation is the most critical factor that affects the local vasomotion. A tissue that is under hypoxic conditions for a few seconds will rapidly induce the precapillary sphincters to dilate. The interesting point about changes in vasomotion is that it is independent on the organization of the capillary arrangements that were shown in Figure 6.2; changes in capillary perfusion can alter the precapillary sphincter of any feeding arteriole. The most common cause for decrease in oxygen concentration is an increase in oxygen usage through a cellular respiration process. Under these conditions, the flow through that particular vascular network will be turned on for the majority of the time and it would only be turned off for very short periods of time, so that other tissue locations do not become hypoxic. This condition would persist until the tissue oxygen concentration returns back to normal levels. Once the oxygen levels within the capillary bed return to a normal level, the flow throughout the network returns to its normal on/off pattern.

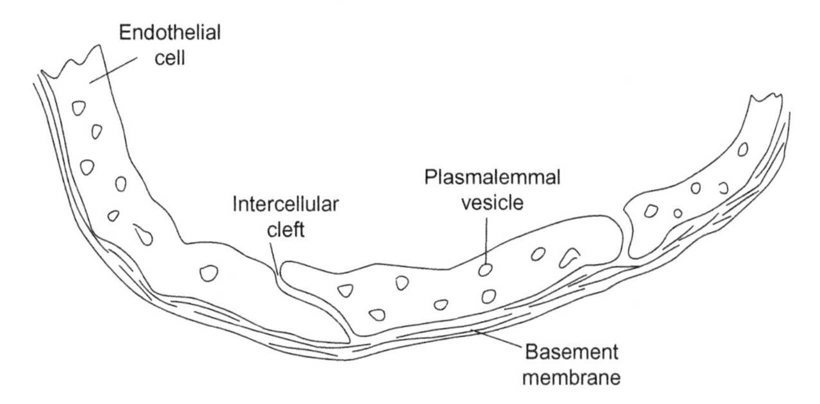

FIGURE 6.3 Cross-section of a capillary, showing possible transport mechanisms across the capillary wall. Recall that the capillary is composed of a single endothelial cell (in thickness) surrounded by a small basement membrane. There are typically two to three endothelial cells that form a capillary in circumference, but these cells do not overlap with each other in thickness. In between endothelial cells, there is an intercellular cleft which is responsible for the majority of transport between the blood and the extravascular space. Plasmalemmal vesicles are responsible for the transport of larger molecules, however, this is a slow transport mechanism. *Adapted from Guyton and Hall (2000).*

As stated above, flow through the capillary is not the only flow that occurs. There is also a flow of molecules across the endothelial cell wall; nutrients typically leave the blood vessel, while metabolic wastes typically enter the blood vessel. The capillary wall is composed of a single endothelial cell and has a thickness on the order of 0.5 μm. The extravascular side of the endothelial cell is directly connected to the extracellular matrix of the tissue. In most capillary beds, there are two major methods for molecular transport across the endothelial cell barrier: the intercellular cleft and plasmalemmal vesicles (Figure 6.3). Both of these methods will be discussed in more detail throughout the remainder of this section. There are some exceptions to these transport mechanisms, in specialized capillary beds, such as the fenestrated capillaries of the glomerulus or the continuous capillaries within the brain microvasculature.

The intercellular cleft is the location where two neighboring endothelial cells adhere to each other. Endothelial cells do not come directly into contact with each other, leaving a gap of about 10–15 nm in thickness. This endothelial cell–endothelial cell adhesion is primarily carried out by gap junctions, cadherins, and other cell membrane adhesion proteins. These clefts represent a very small proportion of the total capillary surface area (about 1/1000 of the total available surface area), but due to the permeability of nutrients across this barrier, some of the important transported molecules can diffuse very rapidly through the intercellular cleft. Furthermore, flow through the intercellular cleft is approximately 80 times faster than blood flow through the capillary itself, and therefore, a rapid equilibrium is reached between the blood and the interstitial fluid. This type of intercellular cleft is present in most tissues. Two exceptions are the brain and the liver, where the intercellular cleft is much smaller and much larger, respectively. In the brain, these tight junctions only allow small molecules to pass (such as O_2, glucose), preventing the entrance of potential dangerous and unwanted compounds into the brain tissue. In the liver, the pores are so large that even plasma proteins can pass through them. Because the liver partially acts to detoxify blood, it requires access to all of the blood components. In general, only metabolically important compounds are transported through the intercellular cleft, except for within the liver capillary and other fenestrated capillary beds (refer to Chapter 13).

The second method of transport across the endothelial cell is via plasmalemmal vesicles. These vesicles form on both the vascular side and the interstitial side of the endothelial cell via a cellular invagination process. Through this process, fluid and molecules are trapped within the vesicle and then are slowly transported to the opposing endothelial cell surface. Once the plasmalemmal vesicles reach the opposing endothelial cell surface, they fuse with the cell membrane and release the contents into the vascular space or the interstitial space. This is clearly not an effective method to transport nutrients/wastes to every cell within the body because it is slow and requires a great amount of energy to (i) form vesicles and (ii) transport vesicles across the cell. However, larger molecules that cannot fit through the intercellular cleft are transported with this mechanism. Additionally, the efficiency of this system is relatively low as compared with transport through the intercellular cleft.

6.2 ENDOTHELIAL CELL AND SMOOTH MUSCLE CELL PHYSIOLOGY

As already discussed, endothelial cells compose the interior lining of all blood vessels within the body. These cells can partially regulate the transport of nutrients across the cell and play a very critical role in coagulation. Endothelial cells join to form a sheet with one of three different structures. Continuous endothelial cells are found in arteries, veins, and capillaries throughout the body. The transport of nutrients across these endothelial cell sheets is regulated by the type and quantity of junctions between neighboring cells. Primarily, cadherin junctions mediate cell–cell adhesion and gap junctions regulate the communication between cells. Endothelial cells are typically 10 μm in diameter and depending on their location within the circulatory system, their length can vary from 50 to 200 μm. Continuous endothelial cell capillary beds are by far the most commonly found in the body. A subset of continuous endothelial cells are endothelial cells that essentially prevent the free diffusion of molecules between the blood vessel and the extravascular space. These types of capillaries are primarily found in the vasculature that surrounds the brain and helps to restrict the movement of molecules into the central nervous system.

The second type of endothelial cell sheets is the discontinuous endothelial sheet. These are endothelial cells that are not tightly connected to each other. This allows for a free diffusion around the endothelial cells and is typically present in the liver and spleen. Most large molecules including proteins can pass through this type of endothelial cell sheet. The third type of endothelial cell sheet is the fenestrated endothelial sheet. These types of endothelial cell sheets are located within the kidney's glomerulus and have similar intercellular connections as continuous endothelial cells. However, there are channels that span throughout the entire cells, which effectively expose the extracellular space to the blood vessel lumen. These channels facilitate the movement of large molecules across the endothelial cell wall. In the glomerular capillaries, the podocytes and the slit membrane are primarily responsible for restricting the diffusion of molecules between the blood vessels and the nephron tubules.

VE-cadherin is the primary cadherin located on the endothelial cell membrane. This protein is expressed as a dimer of two cadherin proteins. Each protein passes through the cell membrane and has an intracellular portion that is connected to the cytoskeleton. This strong cytoskeletal connection allows for force transmittance through the endothelial cell. The association of these proteins is dependent on the local calcium concentration. With calcium concentrations lower than approximately 0.5 mM, each cadherin dimer dissociates and then the cadherins cannot make connections to other cells or the cytoskeleton. When the calcium concentration increases, cadherin monomers associate into dimers and can then adhere to a cadherin dimer from a neighboring cell. The role of cadherin is to connect cells together and to mechanically couple the endothelial cell sheet with the intracellular load-bearing components.

Vascular smooth muscle cells perform work in a similar way that skeletal muscles perform work, except that the loads they transmit are typically orders of magnitude smaller than the skeletal muscle loads. Smooth muscle cells contract when actin and myosin form a cross-bridge. Similar to skeletal muscle, calcium regulates cross-bridge formation in vascular smooth muscle cells. However, the internal arrangement of proteins within the smooth muscle cells is vastly different from the ubiquitous sarcomere structural unit seen in skeletal muscle cells. The size of vascular smooth muscle cells is also smaller than skeletal muscles. They typically range from 1 to 5 μm in diameter and approximately 100 μm in length.

Neighboring vascular smooth cells form one structural unit which contract or dilate as one large muscle "fiber." The cell membranes of these cells are connected to each other via cadherin junctions (similar to endothelial cells) and gap junctions. This allows for the uniform force generation across smooth muscle cell sheets. Gap junctions are also present within the sheet of smooth muscle cells to electrically couple neighboring cells. Gap junctions allow for the free movement of ions between cells. Similar to the cardiac muscle cells, this allows for the rapid transmittance of the action potential between cells. Interestingly, there is a second group of specialized gap junctions that allow for the free communication between endothelial cells and smooth muscle cells.

Briefly, there are many similarities between smooth muscle and skeletal muscle contraction. However, within smooth muscle cells, actin and myosin filaments are able to interact without the troponin/tropomyosin complex, which regulates skeletal muscle contraction. Smooth muscle cell contraction is regulated by calcium and adenosine triphosphate (ATP) concentration in a slightly different way than skeletal muscle. Also, instead of a sarcomere subunit, smooth muscle cells have a more disorganized structure. Actin filaments are directly connected to dense bodies, which are typically located within the cell membrane of the smooth muscle cell (some dense bodies are located sporadically throughout the cytoplasm). Myosin filaments are randomly dispersed throughout the cytoplasm, potentially coming in contact with the actin filaments (Figure 6.4). Cell membrane-associated dense bodies of one smooth muscle cell can adhere to cell membrane-associated dense bodies of a neighboring smooth muscle cell, which allows for force transmittance throughout the vascular smooth muscle sheet. Due to this random arrangement, there is no organized contraction direction within the smooth muscle cell; however, the extent of

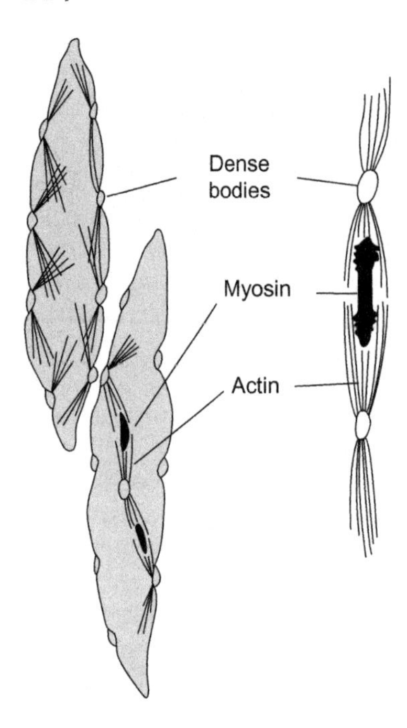

FIGURE 6.4 Structure of a smooth muscle cell, with an inset showing the relationship between actin, myosin, and the dense bodies. Adjacent smooth muscle cells are connected by dense bodies, which are also interspersed throughout the cells cytoplasm. The actin and the myosin have a much more random orientation than skeletal or cardiac muscle cells. *Adapted from Guyton and Hall (2000).*

connections between smooth muscle cells and how those connections are organized can dictate the contraction direction.

To initiate smooth muscle cell contraction, intracellular calcium ion concentration increases, in response to nervous input, humoral signals, hormonal signals, and mechanical stretch. Once there is a sufficient calcium ion concentration within smooth muscle cells, calmodulin activates through the binding/association of calmodulin with four calcium ions. Calmodulin, instead of troponin, initiates myosin–actin cross-bridge formation via activating a separate protein, myosin kinase (calmodulin associates with myosin kinase, forming a protein–protein complex). One of the myosin protein chains becomes phosphorylated, through the action of myosin kinase, which is always active when calmodulin associates with four calcium ions. This allows myosin to constantly form cross-bridges with actin, if sufficient ATP is present. To end this process, a second enzyme, myosin phosphatase, removes the phosphate group from myosin and thus myosin can no longer form cross-bridges with actin. As long as there is a high enough concentration of calcium within the vascular smooth muscle cell, calmodulin and myosin kinase remain active and therefore myosin is always in a phosphorylated state. However, once the calcium concentration drops below a critical value, calmodulin dissociates from calcium and myosin kinase, thus causing myosin kinase to become inactive. Therefore, under these conditions, actin cannot form cross-bridges with myosin because myosin phosphatase is the dominate enzyme. There is a quite complex signaling pathway that occurs to transmit an extracellular signal into increase in intracellular calcium concentration. The following section deals with the control of intracellular calcium in response to various external signals.

6.3 LOCAL CONTROL OF BLOOD FLOW

The basis for the local control of blood flow rests with the needs of the tissue that the capillary network supplies. This means that every tissue in the body can have a slightly different requirement at one instant in time and also that this requirement can change over time based on the tissue's energy usage. For instance, how much blood does a stomach need immediately before eating Thanksgiving dinner (assuming you have not smelled the food cooking yet) as compared to immediately after eating Thanksgiving dinner? Would this be different? In general, the answer to the question is yes, the needs of the stomach would be different before and after any meal. For the example of the gastrointestinal tract, any nutrients that have already been broken down in the stomach and need to be absorbed into the cardiovascular system will occur after eating a meal and not before eating a meal. In the stomach, the motion of the stomach and the secretion of gastric juice accelerates after eating a meal and thus the blood flow would need to accommodate these changes. This is one of the reasons why it is easier to get muscle cramps during a workout immediately after consuming a meal. Under these conditions, a certain percentage of the blood is directed to the digestive system, and therefore, there is potentially not enough blood to supply the muscular system with the nutrients (and cannot remove the metabolic wastes) it needs to function. Blood flow to any particular tissue is normally maintained at a level just above the minimum requirements for that tissue. The most critical nutrient for most tissue is oxygen, and therefore, this minimum requirement would be based on the level of oxygen needed to maintain a normal tissue oxygenation level, but not to exceed the normal tissue oxygenation level (this would be hyperoxic conditions). If all tissues in the body were constantly supplied with blood to significantly exceed its particular tissue oxygenation level, the amount of blood needed would exceed the work capacity of the heart and nutrient delivery would not be effective. Also, as tissues do not store oxygen, most of the excess oxygen would go to "waste" and would be expelled out through the lungs or potentially diffuse out of the skin. Therefore, under normal conditions, the blood supply to all tissues meets the minimum requirements of the tissue and no more than that.

In general, there are two mechanisms to control blood flow to tissues. The first mechanism is an acute local control, which accounts for rapid constriction or dilation of the blood vessels. The second control system is a long-term control, which leads to changes in the number of blood vessels and/or the resting size (diameter/length) of the blood vessel. We will restrict our discussion to the rapid acute control mechanism at this point, since the mechanism(s) involved in the long-term control are fairly complex and there are a significant amount of control systems that are involved with this process.

As mentioned earlier, the first major regulator of blood flow is the tissue oxygenation level. When the quantity of oxygen in the blood or the quantity of oxygen within the tissue reduces, there is a noticeable increase in the blood flow to that tissue. The rate of blood flow increase is almost inversely proportional to the percentage of oxygen within the arterial blood. At a level of 50% of the normal oxygen concentration, the blood flow increases by approximately 2 times. At a level of 25% of the normal oxygen concentration, the blood flow increases by slightly more than 3 times. Be cautioned though that this is not a linear relationship. The first possible mechanism that regulates increased blood flow under decreased arterial oxygen concentration is an increased vascular adenosine concentration to act as a vasodilator (other vasodilators such as carbon dioxide, nitric oxide, and hydrogen ions, among others, can follow a similar mechanism, but currently it is unknown which of these vasodilators is the most important). Due to a decreased availability of oxygen, the surrounding tissues release adenosine (or other dilators) into the bloodstream. These dilators cause the local vascular smooth muscle cells to relax and effectively increase blood flow to that tissue by increasing flow through the capillary network. One problem with this theory is that the vasodilator which is released into the blood capillaries would have to diffuse against the blood flow (i.e., upstream) to the arterioles/precapillary sphincters to have a substantial effect on blood flow changes. Therefore, this mechanism cannot account for all of the vascular changes seen in response to low tissue oxygenation.

A second common thought is that low oxygen concentration itself can directly induce changes in the diameter of blood vessels. There is some evidence for this mechanism as well, but the same problem about downstream sensing receptors and the communication to the upstream vascular smooth muscle cells is associated with this theory. However, a better theory would include some combination of these mechanisms. Some microcirculatory research groups have experimental results that suggest that there is a rapid communication between the endothelial cells that form the capillaries and those that line the metarterioles. This communication exists through the presence of gap junctions between cells. In a similar manner to the above theory, when oxygen (or other nutrients) concentration decreases within the tissue supplied by the capillary beds, by-products of cellular respiration or other vasoactive compounds can be formed within endothelial cells. These compounds can then be transported up through the vascular network to the endothelial cells that comprise the lining of the metarterioles, via intercellular channels such as gap junctions. With this communication mechanism, there would be no problem associated with movement against the blood flow direction. Once these compounds have made it to the metarteriole location, they can communicate with the precapillary sphincters to induce constriction or dilation. It is known that there are communicating junctions between endothelial cells and vascular smooth muscle cells that help to regulate changes in the diameter of the blood vessel. However, we also know that vasoactive compounds are released into the bloodstream. Therefore, neither theory represents the physiology completely, but as stated above, it is likely that some combination of these theories would be responsible for changes to the caliber of the blood vessels.

There are two other theories about local blood flow regulation that are based on the pressure changes that can occur within the blood vessel. Imagine an increase in arterial blood pressure. This would be accompanied with an increase in blood flow to the tissue, causing an increased oxygen/nutrient delivery to the tissue. It has been experimentally shown that within a minute, this increase in blood flow returns back to normal levels in a process termed autoregulation of blood flow. The first theory, the metabolic theory, proposes that an increased oxygen/nutrient delivery causes constriction of the blood vessels in much the same way that was discussed previously. Increased oxygen (or nutrients) causes an increase in cellular by-products to be released within the vascular network (or within endothelial cells) which directly communicates with precapillary sphincters, causing a constriction. The second theory, the myogenic theory, suggests that it is not cellular compounds that regulate blood flow. Instead, the constriction of blood vessels is caused by the pressure-induced stretch of the blood vessels. It has been observed that after blood vessels experience a small sudden stretch, the vascular smooth muscle cells start to constrict to counter the stretch. Also, when the pressure reduces in the blood vessels, the diameter suddenly reduces, causing a relaxation of the vascular smooth muscle cells to counter the decreased pressure forces. This relaxation may be due to mechanical factors to counter the reduced blood vessel diameter and/or changes in the release of local metabolically active compounds. Either of these changes has the net effect of bringing the vessel back to its normal resting state. Again, it is likely that some combination of all of these mechanisms is responsible for the local control of blood vessel diameter.

We will now briefly discuss the humoral regulation of blood vessel diameter. Three of the most potent vasoconstrictors are vasopressin, angiotensin, and norepinephrine. Vasopressin works in conjunction with the kidneys to increase water absorption by decreasing the flow through the circulatory system. Angiotensin acts to increase the vascular resistance of all of the arterioles within a specific tissue upon release. This effectively decreases blood flow through the tissue. During stress situations, norepinephrine is released and can partially regulate blood flow to areas of need by shunting blood away from areas with a lower need. Two of the most potent vasodilators are the superfamily of kinins and histamine. Both of these compounds have direct roles in the inflammatory response by increasing vascular flow to an inflamed region. Also, these compounds increase the permeability of the capillaries to allow for the rapid extravasation of white blood cells.

6.4 PRESSURE DISTRIBUTION THROUGHOUT THE MICROVASCULAR BEDS

The pressure gradient across microvascular beds has been shown to be one of the most critical fluid parameters in regulating the flow throughout these vessels (the fluid velocity is so slow in these vessels that small changes in the pressure can cause large changes in the flow conditions, including velocity, shear stress, and shear rate). The pressure gradient across microvascular beds has been experimentally measured using two flow probes inserted into various branches along the network. By knowing the distance between these probes, and the pressure readings obtained from the probes, a pressure gradient can be calculated. As the diameter of the blood vessel decreases through the vascular network (e.g., arterioles to capillaries) the pressure gradient significantly increases. As the diameter of the blood vessel increases (capillaries to venules) the pressure gradient reduces again. This can be explained by the rapid changes in hydrostatic pressure within the precapillary arterioles, caused by constriction or dilation of the precapillary sphincter.

In contrast, the hydrostatic pressure within the arteriolar side is relatively constant until the blood vessels approach a diameter of approximately 40 μm. Therefore, the pressure gradient throughout these vessels is relatively low (refer to Figure 5.26 which shows that the mean arterial pressure is relatively stable in vessels larger than capillaries on both the arterial and venous side of the circulation). Vessels with a diameter in the range of 40 μm would typically be found one to two bifurcations upstream of the metarteriole/precapillary sphincters. For arterioles in the range of 15–40 μm in diameter, there is a rapid decrease in pressure to approximately 30 mmHg, which is associated with a rapid increase in the pressure gradient throughout the vessel. This rapid decrease occurs so that the blood velocity is slow enough for nutrient and waste exchange to occur within the capillaries at the same time that blood is directed rapidly into the capillaries. Observe that the gauge mean arterial pressure variation in capillaries and venules is much lower than that seen in the precapillary arterioles (see Figure 5.26). Within capillaries (diameter ranging from 5 to 10 μm) the pressure decreases from approximately 25 mmHg to at most 20 mmHg, under normal conditions. However, the pressure gradient driving blood into the capillaries is relatively large so that blood movement through these vessels is efficient. Within the venous circulation, the pressure continually drops (to approximately 0 mmHg in the right atrium), but again, it is much more gradual, taking the entire length of the venous system. Therefore, the pressure gradient is much lower within the venules/venous system. In postcapillary venules (diameter as large as 50 μm), the pressure is no more than 15 mmHg under normal conditions.

To continue the discussion of the pressure gradient throughout microvascular beds, there is approximately an eightfold increase in the pressure gradient within small capillary segments (100–300 μm length) as compared to arterioles and venules (approximately 2000 μm length, approximately 40 μm diameter). For metarterioles and postcapillary venules (approximately 15 μm diameter), the pressure gradient is 50% of the pressure gradient throughout the capillaries. This suggests that the flow is directed into the capillaries and then it slows to allow sufficient time for nutrient exchange. Recall that the gradient may be large, but flow will be diverted into many small capillaries to perfuse the entire vascular bed.

The pressure variation in microvascular beds under hypertensive and hypotensive conditions has also been investigated. Interestingly, under both conditions, the mean hydrostatic pressure within the capillaries as well as the pressure gradient across the capillaries was equivalent to that seen under normal conditions. Also, the pressure (hydrostatic and pressure gradient) within the postcapillary venules was the same under these conditions as under normal conditions. The major change was observed within the arteriolar vasculature where the pressure gradient was significantly greater under hypertensive conditions or significantly lower under hypotensive conditions. This suggests that metarterioles (and the precapillary sphincters) regulate capillary blood flow to maintain it at its normal levels, so that nutrient exchange is maintained at the optimal level. This also suggests that the circulatory system is designed to maintain a constant flow through the microcirculation, independent of the mean arterial pressure. This is fairly significant and can be considered as a mechanism to dampen out any pressure variations prior to the site of exchange within the vasculature.

The last major variation in pressure throughout the microvascular beds is based on the temporal changes from the cardiac pressure pulse and the wave propagation throughout the vasculature. The pressure values discussed so far have been averaged across the cardiac cycle, by taking multiple readings during the entire cardiac cycle. However, during the cardiac cycle, the hydrostatic pressure changes within the microvascular beds oscillating on the order of 2 mmHg. We previously stated that the hydrostatic pressure within the capillaries is approximately 25 mmHg. This means that the pressure would vary between approximately 24 and 26 mmHg during the cardiac cycle. These temporal changes are not that significant when compared with the relatively stable pressure gradient across the blood vessels. Most of the data that has been discussed in this section was collected by B. Zweifach and his research group during the 1970s.

6.5 VELOCITY DISTRIBUTION THROUGHOUT THE MICROVASCULAR BEDS

In large blood vessels, where the vessel diameter exceeds the diameter of cells within the blood by at least three times, the mean blood velocity is approximately equal to the mean velocity of the cells. This is because the cells tend to spread evenly throughout the cross-section of the blood vessel (imagine the cell distribution across a cross-section of the aorta). However, as the blood vessel diameter approaches the diameter of the suspended cells, this is not necessarily true anymore. The mean blood cell velocity within blood vessels that have a diameter in the range of 15–25 μm is higher than the mean blood velocity. This is caused by a phenomenon where the blood cells are pushed to the centerline of the blood vessel. As discussed previously, under single-phase flow the fully developed parabolic flow will have a centerline velocity that is greater than the mean blood velocity. However, in these small blood vessels the flow profile will be blunted (similar to the Casson model velocity profile) because the blood cells account for a significant portion of the cross-sectional area (this will be discussed further in Section 6.7). As the blood vessel diameter reduces to less than 15 μm, the mean blood cell velocity and the mean blood velocity approach the same value again. The blood vessels that have different mean cell velocities and mean plasma velocities are found within the microvascular beds.

The change in the relationship between the velocity of blood cells and the velocity of the blood occurs because the blood cells account for the majority of the cross-sectional area of the blood vessel. In other words, the blood vessel diameter reduces to the cellular diameter, in the smallest of capillaries. When this occurs, the blood cells tend to plug the vessels and dominate the entire velocity profile. What typically occurs is that red cells group together in packets termed rouleaux and pass through the capillary as one. This is then followed by a packet of plasma, which passes through the capillary at the same velocity as the cells. This again would be followed by a red cell packet (which would more than likely be of a different size than the first packet). However, independent of red blood cell group size, the velocity of each cell/fluid packet will be the same. In larger vessels, when the blood vessel diameter is approximately 15 μm, the red blood cells tend to move toward the centerline, as will be discussed in Section 6.7. In these 15 μm vessels, plasma flows along the walls of the blood vessel with a slow velocity, at the same time that the red blood cells flow along the centerline of the vessel. Therefore, the red blood cell velocity would have an overall higher value than (i) the mean blood velocity and (ii) the plasma velocity. As the blood cells enter the smallest capillaries from these 15 μm vessels, the plasma is squeezed against the wall, and eventually, it leaks back through this narrow gap between the cells and the vessel wall. This leak-back plasma is collected at one location, leaving a gap between the stream of red cell packets. Red cells that are following this first packet of cells also push the plasma back between the wall and cells, which again causes red cells to congregate into a single-file packet followed by a plasma packet (there will be some plasma mixing with cells, but this is a relatively minimal percent of the total mass). Another way to think of this is with the no-slip boundary condition.

Because the cells stream toward the centerline, effectively pushing the plasma toward the walls, the plasma will have a velocity that is close to zero. From the cellular point of view, this looks like the plasma is being pushed backward, but in reality the plasma is just holding relatively steady along the wall. Eventually enough of this plasma builds up to break off the wall and cause a break in the stream of red cells, which forms the plasma packet.

The spatial difference in velocity in the radial direction is relatively negligible in capillaries because on average, the entire cell will travel with the same velocity. When a plasma portion moves through the capillary, it does not have sufficient time to fully develop and moves as a plug in-between two plasma packets. To model the flow of blood through these small vessels, it is typical to utilize the assumption that the fluid is invisicid and is composed of elastic particles. However, this will not accurately account for the portion of blood that solely consists of plasma, and the red blood cells do exhibit some viscous properties (the red blood cell membrane is a viscoelastic material). To develop a more accurate model, we will discuss fluid flow at low Reynolds numbers (the Reynolds number will be discussed in more detail in Chapter 14, with numerical methods, but it was developed earlier in the text; see Section 1.6). Without going into the complete details of the Reynolds number here, using typical blood flow properties found in the microcirculation, the Reynolds number is within the range of 0.001 and 0.01. At a Reynolds number in this range, the viscous effects of the blood flow dominate over the inertial effects (i.e., velocity). In this situation, the Navier–Stokes equations simplify to

$$\begin{aligned}
\frac{\partial p}{\partial x} &= \rho g_x + \mu\left(\frac{\partial^2 u}{\partial x^2} + \frac{\partial^2 u}{\partial y^2} + \frac{\partial^2 u}{\partial z^2}\right) \\
\frac{\partial p}{\partial y} &= \rho g_y + \mu\left(\frac{\partial^2 v}{\partial x^2} + \frac{\partial^2 v}{\partial y^2} + \frac{\partial^2 v}{\partial z^2}\right) \\
\frac{\partial p}{\partial z} &= \rho g_z + \mu\left(\frac{\partial^2 w}{\partial x^2} + \frac{\partial^2 w}{\partial y^2} + \frac{\partial^2 w}{\partial z^2}\right)
\end{aligned} \tag{6.1}$$

If gravitational effects can be ignored (which is probably a good assumption within the microvasculature because the length scales are relatively small), then Eq. (6.1) can be simplified even further to

$$\begin{aligned}
\frac{\partial p}{\partial x} &= \mu\left(\frac{\partial^2 u}{\partial x^2} + \frac{\partial^2 u}{\partial y^2} + \frac{\partial^2 u}{\partial z^2}\right) = \mu\nabla^2 u \\
\frac{\partial p}{\partial y} &= \mu\left(\frac{\partial^2 v}{\partial x^2} + \frac{\partial^2 v}{\partial y^2} + \frac{\partial^2 v}{\partial z^2}\right) = \mu\nabla^2 v \\
\frac{\partial p}{\partial z} &= \mu\left(\frac{\partial^2 w}{\partial x^2} + \frac{\partial^2 w}{\partial y^2} + \frac{\partial^2 w}{\partial z^2}\right) = \mu\nabla^2 w
\end{aligned} \tag{6.2}$$

Equation (6.2) shows the famous Stokes equations, which hold for relatively slow, viscous flows that are only pressure driven. It is important to emphasize that the validity of these equations is only for flows where the viscous terms dominate the inertial terms by at least 1 order of magnitude, or that the length scales are so small that the fluid inertial effects do not have time to overcome the viscous effects and that gravitational effects can be neglected. However, we will see in Chapter 9 that many gas flows (e.g., air flow in the lungs) and blood flow within the microcirculation can be characterized as a Stokes flow, because they obey the simplified Navier–Stokes equations shown in Eq. (6.2). The Stokes equations in cylindrical coordinates are

$$\frac{\partial p}{\partial r} = \mu\left[\frac{\partial}{\partial r}\left(\frac{1}{r}\frac{\partial}{\partial r}(rv_r)\right) + \frac{1}{r^2}\frac{\partial^2 v_r}{\partial \theta^2} - \frac{2}{r^2}\frac{\partial v_\theta}{\partial \theta} + \frac{\partial^2 v_r}{\partial z^2}\right]$$

$$\frac{\partial p}{\partial z} = \mu\left[\frac{1}{r}\frac{\partial}{\partial r}\left(r\frac{\partial v_z}{\partial r}\right) + \frac{1}{r^2}\frac{\partial^2 v_z}{\partial \theta^2} + \frac{\partial^2 v_z}{\partial z^2}\right] \tag{6.3}$$

$$\frac{\partial p}{\partial \theta} = r\mu\left[\frac{\partial}{\partial r}\left(\frac{1}{r}\frac{\partial}{\partial r}(rv_\theta)\right) + \frac{1}{r^2}\frac{\partial^2 v_\theta}{\partial \theta^2} + \frac{2}{r^2}\frac{\partial v_r}{\partial \theta} + \frac{\partial^2 v_\theta}{\partial z^2}\right]$$

Example

Consider the flow of a blood cell within the microcirculation (Figure 6.5). Determine the velocity profile within a 16-μm-diameter blood vessel with an 8-μm-diameter red cell (assume that it is cylindrical, Figure 6.5) centered within the blood vessel, flowing with a uniform velocity $v_{z\text{-}max}$. Assume that the viscosity of the blood is μ and that the pressure gradient only acts within the z direction and is $\frac{\partial p}{\partial z}$. Solve for the velocity profile at the blood cells centerline.

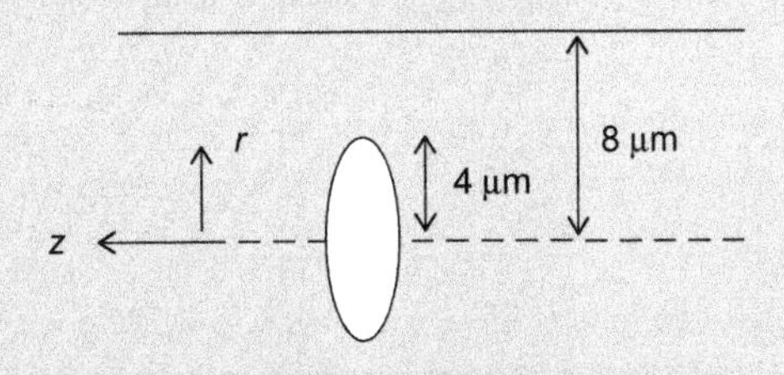

FIGURE 6.5 Schematic of a red blood cell flowing through a centerline. This is associated with the in-text example.

Solution

$$\frac{\partial p}{\partial z} = \mu\left[\frac{1}{r}\frac{\partial}{\partial r}\left(r\frac{\partial v_z}{\partial r}\right) + \frac{1}{r^2}\frac{\partial^2 v_z}{\partial \theta^2} + \frac{\partial^2 v_z}{\partial z^2}\right]$$

$$\frac{\partial p}{\partial z} = \frac{\mu}{r}\frac{\partial}{\partial r}\left(r\frac{\partial v_z}{\partial r}\right)$$

$$\int \frac{r}{\mu}\frac{\partial p}{\partial z} = \int \frac{\partial}{\partial r}\left(r\frac{\partial v_z}{\partial r}\right)$$

$$\frac{r^2}{2\mu}\frac{\partial p}{\partial z} + c_1 = r\frac{\partial v_z}{\partial r}$$

$$\int \frac{r}{2\mu}\frac{\partial p}{\partial z} + \frac{c_1}{r} = \int \frac{\partial v_z}{\partial r}$$

$$v_z(r) = \frac{r^2}{4\mu}\frac{\partial p}{\partial z} + c_1 \ln(r) + c_2$$

$$v_z(4\ \mu m) = v_{z\text{-max}}$$

$$v_z(8\ \mu m) = 0$$

$$v_{z\text{-max}} = \frac{(4\ \mu m)^2}{4\mu}\frac{\partial p}{\partial z} + c_1 \ln(4) + c_2$$

$$v_{z\text{-max}} = \frac{4\ \mu m^2}{\mu}\frac{\partial p}{\partial z} + c_1 \ln(4) + c_2 \tag{1}$$

$$0 = \frac{(8\ \mu m)^2}{4\mu}\frac{\partial p}{\partial z} + c_1 \ln(8) + c_2$$

$$0 = \frac{16\ \mu m^2}{\mu}\frac{\partial p}{\partial z} + c_1 \ln(8) + c_2 \tag{2}$$

Equation (1)–(2)

$$v_{z\text{-max}} = \frac{-12\ \mu m^2}{\mu}\frac{\partial p}{\partial z} - 0.693 c_1$$

$$c_1 = -1.44 v_{z\text{-max}} - \frac{17.31\ \mu m^2}{\mu}\frac{\partial p}{\partial z}$$

$$0 = \frac{16\ \mu m^2}{\mu}\frac{\partial p}{\partial z} + \left(-1.44 v_{z\text{-max}} - \frac{17.31\ \mu m^2}{\mu}\frac{\partial p}{\partial z}\right)\ln(8) + c_2$$

$$c_2 = -\frac{16\ \mu m^2}{\mu}\frac{\partial p}{\partial z} + 3v_{z\text{-max}} + \frac{36\ \mu m^2}{\mu}\frac{\partial p}{\partial z} = 3v_{z\text{-max}} + \frac{20\ \mu m^2}{\mu}\frac{\partial p}{\partial z}$$

$$v_z(r) = \frac{r^2}{4\mu}\frac{\partial p}{\partial z} + \left(-1.44 v_{z\text{-max}} - \frac{17.31\ \mu m^2}{\mu}\frac{\partial p}{\partial z}\right)\ln(r) + 3v_{z\text{-max}} + \frac{20\ \mu m^2}{\mu}\frac{\partial p}{\partial z}$$

$$v_z(r) = \begin{cases} v_{z\text{-max}} & r < 4\ \mu m \\ \dfrac{r^2}{4\mu}\dfrac{\partial p}{\partial z} + \left(-1.44 v_{z\text{-max}} - \dfrac{17.31\ \mu m^2}{\mu}\dfrac{\partial p}{\partial z}\right)\ln(r) + 3v_{z\text{-max}} + \dfrac{20\ \mu m^2}{\mu}\dfrac{\partial p}{\partial z} & 4\ \mu m \le r \le 8\ \mu m \end{cases}$$

This problem simplified to a similar form as many of the Navier–Stokes examples that we have solved previously. It is important to note that we made the assumption that the entire red blood cell will travel with the same velocity and that the flow is axisymmetric and steady. The validity of the red blood cells as a rigid body would need to be verified in the particular flow scenario, but using this assumption may still provide a somewhat accurate reflection of the real flow conditions.

In Stokes flow situations, the viscous forces dominate the flow profile. As a consequence, there is a significant amount of drag on a particle that is within the fluid. To quantify this, we will introduce the drag coefficient (C_D), which relates the viscous stresses on the surface of a particle to the fluid properties. In a general form, the drag coefficient is defined as

$$C_D = \frac{F}{\frac{1}{2}\rho v^2 A} \tag{6.4}$$

where F is the applied force on the particle due to the viscous fluid stresses, ρ is the fluids density, v is the mean fluid velocity, and A is the cross-sectional area of the particle. For a sphere with radius of r, the viscous forces are equal to

$$F = 6\pi\mu r v \tag{6.5}$$

Substituting Eq. (6.5) into Eq. (6.4), the drag coefficient for a sphere is equal to

$$C_D = \frac{6\pi\mu r v}{\frac{1}{2}\rho v^2 \pi r^2} = \frac{12\mu}{\rho v r} \tag{6.6}$$

We will see in a later chapter that this is strongly dependent on the Reynolds number (Re). Incorporating the Re into Eq. (6.6)

$$C_D = \frac{6}{\text{Re}} \tag{6.7}$$

Example

Using the formulation from the previous example, calculate the viscous forces on a red blood cell in the microcirculation as well as the drag coefficient. Assume that the velocity of the red blood cell is 15 mm/s, the viscosity of blood is 5 cP, and the density of the blood is 1100 kg/m^3.

Solution

The average blood velocity is approximately equal to half of the maximum blood velocity of 15 mm/s (if we assume laminar Newtonian steady flow about the red blood cell):

$$F = 6\pi(5\ cP)(4\ \mu m)(7.5\ mm/s) - 2.83\ nN$$

$$C_D = \frac{12(5cP)}{(1100\ kg/m^3)(7.5\ mm/s)(4\ \mu m)} \cong 1820$$

In the microcirculation, it is common for the Reynolds number to range between 0.001 and 0.01. This is associated with a very large drag coefficient for any particle within the fluid. As illustrated in the example above, the actual forces on each red blood cells are not necessarily large (although this is a relatively large force on the cell itself), but due to the high viscous forces compared to the small particle size (as related to the inertial forces), there is a large drag on the cells.

Similar to the pressure distribution within the microcirculation, there is a temporal variation in the velocity distribution based on the cardiac cycle as well as the propagation of the velocity throughout the cardiovascular system. The velocity can change on the order of a few millimeters per second within the capillaries. For most calculations, however, this is a relatively small change and is typically neglected. This allows for a fairly valid assumption of steady flow within the microcirculation. As with the pressure distribution, the relative changes based on vessel diameter are more important to consider than the temporal changes.

6.6 INTERSTITIAL SPACE PRESSURE AND VELOCITY

The interstitial space is the area between cells, which is typically composed of fluid and proteins (such as collagen and/or proteoglycans). Note that some cells, such as inflammatory cells, can reside within the interstitial space, but typically these types of cells account for a very small portion of the total volume of the interstitial space. The proteins that are present within the interstitial space function to "connect" cells together into a uniform tissue that can withstand mechanical loads as one entity. The fluid within the interstitial space has a very similar composition as plasma, except for the much lower concentration of plasma proteins. Plasma proteins are unable to diffuse through the capillary membrane under normal conditions and therefore do not enter the interstitial space. The fluid component of the interstitial space also helps to maintain the structural integrity of tissue under mechanical loads.

In Chapter 7, we will see that the pressure within the interstitial space is critical to determine the transport of fluid and nutrients between the blood vessels and the extravascular space. The flow rate of water across the blood vessel wall is directly proportional to the pressure (osmotic and hydrostatic) within the blood vessel and the interstitial space (Figure 6.6). This is summarized by Starling's theory:

$$\dot{m} = K_p(P_B - P_I - \Pi_B + \Pi_I) \tag{6.8}$$

where $\dot{m}$ is the volumetric flow rate of water. K_p is a permeability constant, which will be discussed later. P_B is the hydrostatic pressure within the capillary, and this tends to force water out of the blood vessel. P_I is the hydrostatic pressure within the interstitial space. This tends to inhibit the movement of water out of the blood vessel (if it is positive), but can aid in fluid movement out of the blood vessel (if it is negative). Π_B is the osmotic pressure within the capillary, and this tends to promote the movement of water into the blood vessel. Π_I is the osmotic pressure within the interstitial space and this

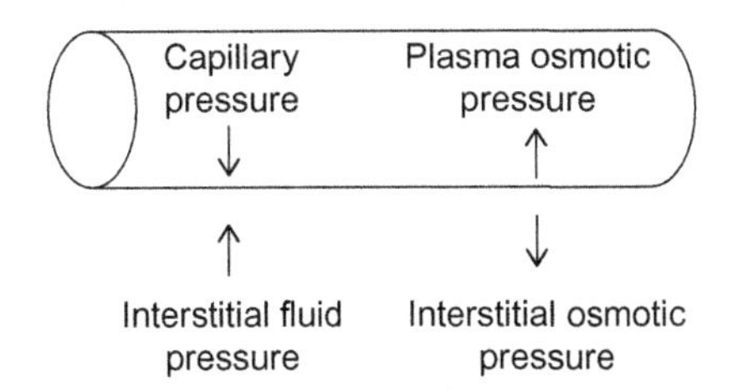

FIGURE 6.6 The capillary and interstitial fluid hydrostatic pressure and the colloidal osmotic pressure (for plasma and interstitial space) affect the movement of fluid within microvascular beds. The capillary hydrostatic pressure and the interstitial osmotic pressure generally aid in water movement out of the capillary. The interstitial hydrostatic pressure and the interstitial plasma pressure generally aid in water movement into the capillary.

tends to promote the movement of water out of the blood vessel. Note that the Starling's theory sometimes includes a reflection coefficient, which relates to the ability of the capillary wall to restrain the movement of particular species; each of the osmotic pressure terms would be multiplied by the reflection coefficient related to transport across the capillary wall in the appropriate direction. As previously discussed, the capillary hydrostatic pressure is within the range of 25 mmHg (on the arterial side) and within the range of 12 mmHg (on the venous side). The interstitial pressure is approximately negative 3 mmHg (this is a gauge pressure), but can increase to positive 4–5 mmHg when the space between the cells constricts or under edema conditions. Cell constriction occurs due to different tissue loading conditions, such as an increased external pressure (e.g., pushing on a region of your body with your hands will locally increase the external pressure). This increase would act to inhibit water movement out of the capillary.

The osmotic pressure within capillaries averages to approximately 28 mmHg. This pressure is developed from all of the compounds within the capillary that cannot diffuse across the endothelial cell barrier. The major contributor to this is albumin, which solely develops an osmotic pressure of approximately 22 mmHg within the capillary. The globulins in the capillary contribute the majority of the remaining portion of the osmotic pressure. The total protein concentration within the interstitial space is greater than the protein concentration within the capillaries. This suggests that the osmotic pressure of the interstitial space should be larger than the plasma osmotic pressure. However, because the volume of the interstitial space is significantly larger than that of the capillary, the total osmotic pressure of the interstitial fluid is approximately 7 mmHg. The osmotic pressure does not change between the arterial and venous side of the capillary because proteins do not leave the blood vessel and the blood volume remains relatively constant, even though there is some transport that occurs.

The total driving force for water out of the capillary is simply the summation of all of these pressure values, according to Starling's theory, if we assume that the reflection coefficient is one for the transport of nutrients across the capillary wall in both directions. On the arterial side of the capillary, the summation of the forces that tend to aid in fluid movement out of the capillary is

$$P_B - P_I + \Pi_I = 25\ mmHg - (3\ mmHg) + 7\ mmHg = 35\ mmHg$$

The only force that tends to prevent fluid movement out of the capillary is the plasma osmotic pressure:

$$\Pi_B = 28\ mmHg$$

Therefore, the total driving force of fluid out of capillaries on the arterial side is approximately 7 mmHg. By using this same analysis for the venous side of the capillary, the balance of the forces shifts to aid in water movement back into the capillary:

$$P_B - P_I - \Pi_B + \Pi_I = 15\ mmHg - (-3\ mmHg) - 28\ mmHg + 7\ mmHg = -3\ mmHg$$

Therefore, along the venous side of the capillary there is a net reabsorption of water that was lost along the arterial side. This change from water loss within the capillary to water gain within the capillary is primarily governed by the change in the capillary hydrostatic pressure. The other pressures (osmotic and interstitial hydrostatic) are relatively constant throughout the capillary length, note that we made the assumption that there is little change in the plasma osmotic pressure, but observe that there is not a balance between water lost on the arterial side and water gained on the venous side. Therefore, there must be some (although it might be minor) change in the capillary osmotic pressure. Also, these hydrostatic and osmotic pressure values (as well as the permeability) are tightly controlled by the body, because slight changes in any of them can cause either edema (increased water within the interstitial space; swelling) or dehydration, by having a significant effect on the movement of water across the capillary wall. This is primarily found in the reflection coefficient and the permeability of the vascular tissue to specific species.

The velocity of fluid within the interstitial space is relatively slow compared to the flow within the cardiovascular system, and most of the fluid does not have the ability to move freely. Due to the high concentration of charged proteins (collagens/proteoglycans) within the interstitial space, water becomes associated with the proteins through hydrogen bonding events. Because the interstitial space is majorly composed of these charged proteins, water does not have the ability to flow through the interstitial, but instead it diffuses via hydrogen bonding to different proteins. For instance, a water molecule must break a hydrogen bond to make a new hydrogen bond. The energetics of these reactions are fairly significant when the scale of the body is considered. Although the vast majority of fluid in the interstitial space is associated with proteins, there are small vesicles of fluid that can flow freely between the proteins. Under normal conditions, less than 1% of the water in the interstitial space can flow freely in this manner and the remaining 99% is loosely bound to proteins through hydrogen bonding events. However, under disease conditions such as edema, the amount of free fluid can increase to approximately 50% of the total fluid in the interstitial space. This is caused by an increase in water volume and not a decrease in protein concentration.

Diffusion of water through the interstitial space occurs at a rate on the order of 50–100 μm/s. This is slightly slower than the free movement of water through the interstitial space. However, because the distance that water (and dissolved solutes) must move in the interstitial space is small, the transport of nutrients is very rapid and can be accounted for by the water associated with charged molecules. Therefore, almost as soon as the water/nutrients enter the interstitial space they are absorbed by cells that are close to the blood vessel. Thus, this is a very effective way to transport nutrients throughout the interstitial space that surrounds the capillary bed.

6.7 HEMATOCRIT/FAHRAEUS–LINDQUIST EFFECT/FAHRAEUS EFFECT

We have already discussed how the blood within the capillary is nonhomogenous because there are packets of red blood cells followed by packets of plasma. These individual packets do not fully mix until they reach the venule. This has an effect on the hematocrit (the percentage of red blood cells in whole blood), which the reader should recall is typically close to 40% under normal conditions. There are a few reasons why the hematocrit changes within the microcirculation. Imagine a bifurcation in which the flow velocity is not the same within the two daughter branches. Because red blood cells tend to move toward the blood vessel centerline (i.e., high-velocity location) prior to the bifurcation, they will follow the higher velocity pathway through the bifurcation. This means that the daughter branch with a higher velocity will tend to have more red blood cells entering it and thus a higher hematocrit as compared to the daughter branch with the slower velocity. Also, the daughter branch with the higher velocity is fed more from the feed vessels centerline as compared to the other slower branch. This ignores branch geometrical considerations which can cause cell skimming into one of the branches.

Mathematical and experimental studies have been conducted in order to determine the relationship between the hematocrit within the daughter branches and the velocity of blood within the branches. In these experiments, it is typical to use the same diameter tubing for the inflow and the two daughter branches, but this is not a necessary condition for these experiments. Regardless of the tube size, this allows for a relationship between the volume flow rate within each of the three tubes as follows:

$$Q_{\text{inflow}} = Q_{\text{branch 1}} + Q_{\text{branch 2}} \tag{6.9}$$

which is developed from the continuity equation assuming that the fluid density within each tube is the same. Again, this equation will hold for any experiment regardless of tube cross-sectional area. To determine how the hematocrit is related to the tube velocities, multiple groups have conducted experiments where a known hematocrit is fed through a tube of known diameter which branches into two daughter tubes with known diameters. The fluid that is discharged from each of the daughter branches is collected and the hematocrit in each collection tube is measured. The velocity of red blood cells as well as the average tube velocity can be quantified with optical means in these experiments (if the tubes are transparent). A relationship has been developed for the quantity of red blood cells within a particular cross-section at an instant in time (tube hematocrit). The quantity of red blood cells was found to be equal to the mean tube speed multiplied by the cross-sectional area of the feed tube multiplied by the discharge hematocrit. This was also found to be equal to the mean red blood cell speed multiplied by the cross-sectional area of the daughter tube multiplied by the daughter tube hematocrit. Mathematically this can be represented as

$$Quantity_{RBC} = v_{feed\text{-}tube} A_{feed} Hct_{discharge} = v_{RBC} A_{daughter} Hct_{daughter} \tag{6.10}$$

If the cross-sectional areas are the same, then the discharge hematocrit is equal to

$$\text{Discharge Hematocrit} = \text{Tube Hematocrit}\left(\frac{\text{Mean Red Blood Cell Speed}}{\text{Mean Tube Flow Speed}}\right)$$
$$Hct_{discharge} = Hct_{daughter}\left(\frac{v_{RBC}}{v_{tube\text{-}feed}}\right) \tag{6.11}$$

Different cross-sectional areas can be incorporated into Eq. (6.11) as well. Substituting Eq. (6.11) into Eq. (6.9)

$$Hct_{inflow} v_{inflow} = Hct_{branch\ 1} v_{branch\ 1} + Hct_{branch\ 2} v_{branch\ 2} \tag{6.12}$$

Again, assuming that the cross-sectional areas are the same, then Eq. (6.9) can be simplified to a relationship between the mean velocities within each tube. Using this to simplify Eq. (6.12), we get

$$Hct_{inflow} = \frac{Hct_{branch\ 1} v_{branch\ 1} + Hct_{branch\ 2} v_{branch\ 2}}{v_{branch\ 1} + v_{branch\ 2}} \tag{6.13}$$

because

$$v_{inflow} = v_{branch\ 1} + v_{branch\ 2}$$

As a word of caution, this formulation is highly dependent on the relationship between the cell diameter and the tube diameter. When the ratio of diameters is close to 1, then Eqs. (6.10)–(6.13) are valid. As the ratio decreases, as occurs within the macrocirculation, these equations are no longer valid because the cell velocity has little to no effect on the tube velocity. It is interesting to note that the ratio of the hematocrit in the two daughter branches has empirically been found to be able to be represented by the following linear relationship:

$$\frac{Hct_1}{Hct_2} - 1 = a\left(\frac{v_1}{v_2} - 1\right) \tag{6.14}$$

In Eq. (6.14), a is a phenomenological parameter, which is dependent on the ratio of the cell diameter to tube diameter, the shape of the cell, the rigidity of the cell, and the feeding tube hematocrit.

The hematocrit in capillaries is lower than in the arterial hematocrit as shown in the mathematical formulations above. This is because the red cells are fairly restricted to the centerline (higher velocity flow) in the capillaries and thus within a particular time frame, there will be a lower chance of "catching" a red blood cell as compared to "catching" the slower plasma along the blood vessel wall (using still photography, for instance). With an inflow hematocrit in the range of 40–50%, it has been shown that the capillary hematocrit can be as low as 10% but averages to around 20%. This reduction in hematocrit begins to be seen in arterioles with a diameter in the range of 100 μm (where the cell diameter/tube diameter $\cong$ 0.1). The hematocrit decreases steadily until the capillaries, where the hematocrit is around 10–20%. The hematocrit in the venules rapidly increases to the inflow hematocrit level within vessels that have a diameter in the range of 50 μm (hematocrit is typically 30–35%). This is caused by the overall slow velocity in these blood vessels, the ability for the cells and plasma to mix and the increase in the cell to tube ratio. It has been suggested that the endothelial cell glycocalyx (a region of glycated membrane bound proteins that extends into the lumen) also plays a role in the decreased hematocrit. This region may partially help to force the flow of red blood cells to the centerline in blood vessels with a smaller diameter, but has little to no effect in blood vessels with larger diameters. A typical glycocalyx extends for approximately 100 nm into the lumen of the blood vessel. Upon removal of the endothelial cell glycocalyx, it has been seen that the capillary hematocrit increases, but not back to the inflow hematocrit levels.

Taken together, blood vessels with a relatively high ratio of blood cell diameter to tube diameter have a lower than normal hematocrit. This is partially caused by the fluid dynamics (velocities, no-slip boundary condition, and diameter relationship), but may also be partially regulated by biological factors (such as the glycocalyx).

In the early 1930s, a number of studies were conducted by Fahraeus and Lindquist to determine the effects of vessel radius on viscosity and hematocrit. For these experiments, the diameter of the tube was restricted to be less than 250 μm and the shear rate was relatively high as compared to the normal microcirculation, approximately 100 s^{-1}. The shear rate was this large so that the flow in the feeding tube would be Newtonian. The pressure difference across the tube and the volumetric flow rate were quantified and fit to the Hagen–Poiseuille equation. Interestingly, the normal flow rate relationship was not dependent on the bulk viscosity of the fluid, but instead it was dependent on an effective fluid viscosity. It has been experimentally shown that the new Hagen–Poiseuille relationship that is valid for flows within the microcirculation is represented as

$$Q = \frac{\pi \Delta P r^4}{8 \mu_{eff} L} \tag{6.15}$$

where μ_{eff} is the effective viscosity of the fluid within the microcirculation. The effective viscosity is highly dependent on the diameter of the blood vessel and the pressure drop across the blood vessel.

In general, the effective viscosity reduces with a decreasing tube radius. For a hematocrit of 40%, the normal blood viscosity is in the range of 3.5 cP. As the diameter of the blood vessels reduces to 1 mm, the effective viscosity will reduce by approximately 5%. However, for blood vessels that are smaller than 500 μm, the reduction in viscosity is much more pronounced, and it approaches 2 cP in blood vessels with a diameter less than 50 μm. This drop can also be accounted for by the overall restriction of red blood cells towards the centerline of the blood vessel as discussed above. Within blood vessels with a diameter in the range of the red blood cell diameter, the viscosity increases rapidly again due to plug flow principles (see Section 6.8). These results were explained by the presence of a plasma layer along the blood vessel wall. This layer is void of red blood cells, consisting solely of plasma, which has a viscosity of approximately 1 cP (i.e., water). Because this low viscosity solution is next to the wall, where the shear stresses would be the highest, the overall viscosity appears to reduce because the force needed to overcome this "low" viscosity solution is far less than what would be needed if cells were present.

To more accurately describe the presence of a plasma layer, one must consider the hemodynamic forces that act on a red blood cell. Because a red blood cell has a size associated with it (i.e., it is not a particle), then the forces that act on the cell must satisfy Newton's law of motion under translational and rotational motions. If the red blood cell is not located directly within the flow centerline (assuming that the flow is fully developed and parabolic), then the forces induced by the fluid velocity will be different on each end of the red blood cell. Due to this imbalance of forces, the red blood cell will tend to rotate toward the region with lower stresses. Recall that the higher fluid velocity has a lower shear stress under normal conditions and therefore, the higher shear stress would be located closer to the vessel wall and the lower stresses would be located along the centerline. This continues until the forces acting on the red cell are balanced and do not cause cell rotation. Therefore, red blood cells move to the lower stress regions which have high velocities in order to balance forces and to minimize the overall loading on the cell. This will only occur when the red cell is centered within the blood vessel (again, if the flow is fully developed and the red blood cell is uniform). This is another factor that affects the location of red blood cells within the blood vessel.

A final factor that promotes the formation of a plasma skimming layer is that red blood cells cannot be located within a certain near-wall region. Most red blood cells traverse through the cardiovascular system face-on and not end-on. This means that a red blood cell cannot be located within one cell radius (approximately 4 μm) of the blood vessel wall. If red blood cells entered this region, then part of the cell would have to be outside of the blood vessel or that part of the cell would need to deform. Deformation is not favorable from an energy standpoint, because this would likely induce a higher energy state and would at least require energy input to cause the deformation. Therefore, from a probability standpoint, it is more likely for a red blood cell to be located close to the centerline and far from the vessel wall. This will effectively promote a plasma layer reducing the blood viscosity.

6.8 PLUG FLOW IN CAPILLARIES

As previously discussed, red blood cells flowing through capillaries tend to plug the capillary, preventing the plasma to flow freely as it does in larger blood vessels. While the red blood cells flow fairly steadily, the plasma in between the cells experiences turbulent eddies and recirculation zones. These eddies help to move material from the centerline of the plasma towards the vessel wall, potentially helping in the transfer of materials towards and across the capillary wall. These turbulent eddies tend to move compounds to the endothelial cell wall, so that they can diffuse across the wall, instead of convect across the blood and then diffuse across the wall. Also, these turbulent eddies would act to maintain a high concentration gradient of diffusing species near the wall. Additionally, there is a very small layer of plasma along the blood vessel wall that experiences very high shear stresses, because it is being squeezed between a red blood cell and an endothelial cell. This can be considered the proverbial rock and a hard place for plasma. Because the shear stress increases significantly, this effectively slows the flow of blood and can be quantified through an increase in the apparent viscosity. The slowing down of material near the endothelial cell wall also increases the residence time of materials near the wall, increasing the overall ability for the material to diffuse into the extravascular space.

To model plug flow, the velocity of the fluid is assumed to be constant across any cross-sectional area of the blood vessel. Normally, to solve a plug flow problem, it is assumed that there is no mixing across the individual plugs, so that there is no transfer of material from upstream to downstream around the plug of red cells. It is also assumed that each fluid plug and each cellular plug are homogeneous, but individual plugs may have different compositions. For instance, the first red blood cell plug may consist of 10 red cells, whereas the second plug may consist of 20 red cells. Within that one plug, however, the density, viscosity, velocity, and all other parameters do not change. The fluid plugs can have different volumes with different compositions, but again, it is uniform across the volume (i.e., perfectly mixed within one plug). To simplify the problem, it is easiest to also assume that the flow is steady and that there is no wave propagation throughout the fluid. As discussed before, this is a relatively good assumption within the microcirculation, where the temporal variation in the pressure and velocity waveform is relatively small.

The first governing equation for this type of problem is the Conservation of Mass throughout the individual plugs. For red blood cells, this is simply the number of red blood cells that enter the region of interest must exit the region of interest. Since under normal conditions, red blood cells cannot pass through the endothelial cell wall and are not produced within the capillary; this is the exact solution. This solution assumes that a red blood cell cannot leave one plug and join another, which can be found to occur under physiological conditions, but it is somewhat rare, since the amount of inertia that would need to be overcome to move between plugs is very significant. Unlike cellular matter, which cannot be generated or consumed within the capillaries, the plasma portion of the fluid, as well as any compound dissolved within the plasma, would obey the Conservation of Mass in its full form. For this example, we will make use of oxygen as our species of interest, but we can use this same formulation to discuss any species of interest. Under these conditions, the Conservation of Mass becomes

$$(O_2)_{\text{accumulated}} = (O_2)_{\text{entering}} - (O_2)_{\text{exiting}} + (O_2)_{\text{generated}} - (O_2)_{\text{consumed}} \tag{6.16}$$

Clearly, the amount of oxygen generated and consumed would be dependent on the kinetics of particular molecular reactions within the plasma components. Summing all of these components together, this can be formulated as

$$\pi r^2 u(C_{O_2}(x) - C_{O_2}(x + dx)) + \pi r^2 R_{O_2} Z dx = 0 \tag{6.17}$$

where r is the radius of the blood vessel, C is the concentration of a particular species at location x (the entrance of the vessel, for our example we are referring to oxygen) or $x + dx$ (the exit of the vessel), u is the fluid velocity, R is the kinetic reaction rate of the specific species (this would be positive if generating species, or negative if consuming species; again, we are referring to oxygen), and Z is a stoichiometric coefficient. The first term on the left-hand side of Eq. (6.17) represents the difference between the amounts of species that enter the capillary versus what leaves the capillary as a function of the flow rate within the capillary. The second term on the left-hand side of the equation represents the overall kinetics of the generation and consumption formulas, within the area of interest. For instance, if there is a net generation, this term would be positive whereas a net consumption would appear as a negative rate. We have made the assumption that no oxygen (or any general species) accumulates within the capillary (e.g., Eq. (6.17) is set equal to zero, and thus any imbalance between the inflow and the outflow is accounted for in the kinetics term). This assumption is reasonable, since there is little to no store for oxygen within the tissue or within the blood. Simplifying Eq. (6.17) into a differential form we obtain

$$u\frac{dC_{O_2}}{dx} = ZR_{O_2} \tag{6.18}$$

Note that $\frac{dC_{O_2}}{dx} = \frac{C_{O_2}(x+dx) - C_{O_2}}{dx}$ and that the cross-sectional area is assumed to be constant along the blood vessel length. In typical kinetic reactions, the kinetic reaction rate is dependent on the concentration of the particular species present. As an example, if the kinetic reaction rate can be modeled with a first-order kinetic reaction, then

$$R_{O_2} = kC_{O_2}$$

Using the first-order kinetic reactions within Eq. (6.18), we can derive the following equation to represent the change in concentration as a function of distance. Note that, k is the rate constant for the particular reaction and typically depends on temperature and the medium under which the reaction is occurring in.

$$\frac{dC_{O_2}}{dx} = \frac{kZC_{O_2}}{u}$$

$$\frac{dC_{O_2}}{C_{O_2}} = \frac{kZ}{u}dx$$

$$\ln(C_{O_2}) = \frac{kZx}{u} + C$$

$$C_{O_2}(x) = C_{\text{inlet}}e^{\frac{kZx}{u}} \tag{6.19}$$

Clearly, depending on the particular conditions and the kinetics of the reaction, this formulation can differ significantly. The assumption of first-order kinetics is a reasonably accurate representation for many of the reactions that occur within capillary blood flow, but if the kinetics change, then Eq. (6.19) is no longer valid. Using the few fairly accurate assumptions made about the species concentration within the fluid phase of the plug flow, the Conservation of Mass simplifies for species within the microcirculation to an exponential decay or rise relationship depending on the kinetics of the reaction.

Example

Consider the change in oxygen concentration and carbon dioxide concentration along a 500 μm blood vessel. Assume that the kinetics for each of these two molecules are described by first-order reactions. The blood velocity within this particular capillary is 10 mm/s. The kinetic rate constant for each reaction is −17 mol/s (for oxygen consumption) and 1.17 mol/s (for carbon dioxide formation). For these particular reactions, one oxygen molecule is consumed or two carbon dioxide molecules are formed. The inlet blood concentration of oxygen and carbon dioxide is 400 and 120 mL, respectively (dissolved gas concentration). Plot the change in concentration as a function of distance along the capillary. What is the outlet concentration of oxygen and carbon dioxide?

Solution

Oxygen concentration at the outlet would be

$$C_{O_2}(0.5\ mm) = 400mLO_2 e^{(0.5\ mm*1*-17\ mol/s)/(10\ mm/s)} = 171mLO_2$$

Carbon dioxide concentration at the outlet would be

$$C_{CO_2}(0.5\ mm) = 120mLCO_2 e^{(0.5\ mm*2*1.17\ mol/s)/(10\ mm/s)} = 135mLCO_2$$

The change in concentration over the distance of the blood vessel is shown in Figure 6.7. Since these reactions can be described by first-order kinetics, Eq. (6.19) was directly applicable. Also, note (i) the generation rate constant is positive and the consumption rate constant is negative and (ii) the stoichiometric constant of the carbon dioxide reaction is 2, since two carbon dioxide molecules were generated during each reaction.

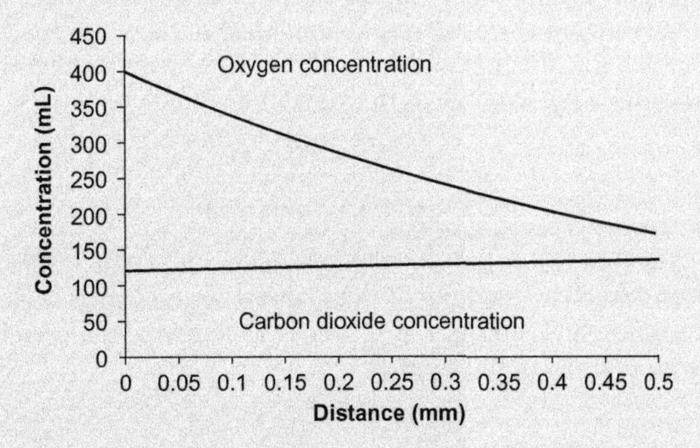

FIGURE 6.7 Change in oxygen and carbon dioxide concentration within a capillary, as modeled with first-order kinetic reactions. This figure is associated with the in-text example.

6.9 CHARACTERISTICS OF TWO-PHASE FLOW

In Chapter 2, we briefly described the nature of two-phase flows. In typical engineering applications, the two phases are either different fluids or the same fluid that exists in two phases (i.e., gas and liquid). Blood is a two-phase flow because the plasma component has different physical properties as compared with the cellular components. Under most conditions, it is not necessary to use two-phase flow principles within the macrocirculation. This is because the cellular component has little to no effect on the overall flow conditions. However, as discussed in this chapter, the cellular component can greatly affect the flow conditions within the microcirculation. For the remainder of this section, we will discuss some of the approaches available to solve two-phase flow problems. For this type of analysis, it is important to know the exact location of each particle within the fluid. Therefore, the Lagrangian approach would be used to solve the two-phase flow problem. However, if average values are more important, then the Eulerian approach should be used.

Using the Lagrangian approach, our goal would be to solve Newton's second law of motion for each particle within the flow field. In this instant, Newton's law takes the form of

$$\vec{F}_p(t) = m_p \frac{d\vec{v}_p}{dt} \tag{6.20}$$

where $\vec{F}_p(t)$ is the summation of all forces acting on one particular particle. To solve this equation, information regarding the exact instantaneous velocity as well as the position of each particle would be required. If at time t, the velocity and the position of a particle is known, then the particle will move to a new position at time δt, according to

$$x(t + \delta t) = x(t) + v_p \delta t \tag{6.21}$$

The velocity of the particle is developed using a known mean velocity component and a fluctuating component, which arises from turbulence, variations in the particles, or variations in the flow field. This is similar to turbulence modeling discussed briefly in Section 5.11. To predict the variation, a number of approaches have been developed which are all related to a Lagrangian correlation function. Most of these approaches use statistical distributions to predict the fluctuations in velocity. The general Lagrangian correlation function is defined as the time average of the velocity fluctuations (u') as

$$R(t) = \frac{\overline{u'(t) * u'(t + \tau)}}{\sqrt{\overline{u'^2(t)}}\sqrt{\overline{u'^2(t + \tau)}}}$$

In most two-phase flow approaches, this correlation function is linearized based on some fluid parameters (such as the time of fluctuations or the length scale of fluctuations). One of the most common and simplest approaches to linearize the correlation function is to use the eddy decay lifetime in the form of

$$R(t) = 1 - \frac{t}{2t_D} \tag{6.22}$$

where t_D is the average eddy decomposition time and t is the time of interest for the fluid within the flow field.

Under the Eulerian approach, the average fluid properties for each phase would be used independently to define the motion of the total fluid. All of the equations that have been developed in previous chapters (for mass conservation, momentum conservation, and Navier–Stokes, among others) apply to each phase individually. There is an extra boundary condition that the interface of the phases must have the same fluid properties (e.g., velocity, pressure, stress), so that there is a balance between the two phases. The difficulty with this approach would be defining where the interface between the two phases occurs within the fluid regime, since the cells are deformable. Also, for a problem where blood is the fluid, there may be more than one interface, which makes the computations very time-consuming by hand. For a 50-μm-diameter blood vessel, there can be a total of five red blood cells aligned within one particular cross-section (if they are flowing face-on). To solve this problem using two-phase flow methods, the exact distance between each cell, at each location, would need to be known. Instead of using the Navier–Stokes equations to solve for the entire flow field within the 50-μm-diameter vessel, each section of the vessel would need to be solved independently and the boundary conditions at each interface would need to be applied for the solution of a different section, which would end up developing a number of coupled differential equations.

6.10 INTERACTIONS BETWEEN CELLS AND THE VESSEL WALL

Within the microcirculation, cell–wall interactions need to be considered to accurately depict the flow field. As discussed previously, red blood cells squeeze through the capillaries, effectively shearing against the endothelial cell wall. In slightly larger vessels (25–50 μm in diameter), red blood cells can rebound off of the wall, changing the flow profile of the cell and the blood. Also, it is common for white blood cells to stick to the vascular wall and roll along it. This changes the cross-sectional area and affects where the no-slip boundary condition would occur. In larger vessels, it is unlikely that any cell–wall interaction will disturb the flow profile and thus is typically not considered within simulations of the macrovasculature.

To model these interactions, we need to make a number of assumptions for the flow around the cells. First, we will assume that there is a no-slip condition at the vessel wall and along the membrane of the particle. This means that the fluid velocity should be matched along the entire length of the cell membrane; however, if the cell is within the free fluid this leads to the imbalance along the cells, which pushes them to the flow centerline. This imbalance also becomes important when one portion of the cell is adhered to the vessel wall and another portion of the wall is exposed to fluid flow. To model these interactions, we will make the assumption that the particles are neutrally buoyant which means that the forces acting on the cell must summate to zero. This allows us to compute a solution by hand instead of using numerical methods (see Chapter 14).

In order to model these conditions, the mechanical properties of the cell and the cell membrane should be developed first. For red blood cells, it has been shown that the bending moment of the cell is very important to allow the cell to squeeze through a small capillary. For instance, if the red blood cell is too stiff to accommodate changes in the capillary cross-sectional area, then the red blood cell seals off the blood flow preventing other red cells from entering that capillary. If the red blood cell is too pliant, then the fluid flow stresses can easily shear the cell, preventing it from functioning properly. For flexible two-dimensional axisymmetric spheres, the principal bending moments can be defined as

$$M_1 = \frac{G(K_1 + \nu K_2)}{\lambda_2}$$
$$M_2 = \frac{G(K_2 + \nu K_1)}{\lambda_1} \tag{6.23}$$

where G is the bending stiffness, ν is Poisson's ratio, λ_n is the extension ratios, and K_n are the changes in curvatures of the membranes. The relationship between the stress and the extension ratios is also needed for these solutions. These are typically defined as

$$T_1 = \frac{E(\lambda_1^3 - \lambda_1)}{2\lambda_2} + T_0$$
$$T_2 = \frac{E(\lambda_2^3 - \lambda_2)}{2\lambda_1} + T_0 \tag{6.24}$$

for the principal directions, where E is the elastic modulus and T_0 is an initial stress. Once the cell membrane mechanical properties are defined (there are many research papers that have investigated these mechanical properties), we need to classify the properties for the interior of the cell. Typically, the interior is assumed to be a uniform Newtonian fluid; that way, forces that act on the cell can be transmitted evenly throughout the cell interior. It is also assumed that the total surface area of each cell remains constant during any deformation. The assumption that the fluid behaves Newtonian is somewhat valid, since the intracellular fluid is primarily composed of water. The second assumption is not very accurate since white blood cells and platelets are known to add and/or remove sections of membrane in response to various stimuli. However, the modeling of these functions is somewhat complex and not typically taken into account.

Using the above assumptions, many groups have studied how red blood cells deform through the microvasculature at different viscosities, pressure differences, blood vessel diameters, and hematocrits. In general, they have all found similar results which suggest that the pressure within the blood vessel depends very strongly on the cell diameter. With greater cellular packing, the pressure across the vessel increases as well, since the inertia of this mass would increase. They have also verified that the cells typically stream toward the centerline and that they have a higher mean flow than the fluid, as was experimentally determined and described above.

The second critical parameter to consider while modeling cell–wall interactions is the adhesion probability between the blood cell and the vessel wall. Adhesion between cells and the blood vessel wall is mediated through cell membrane bound proteins, commonly selectins in the vasculature, but there are many types of adhesion proteins that can account for cell–cell or cell–wall adhesion. There is no easy way to quantify the amount of selectins (or other adhesion molecule) expressed on the cell membrane as well as the location of the specific protein on the cell membrane. To account for this problem, most computational models use probability theory to help predict where and when the molecules will be expressed. Usually, the surface density of the receptor is an assumed parameter and it is assumed that the proteins are evenly distributed across the surface. If the quantity of adhesion molecules is not held constant within the model, then there will be some criterion for expression; for instance, when the shear stress level exceeds 40 dynes/ cm^2, x many adhesion molecules can be expressed. Once this criterion is met, then the model will have a probability statement which states that some percentage of cells will express the adhesion protein at a particular rate. If the criterion for expression persists, then more cells will continually express more of the adhesion molecule. However, if the criterion is transient, then the majority of cells will not express the adhesion molecule, and there are likely statements for the probability of an expressed protein to be recycled within the cell. After the adhesion molecules have been expressed, other probability statements come into play. For instance, they will state if any portion of the cell that is expressing adhesion molecules comes within one cell diameter from the vessel wall, then there is some probability for adhesion to the vessel wall to occur (clearly we are assuming that any protein on the cell membrane is active, although this may not be the case). The adhesion between the cell and the vessel wall is then typically modeled as a spring, so that if the fluid forces exceed the spring's elastic capacity, the bond between the cell and the wall will break. However, once the cell is attached to the wall, there is also some probability that states how likely it will be for other cell membrane adhesion molecules to adhere to other wall adhesion proteins. As more connections are made, the capacity of the system increases. This is similar to a mechanical system, in which more springs are added to the connection. Once the adhesion between the membrane protein and the wall becomes stable (e.g., there are multiple connections), it is then modeled with an on/off probability, which states how likely the cell will maintain the connection. These probabilities can be based on known kinetic constants or other criteria that is defined within the mode. This process is summarized in the following flow chart (Figure 6.8), and examples are discussed in Chapter 14.

6.11 DISEASE CONDITIONS

6.11.1 Shock/Tissue Necrosis

Shock is a condition that can occur after a massive loss of blood due to (i) a significant decrease in the blood pressure and (ii) the decrease in nutrient exchange (due to blood volume loss). In general, shock is defined as a failure of the cardiovascular system to deliver the chemical compounds necessary for cell survival and remove metabolic waste products. This will eventually cause an abnormal membrane structure and irregular metabolism that will eventually lead to cell death. In response to pressure reductions, there are rapid responses that act to increase the blood pressure back to resting conditions. Along the carotid artery and the aortic wall, there are pressure sensors termed baro-receptors. When these receptors sense a significant decrease in blood pressure they instigate a neuronal response that increases cardiac output by increasing the heart rate to over 150 beats per minute and increases the vascular resistance by causing constriction of the blood vessels.

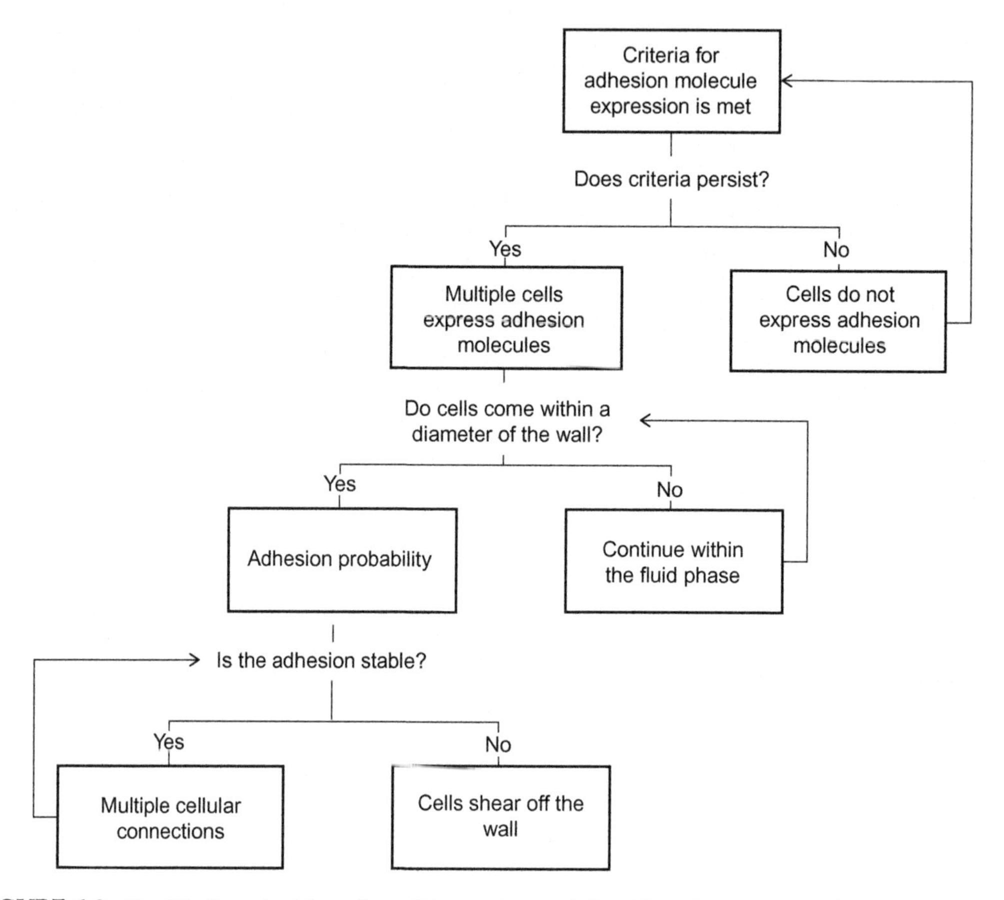

FIGURE 6.8 Possible flow chart for cell–wall interaction modeling. There is a certain predetermined probability that is associated with each step in the flow chart but these probabilities may be adaptive over the time course of interest. Also, the first step in the flow chart can continually change and is reversible, making this flow chart not a direct path from beginning to end. Clearly, this is a somewhat simplified approach to cell wall adhesion, but is intended to provide a general idea for the flow of a probabilistic modeling event.

Combined, these responses have the effect of increasing blood pressure and draw on the venous blood reserve. In the short term, hormones can also be released to aid in vasoconstriction and heart rate increase (epinephrine, antidiuretic hormone, among others). This mechanism typically can work up to a 25% blood volume loss, without showing signs of a decrease in nutrient exchange within the microcirculation.

Medically, there are typically three stages of shock, which if controlled early enough the extent of cellular death can be controlled. Stage 1 is when the patient is presented with low blood pressure (hypotension). There are a number of reason why a patient may experience sever hypotension. The first is a significant loss of blood, as discussed above. In many cases, a significant decrease in cardiac output (maybe due to heart failure) is the initiating factor for shock cases. In septic shock, there is a significant decrease in the peripheral vascular resistance (e.g., an increase in the blood vessel diameter), with little changes to the cardiac output. Regardless of the cause of the decrease in blood pressure, the baroreceptor compensatory mechanism(s) initiate, which aim to restore arterial pressure and blood flow to vital organs.

The second stage of shock is associated with a decreased tissue perfusion. After the patient experiences a massive decrease in blood pressure, the compensatory reactions initiate, but they may not be sufficient to maintain perfusion to vital organs. It is typical for a patient to present with cyanotic, cold and clammy skin, especially at the extremities, if they have progressed into stage 2. At this point, the damage of shock can still be reversed if the patient's cardiac output can be restored by some means. If the cardiac output cannot be restored, the patient typically progress into stage 3 of shock, which is associated with microcirculatory dysfunction. There is typically a significant alteration to cellular metabolism under these conditions, since the cells cannot obtain nutrients or remove wastes at the rate needed to sustain function. Significant vasoconstriction has likely occurred by this point and therefore, blood clots can be found within less vital organs, damage to the GI tract mucosal layer occurs, the glomerular flow and permeability is typically altered, acidosis is found and an overall decrease in cellular function or death of tissue is found. If shock continues to progress unchecked, then the flow through the coronary circulation is diminished, which perpetuates the shock effects.

If the shock progresses to a sever extent but is remedied, then the long-term problem associated with the potential blood loss is the necessary increase of blood volume to make up the loss. A decrease in capillary pressure will increase the movement of water back into the capillary along the venous side. Antidiuretic hormone increases the overall retention of water from the kidneys. Red blood cell maturation rate is increased to make up for the loss of the cellular component of blood. Typically, with a 25% loss of blood volume, the water portion can be returned to a normal volume within a few hours. However, the cellular portion will not return to normal levels for a few days, at best. With larger volumes of blood loss, it would take a longer time to make up the fluid and cellular compartment. In some scenarios, a transfusion may be the best approach to remedy the fluid loss issues, once the cardiac output and the vascular tone (constriction/dilation) has been controlled.

The goal of the combined innate response mechanisms acts to return nutrient exchange to normal levels. However, if any of the mentioned mechanisms fail or there is a more significant loss of blood, it is likely for stage 3 shock to develop. If this condition persists for an extended period of time, there will be local tissue necrosis (tissue death similar to myocardial infarction) in regions where the microvascular beds have been shut down to divert the blood flow to more critical regions (such as the brain). Necrosis is irreversible death or tissue and should be avoided, as much as possible, to maintain the overall health of the patient. However, with the combination of blood volume loss and lowered arterial pressure, it is more difficult for the heart to pump blood throughout the cardiovascular system. Cells that die prematurely though necrotic pathways do not follow the typical planned cell death pathway, termed apoptosis. Cells that undergo necrosis typically release their contents in the surrounding extracellular space, which initiates an inflammatory reaction from cells that are not undergoing necrosis. For the most part, necrotic tissue must be surgically removed, once the shock symptoms have been controlled (or the other symptoms that have initiated the necrotic tissue). Necrosis typically follows one of two pathways, the first initiates through the osmotic swelling of the cells that have a disrupted cell membrane. Under these conditions, the nucleus shrinks and the DNA becomes fragmented and degrades by nucleases. The second pathway involves the fragmentation of the nucleus.

As hinted, there are many causes for necrosis, which include a decrease in blood flow to tissue (as described in the context of shock), damage to blood vessels, frostbite, abnormal innate immune functions (complement system), invading pathogens and venoms.

6.11.2 Edema

Edema is a condition where there is an excessive accumulation of interstitial fluid within the interstitial space. This fluid is not recollected by the capillary networks or the lymphatic system (see Chapter 8). There are many causes of edema, but they all stem from an imbalance between the hydrostatic pressures and the osmotic pressures within the microvascular beds and thus there is a net intake of water that is not met by the excretion of water by the kidneys or by extrarenal mechanisms. The most common cause for edema is an increase in the arterial blood pressure associated with hypertension, which translates to the capillary pressure; this can be observed through changes within Starling's theory. Due to an increase in the capillary hydrostatic pressure at the arterial side, there is a net increase in the movement of water out of the capillary. Additionally, a decrease in the capillary osmotic pressure could aid in net water movement out of the blood vessels. This could arise to defects in plasma protein synthesis, increases in protein degradation or possible transport of plasma proteins out of the vasculature. The inflow of water on the venous side typically cannot make up for the extra loss of water along the arterial side. Therefore, there is a net accumulation of water within the interstitial space.

In some edema cases, the capillary wall begins to breakdown, which allows for the movement of plasma proteins into the interstitial space. This causes an increase in the interstitial osmotic pressure, preventing the movement of water into capillaries along the venous side of the capillary. In a familiar case, bruising, there is swelling because the capillary wall has become damaged. In other conditions, the plasma osmotic pressure can decrease because there is either not enough protein being formed or the proteins are being broken down within the cardiovascular system. In either case, due to the pressure changes that disrupt the delicate pressure balance in the capillaries, fluid has a preference to stay within the interstitial space.

Edema can also occur in specific organs in response to the inflammatory response such as pancreatitis. Particular organs can also develop edema based on certain biochemical and biophysical phenomena that occur within that organ. For instance, cerebral edema commonly occurs when there is an abnormal metabolic state. Corneal edema occurs in conjunction with glaucoma. Cutaneous edema is common with venomous/poisonous insect bites or plant contact such as poison ivy. Finally, the treatment of edema usually includes administering diuretics, but, in general, one must be careful with this approach since diuretics tend to alter the blood pressure. This type of approach can exacerbate the problem as well as remedy the problem.

END OF CHAPTER SUMMARY

6.1 The microcirculation includes all of the capillaries and the smallest arterioles that dictate where blood flows in the vascular beds. Within the microcirculation, there is a transport of nutrients, dissolved species, and water across the vascular wall. Therefore, mass balance can no longer be accounted for by the inflow equaling the outflow. True capillaries have a diameter in the range of 8 μm. Flow through the capillaries is not continuous and is primarily dictated by the need of the surrounding tissue. An intercellular cleft is located between neighboring endothelial cells (typically 10–15 nm thick) and the vast majority of transport is accounted by this channel. The flow through this channel is approximately 80 times faster than the blood flow through the capillary.

6.2 Endothelial cells are typically 10 μm in diameter and range in their lengths. This cell type lines all blood vessels, and they play a role in hemostasis, inflammation, and nutrient transport. Vascular smooth muscle cells can perform work, similar to other muscle cells. However, the regulation of contraction and the organization of the actin and myosin do not mimic other muscle cells. Endothelial cells can communicate with other endothelial cells or with smooth muscle cells via gap junctions.

6.3 There are two primary mechanisms for blood flow control. The first is the acute local control and the second is a long-term control. Acute control is rapid constriction or dilation in response to stimuli. Long-term control is the formation of new blood vessels due to changes in the length/diameter of existing blood vessels. The regulation of these mechanisms is primarily through the need of the tissue although there is some disagreement as to the actual mechanism that controls blood flow. There is no "command center" that continuously controls blood flow through the capillaries.

6.4 The pressure gradient across capillaries is relatively low, not more than 5 mmHg. However, there is a large pressure gradient in the arterioles, which effectively forces the blood into the microvascular beds. However, if one looks at the pressure gradient in small segments of blood vessels, approximately 100-μm lengths, the pressure gradient across the capillaries is much larger than other blood vessels. Under disease conditions, the mean pressure within capillaries remains the same. The changes in pressure under these conditions are largely accounted for by the arterioles. The cardiac pressure pulse is marginally seen within the microvasculature, but this usually does not confound results or analysis techniques.

6.5 The mean velocity in capillaries is the slowest of all blood vessels and can be approximated by the blood cell velocity. Flow through the capillaries is dominated by the viscous fluid effects and not the inertial effects (i.e., there is a low Reynolds number). Under these conditions, the Navier–Stokes solution can be simplified to the Stokes equations, which are

$$\frac{\partial p}{\partial x} = \mu\left(\frac{\partial^2 u}{\partial x^2} + \frac{\partial^2 u}{\partial y^2} + \frac{\partial^2 u}{\partial z^2}\right) = \mu\nabla^2 u$$

$$\frac{\partial p}{\partial y} = \mu\left(\frac{\partial^2 v}{\partial x^2} + \frac{\partial^2 v}{\partial y^2} + \frac{\partial^2 v}{\partial z^2}\right) = \mu\nabla^2 v$$

$$\frac{\partial p}{\partial z} = \mu\left(\frac{\partial^2 w}{\partial x^2} + \frac{\partial^2 w}{\partial y^2} + \frac{\partial^2 w}{\partial z^2}\right) = \mu\nabla^2 w$$

or

$$\frac{\partial p}{\partial r} = \mu\left[\frac{\partial}{\partial r}\left(\frac{1}{r}\frac{\partial}{\partial r}(rv_r)\right) + \frac{1}{r^2}\frac{\partial^2 v_r}{\partial \theta^2} - \frac{2}{r^2}\frac{\partial v_\theta}{\partial \theta} + \frac{\partial^2 v_r}{\partial z^2}\right]$$

$$\frac{\partial p}{\partial z} = \mu\left[\frac{1}{r}\frac{\partial}{\partial r}\left(r\frac{\partial v_z}{\partial r}\right) + \frac{1}{r^2}\frac{\partial^2 v_z}{\partial \theta^2} + \frac{\partial^2 v_z}{\partial z^2}\right]$$

$$\frac{\partial p}{\partial \theta} = r\mu\left[\frac{\partial}{\partial r}\left(\frac{1}{r}\frac{\partial}{\partial r}(rv_\theta)\right) + \frac{1}{r^2}\frac{\partial^2 v_\theta}{\partial \theta^2} + \frac{2}{r^2}\frac{\partial v_r}{\partial \theta} + \frac{\partial^2 v_\theta}{\partial z^2}\right]$$

The Stokes equations are relevant for many gas flows. Due to the cellular matter taking a larger portion of the fluid cross-sectional area, the cellular elements may experience significant drag forces which can be quantified using

$$C_D = \frac{6\pi\mu rv}{\frac{1}{2}\rho v^2 \pi r^2}$$

assuming that the cellular elements are spherical.

6.6 The hydrostatic pressure within the capillaries and the interstitial space, as well as the colloidal osmotic pressure of the plasma and the interstitial space, dictate the net direction of water movement into or out of the capillary. This can be formulated as

$$\dot{m} = K_p(P_B - P_I - \Pi_B + \Pi_I)$$

Along the arterial side of the capillary, the net direction of water movement is generally out of the capillary, with a driving force of approximately 7 mmHg. Along the venous side, the forces tend to balance to aid in water movement back into the capillary with a driving force of approximately 3 mmHg. The fluid velocity within the interstitial space is relatively slow, and the majority of fluid does not have the ability to move freely. This is because there is a high concentration of charged proteins within the interstitial space which attract water molecules.

6.7 The hematocrit within capillaries tend to change based on the bifurcation angles and the splitting of fluid within the microcirculation. Two relationships that can be used to determine the branch hematocrit and branch velocities are

$$Hct_{\text{inflow}} v_{\text{inflow}} = Hct_{\text{branch 1}} v_{\text{branch 1}} + Hct_{\text{branch 2}} v_{\text{branch 2}}$$

$$Hct_{\text{inflow}} = \frac{Hct_{\text{branch 1}} v_{\text{branch 1}} + Hct_{\text{branch 2}} v_{\text{branch 2}}}{v_{\text{branch 1}} + v_{\text{branch 2}}}$$

Within the microcirculation, there is a general reduction in the hematocrit, which reduces the overall effective viscosity of blood. The effective velocity can be used in a modified Hagen–Poiseuille formulation to calculate the flow rate through a capillary as

$$Q = \frac{\pi \Delta P r^4}{8\mu_{eff} L}$$

This reduction in viscosity is primarily caused by a streaming of the blood cellular elements streaming toward the centerline velocity. Under normal conditions, there is an imbalance of forces on the cell, which would push the cell toward the lower shearing force. This would only be balanced when the cell is aligned with the highest fluid velocity.

6.8 In capillaries, red blood cells tend to form packets that act as plugs. Within these red blood cell packets there is very little plasma, and in fact the plasma normally collects in between different red blood cell plugs. Due to this plug formation, we can model the mass conservation of compounds within the fluid by

$$(O_2)_{\text{accumulated}} = (O_2)_{\text{entering}} - (O_2)_{\text{exiting}} + (O_2)_{\text{generated}} - (O_2)_{\text{consumed}}$$

which is shown for oxygen. The rate that a species is generated or consumed would depend on the particular kinetics of that reaction.

6.9 Two-phase flows are flows that have either two phases of matter or one phase of matter with different material properties. In the microcirculation, the plasma phase and the cellular phase must be accounted with different methods when analyzing flow conditions. This would be fairly difficult to do by hand because Newton's second law of motion would need to be solved independently for each particle within the fluid. In general, these types of flows are solved using numerical methods.

6.10 In the microcirculation, the effect of cell–wall interactions must also be accounted for because a cell adhered to the vessel wall may significantly alter the flow profile. Also, it is likely that white blood cells migrate out of the blood vessels and into the extravascular space along the capillary wall. To model these interactions, the mechanical properties of the cells should be developed and included within the solution method. In general, the bending moments and the extension ratios of the cells can be defined as

$$M_1 = \frac{G(K_1 + \nu K_2)}{\lambda_2}$$

$$M_2 = \frac{G(K_2 + \nu K_1)}{\lambda_1}$$

$$T_1 = \frac{E(\lambda_1^3 - \lambda_1)}{2\lambda_2} + T_0$$

$$T_2 = \frac{E(\lambda_2^3 - \lambda_2)}{2\lambda_1} + T_0$$

respectively. To make the modeling of blood cell–wall interactions more complicated, probability functions that describe the expression of cell surface receptors need to be developed. In general, the surface density of these receptors is dependent on various stimuli within the fluid phase.

6.11 Disease conditions that are common within the microcirculation are shock and edema. Shock occurs after there is a massive loss of blood which can lead to a significant decrease in blood pressure and a decrease in nutrient exchange. Edema is a condition in which there is excessive fluid accumulation within the interstitial space, either because the lymphatic system is not collecting the excess fluid or because too much fluid is leaving the blood vessel.

HOMEWORK PROBLEMS

6.1 The blood vessels that play the most important role in regulating the blood flow through microvascular networks are the

- **a.** arteries
- **b.** arterioles
- **c.** true capillaries
- **d.** venules
- **e.** veins

6.2 The exchange of gases, nutrients, and wastes occur only through what type of blood vessels? Why?

6.3 There are four forces that affect the movement of water through the capillary wall. The net osmotic pressure forces water (into, out of) the capillary wall. What is the major constituent of the osmotic force?

6.4 Discuss the primary forces that cause water movement out of the capillary along the arterial end and into the capillary along the venous end.

6.5 There are two major mechanisms that control the blood flow through tissues. Which of these mechanisms is rapid, and which is more long term? Do both mechanisms have the same net effect?

6.6 The myogenic theory is not widely accepted as the only regulator for blood flow through microvascular networks. Prepare a statement that is for the myogenic theory and one that is against the myogenic theory.

6.7 The endothelial cell nucleus protrudes into the bloodstream within the capillary beds. Does this affect blood flow and cellular transport through the microvascular network? Why is the capillary not "smooth"?

6.8 How does the intercellular cleft restrict the diffusion of small uncharged molecules through the capillary wall? How does the intercellular cleft restrict the diffusion of larger charged molecules? Why is there a difference between the diffusion of these two types of molecules?

6.9 List the major differences between vascular smooth muscle cells and skeletal muscle cells. Is mechanical work generated in a similar way between these two cell types?

6.10 Calculate the apparent viscosity for a capillary that has a pressure difference of 4 mmHg, a diameter of 10 μm, and a mean velocity of 12 mm/s, with a length of 100 μm.

6.11 Calculate the apparent viscosity for the same conditions as in Problem 6.10 if the mean velocity reduces to 0.2 mm/s, due to precapillary sphincter constriction.

6.12 Determine the velocity profile within a 12-μm-diameter capillary, with a viscosity of 4 cP, and a pressure gradient of −5 mmHg/500 μm. Assume that gravitational effects can be ignored and that the blood vessel is perfectly cylindrical.

6.13 Determine the velocity profile within a 12-μm-diameter capillary that has a red blood cell (8 μm diameter) flowing down the centerline of the blood vessel. Assume that the centerline velocity is 10 mm/s, that the pressure gradient is −5 mmHg/500 μm, and that the viscosity is 4 cP. The capillary is vertical, so gravitational effects cannot be ignored. Also, assume that the entire red blood cell flows with the same velocity and that the density of blood is 1050 kg/m^3.

6.14 Calculate the drag coefficient and viscous forces on the surface of the red blood cell in Problem 6.13, assuming that the only force which acts on the blood cell is shear stress due to the flowing blood. Also, simplify the geometry of the red blood cell so that the area of interest can be considered a perfect sphere. What is the Reynolds number in this flow scenario?

6.15 Calculate the volumetric flow rate of water out of the capillary under normal conditions where the permeability constant is 0.25 mL/(min*mmHg). How much fluid leaks out of the capillary within 1 day? Is this reasonable?

6.16 A patient is experiencing edema and the permeability constant increases to 1 mL/(min*mmHg). Assuming that the capillary hydrostatic pressure reduces to 23 mmHg and the interstitial pressure increases to −2 mmHg, and that the osmotic pressures have not changed, how much fluid leaks out of the capillary within 1 day?

6.17 What is the hematocrit in each of the daughter branches for a simple one-parent-to-two-daughter-branch network, if the inflow velocity is 50 mm/s (in a tube with a diameter of 60 μm) and the velocity in the first branch is 75 mm/s (in a tube with a diameter of 40 μm)? The diameter of the second branch is 30 μm. The feed hematocrit is 40% and the hematocrit in the second branch is 24%.

6.18 Calculate the concentration of carbon dioxide within a capillary assuming that all reactions can be modeled with first-order kinetics. In this scenario, the blood vessel is 300 μm in length, with a mean blood velocity of 15 mm/s. The kinetic rate constants for carbon dioxide formation and consumption are 1.17 mol/s and −0.2 mol/s, respectively. Assume that the initial concentration of carbon dioxide within the blood vessel is 100 mL.

6.19 Assume that the consumption of oxygen can more accurately be modeled with second-order kinetics ($R_{O_2} = k(C_{O_2})^2$), where the kinetic rate constant for oxygen consumption is 20 mol/s. Calculate the outflow concentration of oxygen if the inflow concentration of oxygen is 200 mL, within a blood vessel that is 50 μm in length, and has a mean velocity of 35 mm/s.

6.20 *Modeling*. Compose a code that can accurately predict the flow through a capillary with consideration on the accumulation of carbon dioxide and depletion of oxygen (using the kinetics discussed) throughout the capillary.

References

[1] J. Aroesty, J.F. Gross, Convection and diffusion in the microcirculation, Microvasc. Res. 2 (1970) 247.
[2] J. Aroesty, J.F. Gross, Pulsatile flow in small blood vessels. I. Casson theory, Biorheology 9 (1972) 33.
[3] J. Aroesty, J.F. Gross, The mathematics of pulsatile flow in small vessels. I. Casson theory, Microvasc. Res. 4 (1972) 1.
[4] R.A. Brace, A.C. Guyton, Interaction of transcapillary starling forces in the isolated dog forelimb, Am. J. Physiol. 233 (1977) H136.
[5] E. Dejana, Endothelial adherens junctions: implications in the control of vascular permeability and angiogenesis, J. Clin. Invest. 98 (1949) (11-1-1996)
[6] C. Desjardins, B.R. Duling, Microvessel hematocrit: measurement and implications for capillary oxygen transport, Am. J. Physiol. 252 (1987) H494.
[7] C. Desjardins, B.R. Duling, Heparinase treatment suggests a role for the endothelial cell glycocalyx in regulation of capillary hematocrit, Am. J. Physiol. 258 (1990) H647.
[8] B.R. Duling, B. Klitzman, Local control of microvascular function: role in tissue oxygen supply, Annu. Rev. Physiol. 42 (1980) 373.

[9] P. Gaehtgens, C. Duhrssen, K.H. Albrecht, Motion, deformation, and interaction of blood cells and plasma during flow through narrow capillary tubes, Blood Cells 6 (1980) 799.
[10] P. Gaehtgens, Flow of blood through narrow capillaries: rheological mechanisms determining capillary hematocrit and apparent viscosity, Biorheology 17 (1980) 183.
[11] B. Klitzman, B.R. Duling, Microvascular hematocrit and red cell flow in resting and contracting striated muscle, Am. J. Physiol. 237 (1979) H481.
[12] B. Klitzman, A.S. Popel, B.R. Duling, Oxygen transport in resting and contracting hamster cremaster muscles: experimental and theoretical microvascular studies, Microvasc. Res. 25 (1983) 108.
[13] A. Krogh, E.M. Landis, A.H. Turner, The movement of fluid through the human capillary wall in relation to venous pressure and to the colloid osmotic pressure of the blood, J. Clin. Invest. 11 (1932) 63.
[14] H.S. Lew, Y.C. Fung, The motion of the plasma between the red cells in the bolus flow, Biorheology 6 (1969) 109.
[15] H.H. Lipowsky, B.W. Zweifach, Network analysis of microcirculation of cat mesentery, Microvasc. Res. 7 (1974) 73.
[16] J. Prothero, A.C. Burton, The physics of blood flow in capillaries. I. The nature of the motion, Biophys. J. 1 (1961) 565.
[17] J.W. Prothero, A.C. Burton, The physics of blood flood in capillaries. II. The capillary resistance to flow, Biophys. J. 2 (1962) 199.
[18] G.W. Schmid-Schonbein, S. Usami, R. Skalak, S. Chien, The interaction of leukocytes and erythrocytes in capillary and postcapillary vessels, Microvasc. Res. 19 (1980) 45.
[19] G.W. Schmid-Schonbein, R. Skalak, S. Usami, S. Chien, Cell distribution in capillary networks, Microvasc. Res. 19 (1980) 18.
[20] G.W. Schmid-Schonbein, Capillary plugging by granulocytes and the no-reflow phenomenon in the microcirculation, Fed. Proc. 46 (5-15-1987) 2397.
[21] G.W. Schmid-Schonbein, Biomechanics of microcirculatory blood perfusion, Annu. Rev. Biomed. Eng. 1 (1999) 73.
[22] H. Schmid-Schonbein, J. Weiss, H. Ludwig, A simple method for measuring red cell deformability in models of the microcirculation, Blut 26 (1973) 369.
[23] T.C. Skalak, R.J. Price, The role of mechanical stresses in microvascular remodeling, Microcirculation 3 (1996) 143.
[24] B.W. Zweifach, Quantitative studies of microcirculatory structure and function. I. Analysis of pressure distribution in the terminal vascular bed in cat mesentery, Circ. Res. 34 (1974) 843.
[25] B.W. Zweifach, Quantitative studies of microcirculatory structure and function. II. Direct measurement of capillary pressure in splanchnic mesenteric vessels, Circ. Res. 34 (1974) 858.
[26] B.W. Zweifach, H.H. Lipowsky, Quantitative studies of microcirculatory structure and function. III. Microvascular hemodynamics of cat mesentery and rabbit omentum, Circ. Res. 41 (1977) 380.

CHAPTER

7

Mass Transport and Heat Transfer in the Microcirculation

LEARNING OUTCOMES

1. Estimate the rate of gas diffusion between the respiratory system and the cardiovascular system
2. Use Fick's laws of diffusion to quantify diffusion flux
3. Determine the rate-limiting steps in gas diffusion into or out of red blood cells
4. Learn important mass transport dimensionless parameters
5. Relate kinetic reactions to diffusion reactions
6. Analyze the tissue oxygenation area with the Krogh model
7. Explain mechanisms for glucose (and other large molecules) transport into the extravascular space
8. Model the vascular permeability for various compounds with different molecular weights, different sizes, and different net charges
9. Consider the energy of adhesion during transport
10. Analyze the strength of various molecular forces, including adhesion, cohesion, and electrostatic forces
11. Calculate the strength of molecular forces
12. Model the organization of water in the vicinity of a charged surface
13. Learn methods to describe transport through porous media
14. Analyze internal forced convection within the microcirculation
15. Describe how white blood cells adhere to the endothelial cell surface and transfer to the extravascular space

© 2015 Elsevier Inc. All rights reserved.

7.1 GAS DIFFUSION

In this section, we will consider the diffusion of oxygen and carbon dioxide from the alveolar space (which is discussed in more detail in Chapter 9) into the pulmonary capillaries and then from the systemic capillaries into the extravascular space/tissue. Chapter 9 also describes the transport of oxygen and carbon dioxide through the cardiovascular system (from the respiratory system standpoint), which is an integral component of gas diffusion, so some of the details may be presented in Chapter 9 instead of here. The pulmonary capillaries are designed as a very efficient and rapid gas exchanger. Under normal conditions, the partial pressure of oxygen within the blood entering the pulmonary capillaries is approximately 40 mmHg. The partial pressure of oxygen in blood leaving the pulmonary capillaries approaches 100 mmHg. Under resting conditions, the partial pressure of oxygen rises from 40 to 100 mmHg along the first one-third of the pulmonary capillary. The remaining distance (two-thirds of the capillary length) is left as a safety reservoir system to allow for the blood oxygen level to increase to 100 mmHg under nonnormal conditions (e.g., disease or strenuous exercise). As we will discuss in detail in Chapter 9, this rapid increase in the partial pressure of blood oxygen is accounted for by at least five different diffusion events. Oxygen must first diffuse through the alveolar space to the respiratory boundary (Figure 7.1). Oxygen must then diffuse across the respiratory boundary, which can be modeled with different components because of the different cells that compose the respiratory boundary (see Chapter 9). Oxygen must then diffuse through the blood and then diffuse across the red blood cell membrane. Lastly, oxygen must diffuse through the cytoplasm of the red blood cell to come into contact with hemoglobin. Carbon dioxide follows a similar process, but in the reverse direction within the pulmonary capillaries. However, a very small portion of carbon dioxide is carried directly by hemoglobin. Therefore, carbon dioxide movement can be modeled as a diffusion process across the red blood cell cytoplasm and/or blood and then the kinetic reactions that describe the conversion of carbon dioxide and water into carbonic acid would need to be considered. The difference in the partial pressure of carbon dioxide within the lung is approximately 40 mmHg in alveolar air and approximately 45 mmHg in the pulmonary capillary.

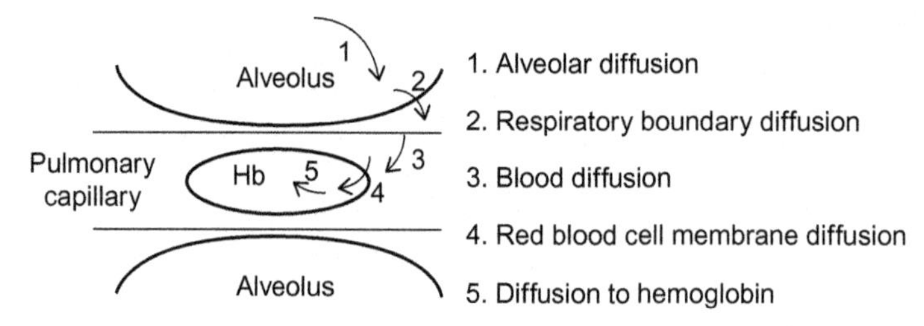

FIGURE 7.1 The five salient diffusion events for red blood cells within the pulmonary capillaries to become oxygenated. By comparing the relative rates of diffusion of these five events, we can see that respiratory boundary diffusion, blood diffusion, and diffusion/association with hemoglobin are the rate-limiting steps of red blood cell oxygenation. Models of red blood cell oxygenation would assume a homogenous distribution of air within the alveolus and a rapid diffusion across the red blood cell membrane.

Diffusion can be characterized by Fick's first and second laws of diffusion. The first law of diffusion relates the flux of molecules across a barrier to the concentration gradient across the same barrier. Mathematically, this can be represented as

$$\vec{J} = -D\nabla\vec{C} \tag{7.1}$$

where J is the diffusive flux, D is the diffusion coefficient, and C is the concentration of species, which may depend on spatial and temporal variables. In one-dimensional diffusion cases, which is what we will typically assume for this textbook, the concentration gradient would be equal to $\frac{\partial\vec{C}}{\partial x}$, because there is no gas movement or volume in the y-and z-directions. Fick's second law of diffusion relates the change in concentration over time to the concentration gradient over the diffusion barrier. This can be represented as

$$\frac{\partial\vec{C}}{\partial t} = D\frac{\partial^2\vec{C}}{\partial x^2} \tag{7.2}$$

for one-dimensional diffusion examples. Fick's second law of diffusion can be directly obtained from Fick's first law of diffusion as

$$\frac{\partial\vec{C}}{\partial t} = -\frac{\partial}{\partial x}(\vec{J}) = \frac{\partial}{\partial x}\left(D\frac{\partial}{\partial x}\vec{C}\right) = D\frac{\partial^2\vec{C}}{\partial x^2}$$

The general form for Fick's second law of diffusion would use the gradient operator for the concentration differences, shown as

$$\frac{\partial\vec{C}}{\partial t} = \nabla\cdot(D\nabla\vec{C})$$

or

$$\frac{\partial\vec{C}}{\partial t} = D\nabla^2\vec{C}$$

if the diffusion coefficient is constant in all directions. Solving the differential equation (7.2) for a one-dimensional diffusion example, the concentration of a species will vary with time and distance, as given by the following function:

$$C(x,t) = C(0)\left[\frac{x}{2\sqrt{Dt}}\right] \tag{7.3}$$

where $C(0)$ is the initial concentration at some point in space at time zero. Note that when we solve these types of scenarios, we typically assume that the feeding concentration (e.g., where oxygen is obtained from) is uniform and that it cannot deplete with time. If we included the decrease in the source oxygen concentration, we would need to couple multiple differential equations in order to solve the diffusion problem.

The denominator of the expression in Eq. (7.3) is termed the diffusion length and is used to provide a relative measure of how far the molecules move within a given time. The diffusion length can be used to compare the diffusion of different molecules; molecules with a larger diffusion length will typically diffuse more freely than those with a smaller diffusion length (under the particular tissue conditions). For an alveolus, which is assumed to be circular with a radius of R_A, the diffusion length will be equal to

$$R_A = 2\sqrt{Dt}$$

$$t = \frac{R_A^2}{4D}$$

assuming that the reference point is the center of the circle (this is derived directly from Eq. (7.3)). Using measured values for the oxygen diffusion coefficient within air (0.2 cm^2/s) and a typical alveolar radius of 100 μm, the average diffusion time for oxygen through the alveolus is 0.125 ms. This is a relatively accurate measurement for the time required for the oxygen concentration to equilibrate within the alveolus, assuming that oxygen enters from the center of a uniform alveolus. Also, this diffusion time is generally very fast as compared to the blood velocity within the pulmonary capillaries, and therefore, it can be assumed that the gas concentration within the alveoli is uniform (i.e., there is no concentration gradient within the alveolar space). Therefore, we typically do not need to consider the oxygen diffusion time throughout the alveoli within the model of red blood cell oxygenation because it is not limiting the diffusion process.

The second diffusion event that we will consider for red blood cell oxygenation is the movement of oxygen across the respiratory boundary, which is primarily composed of the alveolar epithelial cells and the capillary endothelial cells (note that there is a basement membrane that exists between the two cells, but this does not play a significant role in controlling the permeability of the respiratory boundary to oxygen/carbon dioxide). In the simplest case, the diffusion of oxygen across this boundary can be modeled by permeability laws. Assuming that the permeability of oxygen is uniform throughout the entire respiratory boundary, then the permeability can be defined in terms of the diffusion coefficient, the thickness of the boundary, and the partition coefficient for oxygen, as follows:

$$P = \frac{p_c D}{x} \tag{7.4}$$

where p_c is the partition coefficient, D is the diffusion constant, and x is the thickness of the boundary. The partition coefficient is the ratio of the diffusing concentration across the diffusion boundary to the original concentration. In our example, it is the concentration of oxygen in the alveolar space versus the concentration of oxygen within the blood. The average thickness of the respiratory boundary is on the order of 2.2 μm, but can be as small as 0.3 μm. The diffusion coefficient for oxygen across the respiratory boundary is approximately 2.3×10^{-5} cm^2/s. This simplified formula can be extended to the actual respiratory boundary by a summation of the different permeabilities, which can be assumed to be in series (i.e., the thickness and diffusion coefficients for each cell type would be needed for this analysis). An average respiratory permeability can also be obtained through a thickness weighted average for the individual permeabilities (e.g., the thickness of the cell membrane, the thickness of the cytoplasm, among others). Under normal conditions, this process must be considered during red blood cell oxygenation, because the cells that comprise the respiratory boundary can partially regulate the movement of species across the boundary.

The third diffusion event for oxygen is the diffusion of oxygen through the plasma in order to reach the red blood cell membrane. This diffusion is actually a combination of convection and diffusion because of the blood velocity through the pulmonary capillaries will move the dissolved species along the capillary length. Diffusion of oxygen typically occurs tangent to the blood flow direction, and diffusion is the only component that accounts for the movement of oxygen toward the red blood cell wall (this is an assumption that is relatively accurate because the red blood cells typically traverse a capillary in single file, and red blood cells do not have much freedom of movement in the radial direction). The permeability of oxygen through plasma is typically modeled through the use of a mass transfer coefficient, which is related to the Sherwood number (Sh), the diffusion coefficient, and a characteristic length. The Sherwood number is a dimensionless parameter that is the ratio of the convective mass transport to the diffusive mass transport:

$$\mathrm{Sh} = \frac{Kx}{D} \tag{7.5}$$

In Eq. (7.5), K is the mass transfer coefficient, x is the characteristic length, and D is the diffusion coefficient. In plasma, the diffusion coefficient for oxygen is on the order of $2 \times 10^{-5}\ \mathrm{cm^2/s}$. The plasma layer (characteristic length) ranges from 0.35 to 1.4 μm, depending on the exact red blood cell orientation and capillary size. The mass transfer coefficient is based on many physiological parameters, such as blood viscosity, blood speed, and dissolved species within the blood, and can only be calculated from the other parameters and not measured. The Sherwood number has been found to be dependent on the blood hematocrit and can be represented mathematically as

$$\mathrm{Sh} = \mathrm{Sh}_{0.25} + 0.84(\mathrm{Hct} - 0.25) \tag{7.6}$$

In Eq. (7.6), Hct is the hematocrit which is represented as the decimal equivalent of the percent hematocrit. $\mathrm{Sh}_{0.25}$ is the Sherwood number at a hematocrit of 0.25 (25%), which has been experimentally found to be equal to 1.3. Knowing the hematocrit and the characteristic plasma thickness, one can easily calculate the mass transfer coefficient that describes the particular flow conditions and then relate this to the other diffusion/permeability steps. This event must also be considered during normal red blood cell oxygenation because of the length scale that includes.

The fourth diffusion event during red blood cell oxygenation is the diffusion of oxygen across the red blood cell wall. The thickness of the red blood cell membrane is on the order of 10 nm, which is similar for most human cells. The diffusion coefficient for oxygen across the red blood cell membrane has been found to be similar to the diffusion coefficient for oxygen within plasma. Using Eq. (7.4), it can be seen that the permeability of the red blood cell wall will be approximately 100 times greater than the permeability of oxygen across the respiratory boundary (due to thickness differences). This assumes that the partition coefficient is close to 1 for both cases (this is actually a good assumption). Therefore, this diffusion occurs significantly faster than the second diffusion event during blood oxygenation and does not need to be considered in the overall red blood cell oxygenation model; the diffusion of oxygen through two cells (epithelial and endothelial) takes a significantly longer amount of time as compared with the diffusion through one cell membrane.

The fifth diffusion event that occurs for the oxygenation of blood is the diffusion of oxygen within the red blood cell until the oxygen comes into contact with and reacts with hemoglobin. Similar to the convection and diffusion that oxygen experiences while moving within the plasma, once in the red blood cell, oxygen molecules can either react immediately with nearby available hemoglobin or diffuse through the red blood cell to interact with a different available hemoglobin molecule. The Thiele modulus is a dimensionless parameter that relates the kinetic reaction (in this case, oxygen–hemoglobin binding) to the molecular diffusion through the particular medium. The Thiele modulus is defined as

$$\phi = \sqrt{\frac{kx^2}{D}} \tag{7.7}$$

where k is the kinetic association constant, x is a characteristic length, and D is the diffusion coefficient. The oxygen–hemoglobin kinetic association constant within red blood cells is equal to 128 s^{-1} and the diffusion coefficient for oxygen within red blood cells is on the order of 6×10^{-6} cm^2/s. If we take the characteristic length of the red blood cell to be 1.5 μm, which is the average thickness of the red blood cell, then the Thiele modulus is

$$\phi = \sqrt{\frac{128s^{-1}(1.5\mu m)^2}{6_E - 6cm^2/s}} = 0.6928$$

This diffusion/kinetic event must be considered as well, since the kinetic reactions take a significant amount of time as compared with the diffusion events.

Combining the five diffusion events for the oxygenation of red blood cells, we can see that only the diffusion of oxygen across the respiratory boundary, the convection/diffusion of oxygen within plasma, and the association of oxygen with hemoglobin can limit the time it takes for red blood cells to become oxygenated. The remaining two events occur so rapidly that they will have little to no effect on the overall time required for red blood cell oxygenation under normal conditions. The first two diffusion events that we will consider are described by permeability coefficients through a particular medium. To combine these coefficients, we will consider that the barriers to diffusion are in series and that they act to resist the motion. Permeability coefficients are equivalent to conductances in circuit theory, and therefore, the combined oxygen permeability is

$$\frac{1}{P_{barrier\,+\,plasma}} = \frac{1}{P_{barrier}} + \frac{1}{K_{plasma}} \tag{7.8}$$

$$\frac{1}{P_{barrier\,+\,plasma}} = \frac{x_{barrier}}{p_{c\text{-}barrier}D_{barrier}} + \frac{x_{plasma}}{\mathrm{Sh}_{plasma}D_{plasma}}$$

$$\frac{1}{P_{barrier\,+\,plasma}} = \frac{x_{barrier}}{p_{c\text{-}barrier}D_{barrier}} + \frac{x_{plasma}}{[\mathrm{Sh}_{0.25} + 0.84(\mathrm{Hct} - 0.25)]D_{plasma}}$$

$$\frac{1}{P_{barrier\,+\,plasma}} = \frac{2.2\mu m}{1(2.3_E - 5cm^2/s)} + \frac{0.875\mu m}{[1.3 + 0.84(0.4 - 0.25)](2_E - 5cm^2/s)}$$

$$P_{barrier\,+\,plasma} = 792\ \mu m/s$$

assuming that the hematocrit is 40%. To add the resistance of oxygen diffusion through the red blood cell into this combined permeability, a mass transfer Biot number and an effectiveness factor is needed. The Biot number relates the mass transfer of the molecules to the diffusion of the molecules with the following relationship

$$\text{Bi} = \frac{Px}{D} \tag{7.9}$$

For our case the Biot number will equal

$$\text{Bi} = \frac{(792\mu m/s)(1.5\mu m)}{6_E - 6cm^2/s} = 1.98$$

for the mass transfer/diffusion of oxygen within the red blood cell (using the red blood cell characteristics). To combine all of the resistances and diffusion along with the kinetic reaction rates, the effectiveness factor should be used. The effectiveness factor can determine the effect of diffusion on a reaction and whether diffusion or the kinetic reactions are limiting the entire process. A modified effectiveness factor for our case of red blood cell oxygenation within the pulmonary microcirculation can be defined as

$$\eta = \frac{\tanh(\phi)}{\phi\left[\frac{\phi}{\text{Bi}}\tanh(\phi) + 1\right]} \tag{7.10}$$

$$\eta = \frac{\tanh(0.6928)}{0.6928\left[\left(\frac{0.6928}{1.98}\right)\tanh(0.6928) + 1\right]} = 0.7155$$

An effectiveness factor less than 1 suggests that mass transfer is limiting the reaction more significantly than the kinetic reaction rates. However, in this process, the kinetics can change drastically based on percent oxygenation and the partial pressure of oxygen within the blood. Therefore, under normal conditions, oxygen mass transfer across the respiratory boundary and the red blood cell membrane limit the overall oxygenation of red blood cells.

The second case of gas diffusion that we will consider is the delivery of oxygen to the tissues within the systemic circulation. One of the first physiologists to tackle the problem of oxygen delivery to the tissues and to develop a feasible solution to the problem was August Krogh. Similar to the example of blood oxygenation, this process can be simplified into five diffusion events. The first is the dissociation of oxygen from hemoglobin and the red blood cells, then the diffusion and convection of oxygen through plasma, then the diffusion of oxygen across the endothelial cell barrier, then the diffusion of oxygen through the interstitial space, and finally the diffusion of oxygen into cells/mitochondria. Krogh realized that many tissues are arranged with a very regular spacing between the capillaries in three-dimensional space. Assuming that a capillary is perfectly cylindrical, Krogh made the argument that oxygen would be able to diffuse with an equal probability along every radial direction. This also assumes that there is no diffusion restriction in one particular direction, or in other words, the rate of diffusion is the same in each direction. Therefore,

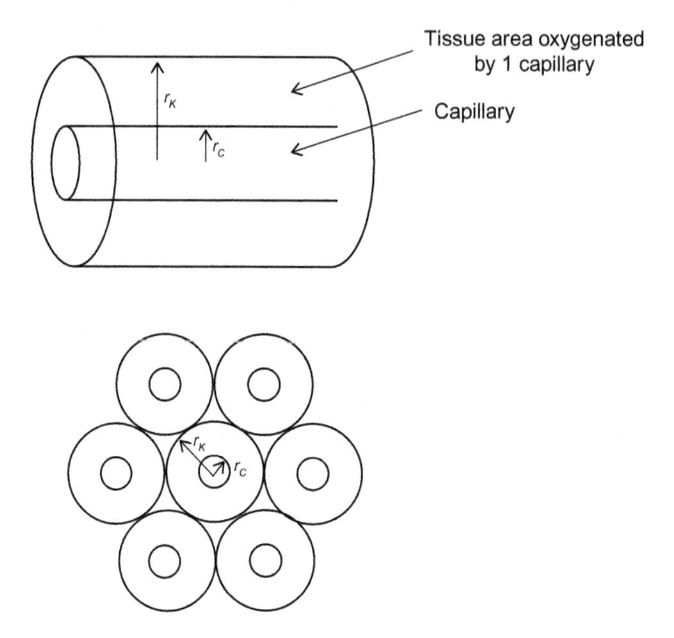

FIGURE 7.2 Krogh model for tissue oxygenation around uniformly spaced capillaries. r_K is the Krogh radius and r_C is the capillary radius. This model assumes that each capillary can supply a uniform area of tissue with oxygen and is very accurate for predicting the oxygenation of tissue with ordered arrays of capillaries (like muscle). However, this model does not take into account the decrease in oxygen along the capillary length, inhomogeneities within the tissue, and nonordered arrays of capillaries (as seen in brain tissue). Even accounting for these limitations, the Krogh model can calculate an accurate estimate of tissue oxygenation.

each capillary would supply a perfect cylindrical area of tissue with oxygen (Figure 7.2). Also, the oxygen concentration at one particular radius would be equal for any angle theta, but the oxygen concentration would decrease as the distance away from the capillary increases. Figure 7.2 illustrates that within this model, there are some areas of tissue that do not receive oxygen (or other nutrients) from the capillaries, if the cylinders do not overlap. Overlap within the supply cylinder could be good for redundancy but could be bad from an energy expenditure point of view. Regardless, at this point, this is a good first approximation for tissue oxygenation. This model is only accurate if the tissue capillaries are arranged in this most efficient organization array. In this model, r_K is the Krogh radius, and its value depicts how effective the capillary is at supplying oxygen to the surrounding tissue. Tissue with a larger r_K can have a lower number of capillaries with a larger spacing between the capillaries, whereas tissues with a smaller r_K need more capillaries to supply the same area as tissues with a larger r_K. This is a similar analysis for the overlapping of blood vessels, because more capillaries would be needed to supply the same area, if overlapping vessels were found.

To develop the mathematics for tissue oxygenation, let us consider that the diffusion of oxygen is steady and only occurs in the radial direction. We can also neglect the oxygen convection because the blood velocity is relatively slow as compared with the oxygen permeability throughout the tissue. A mass balance across this tissue can be obtained by converting the radial component of Eq. (3.62) to a convective/diffusion one-dimensional steady diffusion scenario:

$$\frac{D}{r}\frac{d}{dr}\left(r\frac{dC}{dr}\right) = R \tag{7.11}$$

In Eq. (7.11), D is the diffusion coefficient, r is the radius, C is the concentration of a particular molecule, and R is the reaction rate for that molecule. Integrating Eq. (7.11) twice with respect to r yields a formulation for the concentration of a diffusing molecule with respect to the radial direction:

$$C(r) = \frac{r^2 R}{4D} + C_1 \ln(r) + C_2 \tag{7.12}$$

As with other differential equations, boundary conditions are needed to solve this equation in terms of the particular problem (i.e., to find the two integration constants C_1 and C_2). Let us assume that the plasma concentration of oxygen is uniform and is denoted as C_p. This allows us to formulate the first boundary condition of

$$C(r = r_C) = C_p$$

which states that the concentration of oxygen at the wall of capillary (r_C) is equal to the plasma concentration of oxygen. Our second boundary condition uses the Krogh radius. At r_K, the diffusion of oxygen out of the Krogh cylinder is balanced by the diffusion of oxygen into the Krogh cylinder (from the neighboring cylinder). In other words, the flux of oxygen is equal to zero at the Krogh radius, because this is the farthest point that each capillary can effectively supply. Note that the oxygen concentration is not zero at the Krogh radius; if it was, the tissue would not obtain sufficient oxygen. This second boundary condition states that the movement of oxygen is balanced at the Krogh radius. Therefore, the second boundary condition is

$$\frac{dC}{dr}\Big|_{r=r_K} = 0$$

Solving Eq. (7.12) using the two boundary conditions provides a means to obtain the integration constants, which are

$$C_1 = -\frac{r_K^2 R}{2D}$$

and

$$C_2 = C_p + \frac{r_K^2 R}{2D}\ln(r_C) - \frac{r_C^2 R}{4D}$$

Substituting and simplifying the derived integration constants into Eq. (7.12) develops a mathematical relationship for the change in concentration of a molecule with radial distance. This relationship is normally represented as

$$\frac{C(r)}{C_p} = 1 + \frac{r_K^2 R}{4C_p D}\left[\frac{r^2 - r_C^2}{r_K^2} + 2\ln\left(\frac{r_C}{r}\right)\right] \tag{7.13}$$

Example

Calculate the oxygen concentration at a distance of 10 and 20 μm from a capillary with a radius of 4 μm. Assume that the Krogh radius is 35 μm, the diffusion coefficient for oxygen is 2×10^{-5} cm^2/s, the plasma oxygen concentration is 4×10^{-8} mol/cm^3, and that the oxygen reaction rate is 5×10^{-8} mol/cm^3 s. Also plot the oxygen concentration as a function of radial distance. How does the plot change if the Krogh radius reduces to 25 μm?

Solution

$$C(r) = C_p + \frac{r_K^2 R}{4D}\left[\frac{r^2 - r_C^2}{r_K^2} + 2\ln\left(\frac{r_C}{r}\right)\right]$$

$$C(r) = 4_E - 8\ mol/cm^3 + \frac{(35\mu m)^2(5_E - 8mol/cm^3s)}{4(2_E - 5cm^2/s)}\left[\frac{r^2 - (4\mu m)^2}{(35\mu m)^2} + 2\ln\left(\frac{4\mu m}{r}\right)\right]$$

$$C(10\mu m) = 4_E - 8\ mol/cm^3 + \frac{(35\mu m)^2(5_E - 8mol/cm^3s)}{4(2_E - 5cm^2/s)}\left[\frac{(10\mu m)^2 - (4\mu m)^2}{(35\mu m)^2} + 2\ln\left(\frac{4\mu m}{10\mu m}\right)\right]$$

$$= 2.65_E - 8\ mol/cm^3$$

$$C(20\mu m) = 4_E - 8\ mol/cm^3 + \frac{(35\mu m)^2(5_E - 8mol/cm^3s)}{4(2_E - 5cm^2/s)}\left[\frac{(20\mu m)^2 - (4\mu m)^2}{(35\mu m)^2} + 2\ln\left(\frac{4\mu m}{20\mu m}\right)\right]$$

$$= 1.78_E - 8\ mol/cm^3$$

The oxygen concentration as a function of Krogh radius and distance is plotted in Figure 7.3.

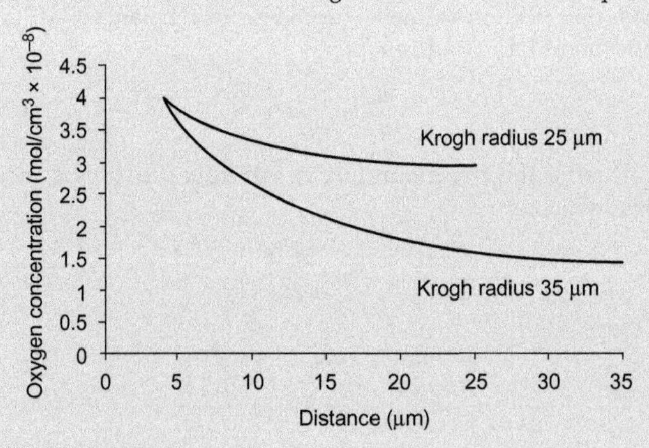

FIGURE 7.3 Oxygen concentration as a function of distance and the Krogh radius.

Although the Krogh method to calculate tissue oxygenation is possibly the most famous and the most often used model, there are some limitations associated with this model. First, Krogh did not consider the reduction in plasma oxygenation concentration along the capillary length. This model also assumes that the capillaries are uniformly distributed in a regular array throughout the tissue. This is fairly accurate for some tissue like skeletal muscle, but for other tissues (especially the brain), the capillary distribution is seemingly random. Krogh also assumed that diffusion only occurs within one direction in both the plasma and the tissue (i.e., radial). If these assumptions are not met, there may be a drastic effect on the movement of oxygen throughout the tissue, and it is likely that the tissue oxygen concentration will not be uniform at one particular radial location.

The same analysis can be conducted for carbon dioxide exchange. The formulations are the same, and it turns out that the same processes are the rate-limiting steps for carbon dioxide movement. The following are experimentally determined diffusion coefficients/kinetic rate constants for carbon dioxide. The diffusion coefficient for carbon dioxide movement through the respiratory boundary is $8.3 \times 10^{-4}\ cm^2/s$. The diffusion coefficient for carbon dioxide through plasma is on the order of $4 \times 10^{-4}\ cm^2/s$. Within the red blood cell, the diffusion coefficient for carbon dioxide is $1.6 \times 10^{-4}\ cm^2/s$ and the kinetic association constant for the formation of bicarbonate within the red blood cell is $0.13\ s^{-1}$ (again, recall that the majority of carbon dioxide does not associate with hemoglobin). One typically does not discuss a Krogh radius for carbon dioxide, since the tissue is supplying the capillary with carbon dioxide and not the reverse.

7.2 GLUCOSE TRANSPORT

Glucose diffusion out of the vascular system is similar to the diffusion of oxygen and carbon dioxide; however, because glucose is not lipid soluble, there are other mechanisms present to efficiently move glucose into the interstitial space. This is typically termed glucose transport. Once glucose passes through the intercellular cleft (see Section 7.3), a carrier protein is required to move glucose into individual cells, so that the glucose can be used as an energy source. In most cases, glucose transport is mediated by a coupled transport protein that moves some molecule in the direction of its electrochemical gradient and at the same time moves glucose against its electrochemical gradient. The energy gained by moving the first molecule down its concentration gradient is used to move glucose against its gradient. Remember that glucose is typically in a higher concentration within the cells than outside of the cell. In the most common transport mechanism for glucose, sodium is moved in the direction of its concentration gradient (i.e., into cells) at the same time that glucose is moved into the cell. The type of transporter that is responsible for this is termed a symporter. The energy that is gained from sodium movement is used to transport glucose against its gradient. This transporter only works after both sodium and glucose have interacted with the protein, which most likely occurs when the protein is open to the extracellular space (because sodium is in a much higher concentration in the extracellular space). Once sodium and glucose bind to the protein, there is a conformational change that closes the protein to the extracellular space and opens it to the intracellular space. The conformational change also causes the release of sodium, and glucose molecules into the cell after the conformational change. Many of these types of transporters exist for various compounds coupled to glucose movement. In the case of sodium–glucose co-transport, the sodium concentration gradient would be maintained through the presence of sodium–potassium ATPase molecules on both the apical and basolateral side of the endothelial cell. The glucose that enters the endothelial cell would then typically leave the endothelial cell via a glucose transporter that is present on the basolateral side of the endothelial cell. Once glucose has entered the extravascular space, it can be transported, by diffusion, to other cells within the system.

In many tissues, this asymmetric distribution is used to move glucose through cells that line a cavity/lumen, for example, the epithelial cells that line the intestines. Within the epithelial cells there is a high concentration of glucose but there is also a low concentration of glucose on either the intestinal side or the basal side of the cells. To effectively transport glucose through the intestinal wall, the sodium–glucose co-transporter brings glucose and sodium from the intestine into the epithelial cells. A carrier protein specific for glucose is also found on the basolateral side of the cell and uses the glucose concentration gradient to move glucose out of the cell. There is a similar mechanism within the kidneys. Glucose transport is difficult to model because of the coupling of the proteins conformational changes and binding kinetics. The affinity of the protein for glucose is dependent on the presence of other ions (such as sodium and its concentration gradient), the pH of the solution, the quantity of proteins within the cell membrane, and the concentration gradient of glucose. Therefore, for most molecules that use this mechanism of transport, many assumptions must be made to conduct some form of analysis.

7.3 VASCULAR PERMEABILITY

We will now move our discussion to generalizations about how molecules move out of and into the vascular system. Diffusion accounts for the majority of exchange between the vascular system and the cells throughout the body. The amount of water and dissolved solutes that move allow for a constant mixing between these two compartments and therefore the constituents of the plasma are very similar to the constituents of the interstitial fluid. As we know, diffusion is a result of random thermal motion of molecules. If the molecule is lipid soluble (i.e., hydrophobic), then it can diffuse directly through the endothelial cell membrane. Physiologically relevant lipid soluble molecules are typically small and uncharged. There are no direct transport mechanisms (e.g., channels, transporters, among others) for these types of molecules present within the cell membrane. The permeability of these molecules is typically much larger than other molecules that are transported within the microvascular networks such as proteins. The permeability of these molecules tends to be very high because the endothelial cells do not restrict the movement of these molecules; it is as if the endothelial barrier is not existent toward these molecules. Examples of these types of molecules are oxygen and carbon dioxide.

Most molecules, however, are not lipid soluble and cannot diffuse readily through the cell membrane. These molecules are typically soluble in water and can only pass through the microvascular wall between endothelial cells at the intercellular clefts or via active transport mechanisms within the endothelial cell membrane. As discussed in the previous chapter, the intercellular cleft accounts for less than 1/1000 of the total capillary surface area. This very small area however accounts for the vast majority of the movement of non-lipid soluble molecules, such as ions, glucose, and water. The diffusion of molecules through the intercellular cleft is so fast that it allows for solutes to rapidly approach equilibrium across the capillary wall (e.g., plasma concentration vs. the interstitial space concentration). In fact, equilibrium of solutes is typically reached within half of the capillary length, due to the overall slow velocity of blood and the very strong diffusional forces across the capillary wall. The remaining portion of the capillary is left as a safety reservoir (this is a similar mechanism that is found within the pulmonary capillaries for blood oxygenation and removal of wastes). The diffusion of molecules through the intercellular cleft occurs at a rate that is approximately 80 times greater than the blood flow through the capillary.

The process of diffusion through the intercellular cleft is principally regulated by the permeability of the pore to the specific molecule. As noted in Chapter 6, the intercellular cleft pore diameter is on the order of 7 nm. Water molecules, ions, glucose, and some other metabolically important molecules all have a molecular radius that is smaller than the 7 nm pore size and therefore can permeate through the pore relatively easily. Plasma proteins typically have a molecular size that is greater than 7 nm and therefore cannot diffuse through the pore under physiological conditions. However, under inflammatory diseases and some pathological scenarios, the size of these pores can increase and they can become permeable to plasma proteins. In the early 1950s, studies were conducted to determine the average permeability of the intercellular cleft to physiologically important molecules. From these studies, it was found that most physiologically important molecules have a permeability that is in the range of water. For instance, sodium chloride in solution has a permeability that is equal to 96% of water's permeability through the intercellular cleft. Glucose, a molecule that is 10 times the weight of water, has a permeability in the range of 60% of water's permeability through the intercellular cleft. Albumin, which has a molecular weight that is approximately 4000 times larger than water, has a permeability of 0.1% of water's permeability through the intercellular cleft and thus is effectively impermeable across the endothelial cell wall.

In most engineering applications, diffusion is principally regulated by the concentration gradient across the membrane. While that is partially true within capillaries, the permeability is much more easily controlled and is therefore the principal regulator of diffusion within the microcirculation (e.g., not the concentration gradient as with most standard engineering applications). For most physiologically important molecules (i.e., glucose, salts, water), the permeability is so large that very small differences in the concentration gradient cause very large changes in the rate of movement across the capillary wall. Also, the energy that would be required to maintain a large concentration gradient for each biologically important molecule would exceed the body's capacity and in some instances would vary from the concentration that is needed for other physiologically relevant processes. For instance, for cells to function in the innate environment, it is required that sodium is present in the extracellular fluid at a concentration that is approximately 100 times that concentration within the cell. However, if concentration gradient was the principle regulator of diffusion, then the concentration in the extracellular fluid would need to be lower than the blood sodium concentration. The blood sodium concentration is variable, is not easily controlled, and is not something that the body would want to maintain at a constant value since digestion/absorption alters the sodium concentration. If we extend this to every metabolically important molecule that diffuses across the capillary wall, each and every molecule would need to be regulated, controlled, and monitored in order to maintain a concentration gradient that favors diffusion in a particular direction. The majority of these metabolically important molecules have a variable concentration with time. Therefore, it is much easier to control the permeability of the diffusing molecule, which is independent of the concentration.

Recall that some molecules, like glucose that was described in the previous section, can also be transported through the endothelial cells that line the capillary walls, instead of being transported in-between the endothelial cells. In this case, specific transport proteins are needed on the apical side (side that faces the blood) and on the basolateral side (side that faces the extravascular space) of the endothelial cell. Most of these transporters derive their energy through the motion of a molecule, such as sodium, down its concentration gradient in order to move a second molecule up its concentration gradient and into the cell. These types of transporters would be present on the apical side of the tissue. On the basolateral side of the tissue, transporters would be present that allow for the motion of the metabolically important molecule down its concentration gradient and out of the cell. To ensure that these proteins function properly, active transport mechanisms, such as the sodium–potassium ATPase, would need to be present to maintain the concentration gradient of the specific molecules used in the first stage of the transport. This requires significant energy and is not a feasible transport mechanism for every metabolically important compound.

An important application of vascular permeability is drug delivery mechanisms. In general, there are two methods to deliver drugs to body: local delivery and systemic delivery. Local delivery is of less interest to biofluid mechanics analysis because the drug is typically directly injected into the site of interest. However, for systemic delivery, the drug can either be given orally, injected into a muscle, injected into the interstitial space, or injected into a vein. For all of these routes, except for intravenous injection, the drug will enter the vascular system via diffusion through microcirculation beds. For all injection methods, the drug will typically bind to plasma proteins and be delivered to all of the organs in the body. In some instances, the metabolites of the drugs are active. Therefore, the drug would need to make it to an organ that can metabolize the drug, usually the liver, and then the metabolites are transported through the vascular system to have some effect. In general, the drug or its metabolites are then removed by the liver or the kidneys. The importance of drug delivery is that the size of drug molecules and the molecular weight of drug molecules is typically larger than the usual molecules transported through these mechanisms. Combined with this, the permeability and diffusion coefficient can be significantly lower than other molecules. Therefore, the design of drugs, including their mode of delivery, need to be carefully considered in order to ensure proper delivery.

It is also important to differentiate between diffusion, active transport, cellular transport, and organ level transport. We discussed diffusion in Section 7.1, and we will relate this to the more physiological transport mechanisms. Active transport requires the presence of a particular receptor within the cell membrane. These receptors are proteins that function to transfer signals, possibly in the form of molecules, from one side of the membrane to the other (for clarification, this also includes transfer across the nuclear membrane and other organelle membranes). Under most conditions, the binding of the ligand, which is the molecule that will be transported, to the receptor, induces a conformational change in the protein structure of the receptor. This change can either cause the ligand to be brought into the cell, a pinocytosis process to cause the molecule to be put into the cell or signal a cascade that may result in a cellular level functional change. Most receptor–ligand events can be described by kinetic reactions that describe the association and the dissociation of the receptor–ligand complex. Depending on the magnitude of the association rate constant as compared to the dissociation constant, the outcome of these kinetic events will be determined. At the cellular level, transport of molecules will either occur via diffusion through the cytoplasm or a coupled diffusion, where the molecule of interest is associated with some other molecule (think of motor proteins). This has been discussed previously, but the only new considerations would be that the diffusion coefficients would differ because the medium that the molecules are moving through are different.

Organ level transport is actually fairly intuitive to model; however, in practice the information that is necessary to complete this type of modeling is typically missing. For instance, the exact blood/lymph flow characteristics would need to be known for the organ that is being modeled. If we assume a Krogh type of model, with uniform or some clearly defined spacing, then this may be known. However, the fluid velocity through each blood and lymph vessel would need to be known, and as we have learned earlier, fluid velocity differs temporally and each capillary is not fully perfused at one instant in time. Therefore, many simplifications/assumptions would be needed to determine the location of blood vessels along with the flow rate of blood through those vessels, over some period of time. The second issue that complicates the modeling of molecules at the organ level is that permeability, diffusion coefficients, and other transport phenomena are generally not known along the entire path. We may be able to develop a formulation that simplifies this transport (recall oxygen transport across the respiratory boundary as discussed in Section 7.1), but then the accuracy of the solution may vary greatly. Also, the diffusion properties depend on the homogeneity of the tissues/organs and as we can imagine, there is a significant amount of inhomogeneities in biological specimens. The last issue with organ level transport, which confounds the solution methods, is that multiple length and time scales are involved in the transport. If we are discussing transport of a species through the digestive tract, we can concern ourselves with the transport across a cell membrane, which is typically 7–8 nm, or we can be concerned with the transport along the tract, which is approximately 30 m in length. This is an astounding difference in length scales, which would need to be coupled if solving an organ transport problem. For a time-scale problem in the digestive tract, molecular transport can occur within a few microseconds or as long as a few days. Therefore, the solutions to these types of problems would need to couple both length and time scales that vary at a few orders of magnitude. Compartmental modeling is a good way to handle these types of problems, and it will be presented in Chapter 12.

7.4 ENERGY CONSIDERATIONS

The transfer of nutrients can be classified as an adhesion reaction between the nutrient and the transporter (e.g., glucose) or as an interaction between the nutrient and water (e.g., ions). In either case, there is some energy that is required to either maintain that adhesion or break the adhesion. In other words, there is a force required to break the adhesive interactions between the two molecules that are involved in the diffusion process so that one molecule can diffuse away freely. Recall that all ions are hydrated in physiological conditions and their transport would require some alteration to the hydrated form. In the case of an adhesion-based transporter, there must be some interaction so that the molecule can associate with the transporter. This adhesion would help to localize the molecule that is being transported but would hinder the overall diffusion of the molecule if the association energy was too high. Adhesion forces are typically small, on the nano- to micro-Newton range, and the adhesive forces increase with a decreasing surface roughness. It is also typical that flexible surfaces have a larger adhesive force than stiffer surfaces because the flexible surface can conform to the surface profile of the material that it is adhering to (therefore there would be a higher contact area). In general, there are four forces that contribute to adhesion: molecular forces, electrostatic forces, capillary forces, and excessive charge forces (which will not be discussed because this is not typically found in common biological systems).

Molecular forces act over the shortest range and are typically the weakest types of forces that play a role in adhesion. These types of attractive forces typically arise due to changes in electronic structure of the atoms that comprise the molecules. This means that electrically neutral molecules can exhibit molecular forces, through transient dipoles created by these fluctuations. Also, any molecule that has a permanent dipole (such as water) will experience molecular forces. This type of interaction is termed a van der Waals interaction. Van der Waals forces are composed of two components: an attractive force and a repulsive force. The attractive force arises from transient dipoles within electrically neutral molecules. The repulsive component arises because two molecules cannot enter the same space, and therefore, no matter how strong the attractive force is, the two molecules cannot exist in one location. The force associated with molecular interactions can be described by

$$\vec{\psi}(d) = \vec{\psi}_0\left[\left(\frac{d'}{d}\right)^{12} - \left(\frac{d'}{d}\right)^{6}\right] \tag{7.14}$$

where Ψ_0 is the strength of the molecular interactions when the molecules are in contact, d is the distance between molecules, and d' is the range of the interactions. Therefore, if $d = d'$, then the molecular forces will be equal to zero.

The second type of force that comprises adhesive forces is electrostatic forces. These can arise from either excess charge on a molecule or an electrical double layer. The force associated with charge should be familiar from a previous classical physics class and is described by Coulomb's law, which states that

$$F = \frac{1}{4\pi\varepsilon_0}\frac{q_1 q_2}{r^2} \tag{7.15}$$

where q is the excess charge on molecule 1 and 2, respectively; r is the distance between the two molecules; and ε_0 is the permittivity of free space. These charges can be removed if the substances are "grounded" or more likely the charges can be transferred to another

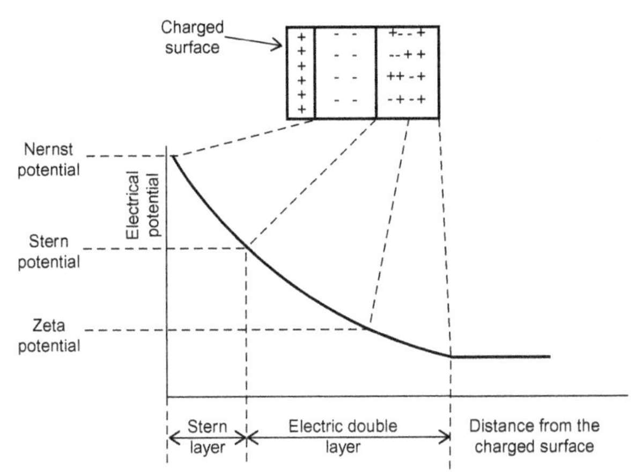

FIGURE 7.4 Electrostatic forces near a charged surface, which tend to organize water (or other ions) close to the boundary. The layer that is closest to the charged boundary and has the highest organization is termed the Stern layer. A second layer, of mostly organized charged species, is termed the electric double layer. Beyond the electric double layer, the charged species would resemble the bulk properties.

molecule. The electrical double layer is more of a concern in biological applications, where you encounter the interaction of a small particle (that may or may not have a charge associated with it) with a larger particle with a surface charge that acts as a boundary. For biological applications, this can be considered as the intercellular cleft (or a protein channel), in which small potentially charged molecules will permeate a larger channel that has a charge associated with it. The charge on the boundary attracts opposite charges within the solution; in the case of biological solutions the majority of the fluid is water. As an example, if the surface has a partial positive charge, then the partial negative charge present on the oxygen atom within water will be attracted to the surface (Figure 7.4; we use the example of water because this is the most likely bathing medium in the body, but this process can be extended to any bathing medium that is a dipole). This suggests that a second, potentially less strong, partial positive layer will be formed from the aligned hydrogen atoms that compose water. A second layer of water will form on top of the original aligned layer, until the electric field generated by the boundary (cell membrane wall or protein backbone) weakens to a point that it can no longer maintain this organized charge structure. The surface potential on the boundary is termed the Nernst potential and the potential at the point that the layer begins to breakup is termed the Stern potential. The entire distance between the boundary and the point where the charges in solution break up is called the Stern layer. The Stern layer is characterized by highly organized solutes, and again in biological applications this is typically water. Directly next to the Stern layer there is a slight organization of the charged molecules until the shearing plane. Between the Stern layer and the shearing plane there is some organization of the bathing medium but one can observe less organization. At the end of the shearing plane, the organization of the bathing medium resembles a solution in contact with an uncharged surface and is therefore organized randomly. This is characterized as the bulk organization of the fluid. In the layer between the Stern layer and the shearing plane, water is partially organized and partially random. The attractive force at the shearing plane is termed the zeta potential. The entire distance from the Stern layer to the point where the electric field equilibrates to the bulk fluid properties is termed the electric double layer. All of these potentials and distances are shown in Figure 7.4.

If there are two charged surfaces that come into close contact with each other, for example, a charged protein and an ion in solution, then the attractive force between these two surfaces can be described by

$$F_{ef}(d) = \frac{2\xi_1\xi_2}{\varepsilon_l\varepsilon_0} Ae^{-\frac{d}{\lambda_D}} \tag{7.16}$$

where ξ is the surface charge densities for the surface and the molecules within the fluid, A is the potential area of contact between the surface and the molecules, ε_0 is the dielectric constant of vacuum, ε_1 is the dielectric constant of the liquid, and λ_D is the Debye length. The Debye length quantifies the electrostatic double layer and is calculated from

$$\lambda_D = \sqrt{\frac{\varepsilon_l\varepsilon_0 k_B T}{e^2 \sum_i^n (c_i \chi_i^2)}} \tag{7.17}$$

where k_B is the Boltzmann constant (1.38×10^{-23} J/K), T is the temperature (in Kelvin), e is the charge of an electron, and c is the concentration of molecule i with a valence of χ. Equation (7.16) is a modification of the electric double-layer formulation, taking into account that the layer from each of the two surfaces will overlap as they approach. The summation in Eq. (7.17) takes into account the valence and concentration of particular species. This tends to be difficult to approximate for biological specimens because it is typical that charged surfaces (e.g., cell membranes) are composed of multiple charged species that can have a time-varying concentration.

Example

Calculate the attractive force between an endothelial cell and a red blood cell that are 3 μm apart. The surface charge density for the endothelial cell is -12 mC/m^2 and that for the red blood cell is -3 mC/m^2. Assume that the dielectric constant of the liquid is 75 and that the blood is flowing at a temperature of 32°C. The summation of all molecular species and valences on these two cell types is 1.7 μm^{-3}. Approximate the potential area of contact between the two cells, using geometric constraints.

Solution

First, calculate the Debye length for this interaction:

$$\lambda_D = \sqrt{\frac{75\left(8.85_E - 12\frac{C^2}{Nm^2}\right)(1.38_E - 23\,J/K)(273+32)K}{(1.6_E - 19C)^2(1.7\mu m^{-3})}} = 8.012\mu m$$

Now calculate the attractive force between the endothelial cell and the red blood cell:

$$F_{ef}(d) = \frac{2(-12\,mC/m^2)(-3\,mC/m^2)}{75\left(8.85_E - 12\frac{C^2}{Nm^2}\right)} Ae^{\frac{-3\mu m}{8.012\mu m}}$$

To approximate the area of interaction, we need to make some assumptions about the cells' contact area. Assume that the endothelial cell can be represented as a stiff plane and the red blood cell as a sphere. This would suggest that the contact area would only be one point on the red blood cell membrane. However, we know that the red blood cell can deform when it comes into contact with the endothelial cell. If we make the assumption that the deformation will encompass a 60° arc length (in three-dimensional space) of the original red blood cell, we get that the potential contact area is

$$A_{60^\circ} = \frac{4\pi(4\mu m)^2}{6} = 33.51\mu m^2$$

For interest, we will show other area calculations as well:

$$A_{30^\circ} = \frac{4\pi(4\mu m)^2}{12} = 16.75\mu m^2$$

$$A_{120^\circ} = \frac{4\pi(4\mu m)^2}{3} = 67.02\mu m^2$$

The effective force of interaction for these three contact areas are

$$F_{ef}(d) = \frac{2(-12\,mC/m^2)(-3\,mC/m^2)}{75(8.85_E - 12\frac{C^2}{Nm^2})}(33.51\mu m^2)e^{\frac{-3\mu m}{8.012\mu m}} = 2.50\mu N$$

$$F_{ef}(d) = \frac{2(-12\,mC/m^2)(-3\,mC/m^2)}{75(8.85_E - 12\frac{C^2}{Nm^2})}(16.75\mu m^2)e^{\frac{-3\mu m}{8.012\mu m}} = 1.25\mu N$$

$$F_{ef}(d) = \frac{2(-12\,mC/m^2)(-3\,mC/m^2)}{75(8.85_E - 12\frac{C^2}{Nm^2})}(67.02\mu m^2)e^{\frac{-3\mu m}{8.012\mu m}} = 5\mu N$$

As we can see, with an increasing contact area, the effective force increases as well.

The third type of force that comprise adhesive forces in biological system is capillary forces, which are subdivided into adhesion, cohesion, and surface tension. Surface tension is a fluid property that causes a surface of liquid to be attracted to another surface (such as a container for wetting). Surface tension arises due to the intermolecular forces within the fluid. For a fluid element within the bulk fluid phase, the intermolecular forces acting on that element are balanced in all directions. However, at the surface of the fluid, there is an interface between the fluid elements and the surrounding medium (e.g., for a glass of water, the fluid at the surface will tend to be pulled into the water because the intermolecular forces for water–water are stronger than the intermolecular forces for water–air). This inward force is balanced by the tendency of the fluid to resist compression until the energy (and hence the fluid surfaces) reaches equilibrium at a low energy state. The surface tension is defined as the force that is required to maintain the fluid in this low energy state per unit length. Surface tension (γ) is quantified by the pressure difference across the surface and the curvature that the surface takes in the low energy state, using the Young-Laplace equation:

$$\Delta p = \gamma\left(\frac{1}{r_x} + \frac{1}{r_y}\right) \tag{7.18}$$

where Δp is the pressure difference across the fluid surface, and r is the radius of curvature measured in the x- and y-directions. This can be normalized to a force by considering the area of contact between the two surfaces, where A_C is the contact area:

$$F_{ST} = \gamma A_C\left(\frac{1}{r_x} + \frac{1}{r_y}\right) \tag{7.19}$$

The surface tension of water is approximately 71 kPa/m, which can be used to approximate the forces or pressures associated with cells coming into contact with other structures. In fact, the surface tension of biomaterials is a common property to alter, in order to promote cell adhesion or inhibit cell adhesion. For instance, for cardiovascular devices, it has become common to make the surface tension very large, so that platelets will not adhere to the device. The reverse is true: if the device was designed to become endothelialized, the surface tension would be reduced to promote endothelial cell adhesion.

Adhesion occurs when two different molecules tend to stick together, due to some attractive force between the molecules. Adhesion is typically quantified by the amount of work that is required to separate the two materials and this is dependent on the surface tension of the materials, as follows:

$$W_A^{12\delta} = \gamma^{1\delta} + \gamma^{2\delta} - \gamma^{12} \tag{7.20}$$

where γ is the surface tension generated between materials 1 and 2 or the medium, δ, within which the materials are bathed. The force associated with adhesion can be calculated from the distance that is required to separate the materials and the work that is input into the system to separate the materials. Cohesion arises due to the interaction of the same molecules within the material. This attraction is caused by intermolecular forces, such as the common hydrogen bonding that occurs between individual water molecules. Dipole interactions can also lead to a cohesive interaction between molecules. Cohesion is also quantified through the surface tension of the material, as follows:

$$W_C^1 = 2\gamma^1 \tag{7.21}$$

where γ is the surface tension between the material phase of material 1 and its equilibrium vapor phase. Again, the force associated with cohesion can be calculated from the distance required to separate two molecules in solution.

Cellular adhesion can be divided into three basic categories, which include cell contact, cell adhesion/attachment, and cell separation. Cell contact is based on random connections that occur due to cells coming into contact with each other. These connections can act over varying ranges and may have a large force associated with them. Cell adhesion/attachment is based on chemical bonds which can be subdivided into the different molecular forces that we have described previously. These typically act over a range of no more than 0.5 nm and are typically in the nano-Newton range. Cell separation is based on the same attachment forces, when there are other forces that are acting to pull the cells apart. Initial adhesion and separation are based on the cell surface properties and geometries. We have hinted at this before: to model these interactions, many simplifying assumptions on these relationships would need to be made because cells are neither homogenous in the expression of surface molecules nor in contact geometry. Furthermore, both of these properties are time varying based on the input biochemical or biomechanical stimuli.

7.5 TRANSPORT THROUGH POROUS MEDIA

Porous media are solid materials that are composed of pore structures, which are typically fluid filled in biological applications. Porous media can have very different topographical and morphological properties, based on the internal organization of the pores. The most common porous media in biological systems is the interstitial space, which is composed of the extracellular matrix and interstitial fluid. As we have learned, the extracellular matrix is mostly composed of proteins and functions as a mechanical scaffolding for cells. This structure is required for the normal functioning of cells, including migration, proliferation, and apoptosis.

It is common to describe porous media based on the porosity, which is defined as

$$\varepsilon = \frac{\text{Volume associated with voids}}{\text{Total volume}} \tag{7.22}$$

The values that comprise porosity may need to be approximated from microscopy images of the porous media, and porosity can be dependent on the loading conditions of the porous media and whether or not cells can interact with and modify the pores. For instance, the extracellular matrix is deformable and thus the porosity may change under particular conditions because both the volume associated with voids and the total volume can change. The porosity of the interstitial space, neglecting blood vessels and cells, approaches 0.9, whereas the porosity of general biological tissues (including blood vessels and cells) would be closer to 0.25.

Porous media can also be classified based on the tortuosity of the pores within the media. Tortuosity is a measure of the random orientation and random spacing of pores throughout the media. Clearly, not all pores will be able to be classified as straight channels through the media, and tortuosity quantifies the degree of bending and/or twisting of the pores (Figure 7.5). Tortuosity is defined as

$$T = \frac{L_{actual}}{L} \tag{7.23}$$

where L_{actual} is the minimum actual distance between two points that are connected via pores and L is the straight-line distance between the two points. Thus, tortuosity can only be equal to or greater than one, and as tortuosity values approach one, the pores more closely resemble a straight pass through the channel. Note that the quantification of tortuosity does not account for the shape of the channel and has no need to, as long as the channel is open to some extent along its entire length.

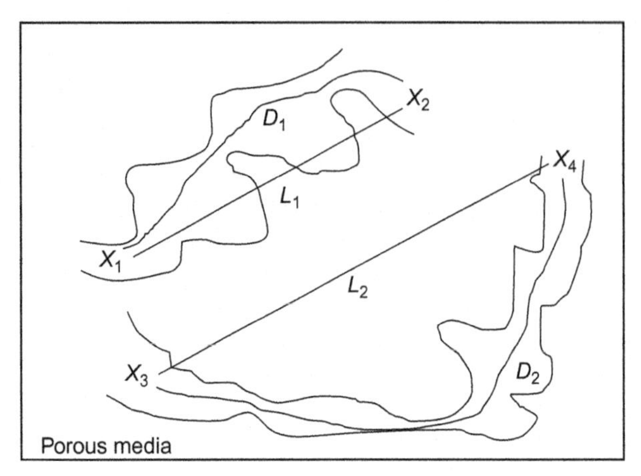

FIGURE 7.5 Tortuosity is defined as the actual minimum distance between two points connected by a pore divided by the straight line path between the points. This value can only be greater than or equal to 1.

$$T_{X_1 \rightarrow X_2} = \frac{D_1}{L_1}, T_{X_3 \rightarrow X_4} = \frac{D_2}{L_2}$$

Fluid flow through porous media is most typically quantified via Darcy's law, which relates the pressure gradient (as the driving force for fluid flow) to the flow rate of fluid through the porous media. Darcy's law is an idealized case that is not applicable for non-Newtonian fluids and for fluids when there is a large momentum transfer between the fluid and the solid particles. Under these conditions, Darcy's law states that the fluid velocity ($\vec{v}$) will be defined by

$$\vec{v} = -K\nabla\vec{p} \tag{7.24}$$

where K is the hydraulic conductivity of the medium and p is the driving pressure. If the porous media is non-isotropic, then the hydraulic conductivity is not constant and must be included within the gradient operator. Also, if blood flow and lymph flow is accounted for in Darcy's law, we get

$$\vec{v}_{blood} - \vec{v}_{lymph} = -\nabla(K\vec{p}) \tag{7.25}$$

which is a more accurate representation of actual biological flows through porous media. The inflow and outflow conditions can be time varying, spatially varying, or some average value of the flow conditions throughout the tissue. Gravitational or other forces can be included within Eq. (7.25) if necessary. The hydraulic conductivity is related to the viscosity of the fluid that is moving through the porous media. The hydraulic conductivity can be defined as

$$\mathrm{K} = \frac{c\varepsilon^3}{\mu\left(\frac{SA}{V}\right)^2} \tag{7.26}$$

where c is a shape factor, ε is the porosity, μ is the viscosity, SA is the interface surface area of the pores, and V is the total volume of the porous media. The hydraulic conductivity is a measure of how readily fluid can move through porous media. The larger the hydraulic conductivity, the more readily the porous media is permeated. The flow through porous media can also be described, in a more general case, by the Brinkman equation:

$$\mu\nabla^2 \vec{v} - \frac{1}{\kappa}\vec{v} - \nabla\vec{p} = 0 \tag{7.27}$$

which can account for transitional flows between the porous media. The Brinkman equation is also used when one would like to account for the flow through the porous media and the flow through particles (e.g., cells) suspended within the media. This formulation is somewhat difficult to use because experimentally or empirically deriving the constants is not easy and is dependent on many factors. Thus, altering the conditions that describe the flow may drastically alter the constants that describe the flow.

7.6 MICROCIRCULATORY HEAT TRANSFER

Heat transfer is an important principle in biological systems. As discussed in an earlier chapter, one of the most important functions of the cardiovascular system is to maintain the temperature of the body. In particular, the cardiovascular system is arranged as a counter-current heat exchanger, so that blood that loses heat as it passes to the extremities can gain heat back as it passes through the venous circulation toward the heart. The role of this arrangement is to ensure that the core body temperature maintains at a near constant temperature under any external conditions. A second example of heat transfer in the body is for air entering the lungs. This air must be warmed (or cooled in some extreme conditions) to body temperature. The purpose of altering the inspired air temperature to body temperature is first, to ensure that the lungs, which are in the core body, maintain a near constant temperature and second, to ensure that there is minimal heat flux from the lungs to other organs within the chest cavity and the blood, with which the oxygen and carbon dioxide exchange with. Major heat flux to these other locations would induce changes in the core body temperature. Cardiovascular heat transfer occurs through all blood vessels, but we will focus on the thinnest walled vessels to make our analysis easier. In thicker vessels, the wall can act as an insulator and heat can be transferred along the length of the wall as well. If we assume that the wall is thin then the analysis of heat transfer is significantly easier. Heat transfer for inspired air occurs primarily within the nasal passages. We will not analyze this situation but similar assumptions would need to be made and an extra simplifying assumption regarding the geometry of the nasal passages would be needed.

There are many different forms of energy in this universe; for this section we will focus on one of these, heat, which is the form of energy that is transferred from two systems as a result of a temperature difference between the two systems. The study of the rate of energy transfer in the form of heat as a function of temperature difference is termed heat transfer. As we stated above, heat transfer is a property of the cardiovascular system, but many other engineering systems that involve temperature differences must consider heat transfer. You may have discussed thermodynamics in a previous course. The distinction between thermodynamics and heat transfer is that thermodynamics only deals with quantities of energy that must be transferred in order to reach a particular state, whereas heat transfer deals with the time involved in the process of a system reaching a different thermodynamic state. The first two laws of thermodynamics lay the foundation for heat transfer analyses, but other principles, some of which will be developed here in the context of the cardiovascular system, are also needed. The first law of thermodynamics requires that the rate of energy transfer from one system equals the rate of energy gain for a second system (assuming a closed system). The second law of thermodynamics requires that heat is transferred from the system with a higher temperature to the system with a lower temperature. The most basic requirement for heat transfer to occur is that a temperature difference exists. This is the driving force for heat transfer, in the same way that pressure differences are the driving force for fluid flow. The rates of heat transfer are dictated by the temperature difference and the resistance between the two systems.

At this point, we will take the time to define some of the common variables that are discussed in heat transfer associated with a flowing fluid. Recall, that the total energy of a system (E) is defined as the summation of all types of energy that are present, such as kinetic energy, thermal energy, potential energy, magnetic energy among others. The internal energy of the system (U) is the sum of all of the energies associated with the molecular structure of the system and the interactions of molecules at the molecular level. The portion of the internal energy that is related to the kinetic energy (or energy of motion) of the molecules is termed the sensible energy. This type of energy is a function of temperature; at higher temperatures the molecules possess a higher kinetic energy and move at faster rates. A second component of the internal energy is based on the interactions of the molecules. These interactions are the strongest in solids and weakest in gases. Thus, a system that is in the gaseous phase must have more internal energy to maintain the system in a gaseous phase and to continually break any bonds that are formed during molecular interactions (e.g., gases are a higher energy state than liquids, which is greater than the energy of solids). The internal energy that is associated with the phase of the material is termed the latent energy. In some instances, heat transfer problems will involve the rearrangement of atoms within molecules or changes within the atomic particles (e.g., protons, neutrons, and electrons). The energy associated with these levels of organization are the chemical bond energy and the nuclear energy, respectively, but these are less common for our examples. Finally, for systems that involve fluid flow, we typically discuss combinations of energy. The energy associated with a stationary fluid would only be composed of the internal energy, whereas a flowing fluid would experience a combination of energy associated with the flow and the internal organization of the fluid. Under most instances, the energy associated with a flowing fluid is represented by the enthalpy (h), which is defined as

$$h = u + p\nu = u + \frac{p}{\rho} \tag{7.28}$$

where u is the internal energy per unit mass, p is the absolute pressure, ν is the specific volume, and ρ is the fluid density.

In heat transfer problems, the specific heat of the material(s) that is involved with the system is an important quantity to know. The specific heat defines the amount of energy that is required to raise the temperature of a unit mass of the material by one degree. This energy is normally dependent on the manner by which the process is carried out, but in general we are interested in and can quantify the specific heat of a substance for a process that is executed under a constant volume or one that is executed under a constant pressure. The specific heat at constant volume (c_v) is the amount of energy that is required to raise the temperature of a unit mass by one degree when the volume is fixed. A similar definition can be made for the specific heat at constant pressure (c_p). The specific heat at constant pressure is always greater than the specific heat at constant volume, because at constant pressure the system can expand and this requires energy that must be supplied by the system. For an ideal gas, the following equations apply

$$\begin{aligned} du &= c_v dT \\ dh &= c_p dT \end{aligned} \tag{7.29}$$

Equation (7.29) can be related to a finite change in the internal energy or enthalpy of the system, via the following

$$\begin{aligned} \Delta u &= c_{v,average} \Delta T \\ \Delta h &= c_{p,average} \Delta T \end{aligned} \tag{7.30}$$

For incompressible materials or materials that can be approximated as incompressible (e.g., the liquids that we have been discussing in the body), the specific heat at constant volume and the specific heat at constant pressure are nearly identical ($c_p \cong c_v \cong c$), because the specific volume of the material does not change significantly during a heat transfer problem, if the material is incompressible (however, we know that no true materials are incompressible). Under these conditions, the specific heat can be defined as

$$\Delta u = c_{average} \Delta T \tag{7.31}$$

There are three basic mechanisms for heat transfer to occur between two mediums. They are conduction, which is the transfer of energy between adjacent non-moving particles; convection, which is the transfer of energy between a solid surface and flowing fluid (liquid or gas); and radiation, which is the transfer of energy through electromagnetic waves (e.g., photons). Conduction can take place in solids, liquids, and gases. In solids, conduction occurs due to the vibrations of the molecules in the molecular structure and energy transport of free energy, whereas in liquids and gases, conduction occurs due to the collisions and random diffusion of the molecules within the material. In general, conduction is dependent on the materials, the thickness of the materials, the geometry of the materials, and the temperature difference between the materials. The differential governing equation for conduction (termed Fourier's Law of Heat Conduction) can be represented as

$$\dot{Q}_{conduction} = -kA \frac{dT}{dx} \tag{7.32}$$

where k is the thermal conductivity of the material and A is the surface area over which conduction can occur. The thermal conductivity is a measure of how readily the material conducts heat. As the thermal conductivity of the material increases, the material can more readily conduct heat and therefore thermal insulators tend to have a very low conductivity. Convection is the mode of heat transfer between a solid surface and the adjacent fluid layer, which must be in motion (if the fluid layer is not in motion, then conduction principles apply). The rate of convection is dependent on the fluid properties, the nature of the fluid motion (e.g., laminar vs. turbulence), fluid velocity, geometric features of the flow (e.g., surface roughness), the area of interaction, and the temperature difference. Although the fluid motion complicates the heat transfer properties, a convenient way to express the rate of heat transfer via convection is through Newton's Law of Cooling, which is

$$\dot{Q}_{convection} = hA_s(T_s - T_\infty) \tag{7.33}$$

In Eq. (7.33), h is the convection heat transfer coefficient, A_s is the surface area over which convection can occur, T_s is the surface temperature, and T_∞ is the free fluid temperature (e.g., the temperature of the fluid sufficiently far from the convecting surface). Unfortunately, the convection heat transfer coefficient is not easily tabulated, since it encompasses all of fluid considerations. The value must be experimental obtained and is only valid within a particular range of fluid properties (e.g., velocity, motions). Radiation, as a mode of heat transfer, does not require the presence of an intervening medium; it can occur through a vacuum. All solids, liquids, and gases emit, absorb, and/or transmit radiation at various rates and it is principally dependent on the surface area of the material and the temperature of the material. The maximum amount of heat transfer that a material can emit by radiation can be represented by the Stefan-Boltzmann Law, which is

$$\dot{Q}_{emit,\ \mathrm{maximum}} = \sigma A_s T_s^4 \tag{7.34}$$

where σ is the Stefan-Boltzmann constant. Under normal nonideal conditions, the maximum amount of radiation is not achieved and thus the rate of heat transfer via emitted radiation can be represented by

$$\dot{Q}_{emit} = \varepsilon\sigma A_s T_s^4 \tag{7.35}$$

where ε is the emissivity if the surface, which ranges from 0 (emits no energy) to 1 (a perfect emitter). All surfaces absorb incident radiation to varying degrees and this can be represented by

$$\dot{Q}_{absorbed} = \alpha \dot{Q}_{incident} \tag{7.36}$$

where α is termed the surface absorptivity. Absorption is principally a function of how much incident energy the material is exposed to and material properties. Finally, in general, more than one mode of heat transfer can occur at a time. It is typical for a body to experience conduction or convection at the same time as radiation (although the heat transfer by radiation is normally small as compared with the other modes of heat transfer). In these instances, one would use a combined heat transfer coefficient, which can be described by the following for a material that experiences both convection and radiation (either heat transfer to or from a surface can be calculated from this equation).

$$\dot{Q}_{total} = \dot{Q}_{convection} + \dot{Q}_{radiation} = h_{convection}A_s(T_s - T_{surrounding}) + \varepsilon\sigma A_s(T_s^4 - T_{surrounding}^4) \tag{7.37}$$

$$\dot{Q}_{total} = h_{combined}A_s(T_s - T_{surrounding})$$

$$h_{combined} = h_{convection} + h_{radiation} = h_{convection} + \varepsilon\sigma(T_s + T_{surrounding})(T_s^2 + T_{surrounding}^2)$$

For reference, the heat transfer by radiation is typically significant as compared with the heat transfer by conduction or natural convection, but is typically very low as compared with forced convection, if the temperature of the involved materials is large and/or the emissivities and absorptivities of the involved materials are close to 1. If these conditions do not hold for the specific problem, then it is typical to assume that the heat transfer by radiation is low.

Each of these heat-transfer mechanisms can be subdivided further into special cases of heat transfer (e.g., we mentioned forced convection as compared with natural convection in the previous paragraph). The analysis methods for each of these particular cases of heat transfer will be slightly different because the modes of heat transfer are different. For this brief discussion of heat transfer, we will focus on internal forced convection. This is the most applicable heat-transfer modality within the vascular system. Internal forced convection is concerned with fluids that are bound on all surfaces (e.g., flow through blood vessels, the respiratory tract, or heat exchangers; this is why it is termed internal) and is driven by some pressure gradient (e.g., the heart, a pump, or a turbine; this is why it is termed forced). For these types of situations, the velocity profile, which has been described in earlier chapters, and the temperature profile are of primary interest. From these two profiles, we can then obtain heat transfer rates and other properties of interest.

We will begin by developing the temperature profile for a fluid in cylindrical coordinates. The temperature within the fluid must satisfy the energy conservation laws, which states that the actual energy transported through a fluid must be equal to the energy transported through a fluid that has a uniform temperature profile throughout the fluid. This is analogous to analyzing fluids as if they are inviscid (and thus have a uniform velocity profile) and making statements about viscous fluids (as if they have a parabolic, or other, velocity profile) from this analysis. Mathematically, the Conservation of Energy equation states that

$$\dot{E}_{fluid} = \dot{m}c_p T_m = \int_{area} \rho c_p T(r)u(r)dA \tag{7.38}$$

where $\dot{m}$ is the fluids mass flow rate, c_p is the fluids specific heat at constant pressure (normally assumed to be a constant), T_m is the mean fluid temperature, ρ is the fluid density (also, normally assumed to be a constant), and $T(r)$ and $u(r)$ are the temperature and velocity profile in radial coordinates, respectively. In this analysis, we assume that the velocity and temperature profiles are only functions of the radial direction (this is analogous to the velocity profile assumptions that we used for the Navier-Stokes equation). Equation (7.38) can be rearranged to calculate the mean temperature of the fluid, if needed, as follows:

$$T_m = \frac{\int_0^R \rho c_p T(r)u(r)2\pi r dr}{\rho v_{average}(\pi R^2)c_p} = \frac{2}{v_{average}R^2}\int_0^R T(r)u(r)rdr \tag{7.39}$$

Solving Eq. (7.38) for a steady two-dimensional (Cartesian coordinates) flow of a fluid with constant fluid properties with negligible stresses, the solution to the energy balance equation is

$$\rho c_p \left(u\frac{\partial T}{\partial x} + v\frac{\partial T}{\partial y} \right) = k\left(\frac{\partial^2 T}{\partial x^2} + \frac{\partial^2 T}{\partial y^2} \right) \tag{7.40}$$

or in radial coordinates

$$u\frac{\partial T}{\partial z} = \frac{k}{\rho r c_p}\frac{\partial}{\partial r}\left(r\frac{\partial T}{\partial r}\right) \tag{7.41}$$

Equation (7.40) or (7.41) can be used to solve for the temperature profile within a fluid as the Navier-Stokes equations were used to determine the velocity profile of the fluid. Note that some information about velocity is needed to solve for the temperature because this is a coupled convection problem. Thus, the components of velocity would need to be known prior to initiating the analysis of the temperature distribution.

Using this same type of energy balance analysis, we can obtain general heat transfer equations when the blood vessel can be assumed to have either a constant surface temperature or a constant surface heat flux. In most biological applications, neither the surface temperature nor the surface heat flux will be constant, but we may be able to use these simplifications to approximate certain conditions and obtain meaningful results. Constant surface temperature (or the approximation of a constant surface temperature) would suggest that the entire surface of the blood vessel is held at a fixed temperature. If the vessel length within the problem is relatively short, this may be a valid approximation; however, as the vessel length increases the probability of obtaining a constant surface temperature along the entire vessel length decreases. Constant surface heat flux (or the approximation of a constant surface heat flux) would suggest that the heat flux into the blood vessel from the blood vessel wall is constant along the entire vessel length. Again with longer vessels, the probability of this approximation being valid decreases. However, in this case, since the heat flux may change along a single cell depending on that cells metabolic activity, one may need to take care when using this assumption. In the absence of any work by the fluid, the Conservation of Energy can be solved to obtain the heat transfer rate, $\dot{Q}$, to or from the fluid. This is represented as

$$\dot{Q} = \dot{m}c_p(T_o - T_i) \tag{7.42}$$

where T_o is the mean outlet temperature and T_i is the mean inlet temperature. You may recall from a previous course in heat transfer that the surface heat flux ($\dot{q}_s$) can be defined as

$$\dot{q}_s = h_x(T_s - T_m) \tag{7.43}$$

where h_x is the local convection heat transfer coefficient, T_s is the surface temperature, and T_m is the mean fluid temperature. Using Eqs. (7.42) and (7.43), we can now analyze our idealized conditions for constant surface heat flux or constant surface temperature. In the case where we can assume that $\dot{q}_s$ is a constant, then the rate of heat transfer can also be expressed as

$$\dot{Q} = \dot{q}_s A_s = \dot{m}c_p(T_o - T_i) \tag{7.44}$$

where A_s is the area available for heat transfer and is equal to πDL for circular blood vessels with a length of L. The mean fluid outflow temperature and the mean surface temperature become

$$T_o = T_i + \frac{\dot{q}_s A_s}{\dot{m} c_p} \tag{7.45}$$

$$T_s = T_m + \frac{\dot{q}_s}{h} \tag{7.46}$$

which both vary linearly with the surface heat flux. By recognizing that $\dot{q}_s$ and h are constants with respect to the z-direction (radial coordinates, along the blood vessel length), we can then obtain

$$T_m = T_i + \frac{2\pi R \dot{q}_s}{\dot{m} c_p} z \tag{7.47}$$

by differentiating and equating the equations that describe the linear variation of the mean outlet temperature and the surface temperature. In Eq. (7.47), R is the blood vessel radius and z is the axial distance along the vessel. Note that for a circular tube,

$$\dot{m} = \rho v_{average} A_c = \rho v_{average}(\pi R^2)$$

Equations (7.44) through (7.47) can be used to solve for various heat transfer properties under the constant surface heat flux assumption.

If the conditions are such that we can make the assumption of a constant surface temperature, instead of a constant surface heat flux, then the heat transfer and temperature conditions will be modified. Here, the rate of heat transfer can be expressed as

$$\dot{Q} = h A_s \Delta T_{average} \tag{7.48}$$

The average temperature difference can be calculated by

$$\Delta T_{average} \approx T_s - \frac{T_i + T_o}{2}$$

where the second term on the right-hand side is the bulk mean fluid temperature, and this is the temperature at which all fluid properties are analyzed for heat transfer problems (e.g., density, viscosity). We can further relate the heat transfer rates to the fluid temperatures by

$$\ln\left[\frac{T_s - T_o}{T_s - T_i}\right] = -\frac{h A_s}{\dot{m} c_p} \tag{7.49}$$

or

$$T_o = T_s - (T_s - T_i) e^{-\frac{h A_s}{\dot{m} c_p}} \tag{7.50}$$

where A_S is the surface area of the tube and can be represented as $A_S = 2\pi RL$. Using the log mean temperature difference approach, which tends to be a bit easier under these conditions, the heat transfer rate can also be calculated from

$$\dot{Q} = h A_s \Delta T_{lm} \tag{7.51}$$

where ΔT_{lm} is the log mean temperature difference, and is defined as

$$\Delta T_{lm} = \frac{T_i - T_o}{\ln\left[\frac{T_s - T_o}{T_s - T_i}\right]} \tag{7.52}$$

Example

Determine the outflow temperature and the surface inlet and outlet temperatures, assuming that there is fully developed blood flow through a 300-μm-long capillary (Figure 7.6). The capillary can be considered to provide the blood with a constant surface heat flux of $-1000\ W/m^2$. The diameter of the capillary is 15 μm, and its convection heat transfer coefficient is 2500 W/m^2K. Blood is flowing with a mass flow rate of 50 g/s at an inlet temperature of 25°C.

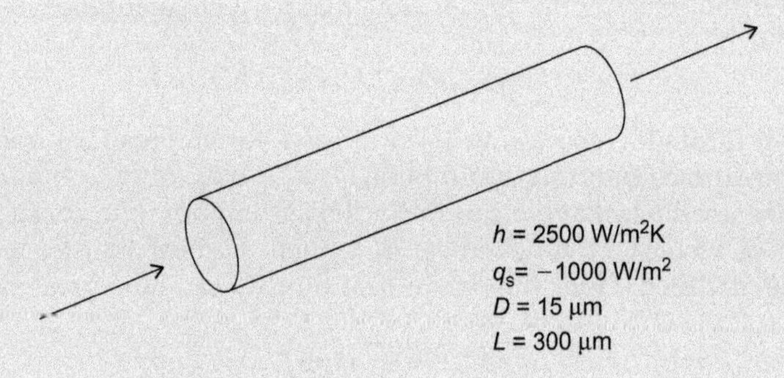

FIGURE 7.6 Heat transfer through a capillary.

Solution

The specific heat at constant pressure for blood is 3.8 kJ/kgK.
First, determine the blood vessel heat transfer rate to obtain the outlet temperature:

$$\dot{Q} = \dot{q}_s A_s = \dot{q}_s \pi DL = -1000\ W/m^2(\pi)(15\mu m)(300\mu m) = -1.414 \times 10^{-5} W$$

$$T_o = T_i + \frac{\dot{Q}}{mc_p} = 25°C + \frac{-1.414 \times 10^{-5} W}{50\ g/s(3.8\ kJ/kgK)} \approx 24.99992°C$$

The surface inlet and outlet temperatures can be found by

$$T_{s,i} = T_i + \frac{\dot{q}_s}{h} = 25°C + \frac{-1000\ W/m^2}{2500\ W/m^2K} = 24.6°C$$

$$T_{s,o} = T_o + \frac{\dot{q}_s}{h} = 24.99992°C + \frac{-1000\ W/m^2}{2500\ W/m^2K} = 24.59992°C$$

This example illustrates that for a relatively short blood vessel length, we do not observe a significant change in the temperature along the blood vessel length.

Example

Determine the outflow temperature, the log mean temperature difference and the heat transfer rate, assuming that there is fully developed blood flow through a 300-μm-long capillary (Figure 7.7). The capillary can be considered to provide the blood with a constant surface temperature of 24°C. The diameter of the capillary is 15 μm, and its convection heat transfer coefficient is 2250 W/m^{2}K. Blood is flowing with a mass flow rate of 50 g/s at an inlet temperature of 25°C.

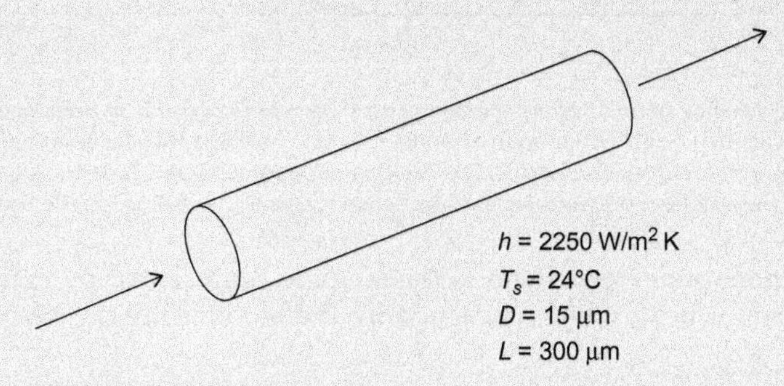

FIGURE 7.7 Heat transfer through a capillary.

Solution

To calculate the outflow temperature, we use

$$T_o = T_s - (T_s - T_i)e^{-\frac{pLh}{\dot{m}c_p}} = 24°C - (24°C - 25°C)e^{-\frac{\pi(15\mu m)(300\mu m)\left(2250\ W/m^2K\right)}{50g/s(3.8\ kJ/kgK)}} = 24.99998°C$$

The log mean temperature difference is

$$\Delta T_{lm} = \frac{T_i - T_o}{\ln\left(\frac{T_s - T_o}{T_s - T_i}\right)} = \frac{(25 - 24.99998)°C}{\ln\left(\frac{24 - 24.99998}{24 - 25}\right)} = -0.9999°C$$

The heat transfer rate will be

$$\dot{Q} = hA_s\Delta T_{\text{lm}} = 2250\ W/m^2K(\pi)(15\mu m)(300\mu m)(-0.9999°C) = -0.000032W$$

Extending this work, we can include the counter-current heat transfer properties of blood vessels within our analysis. Under normal physiological conditions, arteries are always paired with veins, so that any heat that is lost from an artery can be gained back by a vein along the path back to the core body (Figure 7.8). In fact, all warm-blooded animals have this type of arrangement along their cardiovascular system. To begin to analyze this scenario, we assume that blood that is transiting from the core body to the capillaries has a constant rate of heat transfer from the paired artery to the paired vein. Additionally, there is some quantifiable value for the rate of heat transfer from the artery to the environment and a separate rate of heat transfer from the vein to the environment. To begin this analysis, we will use all of the same governing equations as discussed

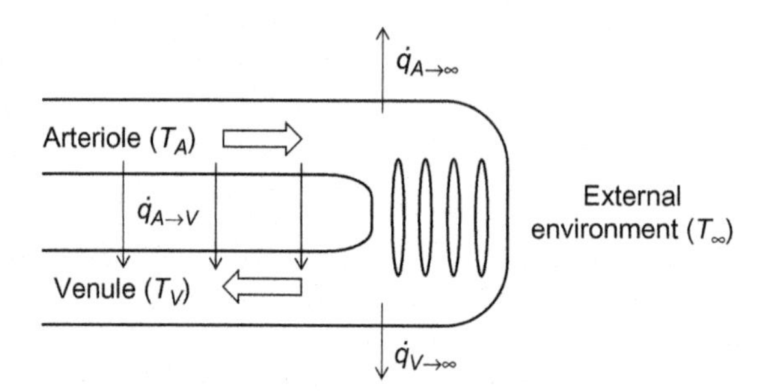

FIGURE 7.8 Representation of an arteriole–venule pair as a counter-current heat exchanger. There are at least three heat transfer events that occur within each arteriole–venule pair; heat transfer from the artery to the external environment, heat transfer from the vein to the external environment, and heat transfer from the artery to vein. Mathematically, this can be represented using convection methods, as outlined in the text.

earlier in this section, but extend them to the scenario represented by Figure 7.8. There is energy, in the form of heat, which is carried by the blood. This can be quantified using the following

$$\dot{Q}_{blood} = \dot{m}\frac{dh}{dx} = \dot{m}c_p\frac{dT}{dx} \tag{7.53}$$

where $\dot{m}$ is the blood mass flow rate and h is the enthalpy of blood. As shown in Figure 7.8, there are also three heat transfer events that occur (artery to vein, artery to environment, and vein to environment). As with most heat exchangers, we begin the analysis of these heat transfer events using a total heat transfer equation, which lumps together all of the parameters/properties of the heat transfer events. The total heat transfer for this heat exchanger can be described by

$$\dot{Q} = UA_S\Delta T_{average} \tag{7.54}$$

where U is the overall heat transfer coefficient, A_s is the surface area and $\Delta T_{average}$ is an appropriate mean temperature difference between the arterial side and the venous side of the blood vessel. Therefore, the total heat transfer for the cardiovascular counter-current system can be represented by combining Eqs. (7.53) and (7.54) as follow

$$\dot{Q}_{blood} + \dot{Q} = \dot{m}c_p\frac{dT}{dx} + UA_S\Delta T_m \tag{7.55}$$

The previous equation would need to be written for the arterial side and the venous side of the heat exchanger and then solved simultaneously. An example of the system that can be written for this heat exchanger scenario is

$$\begin{aligned}
&\dot{m}c_p\frac{dT_A}{dx} + (UA_S)_{AV}(T_A - T_V) + (UA_S)_A(T_A - T_\infty) = 0 \rightarrow \text{Arterial Side}\\
&\dot{m}c_p\frac{dT_V}{dx} + (UA_S)_{AV}(T_A - T_V) - (UA_S)_V(T_V - T_\infty) = 0 \rightarrow \text{Venous Side}
\end{aligned} \tag{7.56}$$

where the subscript A designates arterial side parameters, the subscript V denotes venous side parameters, and the subscript ∞ denotes environmental parameters. We assume that the mass flow rate and the specific heat does not change between the arterial circulation and the venous circulation, in deriving Eq. (7.56). As we have learned, the mass flow rate is likely different between the arterial side and the venous side of the circulation. This equation can be solved in a number of ways to obtain a value for the heat transfer within a cardiovascular counter-current exchanger, using the assumptions that were stated above.

It is important to note that under many conditions the convective heat transfer coefficient cannot be empirically measured due to many of the variables that influence heat transfer. However, under many of the conditions that we have discussed the heat transfer coefficient can be calculated from the Nusselt number. The Nusselt number is a dimensionless number that relates the influence of the convective heat transfer to the conductive heat transfer. The Nusselt number can be defined as

$$Nu = \frac{hd}{k} = \frac{\text{convective heat transfer}}{\text{conductive heat transfer}} \tag{7.57}$$

where h is the convective heat transfer coefficient, d is a characteristic length, and k is the thermal conductivity. Unfortunately, a true thermal conductivity can only be calculated under stagnant flow, which is not the case during a convection problem. Therefore, during the convection scenario, the thermal conductivity is quantified assuming stagnant flow. Regardless of the difficulties in calculating the thermal conductivity, values for k can be tabulated. Luckily, for laminar flow in a circular tube, the Nusselt number is constant if we can apply the constant surface heat flux assumption or the constant surface temperature assumption. Using these assumptions, and the calculated value for the Nusselt number, one can obtain a value for the convective heat transfer coefficient. The appropriate Nusselt numbers are

$$\begin{aligned} Nu &= \frac{hd}{k} = 4.36 \quad \dot{q}_s = \text{constant} \\ Nu &= \frac{hd}{k} = 3.66 \quad T_s = \text{constant} \end{aligned} \tag{7.58}$$

Although we will not go through the entire derivation to obtain Eq. (7.58), it begins by obtaining the temperature profile under either constant surface heat flux or constant surface temperature conditions. For the constant surface heat flux conditions, the temperature profile can be found to be equal to

$$T = T_s - \frac{\dot{q}_s R}{k}\left(\frac{3}{4} - \frac{r^2}{R^2} + \frac{r^4}{4R^4}\right) \tag{7.59}$$

if we assume a radial coordinate system and the following boundary conditions

$$\begin{aligned} \frac{\partial T}{\partial r} &= 0 \quad r = 0 \\ T &= T_s \quad r = R \end{aligned}$$

The mean temperature under these conditions can be found to be

$$T_m = T_s - \frac{11}{24}\frac{\dot{q}_s R}{k} \tag{7.60}$$

If the previous equation is combined with Eq. (7.43), one can obtain the Nusselt number formulation as

$$h = \frac{24}{11}\frac{k}{R} = \frac{48}{11}\frac{k}{D} = 4.36\frac{k}{D}$$

A similar procedure can be completed for the constant surface temperature scenario.

A large area of cardiovascular heat transfer research has been conducted making the assumption that conduction is the primary mode of heat transfer in the body. Within this assumption, one must consider that heat can be transferred (gained or lost) via metabolism and via convection, but that they can be represented in simple heat source (or sink) terms. In a one-dimensional problem, this is typically represented as

$$\frac{d^2T_T}{dx^2} + \frac{\dot{q}_m + \dot{q}_p}{k} = 0 \tag{7.61}$$

where $\dot{q}_m$ is the metabolic heat generation term and $\dot{q}_p$ is the perfusion heat generation term (or convection heat generation term) and T_T is the local tissue temperature. Under these simplifications, any of the heat that is generated due to metabolism is assumed to be able to be incorporated into a modified conduction formula, regardless of the heat transfer modality. Likewise, the convective heat transfer that occurs during blood flow is completely incorporated into the perfusion heat generation term. We also must assume that the thermal conductivity is constant throughout the entire system of interest. In order to make use of this equation, information about the metabolic and the perfusion heat generation term must be known. A significant amount of work, initially by Pennes and colleagues, has been conducted to find an accurate representation for these values. One of the first assumptions that Pennes made was that any heat that is lost within the capillary circulation enters the tissues only and therefore there is no heat lost to the surrounding environment. Although, this is not an accurate representation of the physiology, initially it is true because any heat that exits the capillaries must enter tissue first, before entering the external environment. Pennes also assumed that the blood entering from the arterial side of the capillary had a greater temperature than the blood that was exiting, therefore, heat was always lost along the capillary length. In this scenario, Pennes's formulation for the perfusion heat generation term can be represented as

$$\dot{q}_p = \dot{Q}\rho_b c_p (T_A - T_T) \tag{7.62}$$

where $\dot{Q}$ is the blood volumetric flow rate (assumed to be constant) and c_p is the specific heat for blood. Substituting, Eq. (7.62) into (7.61), we can represent the governing one-dimensional heat transfer equation as

$$\frac{d^2T_T}{dx^2} + \frac{\dot{q}_m + \dot{Q}\rho_b c_p (T_A - T_T)}{k} = 0 \tag{7.63}$$

Eq. (7.63) is a linear second-order differential equation, which can be solved with a variety of techniques. In most cases, it is assumed that all of the properties that are associated with blood and the metabolic heat generation term are all constant with respect to T_T. In practice, the simplifications that were made to analyze this system are not accurate under a wide range of conditions. For instance, it is difficult and impractical, to quantify the thermal properties of biological conditions. It is also likely, that there will be some variance in the thermal properties and fluid properties that were assumed to be constant in this formulation. Additionally, the human body can adjust to the thermal environment in many ways that cannot be quantified in this conduction heat transfer formulation. For instance, sweating and shivering, which are two ways that we compensate for significantly altered heat transfer situations, cannot be included in this type of formulation in a very easy manner. Therefore, while heat transfer and the maintenance of the core body temperature are significant functions for the cardiovascular system, in practice, it is quite difficult to represent these salient features with the classical heat transfer equations. One must make many assumptions about the heat transfer parameters and the conditions in order to apply the formulations that were discussed here.

7.7 CELL TRANSFER DURING INFLAMMATION/WHITE BLOOD CELL ROLLING AND STICKING

One of the most important properties of the endothelium is to allow white blood cells to pass through the endothelial cell wall during inflammatory conditions. From our previous discussion, we have stated that white blood cells, although there are many different types of these cells, are all involved in the removal and/or elimination of foreign particles. This task could not be accomplished if the white blood cells are segregated within the circulatory system because foreign particles can propagate anywhere within the body. One of the earliest responses to an infection is the expression of adhesion molecules on the capillary endothelial cell surface (e.g., E-selectin and various integrins). White blood cells can bind to these receptors to make an initial contact with the endothelial cells. White blood cells tend to roll along the capillary, making and breaking adhesions along the way. At some point, the adhesion between the white blood cell and the endothelial cell wall becomes more stable and the cell does not roll along the blood vessel anymore, and therefore it stays in one location. Once this stable adhesion occurs, it is possible for the white blood cell to cross the endothelial cell barrier, in a process termed transmigration or extravasation. Once this occurs, the white blood cells migrate toward the site of infection along a chemotactic gradient. Again, the initial attachment of white blood cells to the endothelial cell wall is typically mediated through various selectin molecules. These adhesions are characterized by rapid association and dissociation constants. After selectin binding, there is a reorganization of the white blood cell membrane bound integrins, which facilitate a more stable adhesion (the kinetic constants are slower). As more selectins make contact with the endothelial cells, the reorganization of integrins occurs much more rapidly.

Typically, the white blood cell selectin density changes depending on factors such as shear stress, and presence of cytokines, among others. At a selectin density of less than 15 molecules/cm^2, white blood cell adhesion is unstable, even at low fluid velocities. At higher selectin densities, white blood cell adhesion is more stable but transient, and it is likely that integrins will be redistributed to mediate stable adhesion. The force required to break one selectin bond is on the order of 20 pN, and the bond typically acts over a distance of 0.5 Å. As the adhesion becomes stronger, from either more selectin bonds or more integrin bonds, the white blood cell becomes stationary even though the largest fluid shear forces are acting on the bound white blood cell (because the white blood cells would be located along the blood vessel wall). The strength of an integrin adhesion has been reported to be in the range of 50 pN. This process of tight adhesion normally occurs within the venous side of capillaries or within post-capillary venules. At the point of stable adhesion, the cytoskeleton of the white blood cell is reorganized so that the cell flattens out over the endothelial cell wall. This reduces the face on exposure area to the fluid velocity (thus minimizing the face-on pressure). Pseudopods from the leukocyte locate the intercellular cleft between neighboring endothelial cells. Pseudopods typically express many adhesion molecules. The white blood cell passes through the cleft primarily via the action of PECAM (platelet endothelial cell adhesion molecule), which acts to pull the cell through the gap. The strength of a PECAM bond can be as large as 1–10 nN. The leukocytes then migrate through the basement membrane along a chemotactic gradient to reach the site of inflammation. It is important to note that the white blood cell may release vasoactive compounds to increase the diameter of the intercellular cleft. Also, invading pathogens or other inflammatory cells can release cytokines that increase the spacing of the intercellular cleft.

Many groups try to model the adhesion of white blood cells to endothelial cells using the empirical data discussed previously. Binding association and dissociation constants as well as the forces acting on the cell are relatively easy to model. However, the selectin density and actual binding of the white blood cell to the endothelial cell wall are more problematic. Selectin density can be assumed to have some initial value, but this can change drastically based on the conditions described above (shear stress and cytokines, among others). This level of involvement within the model can greatly increase the number of simultaneous equations to solve. Also, probability functions are needed to model the association and dissociation of bonds along with the likelihood for contact. Probability functions are never that accurate and typically depend on the selectin density and the distance between the white blood cell and the endothelial cell. As the reader can imagine, as the selectin density increases and the distance decreases, there should be a higher likelihood for the selectin adhesion to occur within the model. However, the exact probabilities are not known and assumptions have to be made to model this phenomenon. Therefore, the accuracy of these assumptions limits the predictive capability of the model.

It is typical in models of this kind to initiate the adhesion through one receptor–ligand binding event. This event is governed by the receptor density and/or distance between cells, as discussed above. All remaining association and dissociation events are then described by receptor kinetics and the density of the receptors/ligands present. This principal can be used to define the receptor association/dissociation because it is typically assumed that if one bond can be formed, then the other receptors are close enough to bind to other expressed ligands. Also, it is typical to make the time steps in the model small enough that the quantity of receptors–ligand events can only increase by one, decrease by one, or remain the same during each subsequent time step (this significantly simplifies the modeling). In equation form, the quantity of bonds in a time step would be the summation of the amount of bonds in the previous time step, plus the association of a new bond (which is dependent on kinetics) and the dissociation of any one bond (again based on kinetics). This type of modeling will be discussed in more detail in Chapter 14.

END OF CHAPTER SUMMARY

7.1 To model red blood cell oxygenation within the pulmonary capillaries, there are five main diffusion events that we should consider. The first, alveolar diffusion, is governed by Fick's laws of diffusion:

$$\vec{J} = -D\nabla\vec{C}$$

and

$$\frac{\partial C}{\partial t} = D\frac{\partial^2 C}{\partial x^2}$$

Solving this differential equation, we can obtain a measure for the time needed to reach equilibrium. Using this formulation, we can show that alveolar diffusion occurs rapidly and does not limit the red blood cell oxygenation. The second diffusion event is the movement of the gas across the respiratory boundary. This would be modeled with a partition coefficient for each of the boundaries that are included within the model (i.e., alveolus epithelial cells and capillary endothelial cells), as

$$P = \frac{p_c D}{x}$$

The third diffusion event is the convection and diffusion of oxygen to the red blood cell within the pulmonary capillaries. The dimensionless Sherwood number can be used to model this convection–diffusion event, as

$$\text{Sh} = \frac{Kx}{D}$$

and

$$\text{Sh} = \text{Sh}_{0.25} + 0.84(\text{Hct} - 0.25)$$

The fourth diffusion event is the diffusion of oxygen across the red blood cell membrane, which occurs approximately 100 times faster than the diffusion across the respiratory boundary. This event is also not a rate-limiting step in diffusion. The last diffusion event is the diffusion of oxygen to hemoglobin and the kinetic association of oxygen to hemoglobin. This is modeled with a Thiele modulus, which is defined as

$$\phi = \sqrt{\frac{kx^2}{D}}$$

A model of tissue oxygenation should also be formulated and perhaps the most famous model is the Krogh model. Assuming that oxygen diffusion through tissue is homogeneous in all directions and that the capillaries are regularly spaced in three dimensions, the Krogh radius quantifies the area of tissue oxygenation per capillary. The relationship of the capillary radius to the Krogh radius is represented as

$$\frac{C(r)}{C_p} = 1 + \frac{r_K^2 R}{4C_p D}\left[\frac{r^2 - r_C^2}{r_K^2} + 2\ln\left(\frac{r_C}{r}\right)\right]$$

7.2 Glucose transport is also a salient transport mechanism to discuss within the microcirculation. Under most normal conditions, glucose cannot freely diffuse through the capillary wall. Its movement is coupled to the movement of a second ion (such as sodium), where the energy gained by moving this second ion down its electrochemical gradient is used to move glucose up its concentration gradient. Co-transporters are typically used in these processes.

7.3 Lipid soluble molecules can freely permeate the endothelial cells that line the capillaries. Non-lipid soluble molecules cannot diffuse freely through the endothelial cells, but typically they move through the intercellular cleft. It has been experimentally determined that physiologically important molecules generally have a permeability through the intercellular cleft that is in the range of water's permeability through the intercellular cleft. For instance, sodium chloride has a permeability approximately 96% of water's permeability. Large plasma proteins, such as albumin, have a permeability that is significantly lower than the permeability of water.

7.4 The critical adhesive forces that play a role in biological applications are the molecular forces, the electrostatic forces, and the capillary forces. Molecular forces are the weakest, act over the smallest range, and are sometimes transient. The mechanical force associated with these types of adhesive forces can be quantified with

$$\vec{\psi}(d) = \vec{\psi}_0\left[\left(\frac{d'}{d}\right)^{12} - \left(\frac{d'}{d}\right)^{6}\right]$$

Electrostatic forces may arise in biological situations, if there is a net charge on one of the molecules. These forces can be calculated from Coulomb's law:

$$F = \frac{1}{4\pi\varepsilon_0}\frac{q_1 q_2}{r^2}$$

A charged surface would have some potential associated with it, which is termed the Nernst potential. Due to this net charge, the ions in solution or water can organize near the charged surface. This highly organized layer is termed the Stern layer and the potential at which the layer ends is termed the Stern potential. The electric double layer describes the remaining distance that has some organization due to the presence of the charged surface. The Debye length quantifies the electric double layer:

$$\lambda_D = \sqrt{\frac{\varepsilon_l \varepsilon_o k_B T}{e^2 \sum_i^n (c_i \chi_i)}}$$

Capillary forces are composed of adhesion, cohesion, and surface tension. Adhesion occurs when two different molecules tend to stick to each other, due to an attractive force.

Cohesion is the likelihood for two of the same molecules to stick together. Surface tension is a fluid property that arises due to the intermolecular forces within the fluid that occur at the boundary of a fluid and another substance.

7.5 Porous media are solid materials that are composed of pore structures, which are typically fluid filled in biological applications. Porous media can be described by the porosity, which is

$$\varepsilon = \frac{\text{Volume associated with voids}}{\text{Total volume}}$$

The porosity of the interstitial space, neglecting blood vessels and cells, approaches 0.9, whereas the porosity of general biological tissues (including blood vessels and cells) would be closer to 0.25. Porous media can also be classified based on the body's tortuosity, which is a measure of the random orientation and random spacing of pores and can be defined as

$$T = \frac{L_{actual}}{L}$$

Fluid flow through porous media is most typically quantified via Darcy's law, which is

$$\vec{v} = -\mathrm{K}\nabla\vec{p}$$

which can be also be related to the viscous forces present within the media or fluid by

$$\mu\nabla^2\vec{v} - \frac{1}{\kappa}\vec{v} - \nabla\vec{p} = 0$$

7.6 Heat transfer is an integral component to many biological systems. Three basic mechanisms for heat transfer include conduction, convection, and radiation. Most heat transfer formulations begin with the Conservation of Energy principle, which states that

$$\dot{E}_{fluid} = \dot{m}c_pT_m = \int_{area} \rho c_p(r)u(r)dA$$

From this equation, the temperature difference within a fully developed, two-dimensional flow scenario can be solved from

$$\rho c_p\left(u\frac{\partial T}{\partial x} + v\frac{\partial T}{\partial y}\right) = k\left(\frac{\partial^2 T}{\partial x^2} + \frac{\partial^2 T}{\partial y^2}\right)$$

or

$$u\frac{\partial T}{\partial z} = \frac{k}{\rho r c_p}\frac{\partial}{\partial r}\left(r\frac{\partial T}{\partial r}\right)$$

if the viscous forces are negligible. This leads us to two specialized problems, one in which there is a constant surface heat flux and a second where there is a constant surface temperature. In the case where we can assume that $\dot{q}_s$ is a constant, then the rate of heat of transfer can be expressed as

$$\dot{Q} = \dot{q}_s A_s = \dot{m}c_p(T_o - T_i)$$

and the mean fluid outflow temperature and the mean surface temperature become

$$T_o = T_i + \frac{\dot{q}_s A_s}{\dot{m} c_p}$$

$$T_s = T_m + \frac{\dot{q}_s}{h}$$

If the conditions are such that we can make the assumption of a constant surface temperature, then the heat transfer and temperature conditions will be modified as

$$\dot{Q} = h A_s \Delta T_{average}$$

$$T_o = T_s - (T_s - T_i) e^{-\frac{h A_s}{\dot{m} c_p}}$$

Using the log mean temperature difference, the heat transfer rate can also be calculated from

$$\dot{Q} = h A_s \Delta T_{lm}$$

$$\Delta T_{lm} = \frac{T_i - T_o}{\ln\left[\frac{T_s - T_o}{T_s - T_i}\right]}$$

In fact, all arteriole and venule pairs are organized as a counter-current heat exchanger, so that blood that experiences a temperature reduction due to heat transfer to the environment does not reduce the temperature of the core body. The total heat transfer within a counter-current mechanism can be estimated by

$$\dot{Q}_{blood} + \dot{Q} = \dot{m} c_p \frac{dT}{dx} + U A_S \Delta T_m$$

The Nusselt number is a dimensionless number that is used to relate the heat transfer that occurs via convection to the heat transfer that occurs via conduction. In a simplified form, heat transfer in the body can be represented by the bioheat transfer equation, which is a modified conduction equation. Under this assumption, all of the heat that is gained (or lost) by metabolism and perfusion are accounted for in specific heat flux terms, as follows

$$\frac{d^2 T_T}{dx^2} + \frac{\dot{q}_m + \dot{q}_p}{k} = 0$$

or

$$\frac{d^2 T_T}{dx^2} + \frac{\dot{q}_m + \dot{Q} \rho_b c_p (T_A - T_T)}{k} = 0$$

if some assumptions about the fluid properties are made.

7.7 White blood cells express many adhesion molecules that allow them to adhere to the endothelium and roll along the endothelium. Once a more stable adhesion occurs, the white blood cells can stay in one location and transmigrate into the extravascular space. Most of these adhesions are modulated through the actions of selectins that have an average bond strength of 20 pN or integrins that have an average bond strength of 50 pN. Most of these adhesion properties occur during inflammatory reactions, although white blood cells always express a low quantity of adhesion molecules.

HOMEWORK PROBLEMS

7.1 Discuss the steps necessary for red blood cells to become oxygenated. What are the rate-limiting steps of this process?

7.2 Make a mathematical prediction of the changes in oxygen diffusion with either an increased hematocrit or a decreased hematocrit.

7.3 Determine the permeability coefficient (barrier + plasma) for oxygen assuming that the thickness of the barrier increases to 5 μm and the diffusion coefficient decreases to $0.8_E{-}5\ cm^2/s$ (e.g., the patient has edema). Assume that the hematocrit is 38% and that there are no changes in the rate of diffusion characteristics for plasma.

7.4 What is the Biot number and effectiveness factor for homework problem 7.3 (assuming that no changes occur in the kinetics of oxygen saturation)? What does this suggest?

7.5 What is the maximum capillary spacing (using the Krogh model) for a capillary with a radius of 4.5 μm and a plasma oxygen concentration of $5 \times 10^{-8}\ mol/cm^3$? The oxygen reaction rate is $4 \times 10^{-8}\ mol/cm^3\ s$. Assume that at maximum spacing the oxygen concentration at the Krogh radius will be $3 \times 10^{-8}\ mol/cm^3$.

7.6 In the previous example, we used a standard Krogh model to determine the oxygen concentration as a function of distance from the capillary. Extend this model and consider a scenario where there is a differential oxygen usage (as measured by a different oxygen reaction rate) in the extravascular space. Assume that the oxygen reaction rate is $4 \times 10^{-8}\ mol/cm^3\ s$ for $4.5\ \mu m < r < 12\ \mu m$. For r greater than 12 μm, the oxygen reaction rate decreases to $1.5 \times 10^{-8}\ mol/cm^3\ s$. Calculate the maximal spacing under these conditions.

7.7 Under tumorgenic conditions, oxygen usage increases significantly. Assume a two-cylinder Krogh model (as in homework problem 7.6) with an oxygen reaction rate of $2 \times 10^{-7}\ mol/cm^3\ s$ for a radius between 4.5 and 10 μm. For radii greater than 10 μm, the oxygen reaction rate is $4 \times 10^{-8}\ mol/cm^3\ s$. What is the maximal capillary spacing under these conditions? Do new blood vessels need to grow to meet this need?

7.8 (Modeling) Make a model for glucose transport considering the discussion of permeability and transporters needed to move glucose across the vascular wall. Test this model with empirical data.

7.9 A charged element is moving through a protein channel in the microvasculature. The charged element has an excess charge of 50 pC and the protein has an excess charge of 200 pC. If the channel has a mean radius of 100 nm, what is the force associated with these excess charges? Consider that the charged element has a diameter of 5 nm and is centered within the channel, is a half radius from the wall and a quarter radius from the wall.

7.10 Knowing that there are channels that restrict the movement of ions based on charge and size, propose a formulation that can represent this phenomenon.

7.11 Calculate the Debye length for a 120 mM sodium ion ($\sum c_i \chi_i = 5 \mu m^{-3}$) that is being transported through the interstitial space (where ε_l is 60; this is a relative number with no units), at body temperature.

7.12 Calculate the surface tension and the force associated with the surface tension for a red blood cell moving through blood. Assume that the radius of curvature for the red blood cell is 4 μm and the radius of curvature for the blood is 1 cm. The pressure difference across the cell and blood is 25 mmHg. Assume that a red blood cell is a perfect sphere when estimating contact area.

7.13 Determine the velocity of interstitial fluid through the extracellular matrix where the pressure gradient in the fluid direction is equal to −0.10 mmHg/cm. Assume that the porosity of the media is 60%, the surface area to volume ratio is 8/cm and the shape factor is 2.

7.14 Solve for the velocity profile of a uni-directional pressure driven flow, through a small porous channel, in which the hydraulic conductivity can be represented as$K = \frac{k}{\mu}$, where k is a measure of the permeability of the media. Assume that k, μ, and the pressure gradient are constant.

7.15 Blood is flowing through the aorta at a temperature of 37°C. The aortic wall can be considered isothermal with a temperature of 30°C (on a cool day). The aorta has a diameter of 24 mm. If blood enters the aorta at 75 cm/s and leaves the 0.5-m section of the aorta at 33°C, determine the average heat transfer coefficient between the blood and the aorta. Assume that the specific heat for blood is 3.8 kJ/kg°C and that the density of blood is 1050 kg/m^3.

7.16 The trachea is 0.20 m long and 12 mm in diameter and is used to heat air that enters at 25°C ($v = 0.4$ m/s). A uniform heat flux is maintained by the body so that the air enters the lungs at a temperature of 35°C. Assume that the average properties of air to be $\rho = 1000\ kg/m^3$, and $c_p = 4000$ J/kgK, and determine the required surface heat flux.

7.17 Consider that he velocity and temperature profiles for blood flow in a vessel with a diameter of 100 μm can be expressed as

$$u(r) = 0.35\left[1 - \left(\frac{r}{R}\right)^2\right]$$

$$T(r) = 2.1\left[150 + 35\left(\frac{r}{R}\right)^2 - 65\left(\frac{r}{R}\right)^3\right]$$

with units in μm/s and K, respectively. Determine the average velocity and the mean temperature from the given profiles.

References

[1] A.O. Frank, C.J. Chuong, R.L. Johnson, A finite-element model of oxygen diffusion in the pulmonary capillaries, J. Appl. Physiol. 82 (1997) 2036.

[2] J.D. Hellums, P.K. Nair, N.S. Huang, N. Ohshima, Simulation of intraluminal gas transport processes in the microcirculation, Ann. Biomed. Eng. 24 (1996) 1.

[3] R.T. Hepple, A new measurement of tissue capillarity: the capillary-to-fibre perimeter exchange index, Can. J. Appl. Physiol. 22 (1997) 11.

[4] A. Krogh, The rate of diffusion of gases through animal tissues, with some remarks on the coefficient of invasion, J. Physiol. 52 (1919) 391.

[5] A. Krogh, The number and distribution of capillaries in muscles with calculations of the oxygen pressure head necessary for supplying the tissue, J. Physiol. 52 (1919) 409.

[6] A. Krogh, The supply of oxygen to the tissues and the regulation of the capillary circulation, J. Physiol. 52 (1919) 457.

[7] A. Krogh, Studies on the capillariometer mechanism: I. The reaction to stimuli and the innervation of the blood vessels in the tongue of the frog, J. Physiol. 53 (1920) 399.

[8] A. Krogh, Studies on the physiology of capillaries: II. The reactions to local stimuli of the blood-vessels in the skin and web of the frog, J. Physiol. 55 (1921) 412.

[9] A. Krogh, G.A. Harrop, P.B. Rehberg, Studies on the physiology of capillaries: III. The innervation of the blood vessels in the hind legs of the frog, J. Physiol. 56 (1922) 179.
[10] E.M. Landis, L. Jonas, M. Angevine, W. Erb, The passage of fluid and protein through the human capillary wall during venous congestion, J. Clin. Invest. 11 (1932) 717.
[11] E.M. Landis, J.H. Gibbon, The effects of temperature and of tissue pressure on the movement of fluid through the human capillary wall, J. Clin. Invest. 12 (1933) 105.
[12] E.M. Landis, L.E. Sage, Fluid movement rates through walls of single capillaries exposed to hypertonic solutions, Am. J. Physiol. 221 (1971) 520.
[13] H.H. Pennes, Analysis of tissue and arterial blood temperatures in the resting human forearm, J. Appl. Physiol. 1 (1948) 93.
[14] M.F. Perutz, A.J. Wilkinson, M. Paoli, G.G. Dodson, The stereochemical mechanism of the cooperative effects in hemoglobin revisited, Annu. Rev. Biophys. Biomol. Struct. 27 (1998) 1.
[15] A.S. Popel, R.N. Pittman, M.L. Ellsworth, D.P. Weerappuli, Measurements of oxygen flux from arterioles imply high permeability of perfused tissue to oxygen, Adv. Exp. Med. Biol. 248 (1989) 215.
[16] A.S. Popel, Theory of oxygen transport to tissue, Crit. Rev. Biomed. Eng. 17 (1989) 257.
[17] B. Rippe, B. Haraldsson, Transport of macromolecules across microvascular walls: the two-pore theory, Physiol. Rev. 74 (1994) 163.
[18] C. Rotsch, M. Radmacher, Mapping local electrostatical forces with the atomic force microscope, Langmuir 13 (1997) 2825.
[19] M.W. Vaughn, L. Kuo, J.C. Liao, Effective diffusion distance of nitric oxide in the microcirculation, Am. J. Physiol. 274 (1998) H1705.

CHAPTER

8

The Lymphatic System

LEARNING OUTCOMES

1. Identify the components of the lymphatic system
2. Compare the structure of lymphatic vessels to blood vessels
3. Describe the movement of water from the blood capillaries to the lymphatic capillaries
4. Discuss the function of lymph-collecting vessels
5. Identify salient lymphatic organs
6. Explain the function of lymphatic organs
7. Identify the major regulators that act during lymph formation
8. Compare the composition of lymph to the composition of extracellular fluid and blood
9. Model lymph flow rate and the direction of flow through the lymphatic system
10. Describe the lymphatic pump and compare this pump to the venous pump
11. Discuss the lymph node structure
12. Identify the direction of lymph flow through the lymph node
13. Explain cancer metastasis and why the lymphatic system is important during cancer
14. Analyze edema and the role of the lymphatic system during edema

8.1 LYMPHATIC PHYSIOLOGY

The lymphatic system consists of lymphatic vessels, lymphoid tissues and organs, lymph, and a small quantity of lymphocytes. We will discuss the first three components of the lymphatic system in this chapter. The remaining component, lymphocytes, was discussed briefly in Section 5.3 with the other types of white blood cells that circulate in the blood, and will only be reviewed here. The function of lymphatic vessels is to carry lymph away from tissues and back to the venous side of the cardiovascular system or to the right

© 2015 Elsevier Inc. All rights reserved.

side of the heart. Unlike the cardiovascular network, the first vessels in the lymphatic network are the lymphatic capillaries. Lymphatic capillaries are present in most tissues of the body and are only absent from tissues that do not have a direct blood supply. Lymphatic capillaries tend to permeate microvascular beds, to aid in the absorption of interstitial fluid that has not entered cells or returned to the cardiovascular system along the venous side of the capillary (Figure 8.1). If lymphatic capillaries did not collect this extra fluid, then the interstitial space would swell due to continual diffusion of water out of the blood capillaries (recall that overall there was a water loss from the capillaries into the interstitial space). If this was the case, the hydrostatic pressures (capillary and interstitial) and osmotic pressures (capillary and interstitial) would eventually reach equilibrium, so that there would be no net movement of nutrients or water between the blood and the interstitial space; this would severely restrict nutrient delivery and waste removal. Similar to the blood vessels of the cardiovascular network, lymphatic vessels are lined by endothelial cells. However, the endothelial cells in lymphatic capillaries are not bound as tightly as compared to those in blood capillaries. Instead, endothelial cells tend to overlap one another, forming channels that allow for the uni-directional flow of fluids, large molecules (such as proteins), and cells from the interstitial space into the lymphatic capillaries. Lymphatic capillaries are generally larger than blood capillaries with a typical diameter in the range of 20–30 μm. Also, lymphatic vessels are not circular in cross section, because they experience a very low hydrostatic pressure (gauge, approximately −4 to −6 mmHg). In this regard, they mimic the venous blood vessels of the cardiovascular system. As the

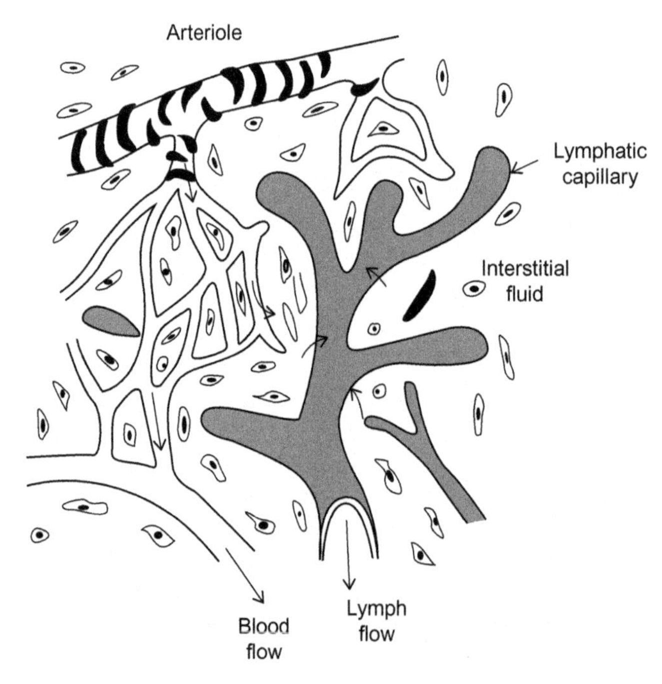

FIGURE 8.1 The association of a lymphatic capillary and a capillary network. Arrows indicate the movement of fluid from the blood vessel into the interstitial space and then into the lymphatic system. The lymphatic system is an open-ended system that starts at the lymphatic capillaries and progresses to larger lymphatic vessels. *Adapted from Martini and Nath (2009).*

reader can imagine, the lymphatic system is not a closed system by itself because the lymphatic capillaries are open-ended.

Multiple lymphatic capillaries converge to form small lymphatic vessels which in turn converge to form large lymphatic vessels. These vessels are comparable in wall structure and size to venules and veins, respectively, and have a similar function as veins. Similar to veins, lymphatic vessels contain valves that are spaced very close together to prevent the backflow of lymph toward the interstitial space. At the valve location, the lymphatic vessel wall bulges to accommodate the opening and closing of the lymphatic valves. Without these bulges, the valves would not form a tight seal and there would be some backflow of lymph. *In vivo*, lymphatic vessels appear to resemble a string of pearls because the diameter changes abruptly and quite often at valve locations. Finally, after multiple convergences, large lymphatic vessels join to form major lymph-collecting vessels that are typically located within the abdomen or chest cavity.

Major lymph-collecting vessels are divided into two different types, the deep lymphatic vessels and the superficial lymphatic vessels. The major difference between these two types of lymph-collecting vessels is their location within the body. Deep lymphatic vessels are located with the deep veins that collect interstitial fluid or blood from the skeletal muscles of the limbs. Superficial lymphatic vessels are primarily located within the skin. Both of these types of lymph-collecting vessels function to transport lymph to the collecting lymphatic trunks which lead to the two main lymph-collecting vessels of the body. The first major lymph-collecting vessel is the thoracic duct. The thoracic duct begins inferior to the diaphragm and then runs upward through the chest cavity parallel to the aorta. The major lymph vessels from the lower body and those from the left side of the upper body transport lymph back to the cardiovascular system via the thoracic duct. The thoracic duct terminates and empties its contents into the left subclavian vein, which is located near the left clavicle. The second major lymph-collecting vessel is the right lymphatic duct. This lymphatic vessel collects all of the lymph from the right side of the upper body. This duct forms as a convergence of the major collecting ducts of the right arm and the right trunk. This occurs near the right clavicle and then the vessel empties into the right subclavian vein. For both major lymph-collecting vessels, the connection with the cardiovascular system forms a closed loop circuit for the fluid that has exited the cardiovascular system, so that the lymph that is collected near the blood capillaries is returned back into the cardiovascular system. This prevents the loss of fluid volume through capillary filtration and cellular absorption. Some argue that in a sense, if you combine the cardiovascular system and the lymphatic system, you have a closed system. This is true but if one isolates the lymphatics, as a contained system, then this is an open-ended system. Be cautious on how you are defining your system of interest.

The lymphatic system is also composed of lymphatic tissue and organs. The lymphatic tissues are composed of many nodules that are associated with a large number of circulating lymphocytes. The two most important lymphatic tissues are the tonsils and the mucosa-associated lymphoid tissue (MALT). The tonsils are large nodules located in the pharynx. The MALT is all of the lymphoid tissues that are associated with the digestive system. The majority of the MALT tissue is found within the lining of the intestines and resembles the tonsils except that they are more spread out along the entire intestinal system. The function of the lymphatic tissue is to provide a centralized location for the production of mature lymphocytes, which can then function to protect the body against inflammation.

Lymphatic organs are more specialized than the lymphatic tissue, and they consist of the spleen, the thymus, and the lymph nodes. The spleen has a number of functions in the body. First, it acts as a blood filter that removes damaged red blood cells, damaged white blood cells, and activated platelets. The spleen can also initiate an immune response, by bringing B-cells and T-cells in close contact with any antigen present in the bloodstream. Lastly, there is a small storage of iron in the spleen obtained from the recycling of damaged red blood cells. This iron store could be used to produce new hemoglobin. The spleen consists of two major parts: the red pulp and the white pulp. The red pulp contains a significant quantity of blood cells and gets its color from red blood cells. The white pulp is similar to lymphatic tissues and also functions as a site for lymphocyte production. It is the red pulp that acts as a blood filter, and it is in that region where an inflammatory response can be initiated. Mature lymphocytes from the white pulp can migrate into the red pulp and enter the blood to circulate throughout the body.

The thymus is located just beneath the sternum and wraps around the trachea. The thymus is divided into a right and left lobe, each consisting of many lobules. The exterior portion of each lobule is composed of dividing/maturing lymphocytes. As the lymphocytes fully mature (primarily T lymphocytes), they migrate into the interior portion of each lobule, where they exit through blood vessels that permeate the entire thymus. In this way, the thymus acts as a lymphatic tissue by producing mature lymphocytes. The thymus also produces hormones that are critical for the immune system development. Upon release, these hormones promote the development and maturation of lymphocytes within other lymphatic tissues/organs. Therefore, the thymus is primarily responsible for the maturation of T-cells and the production of hormones which aid in lymphocyte production in other lymphatic tissues.

The last major lymphatic organs are the lymph nodes. Lymph nodes are relatively small organs (typically do not exceed 25 mm) that resemble the shape of a kidney bean. Each lymph node is innervated and supplied by its own blood vessel. Every lymph node is also directly connected to lymph flow via lymphatic vessels. The lymphatic vessels that bring lymph to the nodes are termed afferent lymphatic vessels, while those that take the flow away from the nodes are termed efferent lymphatic vessels. Lymph nodes function to filter the lymph before it is returned back into the cardiovascular system via the lymphatic ducts. Close to 100% of the antigens within lymph are removed within the lymph nodes by macrophages that are fixed on the lymph node wall. These macrophages engulf and digest the antigens and then present them to lymphocytes within the node. This acts to initiate an immune response and is a warning system for the rest of the body that an antigen has been found within the blood. When there is a major inflammatory process initiated within the body, these glands become swollen because of an increase in the production of lymphocytes, which act to destroy any invading particles found within the blood. Section 8.3 will discuss in more detail the flow of lymph through the lymph nodes, as well as the rest of the lymphatic system.

The last component of the lymphatic system, which has not been discussed previously, is the fluid that circulates throughout the lymphatic system termed lymph. Lymph has a very similar composition to interstitial fluid. One of the primary functions of lymph is to return proteins that have been transported out of the blood vessel or cells back into the cardiovascular system. In this way, there is a very small loss of protein concentration within the cardiovascular system. Lymph derived from the gastrointestinal system provides a second rapid mechanism for nutrient absorption for any nutrients that do not directly enter the cardiovascular system. The remaining sections of this chapter will discuss how lymph is formed and its path through the lymphatic system.

Lymphocytes are the last component of the lymphatic system and are specialized cells that can circulate within the blood or the lymphatic system. Lymphocytes account for approximately 25% of the entire circulating leukocyte cell population. Under normal conditions, only a small portion of the lymphocyte population is circulating throughout the body and the vast majority of these cells are stored in lymphatic tissues/organs. The body contains more than 10^{12} lymphocytes. There are three major classes of lymphocytes circulating within the blood, the T-cells (thymus-dependent), the B-cells (bone marrow-derived), and the natural killer cells (NK cells). Approximately 75% of the lymphocytes are T-cells, which can be sub-divided into the cytotoxic T-cells, the helper T-cells, and the suppressor T-cells. The primary function of the cytotoxic T-cells is to attack foreign cells or destroy the body's cells that have been infected by a virus. The helper T-cells stimulate the activation and production of T-cells and B-cells during an inflammatory response. Suppressor T-cells inhibit the activation of T-cells and B-cells. B-cells account for approximately 15% of the lymphocyte population. Activated B-cells differentiate into plasma cells, which are responsible for mass production and secretion of antibodies in response to an infection. Therefore, T-cells play a critical role during the cell-mediated immunity, while B-cells play a critical role during antibody-mediated immunity. The natural killer cells make up the remaining 10% of the lymphocyte population. These lymphocytes are responsible for attacking foreign cells and innate cells infected with a virus. These cells continuously monitor the quantity of foreign particles, and therefore, their function is termed immunologic surveillance. The life span of lymphocytes varies, but the vast majority of them last for at least 5 years (approximately 80%). Some lymphocytes can last as long as 20 to 30 years in circulation and therefore act as a memory for past foreign particle invasions and can help to stem a secondary attack.

8.2 LYMPH FORMATION

Lymph is composed only of the particular compounds of the interstitial fluid that enters the lymphatic system. Therefore, as stated before, the composition of lymph is very similar to that of the interstitial fluid as it first enters the lymphatic vessels. However, under normal conditions, the protein concentration of lymph is higher than the interstitial fluid protein concentration because a large portion of lymph is derived from the liver. The liver has a protein concentration in the range of 6 g/dL and lymph derived from the liver composes approximately 50% of all lymph within the lymphatic system. The remaining portion of lymph has a protein concentration that ranges from 2–4 g/dL, which is the average interstitial protein concentration (recall that the interstitial fluid protein concentration will vary based on the tissue). This makes the total lymphatic protein concentration in the range of 3–5 g/dL. A second difference between lymph and interstitial fluid is the presence of large bacteria. Large bacterial cells can invade the lymphatic system after they migrate into the interstitial space. These cells will typically be cleared after the lymph flows through the lymph nodes, but appear in a higher concentration than within the interstitial fluid.

Lymph is formed through a process already discussed in Chapter 6, the movement of water and nutrients out of the capillaries into the extracellular space. This means that lymph formation rate is dependent on the hydrostatic pressure of the capillary and the hydrostatic pressure of the interstitial space as well as the osmotic pressure within the capillary and the osmotic pressure within the interstitial space. This is described by the same equation that we discussed in Chapter 6, which is repeated here:

$$\dot{m} = K_p(P_B - P_I - \Pi_B + \Pi_I) \tag{6.8}$$

However, lymph is not solely derived from the blood. As cells perform work, they can uptake substances from the extracellular space (e.g., nutrients) as well as add substances to the extracellular space (e.g., wastes). As such, the composition of lymph changes continually and it is slightly different in each tissue. As lymph enters the lymphatic capillaries, its composition is very similar to the interstitial fluid of that tissue. However, as lymph moves throughout the lymphatic system, it comes into contact with other cells (particularly lymphocytes), other proteins, other molecules and lymph derived from other tissues that change the overall composition of lymph. Therefore, in some sense, lymph is continually being formed until it is returned back to the cardiovascular system via the lymphatic ducts. A modified lymph formation equation can be described by

$$\dot{m} = K_p(P_B - P_I - \Pi_B + \Pi_I) + \dot{m}_t \tag{8.1}$$

where $\dot{m}_t$ is the formation rate of lymph from other cells that may come in contact with lymph within the lymphatic vessels. Note that this can be positive, if the cells are adding lymph or other constituents to the fluid, or negative, if the cells are removing constituents.

8.3 FLOW THROUGH THE LYMPHATIC SYSTEM

The total flow rate throughout the lymphatic system is approximately 125 mL/h. Close to 100 mL of this is returned each hour through the thoracic duct to the cardiovascular system. The remaining portion is returned via the right lymphatic duct. Every day, approximately 3 L of lymph is formed and returned back to the cardiovascular system. If you recall that the total blood volume of an average person is around 5 L and the plasma component of this is close to 60% (or 3 L), this suggests that without the lymphatic system, a person would lose the entire plasma portion of blood each day (if the pressures did not change during this process). If the body had to continually produce plasma to make up for this loss, this would be an inefficient process and the body would not function as we know it, because a significant energy input would be needed to form plasma and this energy would be diverted from other critical functions. Additionally, continual ingestion and absorption of water would be needed and it would be unlikely that the kidneys, if they were present, would produce urine as they currently do (maybe a mechanism that is used in avian would be present). Therefore, the lymphatic system prevents this inefficient process of continually producing blood plasma, which has been filtered within the capillary beds into the interstitial space. Additionally, lymph flow provides a safety mechanism to screen for invading pathogens, since the entire plasma volume can flow through the lymphatics every day.

One of the largest regulators of flow through the lymphatic system is the interstitial hydrostatic pressure. As the hydrostatic pressure increases, the rate of flow within the lymphatic system also increases. At an interstitial pressure of approximately 0 mmHg, the rate of lymph flow increases 10 times over the normal lymphatic flow rate (recall interstitial hydrostatic pressure is typically −3 to −4 mmHg). As the interstitial pressure increases to 2 mmHg, the lymphatic flow rate approaches a maximum of approximately 20 times of the normal lymphatic flow rate. The lymph flow cannot exceed this capacity because at higher interstitial pressures (>3 mmHg), the lymphatic vessels compress, because they cannot withstand the external pressure forces. Lymphatic vessels do not have a thick muscular wall like some of the vessels within the cardiovascular system. The possible compression of lymph vessels effectively increases the resistance to flow through the lymphatic vessels to a point that the driving force cannot overcome. Therefore, for small deviations from the normal interstitial pressure, the rate of lymph flow does not increase significantly. As the interstitial pressure increases significantly above the normal interstitial hydrostatic pressure level, the rate of lymph flow increases to remove the excess fluid within the interstitial space. The major contributors to the interstitial hydrostatic pressure are the capillary hydrostatic pressure, the osmotic pressures of the capillary and the interstitial space, as well as the endothelial cell permeability. Small changes in any of these values can have a significant effect on the hydrostatic pressure of the interstitial space and therefore a major effect on the flow rate of the lymphatic system.

Example

Assume for this example that the osmotic pressure of the blood vessel, the osmotic pressure of the interstitial space, and the hydrostatic pressure of the blood vessel remains the same. Make an approximation of the lymph flow rate at various interstitial pressures (as given in the main text) using a first-order polynomial and a second-order polynomial. Plot the changes in lymph flow rate versus pressure under both conditions.

Solution

Using linear algebra to solve for the least squares regression line through the following data points, we get

$$X = [-4, 0, 2] \text{ mmHg}$$

$$Y = [1, 10, 20] \text{ relative flow rate}$$

First-order Polynomial	Second-order Polynomial
$\begin{bmatrix} 3 & -2 \\ -2 & 20 \end{bmatrix}\begin{bmatrix} a \\ b \end{bmatrix} = \begin{bmatrix} 31 \\ 36 \end{bmatrix}$	$\begin{bmatrix} 3 & -2 & 20 \\ -2 & 20 & -56 \\ 20 & -56 & 272 \end{bmatrix}\begin{bmatrix} a \\ b \\ c \end{bmatrix} = \begin{bmatrix} 31 \\ 36 \\ 96 \end{bmatrix}$
$Y = 3.036X + 12.357$	$Y = 0.4583X^2 + 4.083X + 10$

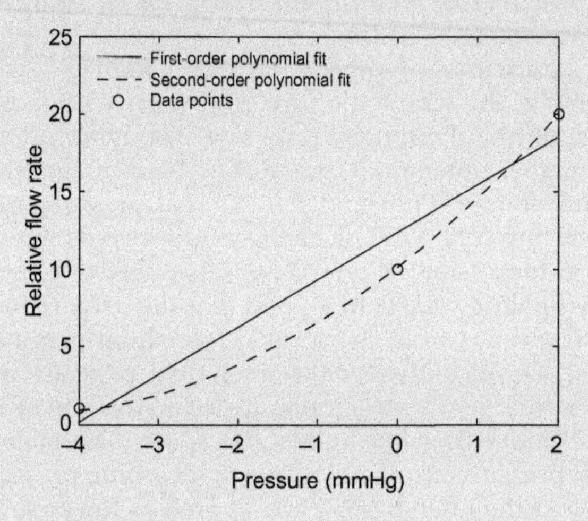

FIGURE 8.2 Regression analysis of data presented in the in-text example.

In these formulations, Y is the relative flow rate and X is the average interstitial pressure in mmHg. With a second-order polynomial, the curve goes through each of the data points and this may be the better approximation for lymph flow rate changes (Y-axis) with changes in interstitial pressure (X-axis), within the range of given values. However, we know that lymph flow rate cannot exceed approximately 20 times its normal flow rate, so outside of this range each approximation will fail (see Figure 8.2). A better approximation should include more data points and possibly better approximation methods.

The lymphatic system has a pump mechanism similar to the one seen in venous flow. As lymphatic vessels fill with lymph, they stretch. This stretch causes an automatic contraction of the smooth muscle cells that surround the lymphatic vessels. The contraction of smooth muscle cells forces lymph through a small lymphatic section (divided by valves) into a downstream lymphatic section. As the fluid fills the next section, the lymphatic vessel stretches again, causing an automatic smooth muscle cell contraction. This pumping mechanism acts to propel the lymph along the entire lymphatic vascular system. Pressure within the lymphatic vessel, due to smooth muscle cell contraction, can be as high as 100 mmHg (in the thoracic duct), but is generally significantly lower than this.

Furthermore, this pumping mechanism is transient and would only experience these high pressure forces for a relatively short duration of time. Smooth muscle cells are not the only mechanism that aids in the contraction of lymphatic vessels. A contraction of neighboring skeletal muscle can propel lymph through the lymphatic system, similar to the venous system. Compression of any tissues that transmits a compressive load to the lymphatic system can also help to move lymph throughout the lymphatic system. Remember that the lymphatic system is a low-pressure system.

Lymph flow through the lymph nodes is also important to discuss (Figure 8.3). Lymph is delivered to the lymph nodes through a few afferent lymphatic vessels (typically in the range of five) per node. Within the lymph node, there are hundreds of lymph node sinuses, which are open passageways throughout the node. Lymph can flow freely through these sinuses. The first set of sinuses that lymph flows through is termed the subcapsular space, which houses a large number of macrophages and other cells to initiate an immune response, if necessary. The subcapsular space is typically on the order of three to five cells in thickness (50–100 μm). After flowing through this space, lymph enters the outer cortex of the lymph node, which contains a large concentration of B-cells. The outer cortex also contains the locations for B-cell division, termed the germinal center. The outer cortex is typically on the order of 10 to 20 cells in thickness (200–400 μm). Lymph then enters the cortex of the lymph node, which surrounds the lymph's artery and vein, connecting the lymph node to the circulatory system. The cortex is composed mainly of T-cells, but many lymphocytes can migrate out of the cardiovascular system into the lymph nodes at this location. Lymph then continues into the medulla of the lymph node, which consists primarily of B-cells and plasma cells. The medulla collects lymph entering from all of the afferent lymphatic vessels and shunts it toward the single efferent lymphatic vessel. At any of these locations within the lymph nodes, lymph can come into contact with immune cells to help

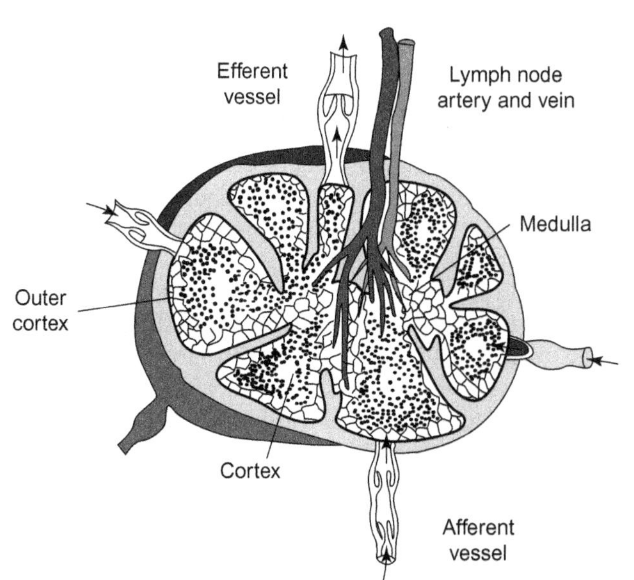

FIGURE 8.3 Structure of a lymph node, showing the internal arrangement of the lymph node sinuses. Each lymph node is associated with one artery and one vein that provide a direct connection between foreign particles within the cardiovascular system and lymphocytes within the lymphatic system. Lymph enters the lymph node from multiple afferent vessels that pass through various chambers of the lymph node. These chambers have multiple lymphocytes to monitor the presence of inflammatory mediators. After passing through these chambers, lymph flows out of the lymph node from one efferent vessel. *Adapted from Martini and Nath (2009).*

initiate an immune response, if needed. Each lymph node is connected to the cardiovascular system via one artery and one vein. The lymph nodes are very efficient at removing antigens from lymph and it has been suggested that over 99% of all antigens are cleared from lymph within lymph nodes. Interestingly, lymph nodes are typically placed along the periphery of the chest cavity to prevent infections from entering the vital organs. The largest collection of lymph nodes are found in the groin region and in the neck, but there is a large concentration of lymph glands associated with the mesenteric flow to protect the body from invading pathogens from the respiratory system and the gastrointestinal system.

Lymphatic flow is a pressure-driven flow. The viscosity of lymph is close to that of water (1 cP). All of the fluid mechanics relationships that we have discussed so far apply to the lymphatic system. In particular, most of the flow scenarios within the lymphatic system will mimic what is seen in the venous system because of the presence of valves, slow pressure-driven flow, a muscular pumping system, and a large elasticity of the vascular walls. We will therefore not reiterate the relationships here because they can be found elsewhere in this textbook.

8.4 DISEASE CONDITIONS

8.4.1 Cancer Metastasis via the Lymphatic System

Metastasis is the spreading of malignant tumor cells from the primary tumor location to another location within the body. Before we discuss the role that the lymphatic system plays in metastasis, it is important to discuss some characteristics of cancer itself. Cancer is a disease that is characterized by uncontrollable cell division. This is typically caused by a destruction and/or modification of some of the genes within the cell. If the rate of cell division exceeds the rate of cell death, then a tumor forms. Tumors are defined as a mass of abnormally dividing cells. Tumors can be benign, which rarely lead to death, or malignant, which are cells that no longer function normally.

The primary tumor is the location where the first cells start to divide uncontrollably. As this process continues, the tissue becomes exceedingly large and it cannot contain all of the tumor cells within that space. Metastasis occurs when tumor cells break away from the primary tumor and invade other nearby locations of the body. To do this, tumor cells migrate through the interstitial space, in a process termed local invasion. Once in the interstitial space, tumor cells can be taken into the lymphatic system or enter the bloodstream, via a process termed intravasation. Intravasation is characterized by cancerous cells passing through the wall of lymphatic capillaries or cardiovascular capillaries. Regardless of which capillary (blood or lymphatic) the tumor cells enter, they can now be carried throughout the body relatively easily. This circulation through the vascular system initiates secondary tumor sites, which are typically more difficult to locate and difficult to treat. At the secondary tumor site, the cancerous cells arrest in the capillaries. This can be caused by a size restriction (e.g., they cannot fit through capillaries at this location) or due to an over-expression of adhesion molecules at a particular location. Regardless of the cause, the cancerous cells begin to extravasate from the lymphatic or cardiovascular capillaries by migrating across the wall. Once the tumor cell has exited the lymphatic system or blood vessels and has reestablished itself in a new location, this location is defined as the secondary tumor site. The cancerous cells can proliferate and induce angiogenesis to this secondary site. If tumor cells have penetrated the lymphatic system, it is also typical for them to amass within lymph nodes, preventing immune cells from recognizing invading particles. It is then likely for a patient to succumb to an infection because the inflammatory process will not proceed normally.

If the cancer cells enter the bloodstream, they can also be carried to other locations within the body to establish secondary tumors. Tumor cells use a great deal of energy at the expense of the surrounding healthy tissue. To accommodate this new demand, cancer cells have the ability to stimulate new blood vessel growth (or angiogenesis). The additional blood supply unfortunately acts to accelerate the rate of cancer cell growth and does not provide the non-cancerous tissue with the nutrients it needs. This leads to the death of non-cancerous tissue and tissue/organ malfunction due to a lack of nutrients. Also, the simple compression of a large mass of tumor cells on a healthy organ causes the organ to not function properly. Cancer is a disease that is intimately associated with the cardiovascular system, and currently there is not a great deal of information available on how tumor cells interact and use the cardiovascular system to its advantage.

The ability of a tumor cell to metastasize, within the lymphatic system or with the cardiovascular system is dependent on a number of properties. First, if the lymphatics or the vascular system is cancerous to begin with, then it is relatively easy for the cells to metastasize based on the mechanical loading conditions of the tissue. If the cancerous tissue is not part of either circulation, then properties of the cancerous cell itself, will allow it to metastasize or not. Surrounding non-cancerous tissue can inhibit or accelerate metastasis based on the responses that it is undergoing. The vasculature (lymphatic or cardiovascular) at the primary tumor site and a potential secondary tumor site can also effect the rate of metastasis. In general, a lot of factors are needed to promote the metastasis of tissue. In addition to all of these factors working together to promote metastasis, a metastatic cancer cell can establish itself at a secondary site, but it may not proliferate for many years. This is part of the reason why locating and treating these secondary sites (which can include multiple secondary sites) is very difficult.

8.4.2 Lymphedema

Lymphedema is very similar to edema associated with microvascular networks. Typically, a lymphatic vessel becomes blocked, and this prevents lymph from passing through the lymphatic system and being returned to the cardiovascular system. In this scenario, lymph collects within the lymphatic vessel and interstitial fluid collects within the interstitial space. The region upstream of the blocked lymphatic vessel becomes swollen because fluid cannot pass through the blockage. The danger of lymphedema is associated with an infectious outbreak due to the decrease in lymph flow. Because a major initiator of the inflammatory response is associated with the lymph tissues and organs, it is likely that during lymphedema, infectious agents never arrive at these lymph tissues and organs. Because the interstitial fluid and lymph is also stagnant, it is unlikely that many white blood cells will be in the vicinity of the infectious agents. Therefore, the infectious agents can overcome the local immune response and reach a very serious level. Also, with any edema, the increased pressure can affect local tissue function and local blood flow.

Lymphedema may be inherited (primary lymphedema) or caused by lymphatic vessel injury (secondary lymphedema). Primary lymphedema can be classified by the time of onset; if it is present at birth it is termed congenital lymphedema, if it appears around puberty it is termed praecox lymphedema, if it appears around age 35 it is termed tarda lymphedema or it can manifest as due to genetic reasons at birth it is termed Milroy's disease. Primary lymphedema is more prevalent in females than males; females account for approximately 90% of the cases. Additionally, the onset of lymphedema manifests prior to the age of 40 in more than 90% of the cases. Secondary lymphedema is most commonly caused by inflammation that follows some injury to lymphatic tissue. It is most frequently seen after lymph node dissection, surgery, and/or radiation therapy, in which damage to the lymphatic system is caused during the treatment itself.

The exact cause of primary lymphedema is unknown, but it generally occurs due to missing lymph nodes or channels between the lymph nodes. Secondary lymphedema affects both men and women, but it occurs in different locations. In women, lymphedema is most commonly found in the upper limbs. In men, it is most commonly found at the location of an injury or the legs. Regardless of the type or location of the lymphedema, it develops in stages. The first stage, Stage 0, is termed the latent stage and is classified by damage to the lymphatic vessels that has not become apparent yet. Transport of interstitial fluid still occurs but it is typically shunted away from the damaged vessels. Therefore, there is not a significant pooling of lymph. Stage 1 is termed the spontaneously reversible stage and is characterized by pooling of lymph that reverses during rest. In most cases, after depressing the affected area, an indentation will remain on the surface for some time. Stage 2 is termed the spontaneously irreversible stage. This stage is characterized by morphological and biochemical changes to the tissue due to the presence of excess interstitial fluid. The affected region is hardened, and there is a general increase in tissue size. The final stage, Stage 3, is termed the lymphostatic elephantiasis stage. The swelling in this stage is irreversible, and the affected area is relatively large. The tissue is very hard and does not depress when indented.

Treatments for lymphedema are somewhat palliative. For primary lymphedema it is most common to elevate the effected limb to prevent fibrosis and/or recurrent infections. In some instances, elastic supporting structures can be worn to maintain an elevated external pressure on the effected limb. There have been a number of studies that have shown that treatment with benzopyrones can increase the degradation of proteins within lymph, to aid in flow through the lymphatic system, however, the majority of these molecules have wide ranging effects and thus care should be used when applying them to patients.

END OF CHAPTER SUMMARY

8.1 The lymphatic system is composed of lymphatic vessels, lymphoid tissues and organs, lymph, and lymphocytes. Lymphatic vessels function to transport lymph away from microvascular beds and into the cardiovascular system. The first vessel within the lymphatic system is the lymphatic capillary, which is composed of loosely bound endothelial cells to facilitate the movement of water into the lymphatic system. Lymphatic capillaries converge into larger lymphatic vessels, which are similar in structure and function to venules. Lymphatic vessels contain valves to prevent lymph backflow. The major function of lymphatic tissues/organs is to produce and control lymphocytes and to bring blood into a close proximity with lymphocytes. Lymphocytes either directly destroy invading pathogens or associate them with antibodies so that other cells can recognize and remove them.

8.2 Lymph is formed as a filtrate of interstitial fluid, except that there is typically a higher protein concentration in lymph than the interstitial fluid. The regulation of lymph formation is therefore governed by the formation of interstitial fluid. Recall that the major players in interstitial fluid formation are the capillary hydrostatic pressure, the interstitial space hydrostatic pressure, the plasma colloidal osmotic pressure, and the interstitial colloidal osmotic pressure. There is an added component that lymph can be continually formed by cells within the lymphatic system. Therefore, lymph formation can be described by

$$\dot{m} = K_p(P_B - P_I - \Pi_B + \Pi_I) + \dot{m}_t$$

8.3 Lymph flows through the lymphatic system at a rate of approximately 125 mL/h for a total of approximately 3 L of lymph formed per day. This is approximately 60% of the average blood volume, and this would account for a significant loss of body fluids per day if it was not returned to the cardiovascular system. The principal regulator of lymph flow is the interstitial hydrostatic pressure, because the lymphatic system is also under a very low pressure throughout the entire system. Small changes in the interstitial pressure can have a significant effect on lymph flow through lymphatic vessels. Lymph is also propelled through the lymphatic vessels by compression of surrounding tissue (e.g., muscle contraction). Lastly, lymph must flow through lymph nodes, prior to being returned to the cardiovascular system. As lymph enters the lymph node, it passes through many sinuses that contain multiple lymphocytes to investigate the cells within the lymph. These cells can initiate an inflammatory response if they come into contact with pathogens.

8.4 Two diseases that are associated with the lymphatic system are cancer and edema. Cancerous cells can easily enter the lymphatic system and then move to other locations within the body. This would possibly initiate a secondary tumor site somewhere else in the body. Interestingly, tumor cells can metastasize through the lymphatic system or the cardiovascular system. Lymphedema is similar to edema and is defined as an increase in extravascular space water volume (and pressure) due to a blockage or a defect of the lymphatic system. This can potentially cause a large inflammatory response because it is possible that pathogens do not come into contact with lymphocytes in the lymphatic tissues.

HOMEWORK PROBLEMS

8.1 What type of blood vessels do the lymphatic vessels most closely match? Are the functions of these two vessels similar?

8.2 Which lymphatic organ removes damaged red blood cells from circulation? What other functions does this organ have?

8.3 Describe the differences between the deep lymphatic vessels and the superficial lymphatic vessels.

8.4 What functions do the T-lymphocytes and the B-lymphocytes perform?

8.5 Why is the protein concentration in lymph typically more concentrated than interstitial fluid?

8.6 Lymphatic vessels have valves that prevent the movement of lymph back toward the capillary beds. What will happen to water movement across the capillary wall if the lymphatic valves are not functioning properly?

8.7 Approximate the flow rate through a lymphatic vessel with a radius of 75 μm, a pressure difference of 5 mmHg, and an overall length of 1 mm. The viscosity of lymph in this section of lymphatic vessels is 1.1 cP. If a muscle surrounding the lymphatic vessel constricts the vessel so that the radius reduces to 25 μm and the pressure difference increases to 25 mmHg, what does the flow rate become under these conditions?

8.8 There are no active pumping mechanisms to propel lymph movement through a lymph node, which is composed of many interconnected chambers. Discuss the movement of lymph through these structures. Is it likely to be laminar, steady, turbulent, and so on? How does the body ensure the proper mixing of lymph within these chambers?

8.9 The permeability constant can be approximated from

$$K_p = \frac{Ar}{8\mu}$$

where A is the vessel area. Discuss and plot variations in this permeability in relation to the mass flow rate through the lymphatic system. Consider that lymph is continually being added to the system at a constant rate.

References

[1] K. Aukland, R.K. Reed, Interstitial-lymphatic mechanisms in the control of extracellular fluid volume, Physiol. Rev. 73 (1993) 1.
[2] M. Foldi, The brain and the lymphatic system (I), Lymphology 29 (1996) 1.
[3] M. Foldi, The brain and the lymphatic system (II), Lymphology 29 (1996) 10.
[4] A.C. Guyton, Pressure-volume relationships in the interstitial spaces, Invest. Ophthalmol. 4 (1965) 1075.
[5] A.C. Guyton, Interstitial fluid pressure. II. Pressure-volume curves of interstitial space, Circ. Res. 16 (1965) 452.
[6] A.C. Guyton, K. Scheel, D. Murphree, Interstitial fluid pressure. 3. Its effect on resistance to tissue fluid mobility, Circ. Res. 19 (1966) 412.
[7] A.C. Guyton, J. Prather, K. Scheel, J. McGehee, Interstitial fluid pressure. IV. Its effect on fluid movement through the capillary wall, Circ. Res. 19 (1966) 1022.
[8] A.C. Guyton, T.G. Coleman, Regulation on interstitial fluid volume and pressure, Ann. N.Y. Acad. Sci. 150 (1968) 537.
[9] A.C. Guyton, H.J. Granger, A.E. Taylor, Interstitial fluid pressure, Physiol. Rev. 51 (1971) 527.
[10] A.C. Guyton, A.E. Taylor, H.J. Granger, W.H. Gibson, Regulation of interstitial fluid volume and pressure, Adv. Exp. Med. Biol. 33 (1972) 111.
[11] A.C. Guyton, A.E. Taylor, R.A. Brace, A synthesis of interstitial fluid regulation and lymph formation, Fed. Proc. 35 (1976) 1881.
[12] A.C. Guyton, Interstitial fluid pressure and dynamics of lymph formation. Introduction, Fed. Proc. 35 (1976) 1861.
[13] A.C. Guyton, B.J. Barber, The energetics of lymph formation, Lymphology 13 (1980) 173.
[14] P.A. Nicoll, A.E. Taylor, Lymph formation and flow, Annu. Rev. Physiol. 39 (1977) 73.
[15] G.W. Schmid-Schonbein, Microlymphatics and lymph flow, Physiol. Rev. 70 (1990) 987.
[16] G.W. Schmid-Schonbein, Mechanisms causing initial lymphatics to expand and compress to promote lymph flow, Arch. Histol. Cytol. 53 (Suppl) (1990) 107.

PART IV

SPECIALITY CIRCULATIONS AND OTHER BIOLOGICAL FLOWS

CHAPTER

9

Flow in the Lungs

LEARNING OUTCOMES

1. Identify the salient components of the respiratory system
2. Describe the physiology of the lungs
3. Compare the branching structure within the respiratory system to that found within the vascular system
4. Explain the alveolar structure
5. Use Boyle's law to predict changes in the lung size
6. Model lung movement during inspiration and expiration
7. Formulate a relationship for lung blood-vessel elasticity
8. Analyze the separation distance between the alveoli and the capillaries
9. Discuss the air volumes associated with normal breathing
10. Describe the lungs' reserve volumes for breathing
11. Analyze the composition of air in the atmosphere and in the lungs
12. Formulate the diffusion rate within the alveoli
13. Understand ventilation perfusion (V/Q) matching
14. Calculate changes in breathing rate associated with changes in atmospheric conditions
15. Model the respiratory boundary as a semipermeable membrane
16. Describe the mechanisms for gas transport within the blood
17. Explain the role of hemoglobin in oxygen/carbon dioxide transport
18. Discuss compressible fluid flow
19. Discuss diseases that affect the respiratory system

© 2015 Elsevier Inc. All rights reserved.

9.1 LUNG PHYSIOLOGY

The respiratory system provides the means for gas exchange between the air within the atmosphere and the cardiovascular system. Before atmospheric air flows into the lungs, it passes through the upper portion of the respiratory system, which is composed of various passageways that deliver air to the lungs. The initial conduits for air to flow are typically the nose, the nasal cavity, the mouth, the pharynx, the larynx, and the trachea (Figure 9.1). All of these passageways, with the exception of the mouth, are lined with a mucous membrane, which act as a first-line defense against toxins or particulate matter within the air. The mucous membrane traps toxins and particulate matter and brings them into close proximity with white blood cells for removal and/or destruction. The other major function of these conducting passages is to warm atmospheric air to body temperature and to humidify atmospheric air to body conditions. Under very dry conditions, it is common for the nasal cavity to lose its moisture and then the blood capillaries in the nose rupture, causing a nosebleed, because all of the moisture has been used to humidify the atmospheric air.

The respiratory system is composed of many branching conducting pathways, to effectively transport and deliver the air into a close proximity with the cardiovascular system. This branching pathway also increases the available surface area for gas exchange at the level of the pulmonary capillaries. The branching respiratory passageways begin with the trachea, which branches into the right and left bronchi (Figure 9.2). Each bronchi divides

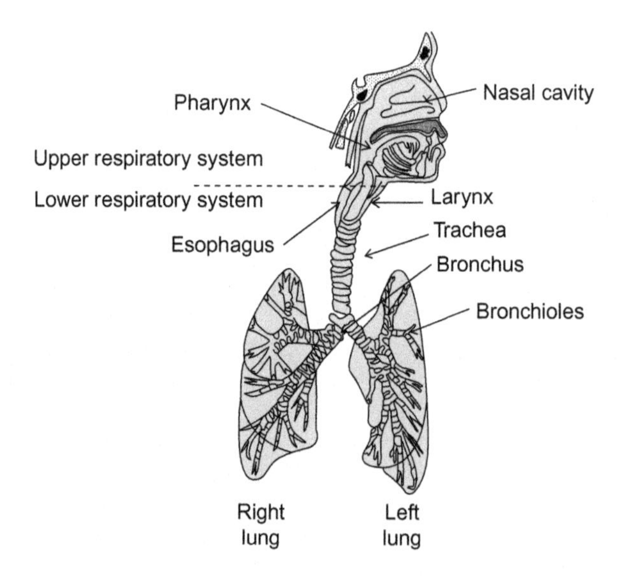

FIGURE 9.1 Anatomical arrangement of the components of the respiratory system. Air enters from either the nose (nasal cavity) or the mouth and passes into the pharynx. From here, air passes through the larynx, the trachea, and the primary bronchi. The bronchioles continually divide until the air reaches the alveoli where gas is exchanged with the cardiovascular system. Only the conducting portion of the respiratory system is shown in this figure.

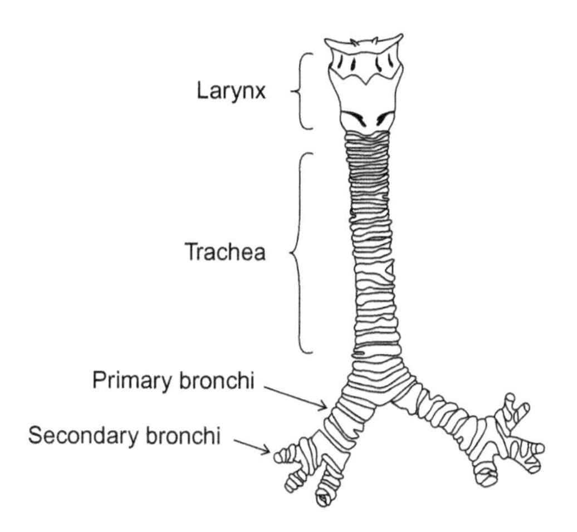

FIGURE 9.2 Schematic of the lower respiratory system, which is composed of the larynx, the trachea, and the bronchi. The trachea and the bronchi are surrounded by cartilage to maintain the rigid shape of this portion of the respiratory system.

into smaller secondary bronchi, which branches into smaller tertiary bronchi. These tertiary bronchi divide approximately three to five times into smaller bronchi within the lungs. Eventually, the bronchi give rise to bronchioles, which themselves divide approximately five times, eventually branching into the terminal bronchioles. Terminal bronchioles have an internal diameter in the range of 500 μm and there are approximately 6000 terminal bronchioles that arise from each tertiary bronchus (there are approximately 20 tertiary bronchioles in the lungs). This branching structure is very similar to the arterial vascular branching pattern between the single large aorta and the many terminal arterioles.

The larger conducting pathways within the lungs are surrounded by cartilage. Cartilage is relatively stiff and acts to protect the airways from collapsing or over-expanding during respiratory pressure changes. The tracheal cartilage wraps around the circumference of the trachea in a C-shape, and the cartilage ends are connected by a smooth muscle cell. The contraction and dilation of this smooth muscle cell changes the diameter of the trachea, which regulates the tracheal air flow by either increasing or decreasing the overall resistance to flow. The exterior walls of each bronchus are also wrapped with cartilage; however, the cartilage coverage is not as regular as seen along the trachea. This cartilage also acts to stiffen the conducting pathways and can partially regulate air flow through this portion of the lungs. The walls of bronchioles do not contain cartilage but do contain a large number of smooth muscle cells to regulate the diameter of the vessel. Bronchioles are similar to the arterioles of the cardiovascular system because they are the major resistance vessels to airflow in the lungs. These vessels are innervated by the autonomic nervous system, which can induce bronchodilation or bronchoconstriction, to change the airflow pattern throughout the lungs.

Each terminal bronchiole is connected to multiple alveolar sacs via small respiratory bronchioles. The alveolar sacs are common chambers that lead to the individual alveolus

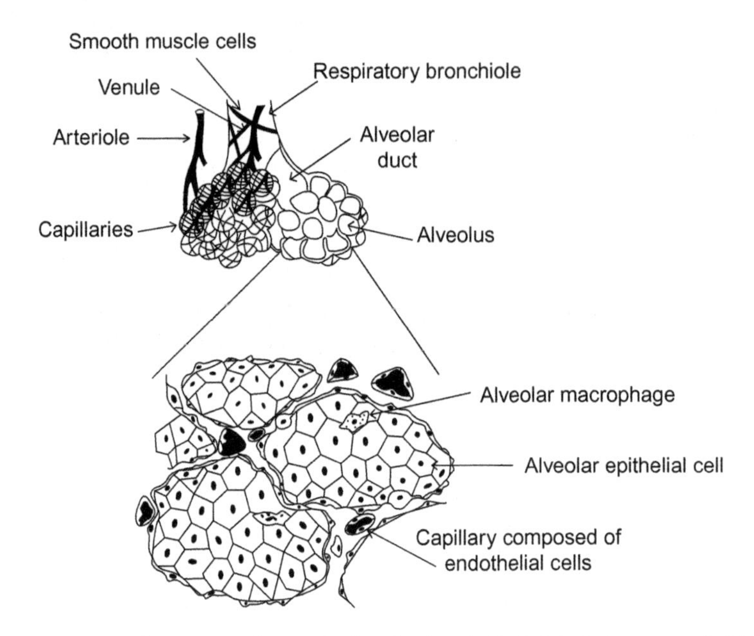

FIGURE 9.3 The anatomical structure of a single alveolar lobule. Each lobule is surrounded by a capillary network to efficiently exchange gas with the alveolar space. Bronchioles are wrapped with smooth muscle cells to change the diameter of the bronchiole and increase or decrease the resistance of air flow into the alveolar sac. This figure also shows an expanded section of an alveolar sac, which is composed of epithelial cells and the surrounding capillaries. *Adapted from Martini and Nath (2009).*

(Figure 9.3). Combined, there are approximately 300 million alveoli in the lungs. Each alveolus is directly associated with blood capillaries, allowing for a large surface area for gas exchange (approximately 70 square meters total). Alveolar walls are composed of thin epithelial cells. Macrophages (termed dust cells) monitor the lung epithelial surface and engulf any foreign particles that make it through the respiratory track and into the alveolar space. A second cell type, the pneumocyte type II cell, is inter-dispersed throughout the epithelial surface. These cells produce and excrete a surfactant, which is a viscous fluid mainly composed of phospholipids. This surfactant covers the entire alveoli surface in a thin layer of fluid, which acts to reduce the surface tension of the alveoli. If the surfactant is not present, the surface tension increases to a level that would tend to collapse the alveoli, called adhesive atelectasis, preventing gas exchange in the lungs. Therefore, the surfactant reduces the surface tension of the respiratory surface to provide a large surface area for gas exchange to occur between the lungs and the cardiovascular system.

As we have stated, gas exchange occurs across the alveoli epithelial surface. Gases must cross the respiratory boundary, which is composed of three layers to enter the blood capillary (Figure 9.4). The first layer of the respiratory boundary is the epithelial cells that line the alveoli. The second layer is termed the fused basal laminae layer which anchors the alveolar epithelial cells to the capillary endothelial cells. The basal laminae are the

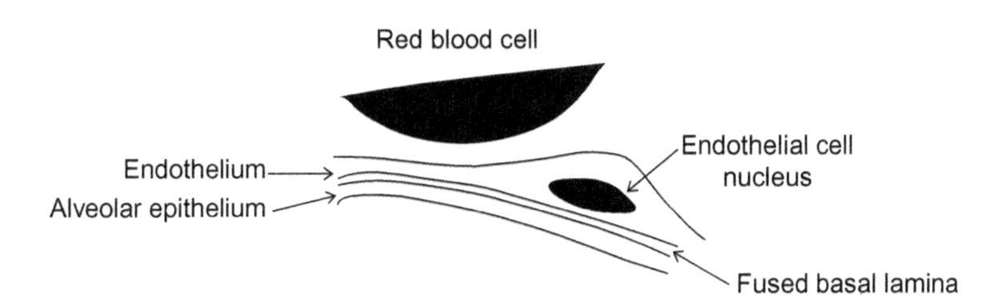

FIGURE 9.4 The respiratory membrane, which is composed of an alveolar epithelial cell, a capillary endothelial cell, and their fused basal lamina layer. In general, a red blood cell would be within one cell diameter of the endothelial cell. As discussed in the text, this boundary can be modeled as a single permeable membrane or a composite permeable membrane (with different diffusion permeabilities). *Adapted from Guyton and Hall (2000).*

extracellular matrix of each cell (epithelial and endothelial) and are primarily composed of elastin and collagen fibers within the respiratory boundary. The third layer is the single endothelial cell that forms the blood capillary wall. Because oxygen and carbon dioxide are uncharged and lipophilic, they easily and rapidly diffuse across the respiratory boundary. Remembering that in the lung circulation, the arterial vessels are deoxygenated, at the arterial end of the blood capillary, there is a large concentration gradient for oxygen to diffuse into the blood and a small concentration gradient for carbon dioxide to diffuse out of the blood. Under normal conditions, at the venous end of the blood capillary, both the oxygen and carbon dioxide concentrations between the alveoli and the blood are in equilibrium (Section 9.3 will discuss oxygen and carbon dioxide exchange in more detail). The distance that separates air within the alveoli and blood within the capillary is approximately 0.5 μm (which is the thickness of the respiratory boundary).

Moving to the gross anatomy of the lungs, each lung is composed of distinct lobes. The right lung has three lobes (the superior lobe, the middle lobe, and the inferior lobe) that are separated by the horizontal and the oblique fissures, respectively. The left lung has two lobes (the superior lobe and the inferior lobe) that are separated by the oblique fissure. The right lung is wider than the left lung because the heart is located within the left thoracic cavity and requires some physical space. The right lung is slightly shorter than the left lung to accommodate the liver, which is directly inferior to the right lung. Each lung is located within a space called the pleural cavity within the chest and is surrounded by a membrane termed the pleura. The pleura membrane is composed of two layers, the parietal pleural and the visceral pleural. The parietal pleural is the exterior layer, and it is connected to the thoracic wall, the diaphragm, and the ribs. The visceral pleural is the inner layer and covers the outer surface of each lung. Between the parietal pleural and the visceral pleural is a thin fluid layer. The fluid within this layer is termed the pleural fluid and is critical during respiration.

In fact, the relationship among the atmospheric pressure, the interpulmonary pressure, and the intrapleural pressure determines the direction of airflow in the lungs. Following simple fluid mechanics laws, if the atmospheric pressure and the interpulmonary pressure are the same, there is no net air movement in the respiratory system. If the interpulmonary pressure drops below the atmospheric pressure, then air will flow into the lungs. When the interpulmonary pressure rises above the atmospheric pressure, then air will flow out of the lungs. However, there is no internal mechanism for the interpulmonary pressure to change by itself because of some demand on the system (e.g., the lungs do not actively change pressure by contracting, relaxing, shape change, etc.). Also, there is clearly no mechanism for the body to alter the atmospheric pressure in relation to the interpulmonary pressure. In order to alter interpulmonary pressure, the body makes use of Boyle's law. Boyle's law states that at a constant temperature the pressure of a system is inversely related to the volume of the system. A constant temperature is a good assumption for most biological systems, and under normal breathing we will typically assume that the temperature within the human body is constant (although the temperature of atmospheric air may differ). Mathematically, Boyle's law is written as

$$P_1V_1 = P_2V_2 \tag{9.1}$$

if we assume that there are negligible changes in temperature and amount of substance between the two different states that are being described. Also, in this case, the two different states that we are analyzing will typically be the difference in the lungs immediately prior to inhaling (1) and immediately after inhaling (2). Using the principle of Boyle's law, the interpulmonary pressure is altered by changing the volume of the lungs. During inhalation, the rib cage moves upward and outward, while the diaphragm moves downward. Because the parietal pleura is attached to the diaphragm and the rib cage, it expands at the same rate as the diaphragm and ribs. This acts to decrease the intrapleural pressure, because the parietal pleura moves outward as the visceral pleural remains in the same location, to approximately negative 5 mmHg (gauge pressure). Due to the surface tension of the pleural fluid, the visceral pleural is pulled toward the parietal pleural. Because the visceral pleura is attached to the lobes of the lungs, the lung tissue gets pulled outward as well. Under normal conditions, this accounts for an approximately 20% change in lung volume, which decreases the interpulmonary pressure by 1–2 mmHg (gauge pressure), allowing for air to flow into the lungs. Before exhalation begins, the lung pressure becomes equilibrated with the atmospheric pressure again because enough atmospheric air has entered the lungs. During exhalation the reverse occurs. The ribs move downward and inward as the diaphragm moves upward, which pushes the parietal pleural into close contact with the visceral pleural. This movement increases the intrapleural pressure slightly (because the pleural fluid is not highly compressible), which acts to push the visceral pleural and hence the lungs. The interpulmonary pressure increases by approximately 1 mmHg above atmospheric pressure which forces air out of the lungs. Again, before the next round of inhalation begins, the atmospheric pressure and the interpulmonary pressure equilibrate. The pleural pressure is a measure of the elastic forces that tend to collapse or expand the lungs.

The lungs are supplied with blood through two pathways. The first supplies blood to the lung tissue, for nutrient delivery and waste exchange with the cells that compose the lungs (via the bronchial arteries). This is similar to the systemic circulation of every other tissue within the body. The second pathway is used for gas exchange to oxygenate blood and is termed the pulmonary circulation. We will not discuss the first pathway here because it is part of the systemic circulation and was covered in Parts 2 and 3 of this textbook.

The pulmonary circulation receives blood from the right side of the heart. The right ventricle pumps deoxygenated blood to the alveolar capillaries via the left and right pulmonary arteries. These arteries enter the lungs at the level of the bronchi and branch at the same locations that the respiratory vessels branch. Thus, the blood vessels follow the pulmonary tree down to the level of the alveoli. Each terminal bronchiole receives blood from one terminal arteriole and blood exits via one post-capillary venule. As stated earlier, each alveolus is surrounded by a dense mesh of capillaries to provide a surface for gas exchange. After the blood becomes re-oxygenated it passes into the post-capillary venule and follows the pulmonary tree back up to the level of the bronchi. The blood is returned to the heart via the pulmonary veins for systemic circulation. Recall that the pulmonary circulation flows at much lower pressures than the systemic circulation. It flows at approximately one-sixth of the systemic circulation, rarely exceeding a systolic pressure of 30 mmHg.

9.2 ELASTICITY OF THE LUNG BLOOD VESSELS AND ALVEOLI

As we have discussed in previous chapters, blood flow through vessels is highly dependent on the cross-sectional area of the blood vessels. We have discussed earlier in this chapter that during each breath, the lungs and all of the tissues and cells within the lungs experience a deformation. Generally during inhalation the lungs experience a 20% increase in volume (a tensile stretch) which is brought back to normal levels after expiration has occurred (a compressive deformation). From this knowledge, we would presume that the blood vessels in the lungs would also be subjected to large strains, and therefore, the cross-sectional area may change significantly, thus altering the blood flow through the pulmonary vasculature.

Because the pulmonary blood vessels are viscoelastic in nature, the stress–strain relationship must take into account the strain rate history. Strain rate history would be a function of loading rates and the applied loads. Because this biological material is under large deformations, the Kirchhoff stress (S_{ij}) and the Green's strain (E_{ij}) are the most applicable stress/strain values to use to describe the loading conditions. Experimentally, it was found that the stress-strain relationship was actually independent of the loading frequency (e.g., breathing rate) and therefore was pseudo-elastic. In these same experiments, the results were fit to a classical two-dimensional strain energy function fairly accurately. From this experimental work, the circumferential stress (S_{xx}) and the longitudinal stress (S_{yy}) can be defined as

$$\begin{aligned} S_{xx} &= a_1 E_{xx} + 0.6429 E_{yy} \\ S_{yy} &= a_2 E_{yy} + 0.6429 E_{xx} \end{aligned} \tag{9.2}$$

where E_{xx} and E_{yy} are the Green's strains in the circumferential and longitudinal direction, respectively, and a_1 and a_2 are material properties. The material property constants change from person to person and can be altered during disease states. Other experimental work has shown that the relationship between pressure (blood pressure minus the pleural pressure) and vessel diameter is linear within the pulmonary blood vessels. In comparison, for systemic vessels this relationship is highly non-linear. More than likely this is due to the fused basal laminae layer which intimately links the pliant blood vessels to the stiffer lung tissue. Therefore, linear elastic relationships can be used when determining the blood vessel cross-sectional area change as a function of hydrostatic pressure, interpulmonary pressure, and intrapleural pressure.

Changes in the interpulmonary pressure and the blood hydrostatic pressure can also affect the size of the alveoli as well as the distance that separates the gas in the alveoli and the gas in the blood capillary. Under normal conditions, the alveolar size remains constant as a function of blood hydrostatic pressure but varies with interpulmonary pressure. In simple terms, the surface area of the alveoli will follow the inverse of Boyle's law, so that when the interpulmonary pressure increases, so does the size of the alveoli. This is a direct relationship so that mass is conserved during lung expansion. The separation distance is a slightly more difficult problem, because it will be affected by the interpulmonary pressure as well as the capillary hydrostatic pressure. Let us define

$$\Delta P = P_{\text{capillary-hydrostatic}} - P_{\text{interpulmonary}} \tag{9.3}$$

This pressure difference is the pressure that effects the separation distance between the alveolar space and the blood. Using Eq. (9.3), we can define the separation distance between the alveolar gas and the capillary gas (this is a measure of the respiratory boundary thickness and not the exact diffusion distance, which may follow random paths). The values for separation distance have been quantified experimentally and agree with the following discussion. If ΔP is negative and less than negative 0.7 mmHg, the separation distance is equal to 0. When ΔP is between negative 0.7 mmHg and 0 mmHg, the separation distance increases from zero to its nominal resting value of approximately 0.5 μm. As ΔP increases from zero, the distance increases linearly as well to some limiting value of approximately 1.3 μm at a pressure of approximately 30 mmHg. Mathematically, this can be represented as a piecewise continuous function:

$$h(\Delta P) = \begin{cases} 0 & \Delta P < -0.7\text{mmHg} \\ 0.5\mu\text{m} + \dfrac{0.5\mu\text{m}}{0.7\text{mmHg}}\Delta P & -0.7\text{mmHg} \leq \Delta P < 0\text{mmHg} \\ 0.5\mu\text{m} + \dfrac{2\mu\text{m}}{75\text{mmHg}}\Delta P & 0 \leq \Delta P < 30\text{mmHg} \\ 1.3\mu\text{m} & \Delta P > 30\text{mmHg} \end{cases} \tag{9.4}$$

The "slope" values that appear before the ΔP in the second and third relationships are known as the compliance coefficients of the alveolar wall. Under disease conditions, these compliance values can change significantly, and this is what effectively inhibits or accelerates gas movement from the atmosphere to blood under pathological conditions. To understand the separation distance as a function of pressure, one must look into which pressure is greater and what phase of matter is associated with the greater pressure. When the blood pressure is greater than the interpulmonary pressure, the blood can exert a net force on the alveolar space. Since the alveolar space is filled with a gas, it would tend to collapse under these conditions and pull away from the capillary. In the reverse scenario, the interpulmonary pressure is greater than the capillary hydrostatic pressure and therefore, the net force pushes the alveolar space closer to the capillary space.

9.3 PRESSURE-VOLUME RELATIONSHIP FOR AIR FLOW IN THE LUNGS

We have briefly discussed in a previous section that air movement into the lungs is caused by changes in the lung volume. Following Boyle's law, the gas pressure must change inversely with the change in volume (assuming that the temperature is constant and there is no change in the amount of gas within the system). Here we will briefly discuss some of the mechanics of breathing that bring about the changes in volume. Recall that we have stated that during inspiration, the rib cage moves outward and upward and the diaphragm moves downward. The movement of the rib cage is controlled by the intercostal muscles and the diaphragm is a skeletal muscle. During inspiration, the diaphragm and the external intercostal muscles contract causing an increase in the thoracic cavity volume. The contraction of the diaphragm accounts for approximately 75% of the air movement during normal breathing. During strenuous activity, the increase in the speed of rib movement and the increase in rib displacement are controlled by the pectoralis minor, the scalene, and the sternocleidomastoid muscles, which completely accounts for the increased volume of the thoracic cavity. During expiration, the contraction of the internal intercostal muscles brings the rib cage back to its normal position and the abdominal muscles contract to assist the internal intercostal muscles and to force the diaphragm upward.

The respiratory system can adapt rapidly to the oxygen demands of the body. The rate of breathing as well as the amount of air moved with each breath can vary significantly. Under extreme conditions, it is possible that the amount of air moved by the respiratory system can exceed 50 times the normal breathing capacity during strenuous exercise. The respiratory rate is defined as the number of breaths that are taken within 1 minute. Under normal resting conditions, this is close to 12 to 15 breaths per minute. The tidal volume is the amount of air that is inhaled during one breath, and this is approximately 500 mL for an average adult. Therefore, at rest, the amount of air moved into the lungs per minute is 6000–7500 mL. Of the 500 mL of air that enters the respiratory system during each breath, only approximately 350 mL of this enters the alveolar space and is used for gas exchange. The remaining 150 mL fills up the dead space (nose, trachea, and bronchi, among others) within the conducting system and is never used for gas exchange. Therefore, the alveolar space is ventilated with 4200–5250 mL of air each minute (clearly this is dependent on the exact size of the lungs and the respiratory rate). Interestingly, the respiratory rate and the tidal volume can be controlled independently of each other. If the respiratory rate increases to 25 breaths per minute, then the tidal volume must drop to 300 mL to maintain the same lung ventilation per minute (7500 mL/min). However, because the dead space of the conducting system does not change, the alveolar ventilation rate drops to 3750 mL/min (as compared with 5250 mL/min) because 150 mL of incoming air stays within the dead space during each breath.

Under extreme conditions, the respiratory system can move 4.8 L of air during each breath. This is termed the vital capacity of the lungs. The lungs never fully deflate because the alveolar wall would adhere to itself (adhesion atelectasis) and then it would be difficult to inflate the alveoli again. There is approximately 1.2 L of air within the lungs that is used to keep the lungs and the alveoli open under the most strenuous breathing conditions. The extra 1.2 L is termed the residual volume of the lungs, which combined with the lung vital

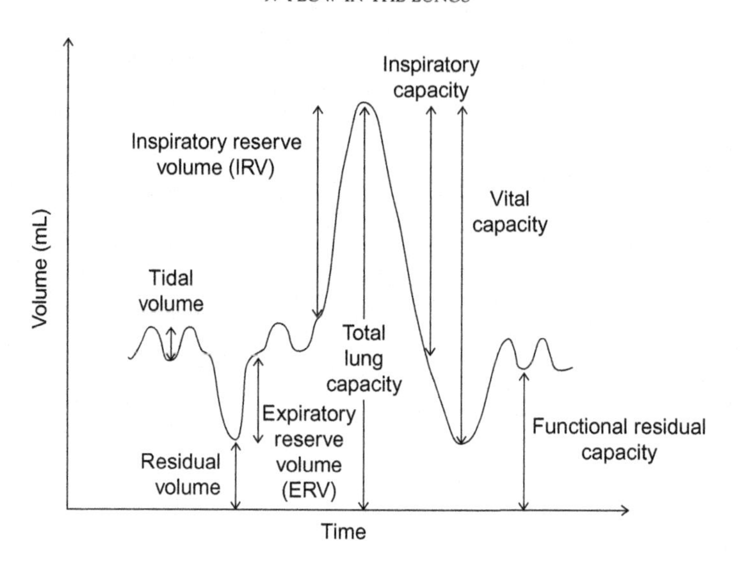

FIGURE 9.5 The volume of the lungs under normal breathing and various other respiratory movements. Under normal conditions, the tidal volume will be approximately 500 mL for each breath and there is an inspiratory reserve volume of approximately 3300 mL and an expiratory reserve volume of approximately 1000 mL. The total lung volume is approximately 6000 mL.

capacity suggests that the total lung volume is 6 L (Figure 9.5). During extremely strenuous activity (when the lung is moving 4.8 L per breath), the respiratory rate can increase to approximately 50 breaths per minute. Under these conditions, the respiratory system would move a total of 240 L of air per minute. In practice, only extremely well-conditioned athletes can approach this theoretical maximum respiratory rate.

We will provide a few more definitions that are associated with Figure 9.5, that relate to two ways to describe gas in the lungs: the volumes and the capacities. The expiratory reserve volume is the amount of air that can voluntarily be forced out of the lungs after the end of a normal expiration of the tidal volume. This is typically on the order of 1000 mL and is used during strenuous activities. The inspiratory reserve volume is the amount of air that can enter the lungs in addition to a normal inspiration of the tidal volume. On average this is close to 3300 mL, and again, this is used during strenuous activities. As defined before, the total lung capacity would be the residual volume plus the expiratory reserve volume plus the tidal volume plus the inspiratory residual volume (1200 mL + 1000 mL + 500 mL + 3300 mL = 6000 mL). The inspiratory capacity is the total amount of air that can be taken into the lungs. This is equal to the tidal volume plus the inspiratory reserve volume (500 mL + 3300 mL = 3800 mL). The vital capacity is the maximum amount of air that the lungs can take in or move out, which is equal to the expiratory reserve volume plus the tidal volume plus the inspiratory reserve volume (1000 mL + 500 mL + 3300 mL = 4800 mL). The functional residual capacity is the amount of air remaining in the lungs after a normal exhalation. This is equal to the residual volume plus the expiratory reserve volume (1200 mL + 1000 mL = 2200 mL).

Assuming that atmospheric air is uniform, then all of the previous discussions on the pressure changes required to move air into or out of the lungs are still valid. Again, primarily Boyle's law will dictate the movement of air into the lungs.

9.4 VENTILATION PERFUSION MATCHING

Adequate respiration means that the ventilation rate and blood flow rate must be matched so that an optimal amount of oxygen is loaded into the erythrocytes of the blood and an optimal amount of carbon dioxide is unloaded into the alveolar gas space. This process is termed ventilation (V) perfusion (Q) matching, and is described by the V/Q ratio. A simple way to think of this ratio is that alveoli without fresh gas cannot supply the blood with oxygen ($V/Q = 0$), and capillaries without flow cannot participate in gas exchange ($V/Q \sim \infty$). The typical normal V/Q ratio is 0.8, using a resting ventilation rate of 4000 mL/min and perfusion rate of 5000 mL/min. At this V/Q ratio, the partial pressure of oxygen will be close to 100 mmHg, and the partial pressure of carbon dioxide will be close to 40 mmHg as blood enters the venous pulmonary circulation.

Due to both the lung anatomy and the force of gravity, the V/Q ratio of 0.8 is not uniform in all alveoli. In fact, in an upright position the upper lobes of the lung are better ventilated than the lower lobes. They receive a larger volume of gas on each breath. This is in part due to the blood vessels that feed the lower lobes being fuller of blood than the upper lobes, due to the force of gravity pulling the weight of the blood down (remember that the overall pressure within the pulmonary circuit is relatively low). Thus the upper lobes of the lungs have better ventilation and less flow; the V/Q ratio is greater than 0.8. The lower lobes have less ventilation and more flow; the V/Q ratio is less than 0.8. One can imagine a point in the middle where $V/Q = 0.8$. This concept is illustrated in

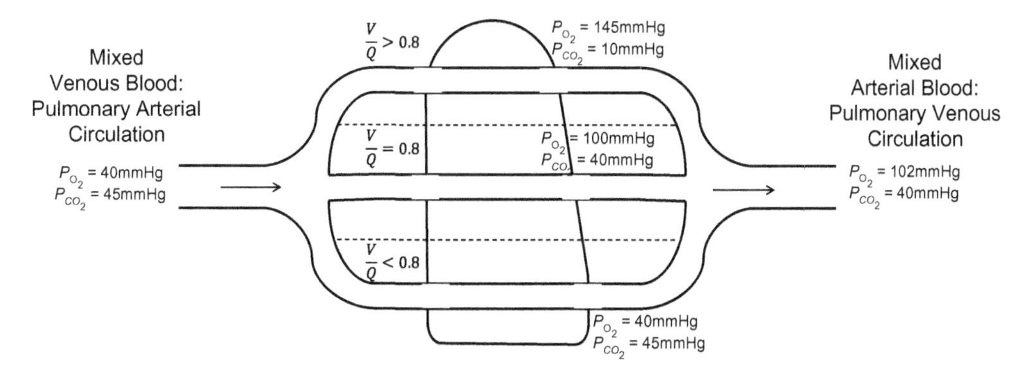

FIGURE 9.6 Ventilation perfusion ratio for lungs in an elevated position. Under normal conditions the upper portion of the lungs has very low flow with a higher ventilation. This allows the blood from this zone to have an elevated oxygen and reduced carbon dioxide concentration as compared with mixed arterial blood. The middle zone has an oxygen and carbon dioxide concentration that is close to that of mixed arterial blood. Finally, the lowest zone has a reduced oxygen and an elevated carbon dioxide concentration as compared with mixed arterial blood, because this portion of the lungs is not ventilated well.

Figure 9.6, which shows one lung that has three different ventilation/perfusion zones. Under normal conditions, in a sitting position the upper zone has a V/Q ratio in the range of 3 and the lower zone has a V/Q ratio in the range of 0.5.

Lung areas that are ventilated, but there is no gas exchange due to no blood flow, are dead space. Likewise lung areas that are not ventilated, but have blood flow, represent a shunt. In various disease states, the V/Q ratio becomes diagnostic and is different from emphysema versus asthma, for instance. This is elaborated on in Section 9.8.

9.5 OXYGEN/CARBON DIOXIDE DIFFUSION

Diffusion is the random thermal motion of molecules, which incurs a net movement of molecules from a high concentration to a lower concentration. Diffusion is the only way that oxygen enters the blood from alveolar air and that carbon dioxide enters the alveolar air from the blood (meaning that there is no active transport mechanism for these molecules). Atmospheric air is composed of a number of molecules and the pressure that each exerts is directly related to the concentration of that particular molecule. The normal atmospheric air contains 21% oxygen; therefore, 21% of atmospheric pressure is contributed by oxygen (i.e., $0.21 * 760\text{mmHg} = 160\text{mmHg}$). This partial pressure is denoted by P_{O_2}, for oxygen. Gases that are dissolved in liquids also exert a pressure. The partial pressure of a gas dissolved within a fluid is determined by its concentration and its solubility coefficient. Henry's law describes the relationship between a gas partial pressure and the factors that affect the pressure. Henry's law takes the form of

$$P = \frac{\text{Gas Concentration}}{\text{Solubility Coefficient}} \tag{9.5}$$

Looking at this relationship, it shows that for a given concentration, if the gas solubility is high, the gas will exert a lower pressure as compared with a gas that has a lower solubility. Gases that are highly soluble exert a lower pressure because there will be an attraction between the liquid molecules and the gas molecules, which prevents the gas from moving randomly and quickly through the fluid. Also, it is easy for gas molecules to enter the liquid, and the molecules would prefer to be associated with the liquid molecules. However, when the solubility is low, there is a very small attraction between the liquid and gas molecules which allows for the gas to move more freely. Also, it is harder for gas molecules to enter the liquid and the molecules would prefer to not be in the liquid. At body temperature, the solubility coefficients for oxygen and carbon dioxide are 0.024 and 0.57, respectively. Therefore, the net movement of gas molecules across the respiration boundary depends on the relationship between the partial pressure of the gas in atmospheric air and the partial pressure of the gas within blood. Under normal conditions, the partial pressure for oxygen is higher in the alveoli, so that there is a net movement of oxygen into the blood. Also, the partial pressure for carbon dioxide is typically higher in the blood, so that there is a net movement of carbon dioxide into the alveoli.

Like diffusion of any other molecule, the rate of diffusion is not solely based on the pressure gradient (for solids/liquids we typically talk about a concentration gradient, regardless of the phase of the molecule there is some driving force for diffusion to occur). The other factors that affect gas diffusion are the temperature of the exchange medium, the distance that the gas must diffuse over, gas solubility, the cross-sectional area of a barrier separating the diffusing molecules, and the molecular weight of the gas. In equation form, the diffusion rate is proportional to

$$D \propto \frac{\Delta PAs}{\Delta x\sqrt{M}} \tag{9.6}$$

where D is the diffusion rate, ΔP is the pressure difference, A is the cross-sectional area, s is the solubility coefficient, Δx is the distance, and M is the molecular weight of the gas. This is intuitive because as the pressure difference, cross-sectional area, and solubility increase, more gas molecules should be able to diffuse into the medium. As the distance and molecular weight increase, then the amount of molecules that can diffuse should decrease. From previous engineering or science courses, you may be familiar with the concept of a diffusion coefficient. For gases in solution, the diffusion coefficient is proportional to the solubility of the gas divided by the square root of the molecular weight of the gas. If we set the relative diffusion coefficient of oxygen to be 1.0 under normal physiological conditions, then the relative diffusion coefficient for other physiologically relevant gases are listed in Table 9.1.

Similar to the diffusion of gases through fluids, the diffusion of gas through the respiratory boundary is regulated by the same factors. Namely, the pressure difference, the thickness of the membrane, the cross-sectional area available for exchange, and the diffusion coefficient (solubility/molecular weight) are the most important factors that determine the diffusion of gas across the respiratory boundary. The pressure difference across the respiratory boundary is determined by the partial pressures of the gas in the alveolar air and in blood (which are listed in Table 9.2). The difference in these two values will provide a means to determine the net direction of movement for gas molecules across the respiratory boundary. The respiratory boundary thickness is a measure of the linear distance that the molecules must traverse in order to be exchanged. As this membrane increases in thickness, which commonly occurs under disease conditions, the rate of diffusion will decrease significantly. The surface area for diffusion determines the available regions for gases to

TABLE 9.1 The Relative Diffusion Coefficients for Important Respiratory Gases

Gas	Relative Diffusion Coefficient
Oxygen	1.0
Carbon Dioxide	20.3
Nitrogen	0.53
Carbon Monoxide	0.81

TABLE 9.2 Respiratory Gas Partial Pressures

Gas	Atmospheric Air (mmHg)	Alveolar Air (mmHg)	Deoxygenated Blood (mmHg)	Expired Air (mmHg)
N_2	597	569	563.3	566
O_2	159	104	40	120
CO_2	0.3	40	45	27
H_2O (as vapor)	3.7	47	Variable	47
Total	760	760	~760	760

be exchanged. As the surface area for exchange decreases, which again is common during disease conditions, the rate of diffusion will also decrease. Lastly, the diffusion coefficient is directly related to the molecular properties of the gas. With a greater solubility and a lower molecular mass the diffusion coefficient will be larger, which in turn leads to a higher rate of diffusion. The changes in the rate of diffusion for gases across the respiratory boundary follow the same principles as discussed for diffusion through a fluid.

The same factors that regulate the diffusion of gases through fluids regulate the diffusion of gases through the respiratory boundary. The diffusing capacity is the ability for a boundary to exchange gas and is defined as the volume of gas that diffuses across a membrane at a pressure difference of 1 mmHg in 1 minute. Under normal resting conditions, the diffusing capacity of oxygen is 21 mL/min/mmHg. As blood enters the alveolar capillaries the pressure difference for oxygen is approximately 60 mmHg, but this reduces to approximately 2 mmHg at the venous (oxygenated) side of the capillary (we are considering the mixed venous and arterial blood levels). If we approximate this to an average of 30 mmHg, this suggests that under normal conditions 630 mL of oxygen diffuses across the respiratory boundary, every minute. During strenuous exercise, the diffusing capacity of oxygen can increase to 65 mL/min/mmHg. The diffusing capacity of carbon dioxide is approximately 20 times greater than oxygen because it is directly related to the diffusion coefficients (see Table 9.1). Therefore, the diffusing capacity of carbon dioxide is approximately 425 mL/min/mmHg under normal conditions or approximately 1300 mL/min/mmHg during strenuous exercise.

Example

Imagine standing on the top of a mountain where the atmospheric pressure is 480 mmHg and the ambient temperature is 10°C. Assume that the percent composition of air is the same as described in Table 9.2 (although the exact partial pressures will be different). Calculate the respiration rate needed to maintain the body's oxygen requirements of 270 mL/min at standard body temperature and pressure (i.e., 760 mmHg, 37°C). Assume that your tidal volume increases to 750 mL and that 35% of the inspired oxygen enters the blood and that all of this would meet the body's oxygen requirements.

Solution

First let us determine the partial pressures of each of the major gases that are in air (calculated in Table 9.3):

TABLE 9.3 Partial Pressure of Gases Calculated from the Given Values

Gas	Percent Composition	Partial Pressure
N_2	78%	374.4 mmHg
O_2	21%	100.8 mmHg
CO_2	0.04%	0.2 mmHg
H_2O	0.5%	2.4 mmHg

Use a more general form of Boyle's law to determine the tidal volume needed at body temperature/pressure (assuming that the amount of substances remains the same):

$$\frac{P_B V_B}{T_B} = \frac{P_M V_M}{T_M}$$

$$V_B = \frac{T_B P_M V_M}{P_B T_M} = \frac{310K(480\text{mmHg})(750\text{ mL})}{760\text{mmHg}(283K)} = 520\text{ mL}$$

The subscript B represents the normal body conditions, whereas the subscript M represents the "mountain top" conditions. Twenty-one percent of this volume is oxygen and of that, your body can only use 35%:

$$520\text{mL} * 0.21 * 0.35 = 38.22\text{ mL}$$

Thus, the body takes in 38.22 mL O_2/breath under normal conditions.

The number of breaths needed on the mountain top would be

$$\frac{270mLO_2/\text{min}}{38.22mLO_2/\text{breath}} = 7\frac{\text{breaths}}{\text{min}}$$

To model the diffusion of oxygen into blood or that of carbon dioxide out of blood, let us consider the respiratory boundary as a semipermeable membrane (Figure 9.7). Blood flows along one side of the membrane and gas flows along the other side. For simplicity,

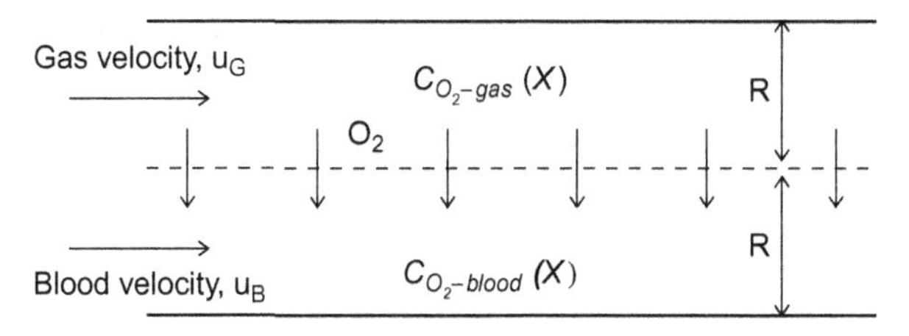

FIGURE 9.7 Respiratory boundary modeled as a semipermeable membrane. Using this type of model, the diffusion of respiratory gases across the respiratory boundary can be calculated if we assume that the total concentration of the gas remains constant along the channel.

let us assume that the membrane is only permeable to one respiratory gas (O_2 for this example). The net movement of oxygen across this channel will be from the gas side to the blood side. If we take the mass balance of the gas within each chamber across this semi-permeable membrane, assuming that the concentration of oxygen is constant (i.e., $C_{O_2-gas} + C_{O_2-blood} = \text{constant}$), we get

$$Ru_G(C_{O_2-gas}(x + \Delta x) - C_{O_2-gas}(x)) + J\Delta x = 0 \tag{9.7}$$

and:

$$Ru_B(C_{O_2-blood}(x + \Delta x) - C_{O_2-blood}(x)) - J\Delta x = 0 \tag{9.8}$$

where J is the flux of oxygen across the membrane defined by

$$J = \frac{D}{h}(C_{O_2-gas} - C_{O_2-blood}) \tag{9.9}$$

where h is the thickness of the respiratory boundary. If Δx approaches zero, Eqs. (9.7) and (9.8) become

$$\frac{dC_{O_2-gas}}{dx} = -\frac{D}{Ru_G h}(C_{O_2-gas} - C_{O_2-blood}) \tag{9.10}$$

and:

$$\frac{dC_{O_2-blood}}{dx} = \frac{D}{Ru_B h}(C_{O_2-gas} - C_{O_2-blood}) \tag{9.11}$$

respectively. To prove the assumption of the oxygen concentration being constant, add Eqs. (9.10) and (9.11) together to get

$$\frac{dC_{O_2-gas}}{dx} + \frac{dC_{O_2-blood}}{dx} = 0$$

which leads to

$$C_{O_2-gas} + C_{O_2-blood} = \text{constant}$$

if you integrate the differential equation with respect to distance. Equations (9.7) through (9.11) can be used to model the gas movement across any semipermeable membrane if the assumption regarding a constant concentration is valid.

9.6 OXYGEN/CARBON DIOXIDE TRANSPORT IN THE BLOOD

Oxygen and carbon dioxide have a very low solubility in plasma, which limits the amount of oxygen and carbon dioxide that blood can carry. Red blood cells, however, can "remove" the dissolved oxygen and carbon dioxide from blood and help blood transport more oxygen and carbon dioxide. In the case of oxygen, hemoglobin directly binds to oxygen molecules. Carbon dioxide, however, is modified into plasma soluble molecules by proteins carried within red blood cells and is therefore indirectly transported by red blood

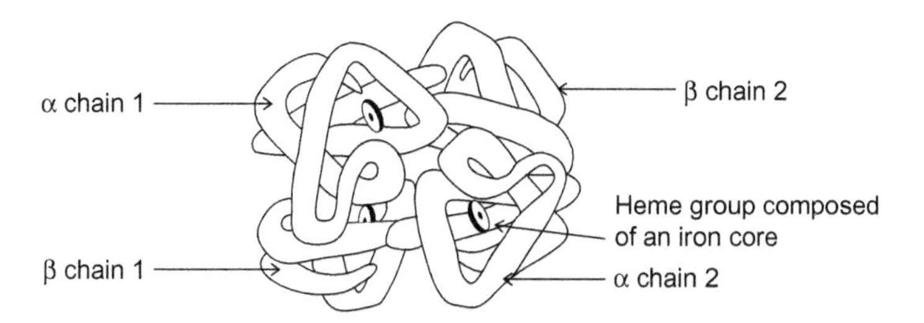

FIGURE 9.8 The protein structure of hemoglobin, which consists of four globular protein subunits. Each subunit contains one heme molecule, which is a non-protein compound that surrounds an iron core. It is this iron molecule that facilitates the transport of oxygen and carbon dioxide. Each red blood cell contains close to 300 million hemoglobin molecules. *Adapted from Martini and Nath (2009).*

cells. These processes solve the problem of the solubility of oxygen and carbon dioxide in plasma. It also allows for the gases to continually diffuse into the bloodstream as long as the red blood cells are not saturated, because the concentration gradient between alveolar air and plasma is never quenched.

Every 100 mL of blood leaving the alveolar capillaries carries approximately 20 mL of oxygen, of which only approximately 0.3 mL is pure oxygen dissolved in blood. The remaining 19.7 mL of oxygen is bound to hemoglobin (Hb). Each hemoglobin molecule consists of four protein subunits with one iron ion per protein subunit (Figure 9.8). As hemoglobin becomes oxygenated, it is termed as oxyhemoglobin (HbO_2). Each red blood cell carries over 250 million hemoglobin molecules. Oxygen-hemoglobin binding is a reversible process that can be described by kinetic rate constants that are dependent on the partial pressure of oxygen within the blood. The hemoglobin-oxygen reaction can be defined with the following kinetic reaction:

$$Hb + O_2 \leftrightarrow HbO_2$$

As the oxygen partial pressure increases, the rate of oxygen-hemoglobin binding increases, such that the equilibrium is shifted to favor more oxyhemoglobin. This is typically quantified by the percent saturation of hemoglobin. At a normal alveolar oxygen partial pressure of approximately 100 mmHg, nearly all of the hemoglobin in the blood becomes saturated with oxygen (approximately 98%). In tissues surrounding systemic capillaries, the partial pressure of oxygen is close to 40 mmHg, which would lead to only 75% hemoglobin saturation. Therefore, under normal conditions, hemoglobin remains fairly saturated with oxygen, even within the capillaries. However, if the tissue is under slightly hypoxic conditions, the percent saturation drops rather quickly to bring the tissue oxygen level back to normal (at an oxygen partial pressure of 20 mmHg the hemoglobin saturation is close to 35%, meaning that a significant portion of oxygen has dissociated from hemoglobin under these conditions).

Carbon dioxide transport is slightly more complicated than oxygen transport within the blood. Upon entering the blood, carbon dioxide can follow one of three pathways at a carbon dioxide partial pressure of approximately 45 mmHg, which is commonly found within the systemic capillaries. (Figure 9.9). The first is the conversion of carbon dioxide to carbonic acid, which accounts for nearly 70% of all of the carbon dioxide transported

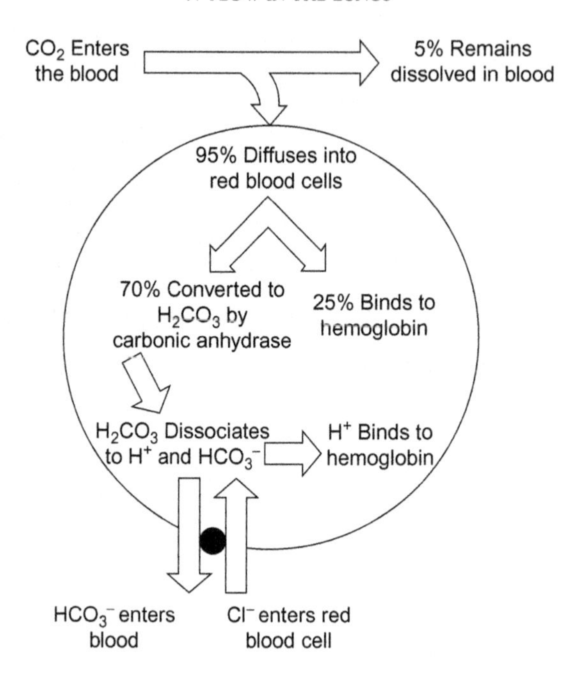

FIGURE 9.9 Pathways for carbon dioxide transportation in blood at a carbon dioxide partial pressure of 45 mmHg. The majority of carbon dioxide is converted to hydrogen and bicarbonate ions. About 25% of the carbon dioxide binds directly to free amino groups in the hemoglobin protein. The remaining carbon dioxide is transported directly within the blood.

through the blood. This reaction proceeds through the action of an enzyme, carbonic anhydrase, which is stored within red blood cells. Carbonic acid immediately dissociates into a hydrogen ion and a bicarbonate ion. This kinetic reaction can be described by

$$CO_2 + H_2O \xleftrightarrow[\text{Carbonic Anhydrase}]{} H_2CO_3 \leftrightarrow H^+ + HCO_3^-$$

Most of the hydrogen ions formed in this way bind to hemoglobin to maintain the pH of the plasma (e.g., if the hydrogen ion was transported into the plasma the pH would decrease). Most of the bicarbonate ions are transported into the plasma through a membrane bound transport protein. For each bicarbonate ion that enters the plasma, one chloride ion is transported into the red blood cell using the same transport protein (Band 3 anion transport protein). No energy is required for this transport, and it results in a large movement of chloride ions into the red blood cells, which is known as the chloride shift. The second possible outcome for carbon dioxide is adhesion to hemoglobin. Close to 25% of carbon dioxide follows this path. Carbon dioxide binds to the hemoglobin protein and not the iron ions like oxygen. Specifically, carbon dioxide binds to free amino groups (NH_2), forming carbaminohemoglobin ($HbCO_2$) in the following reaction:

$$Hb + CO_2 \leftrightarrow HbCO_2$$

The remaining portion of carbon dioxide (approximately 5–7%) is transported as dissolved carbon dioxide in blood. Recall that the solubility of carbon dioxide is approximately 20 times greater than that of oxygen, and therefore, we would anticipate that blood can carry more carbon dioxide as compared with oxygen.

9.7 COMPRESSIBLE FLUID FLOW

One of the most critical differences in compressible fluid flow, as compared with incompressible fluid flow, is that the physical properties of the fluid are dependent on changes in container area, frictional forces along the walls and heat transfer. We will begin our discussion of compressible fluid flow with isentropic flows. For this type of flow, friction and heat transfer are neglected and the only independent variable is the change in the cross-sectional area of the conducting vessel. The equations for isentropic flows are derived from the basic equations developed in Chapter 3 of this textbook, for a fixed volume of interest with steady one-dimensional flow. The properties of interest for an isentropic flow are generally the fluid's temperature, the pressure, the density, the cross-sectional area, and the velocity at any location within the flow field (Figure 9.10).

For a steady isentropic flow, the continuity equation simplifies to

$$\rho_1 v_1 A_1 = \rho_2 v_2 A_2 = \dot{m} \tag{9.12}$$

With the same assumptions, the Conservation of Momentum equation will simplify to

$$R_x + p_1 A_1 - p_2 A_2 = \dot{m} v_2 - \dot{m} v_1 \tag{9.13}$$

Similarly, the first and second law of thermodynamics can be written as

$$h_1 + \frac{v_1^2}{2} = h_2 + \frac{v_2^2}{2} = h_0 \tag{9.14}$$

$$s_1 = s_2 = s \tag{9.15}$$

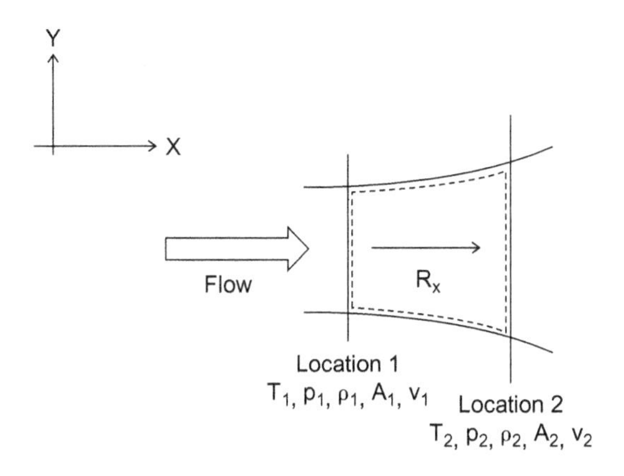

FIGURE 9.10 General isentropic flow in an expanding channel. In general, we would be interested in the temperature, pressure, density, and velocity of the fluid for a given change in area. The fluid mechanics governing equations for compressible flows can be found in the following discussion.

where $h = u + pV$ and h_0 is defined as the stagnation enthalpy (recall that u is the internal energy of the system, p is the pressure and V is the volume of interest). The stagnation enthalpy is the enthalpy that would be obtained if the fluid were decelerated to a zero velocity. For many compressible fluids, the enthalpy and the density are tabulated as functions of entropy and pressure. If the fluid is an ideal gas (or if it is valid to assume that it is an ideal gas) then the following equations can be used during problem solving:

$$p = \rho RT \tag{9.16}$$

$$h_2 - h_1 = c_p(T_2 - T_1) \tag{9.17}$$

$$\frac{p}{\rho^k} = \text{constant} \tag{9.18}$$

where k is the ratio of specific heats $k = c_p/c_v$.

The Mach number (M) is defined as the ratio of the fluid velocity to the speed of sound:

$$M = \frac{v}{c} \tag{9.19}$$

For most, if not all, biological applications, flow will be subsonic in which the Mach number is less than 1. In this case, a decrease in cross-sectional area causes an increase in velocity, while an increase in cross-sectional area causes a decrease in flow velocity. However, in supersonic flows ($M > 1$), the effects are exactly opposite. Flow accelerates in an expanding channel and decelerates in a narrowing channel.

Let us now consider the compressible flow through a channel with friction but no heat transfer. This assumption is valid if the channel is relatively short. For these conditions all of the previous equations remain the same except for Eqs. (9.13) and (9.15). In Eq. (9.13), the reaction force in the x-direction must now include frictional forces (F_x), so that the Conservation of Momentum will become

$$R_x + F_x + p_1A_1 - p_2A_2 = \dot{m}v_2 - \dot{m}v_1 \tag{9.20}$$

The fluid flow is no longer reversible since energy is lost to friction. Therefore, the second law of thermodynamics becomes

$$s_2 - s_1 = c_p \ln\left(\frac{T_2}{T_1}\right) - R \ln\left(\frac{p_2}{p_1}\right) \tag{9.21}$$

If the fluid can be assumed to be an ideal gas, we can also make use of the ideal gas laws, described in Eqs. (9.16) through (9.18).

If there is heat being transferred during the flow conditions but there is no friction, then the equations to describe the flow field will take the following form. Continuity and Conservation of Momentum take the same form as Eqs. (9.12) and (9.13), respectively (although there may be no surface forces acting on the flow). The first law of thermodynamics will simplify to

$$\dot{Q} = \dot{m}\left(h_2 + \frac{v_2^2}{2} - h_1 - \frac{v_1^2}{2}\right) \tag{9.22}$$

The second law of thermodynamics is the same as Eq. (9.21).

In general, the equations just provided are all applicable to the diffusion of gas across the respiratory boundary. Under these conditions, it is possible that there is some exchange of energy during gas flow and potentially friction losses. Because the respiratory boundary is so thin, it is possible to use the more general formula for steady isentropic flows.

9.8 DISEASE CONDITIONS

9.8.1 Emphysema

Emphysema is a chronic disease that is associated with shortness of breath which prevents the patient from participating in strenuous activities. One of the most common causes of this disease is exposure to cigarette smoke, both from mainstream smoke (i.e., the smoker) and from sidestream smoke (e.g., secondhand smoke). This biological problem is associated with the destruction of the alveolar sacs, leaving larger air pockets where several alveoli were. This effectively reduces the surface area for oxygen/carbon dioxide exchange to occur, which reduces the diffusion rate of each molecule. In severe cases of emphysema, alveoli merge, forming large sacs covered with fibrous tissue that ineffectively participate in gas exchange. These larger sacs are fairly rigid and also decrease the lung elasticity, making it more difficult to experience volume changes with the small pressure changes (i.e., Boyle's law) during inspiration and exhalation. Without the proper surface area for gas exchange, the body does not obtain enough oxygen and not enough carbon dioxide leaves the body. The V/Q ratio is well above 0.8, under these conditions, indicating adequate blood flow (often with shunting) and increased ventilation (hyperventilation).

As the alveoli continue to break down, hyperventilation is unable to compensate for the decreased surface area for gas exchange. Therefore, the blood oxygen level begins to decrease and there is an overall constriction of the pulmonary blood vessels (hypoxic pulmonary vasoconstriction), to account for this lowered blood oxygen level. Unchecked, this places a large load on the right ventricle of the heart, and can lead to right heart failure, making it more difficult to return venous blood to the right atria. If this continues, the entire cardiovascular system is placed under high pressure, which places a large load on the left ventricular heart muscle. To reverse this new load from the high blood pressure, the heart muscle begins to thicken, which as we have learned previously will lead to heart failure.

Emphysema is an irreversible condition. The only way to slow its progression is to have the patient stop smoking and avoid all other lung irritants. It is possible for the patient to breath bronchodilators to help ventilate the lungs more, but with significant destruction of alveoli, there is little that physicians can do to reverse the effects of emphysema. Patients are typically also placed on anti-inflammatory agents to decrease the risk of the inflammatory system from further attacking the damaged alveoli.

9.8.2 Asthma

Asthma can be a chronic condition, and/or seen acutely influenced by environmental factors. Common environmental factors include smoke, allergens, cold air and exercise. This biological problem occurs when the irritant causes bronchoconstriction, restricting the flow of air into the lungs. The gas exchange is compromised due to a decreased ventilation rate and diminished volume of fresh air per breath. The ventilation perfusion ratio is lower than 0.8 during an asthma attack, indicating that ventilation has decreased with little effect on the blood flow. Shunt is not present. Drugs that act to dilate the bronchioles are prescribed by physicians to alleviate the bronchoconstriction; they will increase both blood flow and ventilation rate. Untreated asthma spirals into a proinflammatory state in which the bronchioles are inflamed and more reactive to irritants. This leads to thickening of the bronchiolar walls, further, and permanently, restricting ventilation flow rate.

9.8.3 Tuberculosis

Tuberculosis results from an infection of the lungs by a specific bacterium. The bacteria may propagate in the conducting pathways as well as the alveoli. Like most infections, the patient has a fever and general malaise, but during severe cases the lungs can begin to be destroyed. As this occurs, the patient will begin to "cough up" blood. There are some similarities between tuberculosis and any other respiratory infection. In general, the defense mechanisms in the lungs become overwhelmed by the invading pathogens. When the pathogens begin to grow within the lungs there is a large mucus buildup along the conducting pathways. This tends to increase the resistance to air flow and may disrupt gas exchange. If the pathogens destroy lung tissue, the surface area for gas exchange reduces and blood may enter the lung space in extreme conditions.

It has been estimated that one-third of the world's populations (and approximately 5% of the U.S. population) is currently infected with the bacteria that causes tuberculosis: *Mycobacterium tuberculosis*. New tuberculosis infections are occurring at a rate of approximately one person every 3 seconds, but the death rate from tuberculosis is falling. Therefore, the absolute number of people with tuberculosis is increasing. The disease is most prominent in developing countries, but more people in developed countries are contracting the tuberculosis bacteria because their immune systems are compromised by drug use or other immunosuppressant diseases such as AIDS.

Tuberculosis infections manifest when *M. tuberculosis* reach the pulmonary alveoli, where they take over and replicate within alveolar macrophages. It is common for lymphatic vessels and blood vessels to transport bacteria to other locations within the body, but it is the lungs that are most susceptible to degradation from this bacteria. If the tuberculosis bacteria enter the bloodstream, patients have nearly a 100% fatality rate. In most cases of tuberculosis, a granuloma is formed within the lung tissue. The granuloma houses many immune cells, which release cytokines to help destroy the bacteria and call other inflammatory cells to the infection site. However, a good portion of the *M. tuberculosis* can remain dormant within the granuloma and cause an infection later on.

END OF CHAPTER SUMMARY

9.1 The respiratory system provides the means for gas exchange between the atmospheric air and the cardiovascular system. It is divided into a conducting pathway (composed of the nasal cavity, the pharynx, the larynx, the trachea, and the bronchi) and a branching pathway, which eventually diverges into the alveolar sacs. The alveoli are the locations for gas exchange and provide approximately 70 square meters of area for exchange. The respiratory boundary is composed of the alveoli epithelial cell, the capillary endothelial cell, and the fused laminae between these cells. Each lung is surrounded by a pleural membrane that facilitates the changes in lung volume to account for the movement of gases into and out of the lungs. This is based on Boyle's law, which for a constant temperature system is represented as

$$P_1V_1 = P_2V_2$$

9.2 The lungs go through an approximate 20% change in volume, which significantly affects the blood vessel area within the lungs. Due to these large stresses, the Kirchhoff stress and the Green's strain are most applicable to relate the forces to the deformation. This is experimentally found to be represented as

$$s_{xx} = a_1E_{xx} + 0.6429E_{yy}$$
$$s_{yy} = a_2E_{yy} + 0.6429E_{xx}$$

The separation distance for the respiratory boundary is also affected by the pressure of the respiratory and vascular system. This distance can be modeled as

$$h(\Delta P) = \begin{cases} 0 & \Delta P < -0.7\text{mmHg} \\ 0.5\mu\text{m} + \dfrac{0.5\mu\text{m}}{0.7\text{mmHg}}\Delta P & -0.7\text{mmHg} \le \Delta P < 0\text{mmHg} \\ 0.5\mu\text{m} + \dfrac{2\mu\text{m}}{75\text{mmHg}}\Delta P & 0 \le \Delta P < 30\text{mmHg} \\ 1.3\mu\text{m} & \Delta P > 30\text{mmHg} \end{cases}$$

9.3 The pressure and the volume of the lungs are altered based on the movement of the ribs and the diaphragm. As the diaphragm moves downward and the ribs move outward the lungs expand and experience a temporal reduction in pressure. The reverse movements of the diaphragm and lungs cause the lungs to shrink and experience an increase in pressure. Under normal conditions, the tidal volume is approximately 500 mL (of which 350 mL is used for gas exchange), and there are approximately 12 to 15 breaths per minute. In extreme conditions, the lungs can move approximately 4800 mL of air per breath, which is using all of the inspiratory reserve and the expiratory reserve. There is 1200 mL of residual volume which maintains the alveolar space open.

9.4 Adequate respiration requires that the ventilation rate of the lungs be matched with the blood flow rate through the pulmonary circulation. The ventilation rate (V) is a measure of the ability of the lungs to take in and exchange fresh atmospheric air, whereas the perfusion rate (Q) is a measure of the flow through the capillaries. If the V/Q ratio is low, not enough fresh air is entering the lungs. In contrast, if the V/Q ratio is high, there is not enough flow to support gas exchange. Under physiological conditions, the V/Q ratio is close to 0.8, whereas under pathological conditions the V/Q ratio can decrease (e.g. emphysema) or increase (e.g. pulmonary embolism).

9.5 Oxygen and carbon dioxide diffusion is based on the solubility of the gas within blood. Henry's law formulates this as

$$P = \frac{\text{Gas Concentration}}{\text{Solubility Coefficient}}$$

The solubility coefficients for oxygen and carbon dioxide in blood are 0.024 and 0.57, respectively. Gas diffusion can also be modeled as the movement of gases across a semi-permeable boundary. In general, we would write the mass balance equations (for oxygen) as

$$Ru_G(C_{O_2-gas}(x+\Delta x) - C_{O_2-gas}(x)) + J\Delta x = 0$$

and:

$$Ru_B(C_{O_2-blood}(x+\Delta x) - C_{O_2-blood}(x)) - J\Delta x = 0$$

assuming that the concentration of the gas remains constant within the channel.

9.6 Oxygen has a very low solubility in blood and therefore another mechanism for oxygen transport is needed. As we know, oxygen diffuses into the red blood cell and becomes associated with the heme groups of hemoglobin. Carbon dioxide has a much more complicated transport mechanism. Approximately 70% of the carbon dioxide enters the red blood cells and interacts with carbonic anhydrase to form hydrogen and bicarbonate ions; 25% of carbon dioxide associates with free amino groups on the hemoglobin protein. The remaining carbon dioxide is transported directly in the blood.

9.7 Gases are fairly compressible, and it is important to develop the relationships for compressible fluid flow. For a steady isentropic flow, the governing equations are

$$\rho_1 v_1 A_1 = \rho_2 v_2 A_2 = \dot{m}$$

$$R_x + p_1 A_1 - p_2 A_2 = \dot{m} v_2 - \dot{m} v_1$$

$$h_1 + \frac{v_1^2}{2} = h_2 + \frac{v_2^2}{2} = h_0$$

$$s_1 = s_2 = s$$

For an ideal gas, we can make use of the following relationships

$$p = \rho RT$$

$$h_2 - h_1 = c_p(T_2 - T_1)$$

$$\frac{p}{\rho^k} = \text{constant}$$

If there is friction without heat transfer, we need to consider

$$R_x + F_x + p_1 A_1 - p_2 A_2 = \dot{m} v_2 - \dot{m} v_1$$

and:

$$s_2 - s_1 = c_p \ln\left(\frac{T_2}{T_1}\right) - R \ln\left(\frac{p_2}{p_1}\right)$$

If there is heat transfer without friction, then we need to consider

$$\dot{Q} = \dot{m}\left(h_2 + \frac{v_2^2}{2} - h_1 - \frac{v_1^2}{2}\right)$$

9.8 Emphysema, asthma and tuberculosis are common diseases that affect the respiratory system. Emphysema is characterized as a destruction of the respiratory boundary, and therefore, there is an ineffective transport of gases between the alveolar space and the cardiovascular system. Asthma is characterized by a severe constriction of the conducting pathways that reduces the ventilation of the lungs. Tuberculosis is characterized by an infection within the lungs that can either increase the resistance to air movement or break down the respiratory boundary.

HOMEWORK PROBLEMS

9.1 Describe the passage of air from the nose to the alveoli and discuss some of the important structures within the respiratory system. Pay close attention to the structure-function relationship of the organs.

9.2 Describe the mechanism of breathing, paying close attention to the movement of the ribs and the diaphragm.

9.3 The surfactant along the alveolar wall is present to prevent the alveoli from collapsing. How does this fluid perform this function?

9.4 Gas exchange along the alveolar wall occurs rapidly and efficiently because the driving forces are large, oxygen and carbon dioxide are lipid soluble, and the surface area for diffusion is large. Describe the role of each of these during exchange.

9.5 The bronchioles in the respiratory system are analogous to the arterioles in the cardiovascular system. How?

9.6 The differences between the atmospheric, the interpulmonary, and the intrapleural pressures determine the direction of air flow within the lungs. How would this change at different elevations where the atmospheric pressure could change drastically?

9.7 Under a disease condition, the separation distance for the alveolar gas and the capillary gas can be described as

$$h(\Delta P) = \begin{cases} 0 & \Delta P < -0.7\text{mmHg} \\ 0.5\mu\text{m} + \dfrac{0.5\mu\text{m}}{0.5\text{mmHg}}\Delta P & -0.7\text{mmHg} \leq \Delta P < 0\text{mmHg} \\ 0.5\mu\text{m} + \dfrac{2\mu\text{m}}{50\text{mmHg}}\Delta P & 0 \leq \Delta P < 30\text{mmHg} \\ 1.7\mu\text{m} & \Delta P > 30\text{mmHg} \end{cases}$$

Plot separation distance h as a function of pressure and compare this with the normal case. What has happened under these conditions, and what are some of the possible characteristics of the disease?

9.8 Under exercise conditions, predict what the tidal volume would be. Why would this be necessary? Where would the extra volume be taken from: the inspiratory reserve volume or the expiratory reserve volume or both?

9.9 What would happen if there was no residual volume within the lungs?

9.10 The solubility coefficient for carbon dioxide is an order of magnitude larger than the solubility coefficient for oxygen. What does this suggest about gas transport within the blood, and is this accounted for by the known processes of gas transport in the blood?

9.11 Using Table 9.1, calculate the solubility coefficients for nitrogen and carbon monoxide, assuming that the cross-sectional area and diffusion distance is the same.

9.12 In the upper atmosphere, the oxygen concentration drops to 15% of the total composition of air at an air temperature of 5°C. Assume that the nitrogen concentration makes up the difference in air composition. Calculate the respiratory rate needed to maintain the body oxygen requirements of 250 mL/min. Assume that the tidal volume remains the same and that 35% of the tidal volume enters the bloodstream and that all of this would be used to meet the body's oxygen requirements. Also, assume that the pressure drops to 700 mmHg.

9.13 Assume at particular atmospheric conditions, a person's respiratory rate increases to 25 breaths/min because the body needs 350 mL/min of oxygen. If the oxygen concentration in the atmosphere is 21% and that percent usage of oxygen is 40%, calculate the air pressure if the temperature is 32°C.

9.14 The carbon dioxide concentration in the gas at the inlet of a semipermeable oxygenator is 5 mmHg and that in the blood is 45 mmHg. Assume that the radius of the channel is 4 cm and that the thickness of the boundary is 100 μm. Calculate the change in carbon dioxide concentration with distance if the diffusion coefficient is 2×10^{-4} cm^2/s, the gas velocity is 5 cm/s, and the blood velocity is 10 cm/s.

9.15 Air is flowing from the alveoli to the trachea and is following an isentropic flow conditions. Calculate the velocity that air is leaving the alveoli if the temperature within the alveoli is 37°C and that in the trachea is 33°C. The velocity of air within the alveoli is 50 mm/s and the specific heat for air (c_p) is equal to 30.02 mJ/(kg mol K).

9.16 Is there heat being transferred under the conditions in problem 9.15 if you can assume steady flow and the alveoli radius is equal to 100 μm? What is the rate of heat transfer?

9.17 What is the change in entropy under the conditions in problem 9.15, if the pressure is 1.1 atm in the trachea and 1.15 atm in the alveoli?

References

[1] E. Agostoni, E. D'Angelo, G. Roncoroni, The thickness of the pleural liquid, Respir. Physiol. 5 (1968) 1.

[2] E. Agostoni, Mechanics of the pleural space, Physiol. Rev. 52 (1972) 57.

[3] E. Agostoni, E. D'Angelo, Pleural liquid pressure, J. Appl. Physiol. 64 (1988) 1760.

[4] A. Al-Tinawi, J.A. Madden, C.A. Dawson, J.H. Linehan, D.R. Harder, D.A. Rickaby, Distensibility of small arteries of the dog lung, J. Appl. Physiol. 71 (1991) 1714.

[5] H. Bachofen, J. Hildebrandt, M. Bachofen, Pressure-volume curves of air- and liquid-filled excised lungssurface tension in situ, J. Appl. Physiol. 29 (1970) 422.

[6] H. Bachofen, J. Hildebrandt, Area analysis of pressure-volume hysteresis in mammalian lungs, J. Appl. Physiol. 30 (1971) 493.

[7] J.S. Brody, E.J. Stemmler, A.B. DuBois, Longitudinal distribution of vascular resistance in the pulmonary arteries, capillaries, and veins, J. Clin. Invest. 47 (1968) 783.

[8] P.J. Dale, F.L. Matthews, R.C. Schroter, Finite element analysis of lung alveolus, J. Biomech. 13 (1980) 865.
[9] A.P. Fishman, The syndrome of chronic alveolar hypoventilation, Bull. Physiopathol. Respir. (Nancy) 8 (1972) 971.
[10] A.P. Fishman, Pulmonary edema. The water-exchanging function of the lung, Circulation 46 (1972) 390.
[11] E. Flicker, J.S. Lee, Equilibrium of force of subpleural alveoli: implications to lung mechanics, J. Appl. Physiol. 36 (1974) 366.
[12] Y.C. Fung, S.S. Sobin, Theory of sheet flow in lung alveoli, J. Appl. Physiol. 26 (1969) 472.
[13] Y.C. Fung, S.S. Sobin, Elasticity of the pulmonary alveolar sheet, Circ. Res. 30 (1972) 451.
[14] Y.C. Fung, Biorheology in the analysis of the lung, Biorheology 19 (1982) 79.
[15] K.A. Gaar Jr., A.E. Taylor, A.C. Guyton, Effect of lung edema on pulmonary capillary pressure, Am. J. Physiol. 216 (1969) 1370.
[16] A.C. Guyton, A.E. Taylor, R.E. Drake, J.C. Parker, Dynamics of subatmospheric pressure in the pulmonary interstitial fluid, Ciba Found. Symp. (1976) 77.
[17] J. Hildebrandt, H. Bachofen, G. Brandt, Reduction of hysteresis in liquid-sealed bell-type spirometers, J. Appl. Physiol. 28 (1970) 216.
[18] A.A. Merrikh, J.L. Lage, Effect of blood flow on gas transport in a pulmonary capillary, J. Biomech. Eng. 127 (2005) 432.
[19] D.S. Moffatt, A.C. Guyton, T.H. Adair, Functional diagrams of flow and volume for the dog's lung, J. Appl. Physiol. 52 (1982) 1035.
[20] M. Ochs, J.R. Nyengaard, A. Jung, L. Knudsen, M. Voigt, T. Wahlers, et al., The number of alveoli in the human lung, Am. J. Respir. Crit. Care Med. 169 (2004) 120.
[21] A.B. Otis, W.O. Fenn, H. Rahn, Mechanics of breathing in man, J. Appl. Physiol. 2 (1950) 592.
[22] J.C. Parker, A.C. Guyton, A.E. Taylor, Pulmonary interstitial and capillary pressures estimated from intraalveolar fluid pressures, J. Appl. Physiol. 44 (1978) 267.
[23] S. Permutt, B. Bromberger-Barnea, H.N. Bane, Alveolar pressure, pulmonary venous pressure, and the vascular waterfall, Med. Thorac. 19 (1962) 239.
[24] H. Rahn, W.O. Fenn, A.B. Otis, Daily variations of vital capacity, residual air, and expiratory reserve including a study of the residual air method, J. Appl. Physiol. 1 (1949) 725.
[25] H. Rahn, A.B. Otis, Continuous analysis of alveolar gas composition during work, hyperpnea, hypercapnia and anoxia, J. Appl. Physiol. 1 (1949) 717.
[26] A. Roos, L.J. Thomas Jr., E.L. Nagel, D.C. Prommas, Pulmonary vascular resistance as determined by lung inflation and vascular pressures, J. Appl. Physiol. 16 (1961) 77.
[27] A.E. Taylor, A.C. Guyton, V.S. Bishop, Permeability of the alveolar membrane to solutes, Circ. Res. 16 (1965) 353.
[28] E.R. Weibel, D.M. Gomez, Architecture of the human lung. Use of quantitative methods establishes fundamental relations between size and number of lung structures, Science 137 (1962) 577.
[29] E.R. Weibel, Morphological basis of alveolar-capillary gas exchange, Physiol. Rev. 53 (1973) 419.
[30] E.R. Weibel, P. Untersee, J. Gil, M. Zulauf, Morphometric estimation of pulmonary diffusion capacity. VI. Effect of varying positive pressure inflation of air spaces, Respir. Physiol. 18 (1973) 285.
[31] B.M. Wiebe, H. Laursen, Human lung volume, alveolar surface area, and capillary length, Microsc. Res. Tech. 32 (1995) 255.

CHAPTER

10

Intraocular Fluid Flow

LEARNING OUTCOMES

1. Describe the physiology of the eye
2. Identify regions within the eye that are filled with fluid
3. Discuss how the eye senses light intensity and the wavelength of light
4. Evaluate the mechanism for aqueous humor formation
5. Explain aquaporins and their function within the body
6. Model the flow through aquaporin channels
7. Describe the flow of aqueous humor through the eye
8. Calculate the flow of aqueous humor through regions of the eye
9. Evaluate intraocular pressure and describe what regulates intraocular pressure
10. Model the time-dependent changes in intraocular pressure
11. Identify common ways to measure intraocular pressure
12. Discuss common disease conditions that affect the eye

10.1 EYE PHYSIOLOGY

The eye is part of the specialized senses and it is an extremely versatile and sensitive visual instrument. The eye is spherical, with an average diameter of 24 mm, and is located within the eye socket of the skull (also termed the orbit). Also within the orbit are the extrinsic eye muscles, which control the movement of the eye; the optic nerve, which transmits visual information to the brain; the blood vessels, which provide the eye with nutrients and removal of wastes; and fat tissue, which acts to support and cushion the eye (Figure 10.1). The exterior wall of the eye is composed of three layers. The first layer of the eye (most exterior layer) is termed the fibrous tunic, and it consists of the sclera and the cornea. The primary functions of the fibrous tunic are to provide the mechanical support for the eye, to provide a location for extrinsic muscle attachment, and to provide

 © 2015 Elsevier Inc. All rights reserved.

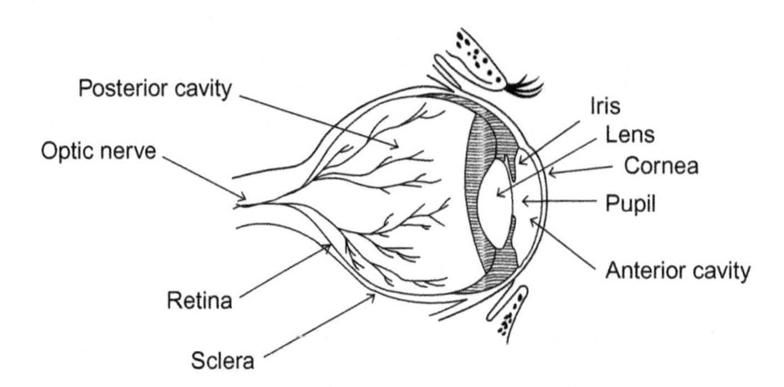

FIGURE 10.1 Anatomical structures of the eye, showing the major optical and sensory components. The anterior cavity and the posterior cavity experience fluid movement. The anterior cavity is continuously forming and circulating aqueous humor, while the posterior cavity experiences a slow flow of aqueous humor.

protection for the interior structures of the eye. The sclera covers most of the eye structure and is composed of connective tissue (mostly collagen and elastin). The sclera is the white portion of the eye, and it gets this color from the high concentration of collagen fibers present within this region. Small blood vessels and nerves are located within the sclera. The cornea is continuous with the sclera and is located on the anterior side of the eye, protecting the eye from mechanical injury and/or contamination. The cornea also acts to protect the iris and lens of the eye. Interestingly, the cornea contains no blood vessels and must obtain enough oxygen and nutrients through diffusion.

The second layer of the eye (middle layer) is termed the vascular tunic (or uvea), and it contains a large quantity of blood vessels and lymphatic vessels. The primary functions of the vascular tunic are to provide a pathway for blood vessels and lymphatic vessels, to secrete and absorb aqueous humor (more on this in Sections 10.3 and 10.5), to control the amount of light that enters the eye, and to control the shape of the lens. The blood and lymphatic vessels are housed with the intrinsic muscle of the eye (these are smooth muscle cells). The intrinsic muscles can partially regulate the shape of the eye through dilatory and contractile events. Also included within this layer are the iris, the ciliary bodies, and the choroid. The iris is primarily composed of smooth muscle cells, which contain blood vessels and pigment cells. When these smooth muscles cells contract, the shape and the size of the pupil change, which effectively regulates the amount of light that enters the eye. There are two types of smooth muscles cells, the constrictor muscles and the dilatory muscles; as the names imply, one constricts the pupil and the other dilates the pupil. Pigment cells, which are also termed melanocytes, partially provide the iris with its color. Like melanocytes in the skin, pigment cells produce melanin at a particular concentration. The concentration of melanin and the light scattering properties of the fluid surrounding the iris determines the overall pigmentation of the eye. The ciliary bodies are a specialized region of the vascular tunic that is composed of a large number of ciliary muscles. The ligaments that are associated with these muscles are attached to the lens, holding it in place to collect and focus all of the light that enters through the pupil. The choroid is the name given to the purely vascular regions of the vascular tunic.

The third layer of the eye (innermost layer) is termed the neural tunic or the retina. The inner layer of the neural tunic is termed the neural part and contains all of the photoreceptors of the eye. The outer layer of the neural tunic is termed the pigmented part and acts to absorb and not reflect light that has passed through the neural part, thus preventing the light to return into the eye structure. There are two types of photoreceptors located within the neural part of the neural tunic. The first type of receptor is termed the rods, which are sensitive to the intensity of light and not the wavelength (color) of the light. The rods allow us to see structures under low light conditions. The second type of photoreceptor is termed the cones, which enable us to distinguish between different wavelengths of light but they are not as sensitive to intensity. There are three types of cones to help determine different colors, but they require a higher intensity light to function properly. There is a higher density of rods along the periphery of the eye, whereas there is a higher density of cones along the center of the eye.

The interior portion of the eye is hollow and is composed of two chambers: the posterior cavity and the anterior cavity, which are divided by the ciliary bodies and the lens. The posterior cavity (also known as the vitreous chamber) contains the vitreous body. This chamber makes up approximately 90% of the hollow portion of the eye. The vitreous body primarily helps to maintain the shape of the eye. The vitreous body is a gel (which can be considered as a very viscous fluid) because cells within this cavity produce and secrete a large quantity of collagen fibers and proteoglycans. The collagens and the proteoglycans are highly charged and can organize and absorb large quantities of water. This water is associated closely with the charged molecules and is not as free to move as compared with water in the bulk phase. The vitreous body is only produced during fetal development and is not replaced during the remaining portion of life.

The anterior cavity is divided into an anterior chamber and a posterior chamber and is filled with the biofluid aqueous humor. The anterior chamber is bounded by the cornea and the iris, while the posterior chamber is bounded by the iris and the ciliary bodies/lens. The aqueous humor also helps to maintain the shape of the eye, but it has other functions as well. The aqueous humor is continually formed and circulates through the anterior cavity as well as along the surface of the retina. Some aqueous humor enters the vitreous chamber, and its movement through this chamber can be described by diffusion laws better than fluid mechanics laws, because of the gel that fills the cavity. The formation of aqueous humor will be discussed extensively in Section 10.3, and the flow of aqueous humor will be discussed in Section 10.5.

The remaining structure of the eye that we will describe here is the lens. The lens is held in place by ligaments attached to the ciliary bodies. The function of the lens is to focus incoming light onto the photoreceptors located within the retina. To accomplish this task, the shape of the lens can be altered. Cells within the lens are organized into concentric circles, layered throughout the lens. These cells are all covered by a fibrous capsule mainly composed of elastic fibers that tend to make the lens less spherical. Elastic fibers along the periphery of the lens are associated with the ciliary body ligaments that mechanically pull on the lens to make it more spherical. Cells within the interior of the lens are termed lens fibers. These cells do not have a nucleus or other organelles, but instead are filled with transparent proteins called crystallins. Crystallins give the lens the ability to focus light onto the retina. In fact the aqueous humor and the vitreous body also function as a liquid lens that helps to further focus the light onto the retina. Light is primarily focused by changing the shape of the lens and small changes to the shape of the eye. When viewing objects that are at a distance, the lens is flat and elongated. However, when the object that is being viewed is close, the lens becomes more spherical. This change in shape is accomplished by a contraction of the ciliary muscles, which pulls the choroid layer of the eye toward the lens, causing the associated ligaments to relax. Due to this reduced tension on the lens, the elastic lens becomes thicker and rounder, which causes the incident light to bend more rapidly.

As stated previously, the retina contains both rod cells and cone cells. There are approximately 125 million rod cells and 6 million cone cells in each eye. Combined, these cells account for more than 70% of all of the sensory receptors within the body. Rods are more sensitive to light intensity; they do not distinguish between colors, and they help us to see at night. Cones can distinguish between different wavelengths of light but are not as sensitive to intensity and therefore do not play an important visual role when it is dark (hence everything looks gray under low-intensity light). Each rod or cone cell is composed of an outer segment that is composed of highly membranous discs. These discs contain the visual pigments that consist of a light-absorbing pigment, retinal, bounded to a membrane protein, opsin. The protein opsin varies in structure based on the photoreceptor type.

The opsin that can be found in rods combines with retinal to form the visual pigment rhodopsin. As rhodopsin absorbs light, it experiences a conformational change that triggers a signal transduction pathway, which ultimately leads to the membrane potential of rod cells to decrease. In fact, when rod cells are in the dark, the cells are in a depolarized state and they release an inhibitory neurotransmitter. When the membrane potential of rod cells decreases (becomes more negative), the rod cells stop releasing the inhibitory neurotransmitter and they begin to send signals through the retina to the brain. Color vision occurs through the presence of three types of cone cells within the retina. Each of these cone cells has their own type of opsin that associates with retinal to form a particular photopsin. Each type of photopsin has an optimal wavelength of light that it can absorb; either yellow (peak wavelength 430 nm); green (peak wavelength 540 nm); or blue (peak wavelength 570 nm) light. Different shades of these colors can be viewed based on the differential stimulation of each of these three types of cone cells.

10.2 EYE BLOOD SUPPLY, CIRCULATION, AND DRAINAGE

What is probably most unique about the circulation of the eye is that all cells must have a readily available supply of blood (e.g., generally within 50–100 μm from a capillary) and that the blood vessels cannot interfere with the transmission of light through the eye and the sensing of light by the optic sensors. In general, the blood vessels of the eye are divided into the uveal circulation and the retinal circulation. The uveal vessels supply the choroid, the ciliary bodies, the iris, and the portion of the retina that contains photoreceptors (via diffusion). The retinal vessels supply all other locations of the eye with blood. Interestingly, during development, blood vessels exist along the cornea and the lens. However, by birth, these vessels have been pruned from the vascular tree (through a process termed rarefaction), so that they do not interfere with light transmission through the eye. These structures, which no longer have a direct blood supply, are supplied with nutrients via the flow of aqueous humor (Sections 10.3/10.5).

Blood is supplied to the eye and some of its surrounding tissue via the ophthalmic artery, which is a branch of the internal carotid artery. Each ophthalmic artery enters the orbit via the optic canal, in parallel with the optic nerve (Figure 10.2). Various vessels branch off from the ophthalmic artery along its length, which supply different regions of the eye with blood. Additionally, some of these branches of the ophthalmic artery end up supplying various tissues associated with the eye. For example, the dorsal nasal artery supplies the lacrimal sac with blood and the medial and lateral palpebral arteries supply the eyelids with blood. In contrast to the one large artery that supplies the eye with blood, two large veins exist to drain blood from the eye and its surrounding structures. Blood leaves the orbit through either the superior ophthalmic vein or the inferior ophthalmic vein, which combine and drain blood into the cavernous sinus which then passes into the internal jugular vein. In general, the superior locations of the eye and the surrounding tissue are drained by the superior ophthalmic vein, whereas the inferior locations are drained by the inferior ophthalmic vein.

The small arteries of the eye play a significant role in the formation of aqueous humor, which will be discussed in the following sections. However, it is important to detail how blood is passed into these important structures. Arterioles that supply the rectus muscle group of the eye are unique, in that they do not terminate in capillary beds. In the rectus muscles, the capillary beds are side branches of the small arterioles and the arterioles

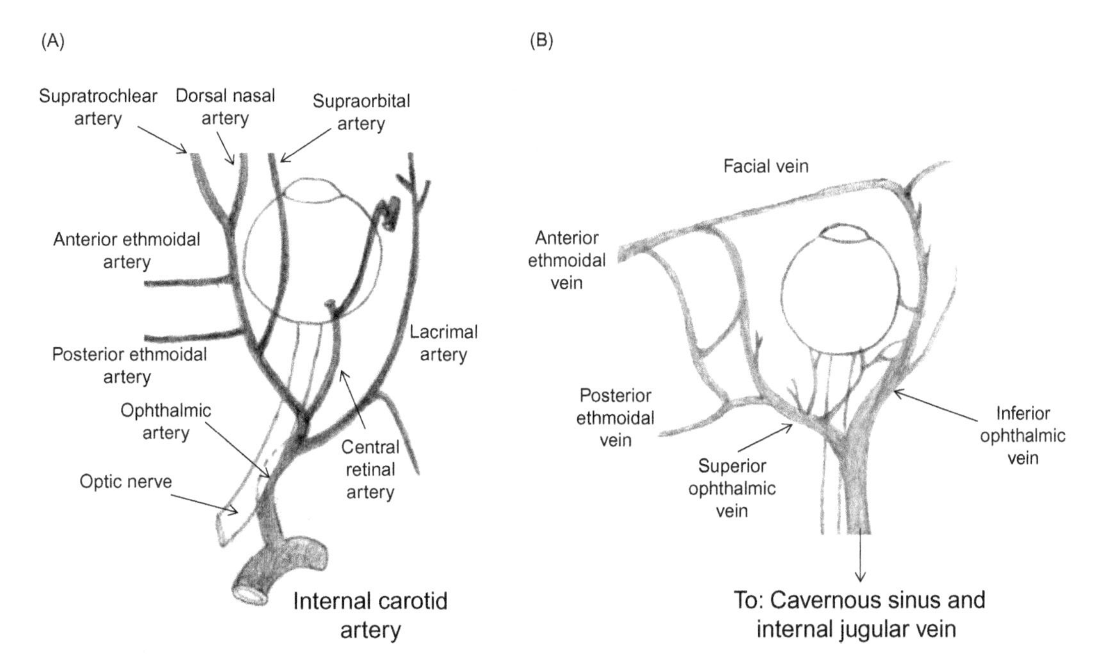

FIGURE 10.2 Blood distribution within the ophthalmic artery system (A) and ophthalmic venous system (B). Some of the major tributaries of the vessels are highlighted on this schematic.

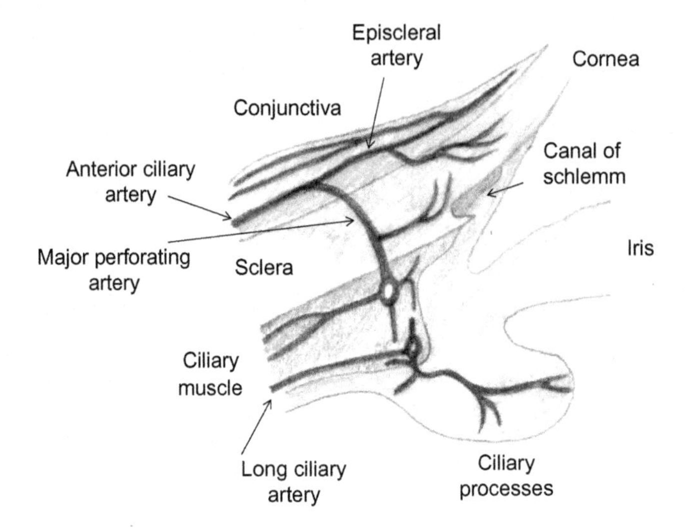

FIGURE 10.3 Branches of the anterior ciliary artery that provide blood to the conjunctiva (episcleral artery) and the ciliary bodies (major perforating artery). This illustrates the redundancy within the optic circulation.

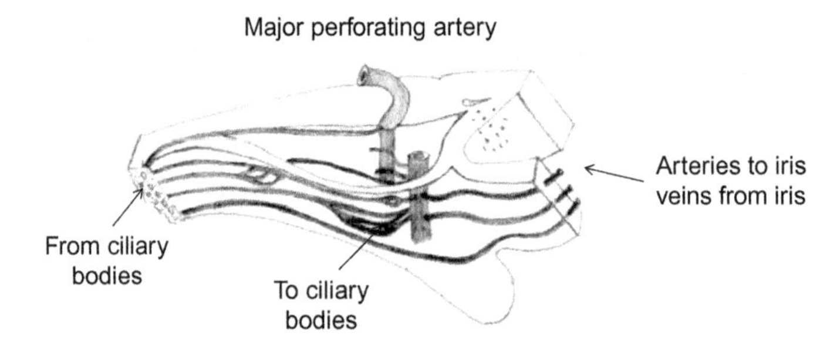

FIGURE 10.4 Detailed view of the blood supply to the ciliary processes, via the major perforating artery and the major arterial circle. The venous circulation from the ciliary processes and the iris converge and drain into the vortex veins.

continue to penetrate the eye and emerge on the orbital surface posterior to the muscle tendons. These arteries continue to progress forward along the eye, at which point their name changes to anterior ciliary arteries. In total, there are typically seven anterior ciliary arteries, which penetrate from each of the different rectus muscles along the eye. Almost immediately distal to the rectus muscle tendons, the ciliary arteries bifurcate; one branch remains along the surface of the eye, whereas the second branch penetrates the sclera to traverse toward the ciliary body (Figure 10.3). The branch that remains along the surface of the eye is named the episcleral artery, which supplies the conjunctiva with blood. The branch of the anterior ciliary artery that penetrates the sclera is renamed the major perforating artery, which branch a number of times as they reach the ciliary bodies. These various branches reconnect forming the anastomosed intramuscular arterial circle. Branches from the major intramuscular arterial circle enter the ciliary processes (Figure 10.4), which contain a very dense intertwined aggregate of small arterioles, capillaries, and venules. This intertwined and interconnected vascular network are surrounded by a small layer (typically two cells thick) of epithelium, which separates blood from the aqueous humor and is responsible for aqueous humor formation. The capillaries of the ciliary processes are somewhat unique in that they typically have a larger diameter than the feeding arterioles with little to no fenestrae ("windows" or openings between or through the cells). This helps to increase the surface area for aqueous humor formation and helps to regulate the formation of aqueous humor. The ciliary processes contain large venules that run in parallel with veins from the iris, where they converge into the choroidal veins. It is interesting to note, that there are typically few episcleral arteries that supply the ciliary processes with blood, but there are significantly more veins coming from the ciliary processes. This is due to the presence of veins that do not collect blood from capillary beds, but instead collect aqueous humor back into the circulation. These veins collect aqueous humor from the canals of Schlemm, which will be discussed in later sections.

The blood–retinal barrier is the critical structure for the separation of blood constituents and eye constituents and the breakdown of this barrier would cause a massive interference with vision. The blood–retinal barrier is composed of the retinal capillary endothelial cells and the epithelium of the retina. The endothelial cells are relatively thick, have extensive tight junctions along the cell-cell-boundaries, and are not fenestrated. It has been shown that the presence of extensive tight junctions is what is responsible for the diffusion restriction in both directions at the blood–retinal barrier. In contrast, the endothelial cells of the choroid capillaries are highly permeable and this has been suggested to play a role in fluid uptake and diffusion of nutrients to the retina. For instance, vitamin A is necessary for retina functions and appears to be obtained via diffusion out of the choroid capillaries and pinocytosis by the retinal epithelium. Similarly, the colloidal tissue oncotic pressure is significantly higher than the retinal colloidal oncotic pressure, which will act to facilitate water movement out of the retina into the choroid capillaries. This water movement will then locally increase the hydrostatic pressure to aid in water movement through the sclera and eventually out of the eye. Overall, the circulation of the eye is efficiently designed to provide nutrients and collect water/wastes from the eye, while not disrupting vision.

10.3 AQUEOUS HUMOR FORMATION

Aqueous humor is continually being formed and reabsorbed within the anterior and posterior chambers. The formation and reabsorption of aqueous humor is balanced by the volume of fluid within the eye as well as the intraocular pressure (Section 10.6). Aqueous humor is formed at a rate of approximately 2 μL/min within the anterior cavity. The majority of this is secreted into the posterior chamber of the anterior cavity from the ciliary processes which are projections from the ciliary bodies. Aqueous humor can flow into the anterior chamber from the posterior chamber by passing through the pupil. Also, some of the aqueous humor can freely diffuse into the posterior cavity to wet the retina. This provides an important route for nutrient delivery and waste removal to nonvascularized structures of the eye. The ciliary processes are located behind the iris where the ciliary muscle attaches to the eye. The ciliary processes have a very large surface area associated with them, approximately 6 cm^2 per eye. Secretory epithelial cells cover the ciliary processes, which are also highly vascularized.

Aqueous humor is actively secreted from the ciliary processes. The epithelia cells that line the ciliary processes regulate the composition of aqueous humor. To form aqueous humor, sodium ions are actively (through the use of ATP) transported out of the ciliary epithelial cells. Chloride and bicarbonate ions diffuse out of the ciliary processes, to maintain the electrical neutrality of the intercellular space. The efflux of these ions into the intercellular space causes an increase in the osmotic pressure of the intercellular space. To counterbalance the increase in intercellular space osmotic pressure, water moves from the blood vessels underlining the ciliary epithelial cells into the intercellular space. As water moves into the intercellular space, it flows across the ciliary processes into the posterior chamber. Other molecules, such as glucose and amino acids, can be transported within the aqueous humor if they migrate or are transported into the intercellular space. Therefore, aqueous humor formation is induced by the active transport of sodium ions and the passive movement of water.

The composition of aqueous humor is similar to arterial plasma and cerebrospinal fluid. However, there is a significantly higher concentration of ascorbate in the aqueous humor than in the blood plasma, and there is a marked reduction in proteins and blood cells (no cells are found suspended in the aqueous humor). Ascorbate is an ester of vitamin C and is used to maintain the health of the lens. With a deficiency in ascorbate, it is common for cataract formation and macular degeneration to proceed. Plasma proteins cannot be filtered by the ciliary bodies, and therefore, proteins account for approximately 0.02% of the total aqueous humor composition (whereas proteins account for nearly 10% of the plasma composition). The major functions of aqueous humor include maintaining intraocular pressure, providing nutrients to the cornea and lens (which are avascular), and removing wastes from the cornea and lens.

10.4 AQUAPORINS

Aquaporins are transmembrane proteins that regulate the flow of water into and out of cells. For many years, it was believed that water movement into and out of the cell was (1) not regulated in any manner and (2) could be accounted for by simple diffusion across the cell membrane. However, the rapid movement of water in aqueous humor formation could not be described by simple diffusion. In the early 1990s, aquaporins were discovered, and it was found that they can selectively control water movement into and out of cells. One of the critical functions of aquaporins is that while they allow the passage of water they prevent the passage of ions. If aquaporins allowed ions through their channels, all ion concentration gradients across the cell membrane would approach zero (i.e., all ions would be in equilibrium) and the cell would not be able to perform many of its critical functions that depend on the concentration gradient of various molecules (e.g., many transport processes that rely on an electrochemical gradient would fail). Also, if ions were allowed to pass through aquaporins, the amount of energy that cells would expend on maintaining the necessary ion concentration gradient, if the cell could maintain a concentration gradient under these conditions, would exceed the amount of energy produced during cellular respiration.

Aquaporins are found in a high concentration in the epithelial cells that produce aqueous humor (as well as other epithelial cells that allow water to move readily across their membrane, e.g., epithelial cells in the kidney). These pores allow water molecules through in a single file. To understand how aquaporins regulate water movement into a single file and prevent the movement of other ions, it is important to understand the three-dimensional structure of the aquaporin protein. An aquaporin channel is composed of six transmembrane α-helices, with both the amino and carboxyl terminal on the cytoplasmic side of the membrane. Two of the five loops that connect the six transmembrane helices are extremely hydrophobic. One of these loops is on the intracellular side of the membrane and the other is on the extracellular side of the membrane. The two hydrophobic loops contain a three amino acid sequence, termed the NPA (Asparagine–Proline–Alanine) motif. The NPA motif folds back into the channel aquaporin created by the six transmembrane helices. In three-dimensional space, the folding back of these two domains resembles an hourglass shape (or a bottle neck for flow). This hourglass constriction restricts water molecules to a single file as they are passing through the channel. Also, the hydrophobic portion "coats" one side of the channel (the reason for this and the effect that his has on ions will be discussed later). The restriction of water most likely occurs due to an electric field created by the charges on the protein structure, inducing the majority of the channel's core to be hydrophobic. This electric field also dictates the direction of the water molecules as the flow through the channel. As water molecules enter the channel, they typically are oriented with the oxygen atom facing the entrance of the channel. As the molecules enter the NPA motif, the water molecules flip, so that the oxygen atom is facing toward the channel's exit. It is believed that the orientation of oxygen changes due to a hydrogen-bonding event with the two asparagine molecules within the NPA motif. Therefore, because each water molecule must be reoriented to pass through the aquaporin channel and it can only be reoriented by interacting with the two asparagine molecules within the NPA motif, only one water molecule can flow through the channel at a time. Through these two restrictions it has been observed that the permeability of aquaporin channels toward water molecules is on the order of $6_E - 14\ \mathrm{cm}^3/\mathrm{s}$, which allows approximately 10^9 water molecules through each pore per second.

There is a second constriction of the aquaporin channel, usually toward the extracellular side of the cell membrane, which acts to restrict the movement of other molecules through the channel. This selectivity filter is termed the aromatic/arginine selectivity filter in aquaporin channels. The selectivity filter is a grouping of amino acids that interact with only water molecules and helps them through the narrowing created by this filter. Other molecules that do not interact with the selectivity filter cannot pass through this narrowing. The aromatic ring weakens the hydrogen bonds between water molecules and then the partial negative charge on the oxygen atom interacts with the positive charge on the arginine. The interaction between the oxygen and the arginine allows water through the channels and prevents the passage of other molecules, especially protons.

All of the restrictions are physically smaller than hydrated ions; remember that all ions that are in the body are hydrated (for instance a sodium ion, which has a positive charge, will typically traverse through the biological substrates, with four water molecules associated with it; the partial negative charge on the water's oxygen weakly hydrogen bonds to the sodium). However, ions in solution can dehydrate and this is how they typically pass through ions channels. To dehydrate an ion, a significant amount of energy would be needed and typically this is counterbalanced by other binding events. In the case of the aquaporin channel, the location where the ions would need to become dehydrated to pass through the pore restriction is associated with hydrophobic amino acid regions. A hydrophobic surface cannot provide a temporary bonding event for a hydrophilic ion. Thus, an ion would need an energy source to break the water hydrogen bonding events and not create new hydrogen bonding events. This significant amount of energy is not readily available, which effectively prevents ions from moving through aquaporin channels.

Four aquaporin channels associate with each other in the membrane, so that in one location there are four possible passageways for water to move through the cell membrane. Each aquaporin channel can have a slightly different protein structure. There are at least four different aquaporins in mammals and upward of 10 aquaporin channels found within plants. The different structures between the aquaporin molecules may allow for the movement of a small quantity of ions or other solutes (such as glycerol and small sugars) through the channels, although most aquaporin channels restrict ion/solute movement. Also, there are some aquaporin channels that respond to stimuli from external hormones or other paracrine molecules. Upon stimulation, the rate of water movement (and possibly the direction of movement) can be altered.

It is important to remember that aquaporins do not actively transport water across the cell membrane; instead they facilitate the diffusion of water across the cell membrane. Due to the slow diffusion of water across the lipid bilayer, aquaporins effectively increase the overall rate of water movement across the cell membrane.

10.5 FLOW OF AQUEOUS HUMOR

Immediately after the formation of aqueous humor (which uses aquaporin channels) within the intercellular space of the ciliary processes, aqueous humor flows between the ciliary muscle ligaments that attach to the lens (Figure 10.5). From here, the aqueous humor flows through the pupil into the anterior chamber of the eye. The fluid flows throughout the anterior chamber and passes through a trabeculae meshwork. The trabeculae act as a filter for the aqueous humor, removing any debris that may enter the aqueous humor. The debris may come from a bacterial infection or hemorrhaging of the vascular system within the ciliary processes. After passing through the trabeculae, the aqueous humor passes into the canals of Schlemm. Each canal of Schlemm is a thin-walled vein that connects to the extraocular veins with the vascular tunic. The canals of Schlemm extend around the entire eye, collecting aqueous humor from all locations. Each canal of Schlemm is lined by highly permeable endothelial cells which allow the passage of red blood cells (if any enter the aqueous humor) and proteins into the vascular network of the eye so they do not accumulate within the eye. Although the canal of Schlemm is a blood vessel, it is typically never filled with blood because the flow rate of aqueous humor is very high. Under normal conditions, the rate of removal of the aqueous humor is equal to

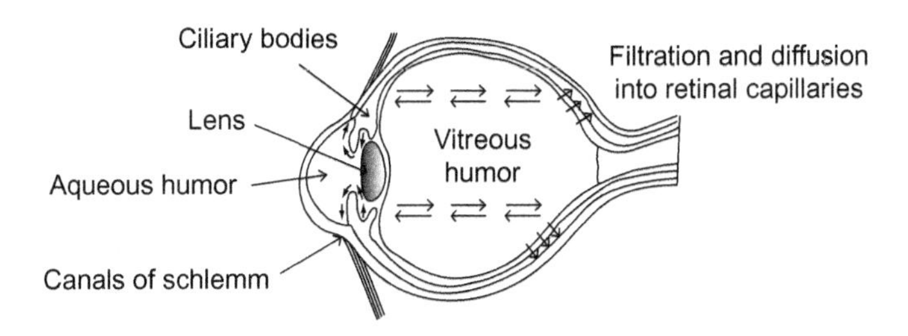

FIGURE 10.5 Anatomical structures salient for the formation of aqueous humor. Aqueous humor is formed at a rate of approximately 2 μL/min within the ciliary bodies. Aqueous humor then flows between the lens and the iris to fill the anterior chamber. Aqueous humor can leave the anterior chamber via the canals of Schlemm or diffuse into the posterior chamber. There is a much slower flow through the posterior chamber, and fluid within this chamber can leave by diffusion into the retinal capillaries. Arrows depict the movement of aqueous humor through the eye. *Source: Adapted from Guyton and Hall (2000).*

the rate of formation (i.e., approximately 2 μL/min). This can be quantified by a mass balance of the aqueous humor:

$$\rho v A_{in} = \rho v A_{out} \tag{10.1}$$

The flow of aqueous humor through many of the portions of the eye can be represented by a simple pressure–flow relationship. In particular, an accurate representation of flow through the canals of Schlemm can be calculated from

$$\frac{dQ}{dx} = \frac{p_{IO} - p(x)}{R} \tag{10.2}$$

where Q is the flow rate along the canals of Schlemm, p_{IO} is the intraocular pressure, $p(x)$ is the local pressure within the canal, and R is the resistance to flow primarily generated from the trabecular mesh. In this formulation, x is the distance along the canal's length. The flow through the canal is a low Reynolds number flow, and the pressure gradient can then be approximated from the flow between two parallel infinite plates, assuming a two-dimensional uniform problem. The pressure gradient can therefore be calculated from

$$\frac{dp}{dx} = \frac{12\mu Q(x)}{zh^3(x)} \tag{10.3}$$

where μ is the aqueous humor viscosity, z is the thickness of the tissue, and $h(x)$ is the local spacing between walls in the channel. Using Eqs. (10.2) and (10.3), the flow rate or pressure gradient through the canals of Schlemm can be obtained.

Very small quantities of aqueous humor enter the posterior cavity of the eye, instead of following the pathway described above. Typically, on the order of a few nanoliters of aqueous humor enter the posterior cavity every hour. As stated above, this portion of the aqueous humor does not "flow" through the posterior cavity because it is filled with the vitreous body, which is a highly charged gel. Remember that the vitreous body does not flow, although it allows for a very slow diffusion of aqueous humor through it. The aqueous humor will typically wet the retina and is eventually collected into the venous flow within the vascular tunic through diffusion processes.

10.6 INTRAOCULAR PRESSURE

Intraocular pressure is a measure of the hydrostatic pressure within the globe (hollow cavities) of the eye. This pressure is essential to maintain the appropriate distance between the lens and the retina. The intraocular pressure is slightly positive as compared with the pressure of neighboring tissue to ensure that the eye ball remains open and does not collapse. The intraocular pressure is maintained by the constant production of the aqueous humor from the ciliary processes. Under normal conditions, the intraocular pressure is close to 15 mmHg. Interestingly, the intraocular pressure varies within an individual by about 2–3 mmHg each day. There is a significant difference between the intraocular pressure during the day and night (or more accurately awake vs. sleeping). During the night, the intraocular pressure decreases because there is a slight reduction in the formation of the aqueous humor. The intraocular pressure is also related to the episcleral venous pressure, which is the hydrostatic pressure of the veins within the sclera. Intraocular pressure is defined by

$$p_{IO} = \frac{F}{C} + p_{EV} \tag{10.4}$$

where p_{IO} is the intraocular pressure, F is the rate of formation of the aqueous humor, C is the drainage rate of the aqueous humor as a function of pressure (or the eye compliance), and p_{EV} is the episcleral venous pressure. The episcleral venous pressure is fairly constant throughout the day and averages close to 14 mmHg under normal conditions.

If the inflow flow rate of the aqueous humor does not balance the outflow flow rate of the aqueous humor, there will be changes in the volume of the eye (this typically can only act for small pressure changes). The volume of the eye is equal to the compliance of the eye multiplied by the pressure of the eye (which may be equal to the intraocular pressure).

$$V_{eye} = Cp_{eye} \tag{10.5}$$

In Eq. (10.5), C is the compliance of the eye. If the intraocular pressure is changing, then the outflow flow rate will not balance the inflow flow rate as described by Eq. (10.1). Instead, the outflow flow rate is equal to

$$Q_{out} = \frac{p_{eye}}{R} \tag{10.6}$$

where R is the resistance to aqueous humor flow. Using the two previous relationships to solve for the Conservation of Mass, we get

$$\frac{dV_{eye}}{dt} = Q_{in} - Q_{out}$$

$$C\frac{dp_{eye}}{dt} = \frac{p_{norm}}{R} - \frac{p_{eye}}{R}$$

$$\frac{dp_{eye}}{dt} + \frac{p_{eye} - p_{norm}}{RC} = 0 \tag{10.7}$$

where p_{norm} is the steady-state pressure that normally acts and regulates the inflow of aqueous humor.

Example

Under a disease condition where the trabeculae become occluded, the intraocular pressure transiently increases by 5 mmHg. If the trabeculae occlusion is removed, which brings the flow resistance back to normal levels, what is the time required to bring the intraocular pressure back to within 5% of the steady-state value? Assume that C is equal to 2.8 μL/mmHg and that R is equal to 5 mmHg min/μL. Assume that the inflow flow rate is 2 μL/min.

Solution

To solve this problem, we will need to solve the differential equation derived for the Conservation of Mass of the eye:

$$\frac{d(p_{eye} - p_{norm})}{dt} = -\frac{p_{eye} - p_{norm}}{RC}$$

$$\int \frac{d(p_{eye} - p_{norm})}{p_{eye} - p_{norm}} = -\int \frac{dt}{RC}$$

$$\ln(p_{eye} - p_{norm}) = \frac{-t}{RC} + K$$

where K is the indefinite integration constant:

$$p_{eye} - p_{norm} = K_0 e^{-t/RC}$$

where K_0 is a different constant from K:

$$p_{eye} = p_{norm} + K_0 e^{-t/RC}$$

To solve for K_0, we know that the initial condition is

$$p_{eye}(t = 0) = p_{norm} + 5\ mmHg$$

$$p_{norm} + 5\ mmHg = p_{norm} + K_0 e^0 = p_{norm} + K_0$$

$$K_0 = 5\ mmHg$$

$$p_{eye} = p_{norm} + 5\ mmHg\left(e^{-t/RC}\right)$$

To solve for the time required to return back to 5% of the steady-state normal pressure

$$1.05 p_{norm} = p_{norm} + 5\ mmHg\left(e^{-t/RC}\right)$$

$$\frac{0.05 p_{norm}}{5\ mmHg} = e^{-t/RC}$$

$$-\frac{t}{RC} = \ln\left(\frac{0.05 p_{norm}}{5\ mmHg}\right) = \ln\left(\frac{0.05 * Q_{in} * R}{5\ mmHg}\right)$$

$$t = -RC\ln\left(\frac{0.05 * Q_{in} * R}{5\ mmHg}\right)$$

$$t = -(5\ mmHg\ min/\mu L)(2.8\ \mu L/mmHg)\ln\left(\frac{0.05 * 2\ \mu L/min * 5\ mmHg\ min/\mu L}{5\ mmHg}\right)$$

$$t = 32.2\ \text{min}$$

Tonometry is a method to directly measure the intraocular pressure. By assuming that the eye is composed of elastic tissue that forms a perfect spherical shell, the effect of increases in external pressure, which causes deformations to the eye, can be used to estimate the intraocular pressure. Increases in intraocular pressure would make it harder to deform the eye. There are two general approaches to conduct tonometry: the first is to measure the force required to deform a fixed area of the eye (Goldmann tonometry, which is the gold standard to measure intraocular pressure), and the second is to measure the area deformed by a fixed force (air-puff tonometry). In this ideal situation, the intraocular pressure can be quantified as

$$p_{IO} = \frac{F}{A} \tag{10.8}$$

where F is either the force applied or measured and A is either the covered area or the measured deformed area. Although this formulation can be used to estimate intraocular pressure, it is important to note that the eye is not a thin-walled, homogeneous, purely elastic, spherical shell (of uniform thickness) and that the deformed area would not be the same as anticipated because the fluid that bathes the exterior portion of the eye can dissipate some of the load applied to the eye.

10.7 DISEASE CONDITIONS

10.7.1 Glaucoma

Glaucoma is one of the leading causes of blindness and occurs when the intraocular pressure rises significantly. Typically the intraocular pressure rises slowly during glaucoma, but any increase can eventually lead to blindness, if the force exerts itself on the eye structures for enough time. An increase in intraocular pressure to about 30 mmHg is common during glaucoma. This level of pressure causes a significant decrease in vision and will eventually lead to blindness. However, under severe conditions, the intraocular pressure can rise to 60 mmHg, which will lead to blindness within weeks. In most cases of glaucoma, the rise in intraocular pressure is caused by a decrease in aqueous humor draining. Commonly, the decrease in drainage is due to a blockage of the trabeculae, and thus, the aqueous humor either does not enter the canals of Schlemm or there is an overall increase in the resistance to flow through the anterior cavity. The intraocular pressure rises because the production of aqueous humor is continuous and the sclera is fairly rigid and cannot accommodate large changes in fluid volume by distending its walls. Although not discussed in previous sections, there is one weak point within the sclera. Along the posterior section of the eye, the optic nerve and the central retinal artery and vein pass through the three layers of the eye. This location is termed the optic disc. As the intraocular pressure increases, the optic nerve is forced outward due to the increased pressure forces. This induces irreversible damage to the nerve fiber. When the intraocular pressure increases to twice of the normal intraocular pressure, the damage to the optic nerve interferes with action potential signal transmittance. Therefore, as the action potential becomes blocked, vision is lost.

Glaucoma is common, affecting over 2% of people over the age 35. In the United States alone, this is over 2 million people. If caught early enough, glaucoma can be treated with a topical drug, but it may eventually need corrective surgery. There have been a number of studies conducted on the cause of glaucoma. According to Eq. (10.2), there can be changes in either the episcleral venous pressure or the inflow/outflow rate of the aqueous humor. In multiple species, it has been shown that the episcleral venous pressure and the inflow flow rate of the aqueous humor remain relatively constant under glaucoma conditions as compared to normal conditions. Therefore, as discussed above, glaucoma is more than likely caused by changes in the outflow flow rate of the aqueous humor.

10.7.2 Cataracts

The transparency of the lens depends on the exact balance between structural characteristics of the lens proteins and the biochemical interactions of the proteins. Once this balance is disturbed, the lens will tend to lose its transparency. Cataract is the condition that is associated with this process. There are many causes of cataracts, including aging, exposure to radiation, drug reactions, but it is most commonly found in the elderly. A cataract is defined as a cloudy or opaque region within the lens. During the early stages of cataract formation, some of the lens proteins become denatured. In later stages, the proteins aggregate together, forming the opaque regions.

Over time, it is natural for the lens to become yellow and lose its transparency. It is then typical for the patient to require brighter lights to read and will tend to have "night blindness." When the lens becomes completely opaque, the patient is functionally blind and will require the removal of the lens. The patient is only functionally blind because the photoreceptors within the retina function properly when stimulated with light. After lens removal, an artificial substitute can be put in its place, and eyeglasses are used to focus light onto the retina properly to allow for vision.

END OF CHAPTER SUMMARY

10.1 The eye is a specialized sensing organ that is sensitive to both the intensity and the wavelength of light. The wall of the eye is composed of three layers, which provides mechanical support for the eye (the fibrous tunic); provides blood to the eye (the vascular tunic); or is composed of sensing cells (the neural tunic). The eye can also be divided into two chambers, each filled with fluids. The vitreous cavity maintains the shape of the eye and is filled with a gel that is composed of many proteins and proteoglycans. The anterior cavity is filled with aqueous humor that wets the cornea and lens, providing these structures with the nutrients it needs to function. The lens is made up of the protein crystallin which helps to focus light on the retina. The retina is composed of two sensing cell types: the rods and the cones, which can be stimulated by changes in the incident light and transmit those changes to the brain.

10.2 The blood supply to the eye is unique that all cells need to have a readily available blood supply but the blood vessels cannot interfere with the transmission of light through the eye. The blood vessels of the eye are divided into the uveal vessels, which supply the choroid, ciliary bodies, the iris and parts of the retina, and the retinal circulation, which supplies all other locations of the eye. Small arteries of the eye play a significant role in the formation of aqueous humor. Arterioles that supply the rectus muscle branch into both capillaries, which feed the muscle, and arterioles, which continue to penetrate the eye tissue. These arteries divide and penetrate the ciliary bodies. These capillaries have a unique structure that helps in the formation of aqueous humor in the ciliary processes. The blood-retinal barrier prevents the mixing of blood constituents with the eye constituents.

10.3 Aqueous humor is formed at a rate of approximately 2 μL/min within the anterior chamber. It is secreted by the ciliary bodies, which are located behind the iris and are composed of secretory epithelial cells. Sodium ions are actively transported out of the ciliary bodies, which causes chloride and bicarbonate ions to diffuse out of the ciliary bodies. The efflux of these ions causes an increase in the osmotic pressure of the fluid, and therefore, water moves out of the ciliary bodies to balance this increase in osmotic pressure.

10.4 Aquaporins are transmembrane proteins that facilitate the movement of water across the cell membrane wall. Under normal conditions, water ions can diffuse across the cell membrane freely, however, this is a relatively slow process. Through the addition of aquaporin channels to the cell membrane, water movement occurs significantly faster. Interestingly, aquaporins contain structures that can regulate the movement of water through the channels. A structure termed the NPA motif forms a bottleneck within the aquaporin channel. This constriction causes water ions to move through the channel in a single file. This is likely caused by an electric field that is generated within the channel. A second filter, termed the aromatic/arginine selectivity filter, interacts with water molecules to facilitate their transport through the channel. Other compounds cannot interact with this filter and cannot pass through the channel.

10.5 Aqueous humor flows from the ciliary processes, between the ciliary ligaments into the anterior chamber. Within the anterior chamber, aqueous humor can either pass through the pupil to wet the lens or remain in the anterior chamber to wet the cornea. After passing through the anterior chamber, aqueous humor passes through the trabeculae that act as a filter for the aqueous humor and then enters the canals of Schlemm. The canals of Schlemm connect back to the vascular tunic so that aqueous humor is reabsorbed into the cardiovascular system. A small portion of aqueous humor diffuses into the posterior chamber. In general, the flow of aqueous humor can be quantified through a mass balance equation:

$$\rho v A_{in} = \rho v A_{out}$$

10.6 Intraocular pressure maintains the proper shape of the eye, helps aqueous humor flow, and is a measure of the hydrostatic pressure within the eye. Intraocular pressure can be calculated from

$$p_{IO} = \frac{F}{C} + p_{EV}$$

which is a function of the formation/drainage rate of the aqueous humor and the cardiovascular venous pressure. When the inflow flow rate of aqueous humor and the outflow flow rate of aqueous humor are not balanced, the volume of the eye can increase and the outflow

flow rate is dependent on the resistance to outflow (this usually increases due to a blockage). This can be quantified with

$$\frac{dV_{eye}}{dt} = Q_{in} - Q_{out}$$

which would measure the time rate of change of the eye volume. There are two typical ways to measure the intraocular pressure, the first with a constant area and the second with a constant force. Each of these methods assumes that the eye is a perfect spherical shell with uniform elastic properties.

10.7 Glaucoma and cataracts are two common pathologies that affect the eye. Glaucoma is one of the leading causes of blindness and is caused by a slow increase in intraocular pressure. Intraocular pressure normally increases due to a blockage of the canals of Schlemm. Cataracts are caused by an imbalance of the biochemical interactions of the lens proteins. Due to this imbalance, the lens proteins either experience a structural change or they do not organize properly. Either way, the lens tends to lose its transparency and light cannot be focused on the retina properly. The retina is not normally damaged during cataracts, but it is damaged during glaucoma.

HOMEWORK PROBLEMS

10.1 Describe the three layers of the eye, with respect to the structure/function relationship.

10.2 There is a blind spot along the retinal wall where the optic nerve attaches to the retina. Why does this not impair vision?

10.3 Aqueous humor circulates at a constant inflow flow rate of 2 μL/min and at steady-state drains from the eye at the same flow rate. This flow rate is dependent on the pressure–volume relationship of the eye which can be described by $V = Cp$, where C is the compliance of the eye and the outflow flow rate is equal to p/R, where R is the resistance to outflow. Calculate the compliance of the eye at steady state, knowing that the resistance to outflow is 4.2 mmHg min/μL and the volume of the eye is 7 cm^3. If the pressure is increased by 10 mmHg due to a compression of the eye which reduces the overall volume to 5.5 cm^3, what is the outflow flow rate in these conditions?

10.4 Approximate how many aquaporin channels are required to maintain an inflow flow rate of 2 μL/min, assuming that the permeability of each aquaporin channel is $6_E{-}14$ cm^3/s and that each channel allows one water molecule to pass at a time.

10.5 Discuss in detail the restrictions of water movement through the aquaporin channels. Is this an effective way to move water into/out of cells?

10.6 Assume that there are 10,000 aquaporin channels that when combined can form aqueous humor at a rate of 2 μL/min. Assume that the diameter of each aquaporin channel is 0.25 nm. Under disease conditions, the aqueous humor drains at a rate of 1.95 μL/min. Assuming that there are 2000 canals of Schlemm present, calculate the average diameter of one canal.

10.7 Calculate the intraocular pressure knowing that the episcleral venous pressure is 14 mmHg and the compliance of the eye is 2.5 μL/mmHg min. What is the absolute pressure of the eye under these conditions?

10.8 If the pressure of the eye is decreased transiently by 10 mmHg, calculate the time required to return to within 10% of the steady-state pressure value, if the eye compliance is equal to 2.7 μL/mmHg, the resistance to outflow is 5 mmHg min/μL, and the inflow flow rate is 2 μL/min (all under normal conditions).

10.9 Discuss some of the reasons why glaucoma can cause an increase in intraocular pressure. How can this be represented mathematically with the inflow flow rate, eye compliance, and so on?

10.10 Conduct a parameter analysis for the effects of the rate of formation of the aqueous humor, the eye compliance, and the episcleral venous pressure on intraocular pressure. Assume that the normal value for the compliance of the eye is 2.8 μL/mmHg min, the formation rate of aqueous humor is 2 μL/min and the episcleral venous pressure is 14 mmHg. Which of these parameters alters the intraocular pressure most significantly.

10.11 Considering what you know about the central dogma of molecular biology (e.g., how long it takes to make, transport, and/or activate a protein), would it be more effective to produce more aquaporin channels or increase the permeability of channels, when cells need to move larger quantities of water. If you choose increase the permeability, how do you propose that this would happen.

References

[1] R.R. Allingham, A.W. de Kater, C.R. Ethier, P.J. Anderson, E. Hertzmark, D.L. Epstein, The relationship between pore density and outflow facility in human eyes, Invest. Ophthalmol. Vis. Sci. 33 (1992) 1661.
[2] R.R. Allingham, A.W. de Kater, C.R. Ethier, Schlemm's canal and primary open angle glaucoma: correlation between Schlemm's canal dimensions and outflow facility, Exp. Eye Res. 62 (1996) 101.
[3] S. Barnes, After transduction: response shaping and control of transmission by ion channels of the photoreceptor inner segments, Neuroscience 58 (1994) 447.
[4] S.P. Bartels, Aqueous humor formation: fluid production by a sodium pump, in: R. Ritch, M.B. Shields, T. Krupin (Eds.), The Glaucomas, CV Mosby Co., St. Louis, MO, 1989.
[5] P.J. Bentley, E. Cruz, The role of Ca21 in maintaining the Na and K content of the amphibian lens, Exp. Eye Res. 27 (1978) 335.
[6] P.J. Bentley, The crystalline lens of the eye: an optical microcosm, News Physiol. Sci. 1 (1986) 195.
[7] A. Bill, Blood circulation and fluid dynamics in the eye, Physiol. Rev. 55 (1975) 383.
[8] R.F. Brubaker, The effect of intraocular pressure on conventional outflow resistance in the enucleated human eye, Invest. Ophthalmol. 14 (1975) 286.
[9] R.F. Brubaker, S. Ezekiel, L. Chin, L. Young, S.A. Johnson, G.W. Beeler, The stress–strain behavior of the corneoscleral envelope of the eye. I. Development of a system for making in vivo measurements using optical interferometry, Exp. Eye Res. 21 (1975) 37.
[10] C.R. Canning, M.J. Greaney, J.N. Dewynne, A.D. Fitt, Fluid flow in the anterior chamber of a human eye, IMA J. Math. Appl. Med. Biol. 19 (2002) 31.
[11] D.F. Cole, Aqueous humor formation, Doc. Ophthalmol. 21 (1966) 116.
[12] J.L. Demer, The orbital pulley system: a revolution in concepts of orbital anatomy, Ann. N.Y. Acad. Sci. 956 (2002) 17.
[13] M.J. Doughty, M.L. Zaman, Human corneal thickness and its impact on intraocular pressure measures: a review and meta-analysis approach, Surv. Ophthalmol. 44 (2000) 367.
[14] C.R. Ethier, R.D. Kamm, B.A. Palaszewski, M.C. Johnson, T.M. Richardson, Calculations of flow resistance in the juxtacanalicular meshwork, Invest. Ophthalmol. Vis. Sci. 27 (1986) 1741.
[15] C.R. Ethier, R.D. Kamm, M. Johnson, A.F. Pavao, P.J. Anderson, Further studies on the flow of aqueous humor through microporous filters, Invest. Ophthalmol. Vis. Sci. 30 (1989) 739.
[16] C.R. Ethier, The inner wall of Schlemm's canal, Exp. Eye Res. 74 (2002) 161.

[17] C.R. Ethier, M. Johnson, J. Ruberti, Ocular biomechanics and biotransport, Annu. Rev. Biomed. Eng. 6 (2004) 249.
[18] C.R. Ethier, A.T. Read, D. Chan, Biomechanics of Schlemm's canal endothelial cells: influence on F-actin architecture, Biophys. J. 87 (2004) 2828.
[19] C.R. Ethier, Scleral biomechanics and glaucoma—a connection? Can. J. Ophthalmol. 41 (2006) 9.
[20] J. Heys, V.H. Barocas, Mechanical characterization of the bovine iris, J. Biomech. 32 (1999) 999.
[21] T.J. Jacob, M.M. Civan, Role of ion channels in aqueous humor formation, Am. J. Physiol. 271 (1996) C703.
[22] G.H. Jacobs, Primate photopigments and primate color vision, Proc. Natl. Acad. Sci. USA 93 (1996) 577.
[23] M. Johnson, K. Erickson, Mechanisms and routes of aqueous humor drainage, in: D.M. Albert, F.A. Jakobiec (Eds.), Principles and Practice of Ophthalmology, WB Saunders, Philadelphia, 2000.
[24] M.C. Johnson, R.D. Kamm, The role of Schlemm's canal in aqueous outflow from the human eye, Invest. Ophthalmol. Vis. Sci. 24 (1983) 320.
[25] O. Maepea, A. Bill, Pressures in the juxtacanalicular tissue and Schlemm's canal in monkeys, Exp. Eye Res. 54 (1992) 879.
[26] R.T. Mathias, J.L. Rae, G.J. Baldo, Physiological properties of the normal lens, Physiol. Rev. 77 (1997) 21.
[27] J.L. Miller, A. Picones, J.I. Korenbrot, Differences in transduction between rod and cone photoreceptors: an exploration of the role of calcium homeostasis, Curr. Opin. Neurobiol. 4 (1994) 488.
[28] W.G. Owen, Ionic conductances in rod photoreceptors, Annu. Rev. Physiol. 49 (1987) 743.
[29] C.D. Phelps, M.F. Armaly, Measurement of episcleral venous pressure, Am. J. Ophthalmol. 85 (1978) 35.
[30] J.L. Rae, R.T. Mathias, R.S. Eisenberg, Physiological role of the membranes and extracellular space with the ocular lens, Exp. Eye Res. 35 (1982) 471.
[31] E.M. Schottenstein, Intraocular Pressure, in: R. Ritch, M.B. Shields, T. Krupin (Eds.), The Glaucomas, CV Mosby Co., St. Louis, MO, 1989, p. 301.
[32] H. Wassle, B.B. Boycott, Functional architecture of the mammalian retina, Physiol. Rev. 71 (1991) 447.
[33] S.M. Wu, Synaptic transmission in the outer retina, Annu. Rev. Physiol. 56 (1994) 141.
[34] S. Zhao, L.J. Rizzolo, C.J. Barnstable, Differentiation and transdifferentiation of the retinal pigment epithelium, Int. Rev. Cytol. 171 (1997) 225.

CHAPTER

11

Lubrication of Joints and Transport in Bone

LEARNING OUTCOMES

1. Explain the structure of bones
2. Describe the molecular composition of bones
3. Differentiate between the four bone cell types and identify the function of each type of bone cell
4. Model the Haversian canal system within long bones
5. Classify different types of joints found within the skeletal system
6. Discuss the function of cartilage
7. Describe the special circulation found within bone
8. Describe the function and formation of synovial fluid
9. Model the flow of synovial fluid through a joint
10. Construct a restricted diffusion coefficient based on the molecular weight of a polymer within solution
11. Calculate the flux of species through the synovial membrane
12. Solve for the forces acting on bones and within joints under particular loading conditions
13. Examine the stress and strain bones can experience under loading conditions
14. Relate loading conditions to the formation of synovial fluid within joints
15. Discuss transport of salient molecules to bone and throughout bone
16. Describe disease conditions that are related to the skeletal system

© 2015 Elsevier Inc. All rights reserved.

11.1 SKELETAL PHYSIOLOGY

In this section, we will briefly describe the structure of bones and then move quickly into the structure of common joints, which use a biological lubricant to facilitate motion. The discussion of the blood vessels that perfuse bone will be highlighted in the following section, although some minor discussion will appear here. All of the bones in the body are divided into two groups. The first group is the axial skeleton, consisting of 80 bones that compose the skull, the rib cage, and the vertebral column. The second group is the appendicular skeleton, which consists of the remaining 126 bones within the body. These bones are found in the pectoral girdle, the pelvic girdle, the upper limbs, and the lower limbs. All 206 bones in the human body function to support and protect the body, store minerals, produce blood cells, and act as levers for motion.

In this textbook, we will restrict ourselves to a discussion of the long bones, which are typically elongated and thin and are located within the limbs of the body (Figure 11.1). All long bones have a tubular shaft that extends for the majority of the length of the bone. This shaft is termed the diaphysis of the bone. The wall of the diaphysis is composed of compact bone, which is very solid and stiff and provides much of the mechanical stiffness of bone. The diaphysis also forms a protective layer that surrounds a central open space termed the medullary cavity (or the marrow cavity). At the longitudinal ends of the diaphysis, the bone expands to a region termed the epiphysis. The epiphysis is composed mostly of spongy bone (or trabecular bone). Spongy bone is an open mesh of bone tissue surrounded by a thin covering of compact bone. This thin layer of compact bone is termed cortical bone. Spongy bone acts to transmit stresses applied to the ends of bone from many different directions, and reorient these stresses into a more uniform direction to the stiffer compact bone. In between the diaphysis and the epiphysis is a small region termed

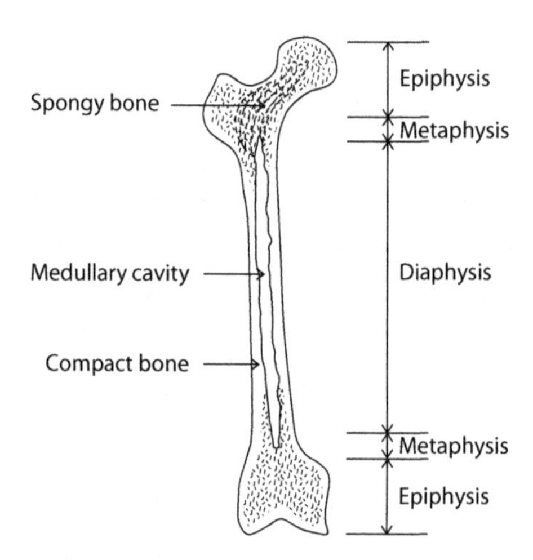

FIGURE 11.1 A representative long bone, showing the various regions within the bone. Long bones are composed of both compact bone and spongy bone. Regions of the bone are classified both by the bone type that composes the majority of the section and by the load direction that the bone normally experiences in that region. *Source: Adapted from Martini and Nath (2009).*

the metaphysis. The metaphysis is composed of both compact bone and spongy bone, with about 50% of each. It is the region of bone that transitions from majorly compact bone in a tube structure to mostly spongy bone.

Bone is a connective tissue that is composed of specialized cells and a stiff matrix that consists of extracellular proteins and a hardened ground substance. The bone matrix is especially stiff due to the deposition of calcium salts within the protein fibers. The major component of bone, by weight, is calcium phosphate, $Ca_3(PO_4)_2$, which accounts for nearly 65% of the total bone mass. In bone tissue, three calcium phosphates interact with one calcium hydroxide to form hydroxyapatite crystals. The molecular formula for hydroxyapatite is $Ca_{10}(PO_4)_6(OH)_2$ and the chemical reaction for hydroxyapatite formation can be described by

$$3Ca_3(PO_4)_2 + Ca(OH)_2 \rightarrow Ca_{10}(PO_4)_6(OH)_2$$

As hydroxyapatite forms, other ions (such as sodium and fluoride) and other calcium compounds (e.g., calcium carbonate) incorporate into the crystal lattice structure of hydroxyapatite. Through the incorporation of these secondary components, these crystals become very stiff but remain quite brittle. Similar to ceramic materials, these crystalline compounds can withstand intense compressive forces but cannot withstand torsion, bending, or forces in tension.

Roughly 33% of bone mass is accounted for by collagen fibers. Different types of collagen can be found in bone matrix and each collagen type forms into a different three-dimensional structure. For instance, collagen type I forms into fibers, whereas collagen type IV forms a more mesh-like structure. Most collagen fibers are long and unbranched proteins and are composed of individual protein subunits that are wound together. Collagen fibers are very strong, especially when subjected to forces in tension. Structurally, collagen resembles a rope and as such does not withstand any forces in compression. Collagen can withstand some pure torsional and bending stresses, but if there is any compressive loading on the collagen fiber, it will tend to fail.

Ground substance fills the spaces between the extracellular proteins, collagen, and the cells within bone tissue. Ground substance is typically clear and viscous due to the presence of proteoglycans. Ground substance is so viscous that most cells have a difficult time migrating through the fluid. Therefore, it is uncommon to see a significant amount of transport occurring within this portion of bone. However, if a foreign cell or a foreign particle enters the ground substance, then the highly viscous fluid aids in the inflammatory response by increasing the amount of time that the foreign cell/particle is visible to white blood cells. Also, the amount of time needed to enter the vascular system is increased, and therefore, the likelihood for spreading to other regions of the body is diminished. The danger of inflammation arises when foreign cells/particles overwhelm the body's defenses in the bone, since it is also difficult for inflammatory cells to migrate through the ground substance. New inflammatory cells can slowly migrate into the ground substance to combat the infection, which may have become very strong and stable.

The remaining 2% of bone mass is accounted for by the cells within the bone matrix. Bone cells are divided into four cell types, which are classified based on their function and appearance. The first type of bone cells are the osteocytes. These are mature cells that make up the vast majority of all of the bone cells. Osteocytes can be found within void spaces throughout the bone matrix. These void spaces are termed lacunae. Each lacuna is occupied by one osteocyte that remains in this location during its entire life span (unless the bone is broken down to such an extent that the lacunae open to the surrounding interstitial space). Lacunae are connected to each other via narrow channels that infiltrate the entire bone matrix. These channels are termed canaliculi, and their primary function is to provide nutrients to osteocytes as well as to allow osteocytes to communicate with each other. In fact, osteocyte cellular extensions fill the canaliculi, and these cellular extensions directly connect to neighboring osteocytes via gap junctions among other junctional proteins. Similar to endothelial cell gap junctions, osteocyte gap junctions allow for the transfer of ions and other signaling molecules to facilitate cell–cell communications.

Osteocytes perform two major functions for bone tissue. The first function is to maintain the mineral content and protein composition of the bone matrix. Bone formation is a process that is continually occurring, depending on the particular loading conditions that the bone is subjected to. This is interesting because if a person starts to run daily as an exercise, this will be accompanied by an increased bone density (among other physiological adaptations). After bed rest or space travel (reduced gravity) the bone density decreases (again among other physiological adaptations). Therefore, bone is a very adaptive tissue. Osteocytes can release compounds to degrade the surrounding matrix or they can aid in the deposition of new hydroxyapatite. Osteocytes also have a critical function in bone repair. When osteocytes are released from a lacuna, they differentiate to osteoblasts or osteoprogenitor cells that function to build new bone.

The second type of bone cell is the osteoblast. Osteoblasts are typically found on the exterior surfaces of bone and function to produce new bone matrix. To produce new bone matrix, osteoblasts secrete the proteins that constitute the bone matrix. Osteoblasts also release calcium phosphate to help calcify the proteins within the matrix; effectively increasing the stiffness of the newly formed matrix. The compounds that osteoblasts release do not become bone until the calcium salts interact with calcium hydroxide to form hydroxyapatite. Prior to this reaction, the organic matrix is termed osteoid. If an osteoblast becomes completely surrounded by bone matrix, the osteoblast differentiates into an osteocyte, and it will stay within that particular lacuna, as described previously.

The third type of bone cell is the osteoclast. Osteoclasts are also typically found on the exterior surfaces of the bone tissue. Osteoclasts function to degrade and recycle the bone matrix. Osteoclasts are multinucleated cells (typically approximately 50 nuclei per cell) that are derived from the same stem cells that produce macrophages. Osteocytes secrete proteolytic enzymes to dissolve the bone matrix in a process termed osteolysis (or resorption). This process is very critical in the regulation of free calcium and phosphate within the body.

Bone is continually being formed and degraded through the actions of osteoblasts and osteoclasts. The balance between these two processes is very critical to the normal function of the particular bone as a tissue. Under normal conditions, if osteoblasts are depositing more matrix than the osteoclasts are removing, the bone itself will become stiffer in compression and larger. The downside of this is that the bone typically becomes more brittle than it should be, and it might exceed the work capacity of the muscles, because the bone mass increases. Therefore, the effort needed to move the bones might be too much for the other organ systems to handle. On the other hand, if osteoclasts remove more bone matrix than is being produced by the osteoblasts, then the bone becomes weaker and thinner. The downside of this process is that the bone cannot withstand the normal loading condition that it would be subjected to, and it would more than likely fail during normal activities. Under normal loading conditions, these processes can occur in everyday life, but they are typically coupled to changes in other systems. For example, individuals who do a lot of strength training increase their muscle mass as well as their bone mass and increase the number of blood vessels perfusing these systems. For someone on bed rest, the loading conditions on their bones have decreased and the muscles begin to atrophy. Therefore, the bone mass and the muscle mass decrease, and the loads that the bones typically withstand are diminished.

The last bone cell is the osteoprogenitor cells. These are mesenchymal cells that can divide and differentiate into osteoblasts. The primary function of osteoprogenitor cells is to maintain the quantity of osteoblasts within bone tissue. These cells play a very important role in fracture healing when it would be necessary to rapidly increase the number of osteoblasts within the bone to repair and produce new bone matrix.

We will now briefly describe the structure of compact and trabecular bone. The most basic structural unit of compact bone is termed the Haversian system (Figure 11.2). In each Haversian system, concentric lamellae surround a central canal, termed the Haversian canal. Lamellae are individual layers of bone matrix that are similar to the concentric circles found within a tree trunk. Lacunae are found in between individual lamellae and not within one lamella, since the bone cells that form/destroy bone are only found on the exterior regions of the bone. Within each Haversian canal are the blood vessels that supply the bone with nutrients. Typically, there is one capillary (or arteriole) and one venule within each canal. Canaliculi run perpendicular to the Haversian canal and connect the lacunae to the blood supply and other lacunae. The canaliculi can penetrate through lamellae to connect bone cells, on neighboring concentric circles, and to aid in nutrient transport. Because the Haversian system is fairly circular, neighboring systems do not

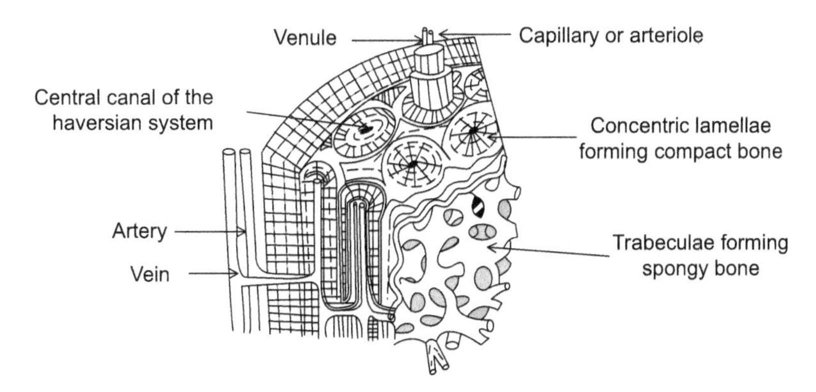

FIGURE 11.2 Schematic of the structure of compact bone. Compact bone is formed from multiple Haversian canal systems that are composed of concentric lamellae. Each Haversian canal has its own venule and capillary (possible arteriole) to provide the osteocytes with nutrients and oxygen. On the interior surface of the compact bone, trabeculae can be found to form a spongy bone marrow within long bones. Osteocytes can be found between the lamellae (black ovals on figure), within a lacuna. *Source: Adapted from Martini and Nath (2009).*

align perfectly with one another. This open space is filled with interstitial lamellae, which is leftover bone matrix from older Haversian systems that have been degraded.

The outer surface of bone is covered by a structure called the periosteum, and the inner surface of bone is covered by the endosteum. The periosteum is a fibrous membrane that isolates bone from other tissue. Within joints, the periosteum is continuous with the surrounding joint connective tissue. The endosteum is a cellular layer that lines the marrow cavity of bone. This layer participates in bone growth and repair and consists of osteoprogenitor cells. In some locations, the endosteum layer is not complete. Here the bone matrix is exposed, and osteoblasts and/or osteoclasts can be found that are actively forming or degrading bone.

In trabecular bone, lamellae do not form a structural unit similar to the Haversian system seen in compact bone. Instead the bone is composed of many thin tubes termed trabeculae, which resemble a branching network. There is no direct blood supply to the cells within these trabeculae. Nutrients must be transported through the canaliculi system that is connected to another blood supply that can be far from the bone cells. Trabecular bone is adapted to withstand stresses in many orientations, but the majority of the trabeculae are oriented along the major stress-loading directions. The trabeculae are branched and interconnected to distribute and transmit the loads and to be able to withstand stresses in non-major directions. Trabecular bone is lighter than compact bone and therefore reduces the overall weight of the bone. Also, some trabecular bone houses the red and the yellow marrow. Red marrow produces blood cells, while the yellow marrow is a store of adipose tissue for future energy needs.

Bone is relatively inflexible, and the only way that movement can occur is at the joints between the bones. There are three types of joints in the body that are each classified based on the range of movement that they permit. The first type of joint allows no movement between the neighboring bones and are termed synarthroses joints. The bones do not rotate about each other because the edges are typically in contact and may even be fused together. These joints are very strong and can be considered an extension of the bone itself. This type of joint is commonly found between the bones that compose the skull. In some sense, the bones are not individual anymore because they fuse to one another. The second type of joint, the amphiarthroses joint, allows some movement between the neighboring bones. These joints are still relatively strong and are typically connected by collagen fibers or cartilage. This type of joint can be found between the radius and the ulna or the tibia and the fibula within the limbs. These bones rotate slightly around each other but do not have a great deal of free motion.

The third type of joint, the diarthroses joint, allows a wide range of motion depending on the structure of the joint. This type of joint is surrounded by an articular capsule and a synovial membrane. The synovial membrane runs along the interior portion of the articular capsule. Synovial fluid (which will be discussed in Sections 11.3 and 11.4) fills the space in between the two bones. There are six types of diarthroses joints, depending on the exact nature of the movement that the joint allows. The first joint type is the gliding joint, which typically allows translation in one or more directions. The bones that are in contact with each other are typically flat or slightly rounded, allowing for translational motion without rotation. Movement within these joints (as well as rotation) is restricted by the ligaments that anchor the bones. An example of this type of joint is the joint that is formed from the connection between the clavicle and the manubrium. The second type of joint is the pivot joint, which typically only allows rotation about one axis. It is normal for the bones in these joints to be touching each other and for a small piece of ligament connected to the first bone to wrap around the second bone. An example of this type of joint is between the atlas and the axis within the neck. The last joint that allows motion in one direction is a hinge joint. This joint also allows rotation about the joint that connects the bones but not translation. The elbow and the knee are two examples of a hinge joint.

The remaining types of diarthroses joints allow more degree of freedom in their movement than the first three described. The first of these joints is the ellipsoid joint, which allows rotation in two planes. Typically, one bone has a convex protrusion which fits within a concave depression of a second bone. An example of this type of joint is found between the metacarpal and phalanges bones within the hand (i.e., finger joints). The next type of joint, the saddle joint, also allows rotation within two planes. One of the bones is shaped like a saddle and the second bone is shaped like a rider's lower body surrounding a saddle. This is also a concave surface facing a convex surface. This type of joint allows the movement of the thumb about the trapezium. The last type of joint is the ball-and-socket joint, which allows for three-dimensional rotation about the joint. In this type of joint, the end of one bone is shaped as a ball and the other bone is concave. The ball fits within the concave depression and is held in place by ligaments. The shoulder and hip joints are examples of ball-and-socket joints.

The surface of bone is rough and the coefficient of friction between two bones is very high. If bones were able to come into contact with each other, to rotate about a joint, the bone itself would become damaged due to the high frictional forces. To prevent bone from coming into contact with another bone, a thin layer of cartilage is present on the ends of the bone. This cartilage is termed articular cartilage and it partially acts to reduce the coefficient of friction between two bones. Under normal conditions, the two cartilage surfaces do not come into contact with each other either. Instead, there is a small fluid layer between the cartilage surfaces. The fluid is termed synovial fluid, and we will discuss this fluid and how it forms in Sections 11.3 and 11.4. The cartilage acts as a safety mechanism for bone–bone contact but it also helps to maintain a layer of synovial fluid within the joint.

Ligaments provide support and strength to joints, by helping to restrict bone movement and to directly connect the bones to one another. Ligaments are typically made of collagen fibers. It takes a great deal of force to tear a ligament because typically they are stiffer than the bones. Therefore, the bone will fracture prior to ligament rupture. However, in a sprain there is some damage to collagen fibers within the ligament. Tendons are not part of the joints themselves, but pass over the joint. Tendons connect bone and muscle together. In relation to the joint, tendons help to hold the joint in place but may restrict the joint motion. However, tendons do provide mechanical support for the joints by connecting bones to other muscles. For instance, many of the muscles that move the humerus are located within the torso and are connected to the axial skeleton. Therefore, the shoulder joint maintains the appropriate orientation because of this connection between the axial skeleton, the muscles, and the humerus.

11.2 BONE VASCULAR ANATOMY AND FLUID PHASES

Bone is perfused by a sparse capillary network that allows for the exchange of nutrients/wastes between blood and the bone cells. As mentioned above, the net movement of calcium ions (Ca^{2+}) to and from bone is critical for the proper functioning of nearly every organ, since calcium ions are necessary for many reactions to occur (e.g., the contraction of muscle, the release of neurotransmitters, coagulation reactions). The vascular tree that is specific to bone can be subdivided into four general categories that are classified based on their size and function. These categories include the conduit vessels, the capillaries of bone, the lacunar–canalicular system and finally the microcanalicular system. Again for this discussion we will restrict ourselves to a discussion on long bone vasculature.

In the vast majority of long bones, a single main artery enters the bone at the region of the diaphysis. This large artery bifurcates and one branch ascends and the other descends along the marrow cavity (Figure 11.3). Along the entire length of this artery, it contains lateral branches that pass to the cortex. As with other tissue small afferent arteries supply the capillary beds of the cortex with blood and regulate the perfusion of these capillary beds. These afferent arteries and each of the feeding arteries are considered part of the conducting or conduit vascular system, which feed the second level, the bone capillaries. While much work has been conducted on the vascular anatomy of bone, it has become apparent that each bone has a slightly different vascular plexus and even the arrangement within analogous bones can vary. For instance, the canine tibial artery has been shown to supply capillary beds using parallel arrangements, whereas the analogous artery in rodents uses a series arrangement. In cortical bone, the arrangement of blood vessels is slightly different; the afferent vessels penetrate along the entire tissue ending with small capillary loops beneath the cartilage layer. The venous drainage system in bones is typical of most vascular beds. Small venules exist that penetrate through the cortex of bone, eventually converging to the periosteal vein. Large emissary veins also exist that penetrate and drain blood from the cortex.

As described above, the Haversian canal system is the location where nutrient exchange occurs between the blood and the bone tissue. Within each Haversian canal system, a small arteriole (which is capable of exchange) or a capillary is present. These small blood vessels are lined first by cells that form the endosteum of bone and then by a continuous basement membrane. It has been observed that the endothelial cell wall that forms bone capillaries restricts the movement of many molecules into bone and the majority of

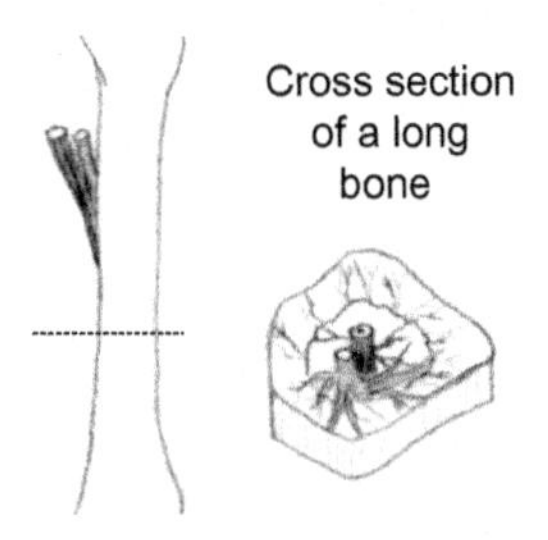

FIGURE 11.3 Blood supply of a long bone, illustrating that an artery and vein are paired along the medullary cavity. Branches of these vessels penetrate into the bone at various levels (see Figure 11.8).

transport occurs through pinocytosis and vesicle transport mechanisms. Some passive diffusion, around the endothelial cell junctions, and active transport can occur as well. It appears that molecular sieving is a property of bone capillaries. The intercellular clefts of bone capillaries appear to be in the range of 20 nm, whereas endothelial transport vesicles appear to be in the range of 100 nm in diameter. While these transport mechanisms are very effective for molecules, the transport of cells through this endothelial cell layer would not occur. In the bone marrow, the endothelial cell layer appears to be discontinuous to allow for the passage of forming blood cells (note that this is a second specialty circulation that is found in bone that will not be discussed in detail here).

The percentage of water found within bone has been shown to be related to the extent of bone mineralization. Higher bone mineralization values are accompanied by lower percentages of water. This can be quantified by an increase in bone density. Bone mineralization increases primarily because there is a higher quantity of hydroxyapatite, which fills the water space in forming bone. Under these conditions, the water is displaced from pores formed by protein fibers to the extracellular fluid, lymphatics, or blood. While it has proven difficult to quantify the water percentage within different sections of bones, the values of water do not appear to vary significantly for different types of bone. However, the vast majority of water in bone is not found within the bone, per say, instead it is found in the blood and the extravascular spaces.

11.3 FORMATION OF SYNOVIAL FLUID

As we have stated in Section 11.1, synovial fluid fills the space between joints to reduce the coefficient of friction between the articular cartilage adhered to the ends of the bones that compose the joint. Synovial fluid is produced by the synovial membrane which is located on the interior side of the fibrous capsule within the joint. The synovial membrane is composed of many interwoven matrix proteins, especially proteoglycans and collagens. Within the joints, this membrane also includes macrophages and fibroblasts to prevent foreign cellular matter from entering the joint space. The combined actions of the woven matrix fibers and the inflammatory cells within the membrane regulate the composition of synovial fluid.

Synovial fluid is produced as a filtrate of interstitial fluid. The composition of synovial fluid is very similar to interstitial fluid, except that there is a significantly higher concentration of proteoglycans, within synovial fluid. The proteoglycans are secreted by fibroblasts within the synovial membrane and thus cannot be found within plasma or the interstitial space. The purpose of the proteoglycans is to increase the organization of water within synovial fluid, so that it can withstand forces better than unorganized water (i.e., the zeta potential and the electric double layer increase, as discussed in Chapter 7). The water becomes organized due to hydrogen bonding between water molecules and the charged amino acids that make up the protein portion of the proteoglycans. Due to this increased arrangement, there are more bonds present within the fluid phase, which strengthens the fluid. The other major function of the proteoglycans is to increase the overall viscosity of the synovial fluid, again due to increased molecular actions. A higher viscous fluid requires more force to cause the same displacement.

The formation of synovial fluid can be described by a sieving coefficient of the synovial membrane. In general, the sieving coefficient is a dimensionless number that quantifies the equilibration potential between a receiving system and a donating system. In the case of the synovial membrane the sieving coefficient can be defined as

$$S = \frac{C_J}{C_I} \tag{11.1}$$

where S is the sieving coefficient, C_J is the mean concentration of a solute in the joint space, and C_I is the mean concentration of a solute in the interstitial space. The mean concentration for the different spaces is quantified immediately prior to and after solute transfer. A sieving coefficient of one would suggest that the solute concentration within the two compartments equilibrates. In most instances the sieving coefficient is less than one, which would suggest that the compartments have not equilibrated because (1) there has not been a sufficient time for exchange, (2) there are other physical considerations effecting the equilibrium concentrations (e.g., an electro-chemical gradient does not reach chemical equilibrium only), or (3) the products are continually consumed, produced, used, or modified so that chemical equilibrium cannot be reached. It is theoretically possible for a sieving coefficient to be greater than one; this would suggest that there is some external forces that are effecting the transport or the solutes are being produced on only one side of the membrane. In the case of the synovial membrane, the sieving coefficient of molecules obtained from the interstitial space would be less than one, whereas the sieving coefficient for proteoglycans would be greater than one.

As we have discussed briefly before, one of the major functions of synovial fluid is the lubrication of the joint. The articular cartilage becomes filled with synovial fluid and the cartilage helps to hold the fluid in place. When the joint is under compression, the two articular cartilage surfaces come into contact with each other, compress each other, and force some of the synovial fluid out of the cartilage into the joint space. The "free" synovial fluid lubricates the two articular cartilage surfaces and significantly reduces the coefficient of friction between these surfaces. When the compressive forces are removed from the joint, the articular cartilage expands again to its normal size. At this point, there is some "free space" within the cartilage and some of the synovial fluid that was forced out of this space may re-enter the cartilage. Therefore, the cartilage effectively holds the synovial fluid within the joint and releases it as needed. It is very devastating when cartilage is torn off the bone or degenerates significantly. When this occurs, the joint can no longer hold significant quantities of synovial fluid, and therefore, there is no added lubrication. Also, the coefficient of friction increases significantly because now the contact surfaces become bone–bone or bone–cartilage. Cartilage can repair itself naturally albeit it a slow process. Additionally, as the cartilage is repairing itself, more damage may occur due to increased coefficient of friction and decreased ability to absorb the loading conditions on bone. Chondrocytes, which are the cells responsible for forming and maintaining cartilage, are bound within lacunae and thus does not necessarily have access to the damaged regions. Additionally, most cartilage does not have a significant blood supply, so supporting cells, growth factors, etc. may not be able to diffuse into the damaged region. There are new techniques that are being developed to repair cartilage, which includes regenerative medicine procedures or injection of juvenile cartilage.

Another mechanical property of the synovial fluid is that it acts as a shock absorber during loading. Because synovial fluid is highly viscous, during compressive loading, there is a time-dependent strain that tends to distribute the load over a larger surface area and to diminish the overall stress within the joint. Synovial fluid is also a dilatant fluid, meaning that for a large shear stress, the amount of strain needed to move the fluid becomes extremely large. This helps to prevent an over-loading condition on the bone, which would effectively damage the cartilage and possibly the bone itself. One of the major proteins that is responsible for these functions is lubricin (proteoglycan 4), which is a fairly large protein produced by both chondrocytes and synovial membrane lining cells.

The last major function of synovial fluid is to bring nutrients into the joint space. The joint space themselves are avascular, but require nutrients so that chondrocytes and the other cells within that comprise the joint space can maintain their proper functions. To perform this task, the synovial fluid must continually circulate around the joint space. This provides nutrients to all of the cells and removes the cellular wastes. During each joint compression, the synovial fluid is circulated throughout the joint space. As the fluid comes into contact with the synovial membrane, wastes are removed and fresh nutrients diffuse into the fluid from the interstitial fluid. The removed wastes enter the interstitial joint space and then diffuse into capillaries that are located outside of the joint capsule. From these three functions, it should be apparent that synovial fluid is critical for the proper functioning of movable joints.

11.4 SYNOVIAL FLUID FLOW

In order to enter the joint space, synovial fluid traverses a fairly complex path starting from blood vessels and ending with re-entering the vascular system. Water, ions, and other small solutes have the ability to cross the endothelial cell boundary that comprises capillaries, as described in Chapter 6. For most diarthrodial joints, the blood vessels that directly supply the joint form an anastomosis around the joint space. This helps to provide redundant circulation to the joint space, which is critical since the nutrient exchange is entirely dependent on diffusion. Capillary exchange through these vessels forms the interstitial fluid, which has a slightly different composition depending on the particular tissue that it is formed within. Synovial fluid is then formed as a filtrate of the interstitial fluid within the joint capsule space (refer to the discussion on the sieving coefficient for some details regarding the formation of synovial fluid). As described previously, the synovial membrane is the tissue that performs the filtering function. This membrane is composed of two layers, the synovium, which is approximately 50 μm thick and the sub-synovium, which is approximately 100 μm thick. The synovium is the layer that is composed of an incomplete layer of macrophages and fibroblasts, which prevents foreign particles from entering the joint space and it produces some of the molecules that comprise synovial fluid. The sub-synovium is composed of extracellular fibers that filter the interstitial fluid. The primary extracellular fibers within this layer are various collagens, fibronectin, hyaluronan, and various proteoglycans. What is interesting about the sub-synovium is that it is permeable enough to allow fluid to enter the joint space but exhibits a high enough resistance to prevent fluid from exiting the joint space (i.e., it acts as a unidirectional valve).

Once the synovial fluid enters the joint space, it is associated with the charged proteins within the cartilage and only "flows" under a compressive load of the joint. The modeling of this flow is quite complex and will be described in the remainder of this section.

The synovial membrane must be modeled as a semipermeable membrane that has a different permeability toward different molecules within the synovial fluid. In some cases, the sieving coefficient can be used as a means to characterize the different permeabilities. Also, cartilage would need to be modeled as a store for the lubricant and molecules, which may become available during loading conditions. The time rate of change of molecule i within the synovial fluid would be represented as

$$\frac{\partial(Vc)_i}{\partial t} = (r_s A_s + r_c A_c - d - J)_i \tag{11.2}$$

where V is the volume of the synovial fluid, c is the concentration of molecule i, r is the formation rate of i by the synovial membrane (denoted with a subscript s) or the cartilage (denoted with a subscript c), d is the degradation rate of molecule i, and J is the flux of i into/out of the joint space. The degradation rate is typically dependent on the concentration and kinetic rate constants of degradation enzymes within the synovial fluid. The flux of molecules within the joint space is typically dependent on the synovial membrane, the permeability of the membrane, the concentration gradient, the cross-sectional area of the membrane, and the ratio of restricted diffusion to free diffusion. For this type of diffusion, we would need to develop a restricted diffusion coefficient, because of the interaction of the molecules with the charged proteins within synovial fluid and the cartilage. The restricted diffusion coefficient can be defined as

$$D_i = De^{(-\sqrt{\theta})\left(1+\frac{a_i}{a}\right)} \tag{11.3}$$

where D is the free diffusion coefficient of species i, θ is the volume fraction of the protein molecules (e.g., proteoglycans), a_i is the effective radius of the molecule i, and a is the effective radius of the protein. The effective radii can be calculated from the Stokes-Einstein formulation (which is valid for any molecule):

$$a_x = \left(\frac{3\text{MW}}{4\pi\rho N_A}\right)^{\frac{1}{3}} \tag{11.4}$$

where MW is the molecular weight, ρ is the density, and N_A is Avogadro's number. From Eqs. (11.2)–(11.4), it is possible to model the changes in the concentration of various species over time. As we have stated before, this is not a typical biofluid flow problem, but instead it relies more on convection and diffusion of species out of a particular compartment.

Example

Calculate the time rate of change of proteoglycans within the synovial fluid, if their diffusion is only restricted by the presence of glycosaminoglycans (GAGs) within the synovial space. Assume that the molecular weight for the proteoglycan of interest is 300 kDa and its density is 1.45 g/mL. The effective radius for the GAGs is 0.5 nm and the volume fraction is 0.008. The free diffusion of proteoglycan in water is $1.1_E - 7\ \text{cm}^2/\text{s}$. The permeability of the proteoglycan through the synovial membrane is $1_E - 6$ cm/s. Assume that the degradation of the proteoglycan is negligible. The formation rate of proteoglycans within the synovial membrane and the cartilage are $6_E - 9$ mg/(cm^{2}s) and $3_E - 7$ mg/(cm^{2}s), respectively. The cross-sectional areas of the synovial membrane and the cartilage are 29.5 mm^2 and 13.5 mm^2, respectively. The concentration gradient of the proteoglycan is 0.13 mg/mL.

Solution

First it will be necessary to calculate the effective radius of the proteoglycan:

$$a = \left(\frac{3 * 300\ kDa}{4\pi * (1.45\ g/mL)(6.02_E 23)}\right)^{\frac{1}{3}} = 4.35\ nm$$

As a side note, Avogadro's constant converts Daltons into grams. Now we can calculate the restricted diffusion coefficient:

$$D_i = (1.1_E - 7\ cm^2/s)e^{(-\sqrt{0.008})(1+\frac{4.35nm}{0.5nm})} = 4.62_E - 8\ cm^2/s$$

To calculate the time rate of change of the proteoglycan, we need to know the flux of the proteoglycans out of the synovial membrane. The flux can be formulated from

$$J = p\Delta cA\frac{D_i}{D} = (1_E - 6\ cm/s)(0.13\ mg/mL)(29.5mm^2)\left(\frac{4.62_E - 8\ cm^2/s}{1.1_E - 7\ cm^2/s}\right) = 0.000058\ mg/h$$

which quantifies the permeation of the proteoglycan as a function of the area of exchange, the concentration gradient, and the percent reduction in diffusion.

Therefore, the time rate of change of proteoglycans within the synovial fluid can be calculated

$$\frac{\partial(Vc)}{\partial t} = r_s A_s + r_c A_c - d - J$$
$$= (6_E - 9\ mg/cm^2s)(29.5\ mm^2) + (3_E - 7\ mg/cm^2s)(13.5\ mm^2) - 0 - 0.000058\ mg/h$$
$$= 0.000094\ mg/h$$

Note that the flux is negative because the concentration gradient for the proteoglycan would act to remove the proteoglycan from synovial fluid.

11.5 MECHANICAL FORCES WITHIN JOINTS

For this section, we will briefly review how to calculate the mechanical forces that may arise in joints, because as we have discussed, under different types of loading conditions, the formation rate of particular molecular species may increase or decrease within cartilage. This material may be familiar from a bio-solid mechanics course or the more traditional engineering statics, dynamics, or solid mechanics course sequence. To start this discussion, we will first restrict ourselves to the standard statics/dynamics assumption that there are no changes in the orientation of the molecules within the joint (i.e., there are no deformations that alter that chemical bonding arrangement). When this assumption is made, the governing equations become

$$\sum \vec{F} = m\,\vec{a} \tag{11.5}$$

and:

$$\sum \vec{M} = I\vec{\alpha} \tag{11.6}$$

where $\vec{F}$ is the forces acting on a body, m is the mass of the body, $\vec{a}$ is the acceleration of the body, $\vec{M}$ is the moments acting on a body, I is the mass moment of inertia of the body, and $\vec{\alpha}$ is the angular acceleration of the body (remember that the moments and the mass moment of inertia were typically calculated about the bodies center of mass). Also we assume that mass and the mass moment of inertia are not functions of time. If you recall from a course in statics, $\vec{a}$ and $\vec{\alpha}$ were always assumed to be zero, allowing you to simplify Eqs. (11.5) and (11.6) to the summation of the forces is equal to zero and the summation of the moments is equal to zero, respectively. In dynamics, we generally include the acceleration (both linear and angular) of the body. Recall from either class that these equations could be broken up into component form (e.g., Cartesian directions) to solve more relevant scenarios. Using Eqs. (11.5) and (11.6), one can calculate the forces required to maintain a joint in place during normal physical activities.

Example

Calculate the force on the shoulder joint for an athlete who is holding a weight with the arm perfectly horizontal (Figure 11.4). To calculate the reaction forces within the shoulder joint, consider that the weight of the arm is equal to 10 lbf and is located at a distance 1 ft from the shoulder joint. The weight that the athlete is holding is equal to 30 lbf, and it is held at a distance of 2.1 ft from the shoulder joint. To simplify the problem, let us consider that the deltoid muscle is the primary muscle holding the arm in place and that it attaches to the humerus at a distance of 0.45 ft from the shoulder joint at an angle of 17°.

Solution

First, let us draw a free-body diagram of the arm.

In Figure 11.5, W_A is the weight of the arm, W is the weight that is being held, F_D is the force of the deltoid muscle, and F_s is the force exerted by the shoulder joint.

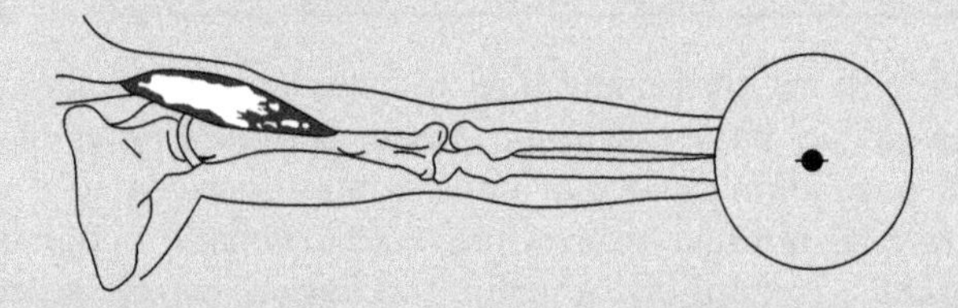

FIGURE 11.4 Diagram of a shoulder joint for the in-text example. *Source: Adapted from Ozkaya and Nordin (1999).*

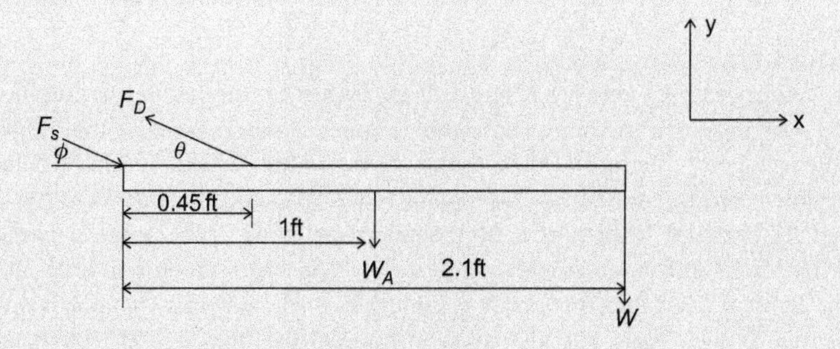

FIGURE 11.5 Free-body diagram of a shoulder joint.

Writing the equations of motion in vector form, we have

$$\sum F_x = 0 = F_s\cos(\phi) - F_D\cos(\theta)$$

$$\sum F_y = 0 = -F_s\sin(\phi) + F_D\cos(\theta) - W_A - W$$

$$\sum M_0 = 0 = (0.45ft)F_D\sin(\theta) - (1ft)W_A - (2.1ft)W$$

if we assume that there is no motion of this join. Solving the third equation for F_D

$$F_D = \frac{(1ft)(10\ lbf) + (2.1ft)(30\ lbf)}{(0.45ft)\sin(17°)} = 555\ lbf$$

Solving the first and second equations

$$F_s\cos(\phi) = F_D\cos(\theta) = (555\ lbf)\cos(17°) = 530\ lbf$$

$$F_s\sin(\phi) = F_D\sin(\theta) - W_A - W = (555\ lbf)\sin(17°) - 10\ lbf - 30\ lbf = 122\ lbf$$

Solving for F_s,

$$F_s = \sqrt{(F_s\cos(\phi))^2 + (F_s\sin(\phi))^2} = \sqrt{(530\ lbf)^2 + (122\ lbf)^2} = 545\ lbf$$

To solve for ϕ,

$$\phi = \tan^{-1}\left(\frac{F_s\sin(\phi)}{F_s\cos(\phi)}\right) = \tan^{-1}\left(\frac{122\ lbf}{530\ lbf}\right) = 13°$$

The force that is generated in the deltoid muscle is very large, which suggests that this is not a stable position for the arm or that other muscles are needed to accommodate this load. Only considering the load that is being held (and not the weight of the arm), there is an approximate 20-fold amplification of the loading conditions in the deltoid muscle.

Example

During normal motion (walking, jogging, or running), for some instant in time, all of our body weight is supported by one leg (Figure 11.6). Typically, the leg is not completely vertical because of the way in which the femur connects to other bones, especially the hip joint (refer to Figure 11.7, the free-body diagram). Using the following values, calculate the reaction forces that act in the hip joint during running at the instant when the body is supported by one leg. The angle between the femoral head and a horizontal plane is 40°. The angle between the distal extremity of the femur and a horizontal plane is 75°. The hip abductor muscle attaches at the junction of the femoral head and the femur at an angle of 75°. Assume that the weight of the leg is 15% of the total body weight and the reaction force at the floor is 340% of the body weight (because of running). Assume that the direct length between the point O (refer to the free-body diagram) and the femoral head is 9 cm. Assume that the weight of the leg acts a distance of 35 cm from point O and that the reaction force acts at a distance of 88 cm from point O. The person has a weight of 800 N.

Solution

First, let us draw a free-body diagram of the leg.

In Figure 11.7, W_L is the weight of the leg, W is the reaction weight from the floor, F_A is the force of the abductor muscle (at an angle of 75° off the horizontal direction), and F_H is the force exerted by the hip joint (at an unknown angle). The angles that are related to the leg geometry are not shown in this figure.

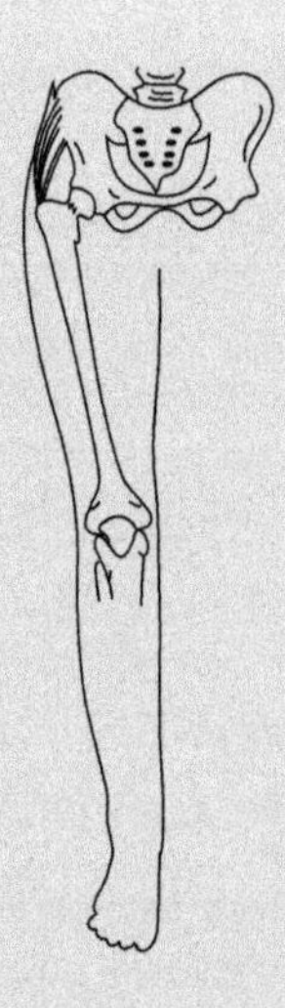

FIGURE 11.6 Diagram of a hip joint during running. *Source: Adapted from Ozkaya and Nordin (1999).*

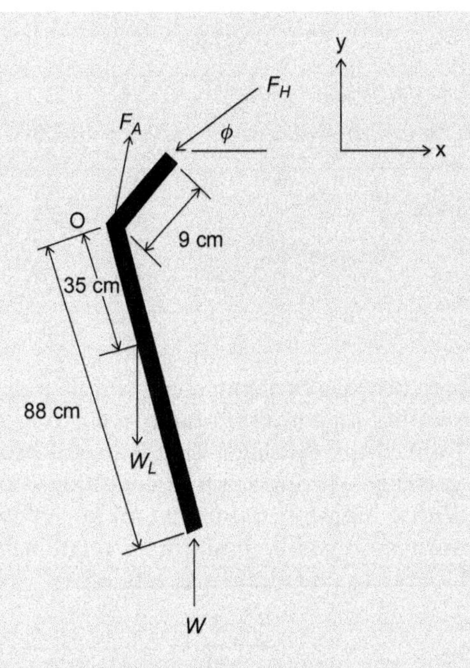

FIGURE 11.7 Free-body diagram for a hip joint.

Writing the equations of motion in vector form, we have

$$\sum F_x = 0 = F_A\cos(75°) - F_H\cos(\phi)$$
$$\sum F_y = 0 = -F_H\sin(\phi) + F_A\sin(75°) - W_L + W$$

$$\sum M_0 = 0 = ((9\ cm)\sin(40°))(F_H\cos(\phi)) - ((9\ cm)\cos(40°))(F_H\sin(\phi)) - ((35\ cm)\cos(75°))W_L + ((88\ cm)\cos(75°))W$$

if we assume that there is no acceleration under these conditions. Using the third equation,

$$(5.785\ cm)F_H\cos(\phi) - (6.89\ cm)F_H\sin(\phi) = 9.059\ cm(0.15 * 800\ N) - 22.77\ cm(3.4 * 800N) = -60,864\ Ncm$$

From the first and second equations,

$$F_H\cos(\phi) = 0.259F_A$$

$$F_H\sin(\phi) = 0.966F_A - 0.15 * 800N + 3.4 * 800N = 0.966F_A + 2600N$$

Substituting these two relationships into the third equation,

$$(5.785\ cm)F_H\cos(\phi) - (6.89\ cm)F_H\sin(\phi) = -60,864\ Ncm$$

$$5.785\ cm(0.259F_A) - 6.89\ cm(0.966F_A + 2600\ N) = -60,864\ Ncm$$

$$(1.5\ cm)F_A - (6.566\ cm)F_A - 17,914\ Ncm = -60,864\ Ncm$$

$$(-5.066\ cm)F_A = -42,949\ Ncm$$

$$F_A = 8478\ N$$

$$F_H\cos(\phi) = 0.259(8478\ N) = 2196\ N$$

$$F_H\sin(\phi) = 0.966(8478\ N) + 2600\ N = 10,790\ N$$

$$F_H = \sqrt{(F_H\cos(\phi))^2 + (F_H\sin(\phi))^2} = \sqrt{(2196\ N)^2 + (10,790\ N)^2} = 11,011\ N$$

$$\phi = \tan^{-1}\left(\frac{10,790\ N}{2196\ N}\right) = 78.5^\circ$$

We show this analysis here because by using some simple assumptions, it is easy to calculate the forces that arise within a joint. From force, we can calculate the stress that acts on the cartilage and the strain of the cartilage. This is useful because the rate of synovial fluid formation from the cartilage, during loading conditions, is dependent on the compression of the cartilage. With a higher compression, more synovial fluid will be released from the cartilage. This information can be built into a relationship for the formation rate of cartilage used in Eq. (11.2). Stress and strain are defined as

$$\sigma = \frac{F}{A} \tag{11.7}$$

and:

$$\varepsilon = \frac{\upsilon}{E} = \frac{\Delta l}{l} \tag{11.8}$$

respectively.

Example

Using the results obtained from the previous example, calculate the cartilage stress and the change in length of cartilage on the femoral head. Assume that the cross-sectional area of the cartilage is 150 cm^2, that the elastic modulus for cartilage is 25 MPa, and that the cartilage thickness is 1.5 mm. Also assume that all of the forces acting on the femoral head are transmitted through the cartilage.

Solution

$$\sigma = \frac{F}{A} = \frac{11,011\ N}{150\ cm^2} = 0.734\ MPa$$

$$\varepsilon = \frac{\sigma}{E} = \frac{0.734\ MPa}{25\ MPa} = 0.029$$

$$\Delta l = \varepsilon l = 0.029(1.5\ mm) = 0.044\ mm$$

This suggests that the thickness of the cartilage changes from 1.5 to 1.456 mm in thickness.

11.6 TRANSPORT OF MOLECULES IN BONE

Fluid transport within the lacunar–canalicular system is key for the passage of nutrients, ions, and wastes to/from the cells that comprise bone tissue. As we have discussed earlier osteoblasts reside on the surfaces of bone tissue. These cells can produce bone matrix and can become embedded within the calcified matrix. Cells that become embedded within bone matrix are termed osteocytes, are located within lacunae, and are highly connected to each other via canalicular passages (Figure 11.8). The lacunar–canalicular passage system provides for a rapid pathway for movement of solutes throughout calcified bone tissue. Previous work has shown that within 30 min of an intravenous injection of a solute with a molecular weight in the range of 40,000 Da, the solute was found throughout the canalicular system and within the lacunae of bone tissue. This suggests that transport through this system is highly efficient and quick.

Many groups have worked on identifying the mode of transport from capillaries within the Haversian canal system to cells on the periphery of bone tissue. These groups have shown that first, molecules are transported across the endothelial cell barrier using typical methods previously described in this textbook (e.g., hydrophilic molecules diffuse through the intercellular cleft, lipophilic molecules are transported with a vesicle mechanism). Once these nutrients enter the interstitial fluid, the fluid flows through the lacunar–canalicular complex, primarily using a diffusion based process. Through similar transport processes, nutrients can enter cells that have entered the canalicular system and wastes can enter the interstitial fluid. However, this transport system suggests the necessity of a lymphatics system, since the fluid flow is largely unidirectional within the lacunar–canalicular complex. It appears that the lymphatic system does not enter the lacunar–canalicular complex, but instead resides on the periosteal surface (outer surface of bone), in close proximity to the periosteal capillaries and the peripheral lacunae. The presence of lymphatics at this one location also supports the notion of unidirectional

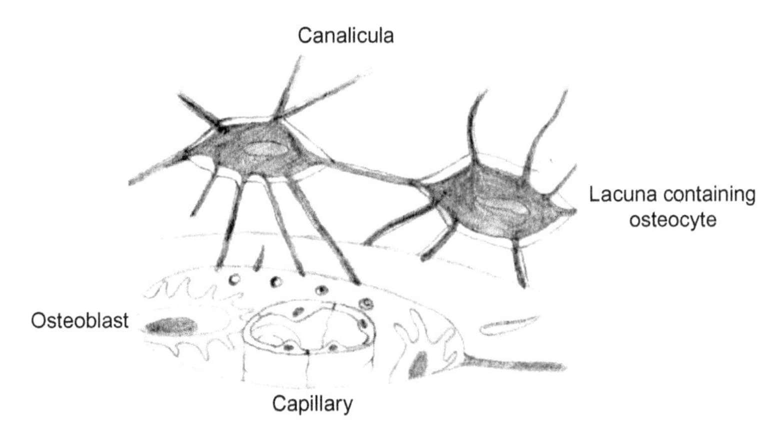

FIGURE 11.8 Relationship between capillaries and the osteoblasts/osteocytes within a lacunae. As illustrated in this figure, canalicula penetrate toward the central canal and help nutrient delivery throughout the entire bone. *Source: Adapted from Handbook of Physiology—The Cardiovascular System Volume 3.*

fluid flow through the lacunar–canalicular system, since the re-uptake of interstitial fluid within the venous portion of the Haversian canal system cannot accommodate all of the fluid that is lost along the arterial portion of the Haversian canal system (recall Starling's relationship).

This leads to some important generalizations regarding the movement of solutes from the capillary fluid into the interstitial fluid of bone. There are three broad categories of substances that can be transported across the capillary wall: (1) small hydrophilic molecules, (2) small lipophilic molecules, and (3) large molecules (such as plasma proteins). As we have learned previously, the majority of large molecules do not cross the capillary wall and thus they establish an osmotic gradient between plasma and interstitial fluid. There are some instances when these molecules are transported; primarily during pathological conditions such as an inflammatory reaction but they can be transported under physiological conditions using vesicular trafficking methods across the endothelial cell wall. In bone, it appears that these large proteins are transported at a higher rate than most systemic vascular beds and it appears that these proteins play three important roles within bone tissue. First, plasma proteins that enter the interstitial space of bone appear to act as carrier molecules for hormones and vitamins. Thus the transport of proteins into the bone interstitial space acts to regulate bone cell activity. Second, the transport of plasma proteins into bone interstitial space appears to dictate the osmotic pressure of bone interstitial fluid and thus the transport of all molecules into bone interstitial space. As compared with other interstitial fluids, the concentration of solutes seems to be highly variable and thus this movement may provide a mechanism to stabilize the osmotic pressure. A third, effect of transport of plasma proteins in bone interstitial fluid is that they can help to buffer the pH of the fluid that is transported through the lacunar–canalicular system. Since there is a reduced availability of a direct blood supply, the fluid within the lacunar–canalicular system would be very highly susceptible to changes in any solute concentration based on the actions of cells within a localized region. Plasma proteins may help to reduce this susceptibility to changes in pH. Remember however, that any changes to the bone interstitial fluid compartment may have significant implications for bone homeostasis since, the clearance of these changes occurs over a very large length scale (e.g., the bone lymphatic system is only present on the periosteal surface).

The transport of hydrophilic and lipophilic molecules through bone is less complex and appears to be under less control than the transport of large plasma proteins. It is typical for small hydrophilic solutes to diffuse out of the capillary through the intercellular cleft and become evenly distributed throughout bone tissue. Local concentration gradients would account for the movement of solutes into or out of the osteocytes and toward the lymphatic system, which always acts as a sink for dissolved solutes. The transport of lipophilic solutes would follow an analogous path, except, this type of molecule would be transported through endothelial cells, which comprise the vascular wall. These molecules would be transported through the bone interstitial space with varying mechanisms that are dependent on the molecular structure and whether or not specific transport mechanisms exist for that molecule.

This leads us to the question of how is efflux of nutrients from the bone interstitial space into the periosteal capillaries or periosteal lymphatics regulated. We can imagine two scenarios for efflux of nutrients. The first is the case when the movement of solute is limited by the flow of bone interstitial fluid and the second is when the movement of solutes is limited by a barrier. For most solutes that have been studied it appears that the efflux of solutes out of the lacunar–canalicular system is limited by the transport through a barrier and cannot be completely described by clearance relationships. It is currently thought that in these cases, efflux occurs due to an energy-dependent mechanism, which functions based on ligand binding relationships. The efflux of potassium and calcium from the bone interstitial fluid follows these barrier-limited relationships.

11.7 DISEASE CONDITIONS

11.7.1 Synovitis

Synovitis is a condition that is characterized by an inflammation of the synovial membrane. As discussed above, the synovial fluid is secreted by the synovial membrane, and it is this process that is disrupted during inflammation of the synovial membrane. During synovitis, the permeability of the synovial membrane increases, allowing white blood cells into the joint space to defend against the infection. Under most instances of synovitis, an increase in the amount of polymorphonuclear leukocytes are the principle cells that are observed in a higher concentration within the synovial fluid. White blood cells in the joint space release enzymes that can degrade the cartilage and the proteoglycan core if the condition persists for an extended period of time. This will cause a great deal of pain and the overall lubrication of the joint will be reduced because (1) the cartilage will not store as much synovial fluid as normal, (2) the shock-absorbing function of the cartilage/synovial fluid will be reduced, and (3) bone may come into contact with bone. A secondary effect associated with synovial membrane breakdown is an increase in interstitial fluid movement into the joint space. Increases in interstitial fluid within the joint space will decrease the viscosity of the synovial fluid and increase the hydrostatic pressure within the joint space. A decreased synovial fluid viscosity will prevent the synovial fluid from protecting the joint during compressive loading. An increased hydrostatic pressure within the joint space acts to decrease the mobility of the joint. Combined, the overall effects of synovitis are a loss of the synovial fluid properties, a possible degradation of the cartilage, and a decreased joint mobility. Synovitis can be recognized by a swelling of the joints and is diagnosed by taking a sample of the synovial fluid from the patient. Minor synovitis is common during arthritis and is generally treated with anti-inflammatory medications.

In severe cases of synovitis, inflammatory cells and/or the products they release can burrow beneath the articular cartilage layer. Under these cases, a significant increase in the presence of collagenase and elastase proteins has been observed. In most instances, the subchondral bone density decreases through an enhancement of resorption processes. If the inflammatory processes are quelled it is possible for the damage to bone to be repaired, given sufficient time. Again, however, any damage to the articular cartilage persists due to the limited capability of chondrocytes to reform the specific proteins needed to maintain cartilage.

11.7.2 Bursitis/Tenosynovitis

Bursitis and tenosynovitis are a group of non-articular rheumatism, which result from the inflammation of the bursa or tendons that surround the joint capsule. In these cases, impairment to the joint and some of the fluid properties can be observed.

Bursae are closed sacs that are filled with synovial like fluid. These sacs can be found at sites of interaction between skin, ligaments, tendons, bone, and muscles. Inflammatory cells and/or mediators found during synovitis can migrate to the bursa inducing secondary inflammatory sites. It is also possible for bursitis to form under direct trauma to the bursa. In general, as the bursa becomes inflamed, its size will increase. This makes the bursa more susceptible to wear by friction and can induce a tear in the lining of the bursa. In most instances, bursitis is treated with mild anti-inflammatory mediators, either systemically or locally.

Tenosynovitis is an inflammation of the sheath that surrounds the tendons. These linings, much like bursa, are comprised of synovial linings and can experience inflammation similar to synovitis. When this sheath becomes inflamed it is common for the joint to "lock-up" and cause significant pain and decreases in locomotion. In some instances, the inflammation of the tendon is so severe that the tendon sheath must be removed surgically.

END OF CHAPTER SUMMARY

11.1 There are 206 bones in the human body that are divided into bones that compose the axial skeleton and those that compose the appendicular skeleton. All bones have the function to support and protect the body, store minerals, produce blood cells, and act as levers for motion. Long bones are divided into three major sections: the epiphysis, the metaphysis, and the diaphysis. The epiphysis is composed of spongy bone and can withstand loads from many different directions. The diaphysis is composed of compact bone and can withstand loads in one main direction. The metaphysis is the interface of these two regions. The bone extracellular matrix is composed of a hardened ground substance that is stiff due to the deposition of calcium salts. Hydroxyapatite crystals are a major component of the ground substance. Thirty-three percent of bone mass is accounted for by collagen fibers. There are four types of bone cells that can be identified. These are the osteocytes, the osteoblasts, the osteoclasts, and the osteoprogenitor cells. The action of these four cell types determines the chemical makeup of bone. Bone is composed of a Haversian canal system structural unit. Each of these units has a central canal that is filled by an artery and vein. Concentric circles of bone surround this canal and are termed lamellae. Canaliculi run perpendicular to the Haversian canal and function to connect bone cells to each other. The junctions between bones form joints where bones can move around each other. The coefficient of friction between bones is relatively high, and in fact, bones can wear on each other. The coefficient of friction is lowered through the presence of cartilage and synovial fluid.

11.2 Bone is perfused by a sparse capillary network that allows for the exchange of nutrients/wastes between blood and the bone cells. The vascular tree that is specific to bone can be subdivided into four general categories: the conduit vessels, the capillaries of bone, the lacunar–canalicular system, and finally the microcanalicular system. Within each Haversian canal system, a small arteriole (which is capable of exchange) or a capillary is present. The percentage of water found within bone has been shown to be related to the extent of bone mineralization. Higher bone mineralization values are accompanied by lower percentages of water. This can be quantified by an increase in bone density. Bone mineralization increases primarily because there is a higher quantity of hydroxyapatite, which fills the water space in forming bone. The vast majority of water in bone is not found within the bone, per say, instead it is found in the blood and the extravascular spaces.

11.3 Synovial fluid fills the space between joints and is produced as an ultra-filtrate of interstitial fluid. The interstitial fluid is filtered by the synovial membrane and is very similar in composition to interstitial fluid, except that there is a significantly higher concentration of proteoglycans. These proteoglycans act to organize water (because they are highly charged) so that the overall synovial fluid can withstand a higher load. Cartilage also has a property of organizing and holding synovial fluid within the joint.

11.4 Synovial fluid enters the joint space as a filtrate through the synovial membrane. This membrane is composed of two layers: the synovium and the sub-synovium. The synovium functions to prevent foreign particles from entering the joint space, while the sub-synovium is composed of an extensive extracellular fiber mesh that acts to filter the interstitial fluid. The synovial membrane is actually a unidirectional valve because fluid can enter the joint space through the membrane, but the fluid does not exit the joint space through the synovial membrane. To model flow through the synovial membrane, we use a modified flux equation that accounts for the degradation rate and formation rate of species within synovial fluid. This can be represented as

$$\frac{\partial(Vc)_i}{\partial t} = (r_s A_s + r_c A_c - d - J)_i$$

The flow of species through the synovial space must also take into account a restricted diffusion coefficient that is based on the restricted movement of ions through the joint space

$$D_i = De^{(-\sqrt{\theta})\left(1+\frac{a_i}{a}\right)}$$

11.5 Using standard engineering statics and dynamics courses, the forces within joints can be solved by a direct summation of the forces that act within the joint and a summation of the moments about a certain location within the joint:

$$\sum \vec{F} = m\,\vec{a}$$

and:

$$\sum \vec{M} = I\vec{\alpha}$$

Recall from a mechanics of materials course that the deformation of a material can be described by the stress, strain, and elastic modulus. This is formulated as

$$\sigma = \frac{F}{A}$$

and:

$$\varepsilon = \frac{\sigma}{E} = \frac{\Delta l}{l}$$

11.6 Fluid transport within the lacunar–canalicular system is key for the passage of nutrients, ions, and wastes to/from the cells that comprise bone tissue. Molecules are first transported across the endothelial cell barrier using typical methods. Once these nutrients enter the interstitial fluid, the fluid flows through the lacunar–canalicular complex, primarily using a diffusion based process. It appears that the lymphatic system does not enter the lacunar–canalicular complex, but instead resides on the periosteal surface, in close proximity to the periosteal capillaries and the peripheral lacunae. Large proteins are transported through bone capillaries at a higher rate than most systemic vascular beds. Small hydrophilic solutes to diffuse out of the capillary through the intercellular cleft and become evenly distributed throughout bone tissue.

11.7 Synovitis is a condition that is characterized by an inflammation within the synovial membrane. This process alters the formation of synovial fluid, and therefore, the overall lubrication of the joint space will be diminished. This would eventually lead to an abrasive wear of the cartilage and possible the bone itself. A second possible outcome of synovitis is that the quantity of synovial fluid that is produced remains the same but the composition of the fluid changes. Typically, there would be a decrease in the viscosity of the synovial fluid, which would tend to decrease its lubricating efficiency (and the fluid would not withstand the same amount of loads as under normal conditions). Bursitis and tenosynovitis are similar pathologies that effect the bursa and tendon sheaths, respectively.

HOMEWORK PROBLEMS

11.1 The formation of blood cells occurs within what component of the bones?

11.2 There are four critical functions of bones which include the protection of internal organs, support of the body, blood cell production, and regulation of internal calcium concentration. Discuss how bones conduct each of these four tasks.

11.3 What are the structural and functional differences between the diaphysis and the epiphysis?

11.4 Spongy bone can withstand stresses that arrive from multiple directions, whereas compact bone can normally withstand stresses in only one direction. Why is there this difference and what structures account for this difference?

11.5 What are the functions of the four types of bone cells?

11.6 We discussed some of the differences that are observed in the microvasculature (e.g., feeding arterioles and capillary beds) found in different bones and even within different regions of bone. Why do you suppose these differences are found? Does the anatomical organization of the bone and/or blood vessels solely account for these differences.

11.7 If the synovial fluid is not formed properly, there is the possibility for the cartilage at the ends of the bone to come into contact with each other. How can synovial flow be re-established and what would happen if it were not re-established?

11.8 Calculate the change in molecular volume of the synovial fluid if the formation rate of synovial fluid by the synovial membrane is 6.5 μg/h/cm^2 and by the cartilage is 73 ng/h/cm^2. Assume that the synovial membrane area is 15 mm^2 and that of the cartilage is 25 mm^2. The degradation rate of the synovial fluid is 23 ng/h, and the flux of synovial fluid out of the membrane is 0.97 μg/h.

11.9 Calculate the restricted diffusion coefficient for proteoglycans within the synovial fluid if the time rate of change of proteoglycans within synovial fluid is 0.095 mg/h. The free diffusion for proteoglycans in water is $1.25_E{-}7$ cm^2/s. Assume that the degradation rate for proteoglycans is negligible and that the formation rate for proteoglycans within the synovial membrane and the cartilage is 2.2 μg/h/cm^2 and 31 ng/h/cm^2, respectively. The area of the synovial membrane is 15 mm^2 and that of the cartilage is 25 mm^2. The permeability of this proteoglycan through the synovial membrane is $1.23_E{-}6$ cm/s, and the concentration gradient for this proteoglycan is 0.07 mg/mL.

11.10 Consider a person that is standing on their tiptoes on one foot. The forces acting on this joint are their weight (175 lb), the force of the muscle (F_M) acting through the Achilles tendon, and the joint reaction force (F_J). Figure 11.9 depicts the location and directions that the forces act on the ankle joint. The angle that the Achilles tendon makes in this position is 40° and the angle that the joint reaction force acts on the ankle is 75°. Calculate the muscle force and the joint reaction force needed to maintain this position. Ignore the weight of the foot itself and geometric considerations.

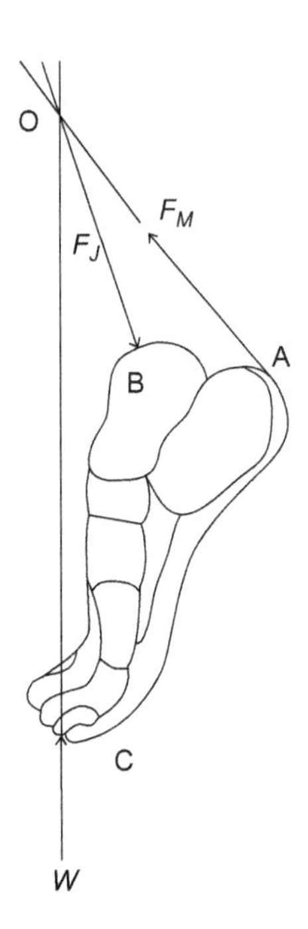

FIGURE 11.9 Diagram of a foot for homework problem 11.9. *Source: Adapted from Ozkaya and Nordin (1999).*

11.11 Calculate the stress on the cartilage and the change in length of cartilage, assuming that the force on the cartilage is 9875 N and that the diameter of the cartilage is 2 cm (assume that the cartilage has a circular area). The cartilage has a thickness of 1.5 mm and an elastic modulus of 250 MPa.

11.12 Using the results from the previous example, calculate the change in molecular volume of synovial fluid if all of the criteria are the same as in homework problem 11.7, except that the formation of synovial fluid from cartilage now includes 142 ng/h/mm from compression.

11.13 Discuss the movement of solutes within and to bone cells. Why is a robust lymphatic system missing from the Haversian canal system? What do you think is the primary mode of delivery and re-up-take of solutes in bone?

11.14 We described some of the mechanical properties of bone using elastic deformation principles. In your opinion, do these calculations suitably account for the mechanical deformations/loading conditions that bone experiences. If so, why? If not, propose some other variables that should be considered when quantifying loading dynamics on bones and discuss why you believe these parameters are important.

References

[1] S. Akizuki, V.C. Mow, F. Muller, J.C. Pita, D.S. Howell, D.H. Manicourt, Tensile properties of human knee joint cartilage: I. Influence of ionic conditions, weight bearing, and fibrillation on the tensile modulus, J. Orthop. Res. 4 (1986) 379.

[2] S. Akizuki, V.C. Mow, F. Muller, J.C. Pita, D.S. Howell, Tensile properties of human knee joint cartilage. II. Correlations between weight bearing and tissue pathology and the kinetics of swelling, J. Orthop. Res. 5 (1987) 173.

[3] E. Amtmann, The distribution of breaking strength in the human femur shaft, J. Biomech. 1 (1968) 271.

[4] E. Amtmann, Mechanical stress, functional adaptation and the variation structure of the human femur diaphysis, Ergeb. Anat. Entwicklungsgesch 44 (1971) 1.

[5] E. Amtmann, On functional adaptation of long bones. Investigations on human femora, Gegenbaurs. Morphol. Jahrb 117 (1972) 224.

[6] G.A. Ateshian, V.C. Mow, Friction, lubrication and wear of articular cartilage and diarthrodial joints, in: G.A. Ateshian, V.C. Mow (Eds.), Basic Orthopaedic Biomechanics and Mechano-biology, third ed., Lippincott Williams & Wilkins, Philadelphia, PA, 2005, p. 447.

[7] C.A. Baud, Submicroscopic structure and functional aspects of the osteocyte, Clin. Orthop. 56 (1968) 227.

[8] D.R. Carter, W.C. Hayes, D.J. Schurman, Fatigue life of compact bone—II. Effects of microstructure and density, J. Biomech. 9 (1976) 211.

[9] D.R. Carter, W.C. Hayes, Fatigue life of compact bone—I. Effects of stress amplitude, temperature and density, J. Biomech. 9 (1976) 27.

[10] D.R. Carter, W.C. Hayes, Bone compressive strength: the influence of density and strain rate, Science 194 (1976) 1174.

[11] D.R. Carter, D.M. Spengler, Mechanical properties and composition of cortical bone, Clin. Orthop. Relat. Res (1978) 192.

[12] R.R. Cooper, J.W. Milgram, R.A. Robinson, Morphology of the osteon: an electron microscopic study, J. Bone Joint Surg. Am. 48 (1966) 1239.

[13] S.C. Cowin, The mechanical and stress adaptive properties of bone, Ann. Biomed. Eng. 11 (1983) 263.

[14] S.C. Cowin, Mechanical modeling of the stress adaptation process in bone, Calcif. Tissue Int. 36 (Suppl. 1) (1984) S98.

[15] S.C. Cowin, Wolff's law of trabecular architecture at remodeling equilibrium, J. Biomech. Eng. 108 (1986) 83.

[16] S.C. Cowin, A.M. Sadegh, G.M. Luo, An evolutionary Wolff's law for trabecular architecture, J. Biomech. Eng. 114 (1992) 129.

[17] J.D. Currey, Metabolic starvation as a factor in bone reconstruction, Acta Anat. (Basel) 59 (1964) 77.
[18] J.D. Currey, Some effects of ageing in human Haversian systems, J. Anat. 98 (1964) 69.
[19] P.P De Bruyn, H.P.C. Breen, T.B. Thomas, The microcirculation of the bone marrow, Anat. Rec. 168 (1970) 55.
[20] L. Dintenfass, Lubrication in synovial joints, Nature 197 (1963) 496.
[21] S.B. Doty, B.H. Schofield, F.H. Chen, Metabolic and structural changes within osteocytes of rat bone, in: R.V. Talmage, P.L. Munson (Eds.), Calcium, Parathyroid Hormone and the Calcitonins, Excerpta Medica, Amsterdam, 1972, pp. 353–364.
[22] F.G. Evans, The mechanical properties of bone, Artif. Limbs 13 (1969) 37.
[23] X.E. Guo, L.C. Liang, S.A. Goldstein, Micromechanics of osteonal cortical bone fracture, J. Biomech. Eng. 120 (1998) 112.
[24] X.E. Guo, S.A. Goldstein, Vertebral trabecular bone microscopic tissue elastic modulus and hardness do not change in ovariectomized rats, J. Orthop. Res. 18 (2000) 333.
[25] A. Keith, Hunterian lecture on Wolff's law of bone transformation, Lancet 16 (1918) 250.
[26] J.A. Lopez-Curto, J.B. Bassingthwaighte, P.J. Kelly, Anatomy of the microvasculature of the tibial diaphysis of the adult dog, J. Bone. Joint Surg. Am. 62 (1980) 1362.
[27] C.W. McCutchen, A note upon tensile stresses in the collagen fibers of articular cartilage, Med. Electron. Biol. Eng. 3 (1965) 447.
[28] C.W. McCutchen, Cartilage is poroelastic, not viscoelastic (including an exact theorem about strain energy and viscous loss, and an order of magnitude relation for equilibration time), J. Biomech. 15 (1982) 325.
[29] C.W. McCutchen, Joint lubrication, Bull. Hosp. Joint Dis. Orthop. Inst. 43 (1983) 118.
[30] V.C. Mow, W.M. Lai, Some surface characteristics of articular cartilage. I. A scanning electron microscopy study and a theoretical model for the dynamic interaction of synovial fluid and articular cartilage, J. Biomech. 7 (1974) 449.
[31] V.C. Mow, W.M. Lai, J. Eisenfeld, I. Redler, Some surface characteristics of articular cartilage. II. On the stability of articular surface and a possible biomechanical factor in etiology of chondrodegeneration, J. Biomech. 7 (1974) 457.
[32] V.C. Mow, W.Y. Gu, F.H. Chen, Structure and function of articular cartilage and meniscus, in: G.A. Ateshian, V.C. Mow (Eds.), Basic Orthopaedic Biomechanics and Mechano-Biology, third ed., Lippincott Williams & Wilkins, Philadelphia, PA, 2005, p. 181.
[33] A.G. Ogston, J.E. Stanier, On the state of hyaluronic acid in synovial fluid, Biochem. J. 46 (1950) 364.
[34] A.G. Ogston, J.E. Stanier, The physiological function of hyaluronic acid in synovial fluid; viscous, elastic and lubricant properties, J. Physiol. 119 (1953) 244.
[35] N. Ozkaya, M. Nordin, Fundamentals of Biomechanics: Equilibrium, Motion, and Deformation, second ed., Springer, New York, NY, 1999.
[36] H. Roesler, The history of some fundamental concepts in bone biomechanics, J. Biomech. 20 (1987) 1025.
[37] A. Unsworth, D. Dowson, V. Wright, Some new evidence on human joint lubrication, Ann. Rheum. Dis. 34 (1975) 277.
[38] P.S. Walker, D. Dowson, M.D. Longfield, V. Wright, "Boosted lubrication" in synovial joints by fluid entrapment and enrichment, Ann. Rheum. Dis. 27 (1968) 512.
[39] P.S. Walker, D. Dowson, M.D. Longfield, V. Wright, Lubrication of human joints, Ann. Rheum. Dis. 28 (1969) 194.
[40] P.S. Walker, J. Sikorski, D. Dowson, M.D. Longfield, V. Wright, T. Buckley, Behaviour of synovial fluid on surfaces of articular cartilage. A scanning electron microscope study, Ann. Rheum. Dis. 28 (1969) 1.
[41] P.S. Walker, J. Sikorski, D. Dowson, D. Longfield, V. Wright, Behaviour of synovial fluid on articular cartilage, Ann. Rheum. Dis. 28 (1969) 326.
[42] P.S. Walker, J. Sikorski, D. Dowson, M.D. Longfield, V. Wright, Features of the synovial fluid film in human joint lubrication, Nature 225 (1970) 956.

CHAPTER

12

Flow Through the Kidney

LEARNING OUTCOMES

1. Describe the physiology and the anatomy of the kidneys
2. Explain the unique characteristics of the renal circulation
3. Identify the functional unit of the kidney
4. Detail the flow of blood and its distribution throughout the kidney and to the nephron
5. Evaluate the flow through a nephron
6. Model the forces that alter glomerular filtration
7. Examine the feedback mechanisms within glomerular filtration
8. Calculate the filtered load for various metabolically important compounds
9. Distinguish between tubule reabsorption and secretion
10. Examine the mechanisms for tubule reabsorption
11. Formulate a function for the amount of a particular substance excreted
12. Describe the filtration rate of a single nephron and how this relates to the overall kidney function
13. Explain the flow of blood through the peritubular capillary system
14. Compare feedback mechanisms for sodium absorption
15. Discuss the autoregulation of blood flow and urine formation within the kidney
16. Use compartmental analysis to describe the movement of solutes throughout the nephron
17. Model multicompartment systems
18. Describe extracorporeal flows
19. Discuss disease conditions that commonly affect the kidney

12.1 KIDNEY PHYSIOLOGY

The kidneys perform two critical functions for the body. The first is the removal of metabolic waste products, foreign particles, and metabolites from the body. If the kidneys are functioning properly, the removal of these waste products should occur as fast as they are

 © 2015 Elsevier Inc. All rights reserved.

being generated through cellular metabolism, so that there is no buildup of toxins within the body. Under normal conditions, the processes that occur within the kidneys occur at about this rate. The second critical function that the kidneys perform is the maintenance of the blood volume within the body and the regulation of the concentration of physiologically important molecules within blood. This is maintained as a balance between what is ingested versus what is excreted from the body (e.g., after a person eats five bananas there may be a large excess of potassium ions within the body, and this excess would normally be removed by the kidneys). The first function occurs through a filtration of the blood plasma. The second function occurs through the reabsorption of molecules from the filtrate at a rate that is dependent on the need of the body combined with an added secretion of unnecessary compounds from blood into the filtrate of blood. Following the banana example, when potassium ions are in excess, the reabsorption of potassium will be low; however, if the potassium ion concentration is depleted within the blood plasma, then the reabsorption of potassium will be higher. At the same time, the secretion of potassium may be altered to dictate the potassium ion concentration.

The kidneys are located along the posterior wall of the abdomen and are approximately the size of a clenched fist. Each kidney is indented at its interior medial location, and this location is termed the hilum (Figure 12.1). At the hilum, the renal artery, which branches directly from the aorta, and the renal nerve enter the kidney. Also at this location, the renal vein, which converges with the inferior vena cava, the renal lymphatic vessels, and the ureter leave the kidney. The kidney is then divided into two main sections: the outer cortex and the medulla. The outer cortex is the location where plasma is filtered, while reabsorption and excretion generally occurs within the medulla (note that there are two main types of neurons, the cortical nephron and the juxtamedullary nephron, which filter/secrete nutrients and wastes in slightly different locations and with slightly different mechanisms). Each renal artery enters the kidney at the hilum and branches to form smaller arteries until they penetrate into the entire outer cortex of the kidney (the kidney has a unique blood supply that will be described in Section 12.2). At this point, the vessels are the size of arterioles. The smallest arterial blood vessel within the kidney circulation is termed the afferent arteriole, which supplies each nephron (the functional unit of the kidney) with blood. The afferent

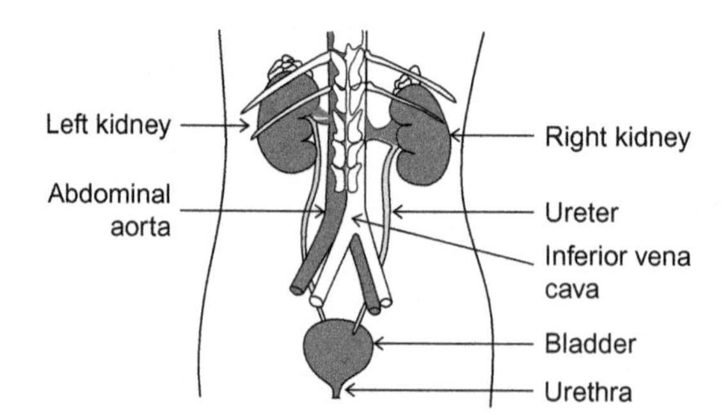

FIGURE 12.1 Anatomical location (posterior view) of the kidneys, showing their location in relation to other structures in the abdomen. The kidneys are protected by the eleventh and twelfth ribs (both shown). Urine formed in the kidneys passes through the left and right ureter into the bladder. As the bladder fills, urine can be removed from the body via the urethra. *Adapted from Martini and Nath (2009).*

arterioles branch into the glomerular capillaries, which then converge to an efferent arteriole. A second capillary network, termed the peritubular capillaries (feeding the cortical nephrons) or vasa recta (feeding the juxtamedullary nephron), surrounds the nephron and then converges to venules. These venules converge to larger veins and eventually the renal vein, which exits the kidney at the hilum. Interestingly, slightly more than 20% of the cardiac output is directed to both kidneys under normal resting conditions. This supports the important role for the kidneys in maintaining homeostasis.

A unique characteristic of the renal circulation is that it has two capillary beds, which are separated by an arteriole. Interestingly, the efferent arteriole primarily regulates the pressure in both capillary beds and hence the flow rate throughout each capillary network. The afferent arteriole, to a lesser extent, regulates the pressure and flow throughout both capillary networks. Unlike the systemic and respiratory circulatory systems, the hydrostatic pressure within the glomerular capillaries (the first capillary bed) is very high, around 60 mmHg. This allows for a rapid filtration of fluid and small molecules from plasma (termed the plasma filtrate), because there is a very high hydrostatic pressure gradient between the blood and the filtrate in the nephrons (this filtration can be described by Starling's law, as well). The pressure within the peritubular capillaries (the second capillary bed) is much lower, close to 15 mmHg (which is similar to previously discussed capillary beds), which allows for rapid absorption of any filtered molecules. Changes within the cross-sectional area of the afferent

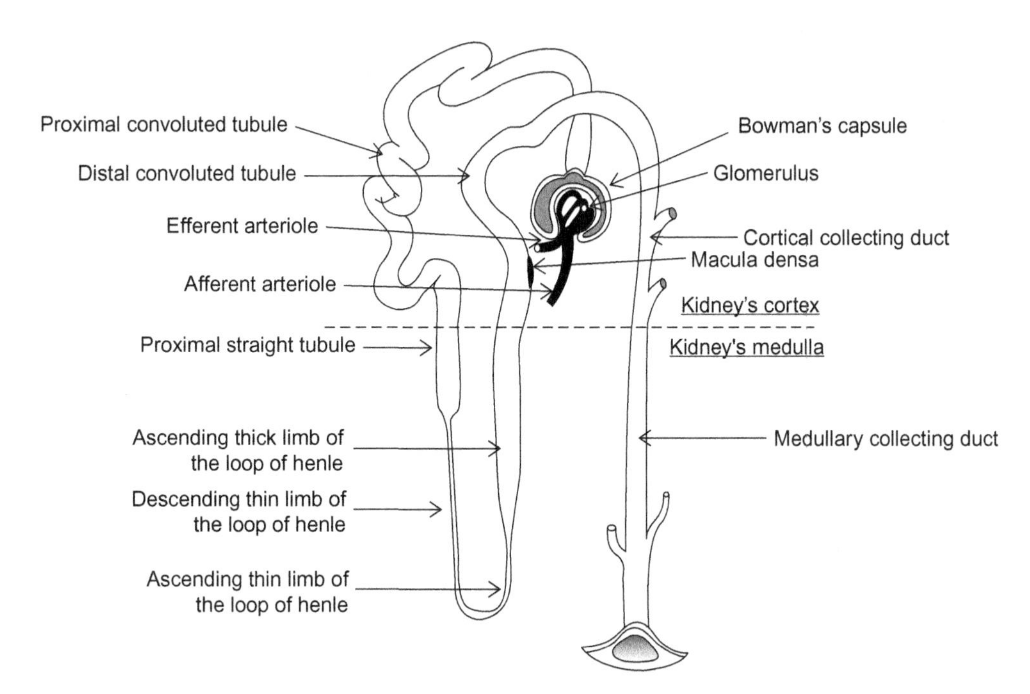

FIGURE 12.2 Schematic of the functional unit of a kidney: the nephron. This figure illustrates the anatomical arrangement of the nephron. Notice that the macula densa is a grouping of cells and not a particular region of the kidney. Blood is filtered in the glomerulus and the filtrate passes through the nephron (proximal convoluted tubule, Loop of Henle, distal convoluted tubule, and collecting duct) to form urine. *Adapted from Martini and Nath (2009).*

arteriole and the efferent arteriole can drastically affect the pressure within the capillary beds, acting as a regulator for the filtration and absorption rates.

The nephrons are located within the cortex and medulla of the kidney (Figure 12.2). There are nearly 2 million nephrons within the adult body, and each one has the ability to form urine. Nephrons do not regenerate after birth; therefore, with injury, disease, and aging there is a loss of the quantity of nephrons within the body (to about 50% of the total quantity at the age of 80). The kidneys and the remaining functioning nephrons can adapt to this decrease in functional unit quantity and still produce enough urine properly to remove wastes/toxins. If the kidneys do not function properly and wastes begin to accumulate in the body, dialysis or surgical procedures may be recommended (see Section 12.10/12.11). Each nephron is composed of two major parts. The first is the glomerulus, which consists of the glomerular capillaries and Bowman's capsule. The glomerulus functions to filter fluid and compounds from blood. The second major component of the nephrons is a tortuous tube which modifies the filtered fluid into urine through differential reabsorption and secretion of metabolically important molecules.

The glomerular capillaries are very similar to all other capillaries within the body, except that they are under a very high hydrostatic pressure (approximately 60 mmHg). Also, unlike other capillary beds, the glomerular capillaries are surrounded by epithelial cells (termed podocytes). In one sense, these capillaries are not only composed of a single endothelial cell wall, but they also have a second layer composed of epithelial cells. However, the epithelial layer does not affect the flow within the blood vessel but functions to partially regulate the blood filtration (see Section 12.3). The glomerular capillary epithelial cells are surrounded by Bowman's capsule, which is connected to the nephron tubule system. All fluid and compounds that are filtered from the glomerular capillaries must pass through the endothelial cell barrier and the epithelial cell barrier before entering Bowman's capsule. The entire glomerulus structure lies within the cortex of the kidney. A term given to the glomerulus and Bowman's capsule is the renal corpuscle.

Any fluid that enters Bowman's capsule is passed directly into the continuous renal tubule. The renal tubule is a narrow-diameter hollow cylinder that is composed of a single layer of epithelial cells, which are attached to a basement membrane. There are 10 different types of epithelial cells that are found within the nephron. Each of these epithelial cell types form a continuous segment of the nephron, each with a slightly different function during urine formation. These 10 segments are typically grouped into 5 tubule systems, each with their own common function. The first epithelial cell type is found within Bowman's capsule (the first tubule system) and functions to shunt the plasma filtrate into the proximal tubule. These epithelial cells have little to no role in reabsorption/secretion. The proximal tubule (the second tubule system) is broken up into two segments, each with its own specific epithelial cell composition. The first segment is the proximal convoluted tubule, and the second segment is the proximal straight tubule. The third tubule system is termed the Loop of Henle, which forms a very sharp 180-degree turn within the nephron. The Loop of Henle is broken up into the thin descending limb, the thin ascending limb, and the thick ascending limb. Very specialized cells, termed the macula densa cells, are located within the thick ascending limb of the Loop of Henle (more details regarding this specialized location will be discussed later). The fourth tubule system is termed the distal convoluted tubule, which is only composed of one segment given the same name. The last tubule system is termed the collecting duct system and is composed of the connecting tubule, the cortical collecting tubule, and the medullary collecting tubule. As the reader can imagine, each epithelial cell type has a different permeability toward compounds within the plasma filtrate, which is termed glomerular filtrate, within the nephron tubule system, thus altering the particular segment's function from the preceding segment and the following segment.

In subsequent discussions, it will become apparent that there are important differences in nephron function based on the regional location within the kidney. For reference, the renal corpuscle, the proximal convoluted tubule, the distal portion of the thick ascending Loop of Henle, the distal convoluted tubule, the connecting tubule, and the cortical collecting tubule are all located within the kidney's cortex. The proximal convoluted tubule, the thin descending limb, the thin ascending limb, the proximal portion of the thick ascending Loop of Henle, and the medullary collecting tubule are all located within the kidney's medulla. It is also important to note that the entire tubule system is surrounded by the peritubular capillaries or the vasa recta (although these are not shown in Figure 12.2). For reference, in humans approximately 85% of the nephrons are cortical nephrons surrounded by the peritubular capillary system and 15% of the nephrons are juxtamedullary nephrons surrounded by the vasa recta.

One of the most unique locations within the nephron unit is termed the juxtaglomerular apparatus (Figure 12.3). At this location, the thick ascending Loop of Henle passes in between the afferent arteriole and the efferent arteriole, which feeds the glomerular capillaries and the peritubular capillaries of the same nephron. In the thick ascending Loop of Henle, there are specialized epithelial cells termed the macula densa cells, which are in direct contact with specialized cells present within the afferent arteriole. The afferent arteriole specialized cells are termed the juxtaglomerular cells. This location provides a feedback mechanism within each nephron to control the filtration rate of the glomerular capillaries of that nephron. The exact details of the feedback mechanism are unknown, but what is clear is that when the flow decreases within the Loop of Henle, there is an increased reabsorption of sodium and chloride ions into the peritubular capillaries (more details regarding reabsorption can be found in Section 12.4). An increased reabsorption of these ions within the early tubule sections decreases the concentration of these ions within the plasma filtrate passing through the ascending Loop of Henle. The macula densa cells are sensitive to the

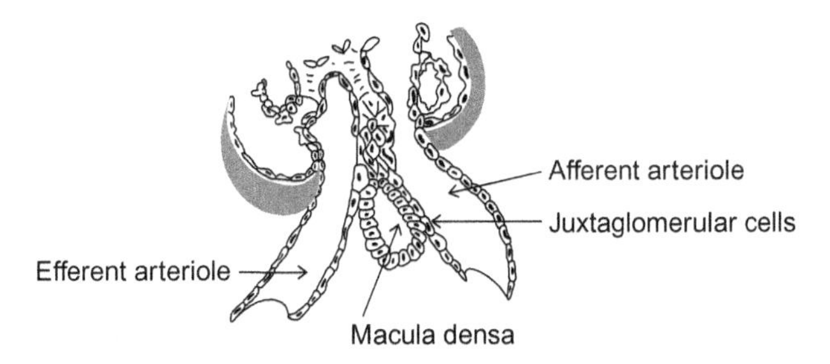

FIGURE 12.3 Anatomical arrangement of the juxtaglomerular apparatus, which can monitor and regulate the glomerular filtration rate, through the production of renin. The thick ascending Loop of Henle passes between the afferent and efferent arterioles of the same nephron to form the juxtaglomerular apparatus. Depending on the concentration of sodium ions and chloride ions, the macula densa cells can either increase or decrease the production of renin. *Adapted from Martini and Nath (2009).*

sodium and chloride concentrations, and when either concentration reduces, a signal is sent from the macula densa that has two effects on nephron function. The first effect is that the resistance of the afferent arteriole increases (e.g., the diameter is reduced), thereby decreasing the amount of filtrate within the tubule system. The second effect is the production of renin from the juxtaglomerular cells. Renin is an enzyme that causes an increased production of angiotensin I, from angiotensinogen, which is then rapidly converted to angiotensin II in the kidney/lungs. Angiotensin II formation is mediated by the angiotensin-converting enzyme (ACE), which is a common target for cardiovascular disease drugs. Angiotensin II constricts the systemic arterioles, which increases the total peripheral resistance and thus decreases blood flow throughout the systemic circulation. The net effect of this is to decrease the filtration of sodium into the kidney and thus increase the retention of sodium in plasma. Angiotensin II has other functions throughout the body, such as an increased release of aldosterone from the adrenal glands. If the processes that control the flow through the glomerular capillaries are functioning properly, the glomerular filtration rate changes marginally with large changes in systemic arterial pressure, and therefore, the kidneys continue to produce urine at a normal rate. Molecular concentrations, which partially dictate osmotic pressure, are the more important contributor to glomerular filtration rate.

12.2 DISTRIBUTION OF BLOOD IN THE KIDNEY

As we briefly mentioned in the previous section, the renal artery brings a direct supply of blood into the kidneys. This major artery branches directly from the aorta and receives close to 20% of the cardiac output under physiological conditions. Immediately after the renal artery passes through the hilum, the artery branches into interlobar arteries, which pass between the lobes of the kidney (Figure 12.4). The interlobar arteries pass through the kidney and at the border between the medulla and the cortex, the arteries bifurcate and

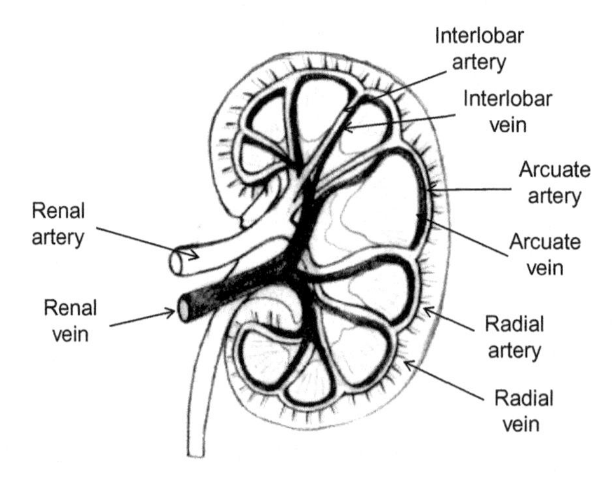

FIGURE 12.4 Schematic of the renal vasculature, which comprises of one renal artery that bifurcates into the interlobar arteries. The interlobar arteries bend over the renal pyramids to form the arcuate arteries, which have the radial arteries coming off of them. The afferent arterioles to each glomerulus form from the radial arteries. The venous system follows a similar path as the arterial system.

bend, connecting to neighboring, bifurcated, and bent interlobar arteries. The vessels, which are now termed arcuate arteries, follow the boundary between the medulla and cortex of the kidney. Many small arteries branch off of the arcuate arteries, which bring blood towards the outer cortex of the kidney. These small arteries branch perpendicularly off of the arcuate arteries and are now termed radial arteries. The afferent arterioles that enter the nephron units branch off of the arcuate arteries.

The venous circulation through the kidney follows a similar path, but in reverse to the artery path. The venous portion of the peritubular capillaries and the vasa recta converge into radial veins, which converge into one larger arcuate vein. The arcuate veins run parallel to the arcuate arteries and converge in between the renal lobules into the interlobar veins. Near the hilum of the kidney, the interlobar veins converge into the renal vein. All of these vessels can be found on Figure 12.4 as well.

We highlighted briefly the difference between cortical nephrons and juxtamedullary nephrons. One of the principle differences is that the peritubular capillaries of the cortical nephrons do not penetrate as deep into the medulla as the vasa recta capillaries do for the juxtamedullary nephrons (the peritubular capillaries and the vasa recta capillaries are the second capillary network for the different nephron types). For cortical nephrons, the efferent arteriole diverges into an anastomosing capillary network. This network, termed the peritubular capillaries, is primarily used as a source of exchange for nutrients/wastes through reabsorption/secretion processes. To a lesser extent these capillaries supply nutrients to the renal tissue. In contrast, the vasa recta of juxtamedullary nephrons are primarily used to (i) supply deep renal tissue with nutrients, (ii) maintain the osmotic concentration gradients necessary for reabsorption/secretion of all the nephrons. Note that these important functions are accomplished by approximately 15% of the nephrons reserving the vast majority of the nephrons for blood filtration, absorption of metabolically important nutrients, and removal of wastes/toxins. This uneven distribution of blood flow within the kidneys has been observed experimentally and some have suggested that during pathological events the distribution of blood between the cortex and the medulla is altered. It is perhaps even more important to recognize that not all nephrons, even those defined as cortical nephrons or juxtamedullary nephrons, function the same and part of this functional difference is possibly due to the difference in blood flow distribution to each of the nephrons.

Originally, researchers speculated that blood flow to different nephrons was determined based on the size of the glomerular capillaries; the glomerular capillary beds of juxtamedullary cortex nephrons are larger in size than the glomerular capillaries of the cortical nephrons. In parallel with this, it had been observed that the nephron filtration rates are significantly greater for the juxtaglomerular nephrons as compared with the cortical nephrons. It has been speculated that part of this uneven distribution of blood flow is due to the longer length of the tubule system that is accompanied by a longer length of capillaries (e.g., the vasa recta tend to have a longer length than the peritubular capillaries). However, this has been difficult to observe under physiological conditions. A second factor that may play a role in this uneven distribution is that the preglomerular vessels (e.g., afferent arterioles) have different size distributions under resting conditions. As we have learned in previous sections, the diameter of the blood vessels is the most effective way that the cardiovascular system controls the perfusion rate through that same vessel; therefore it follows that vascular beds with different perfusion rates should have different resting diameter values. This has also been somewhat difficult to observe, although some groups have used various size microspheres (in the range of 15–50 μm in diameter) to observe where they get trapped within the renal circulation.

The blood flow through the medulla of the kidney has some added functions that were not considered for cortical blood flow. Not only does medullary blood flow provide nutrients and remove wastes from this tissue source, medullary blood flow must remove excess water that has been reabsorbed along the collecting duct length and it must accomplish this task without disturbing the osmotic pressure gradient of the medulla interstitial space. Recall that the osmotic pressure gradient is used to maintain the countercurrent exchange gradient to absorb/secrete materials from each nephron. These apparently contradictory roles absorb water and maintain the osmotic gradient, are primarily completed by the vastly different vasculature of the outer medulla as compared with the inner medulla. To orient yourself to the anatomical arrangement of the nephron within the medulla, the outer medulla primarily consists of the thick ascending limb of the Loop of Henle, the proximal straight tubule and portions of the medullary collecting duct, whereas the inner medulla primarily consists of the thin descending and thin ascending limb of the Loop of Henle and the distal portions of the medullary collecting duct (refer to Figure 12.2). All of the medullary vasculature arises from the efferent arteriole (primarily from the juxtamedullary nephrons). In the outer medulla these capillary beds form extensive anastomosing networks that surround the nephron tubule system. These vessels converge into larger veins that ascend along the tubular section that join with the venous system of peritubular capillaries within the cortex to form the radial veins. In contrast, the vasculature of the inner medulla is composed of a few small vascular plexuses that surround the different levels of the thin limb sections of the Loop of Henle and the distal medullary collecting duct. These vessels arise from the vessels distal to the efferent arteriole, but have a very different organization, hence the differentiation between the peritubular capillaries (which are more extensive, more branching and surround a large section of the nephron tubule) and the vasa recta (which are less branching, less extensive and surround well-defined sections of the tubule system, in well-defined arrangements; note that vasa recta is a Latin word that translates to straight vessel). The venous flow from the vasa recta converges directly with arcuate veins.

Many groups have investigated the hydrostatic and oncotic pressure differences between the descending and ascending sections of the vasa recta. The hydrostatic pressure difference between the descending and ascending vasa recta is in the range of 2–4 mmHg and is in the range of 10–15 mmHg. The oncotic pressure difference is typically in the range of 10 mmHg and varies from 20 to 30 mmHg under normal conditions. The large difference in oncotic pressure has been attributed to the reabsorption of water in the vasa recta along the ascending limb. As described above, there is also a large difference in flow between the vasa recta (and hence the medulla) and the peritubular capillaries (the cortex). This has experimentally been shown to be caused by the many parallel branches of capillaries that arise from one efferent vessel; it is typical for more than 20 vessels to diverge from the efferent arteriole). Recall though that the total perfusion of the kidney is quite high and we are making the point that the medullary blood flow is slow as compared with the cortical blood flow. The cortical blood flow rate is in the range of 0.5 mL min^{-1} $(\text{g tissue})^{-1}$, which is similar to brain perfusion rates and considerably higher than most tissues under resting conditions. The transit time for a single red blood cell through the vasa recta is on the order of 30–40 s, whereas the transit time for a single red blood cell in the peritubular capillaries is on the order of 1–2 s. This slow transit times allows for the equilibration of materials within the descending and ascending limbs of the vasa recta, preventing the loss of the osmotic gradient within the kidney interstitial space. Thus, the microcirculation of the kidney is highly specialized for the salient functions that the kidney accomplishes.

12.3 GLOMERULAR FILTRATION/DYNAMICS

The basic function of the nephron can be divided into three processes: filtration, reabsorption, and secretion. Filtration occurs within the glomerulus of all nephrons and will be the focus of this section. Reabsorption and secretion occur within the tubule section of the nephron and will be discussed in Section 12.4. Filtration is the process where some components of plasma enter the nephron tubule system. Reabsorption is when compounds are removed from the filtrate and re-enter the peritubular capillaries or vasa recta capillary system. Secretion is when compounds are again removed from plasma of the peritubular or vasa recta capillary system, and then enter the tubule system of the nephron. It is important to note that for each substance in the blood (except for cells and proteins), there is a particular combination of filtration, tubule reabsorption, and tubule secretion that occurs to remove the compound from the body or maintain the compound in the body (Figure 12.5). Under normal conditions, cells and proteins should not enter the nephron tubule system from the glomerular filtrate.

The glomerulus acts as a filter for all of the substances within plasma. The glomerular filtrate contains all of the compounds in plasma, except for proteins, at nearly the same concentration as found in plasma. Glomerular filtration is a bulk flow process, where

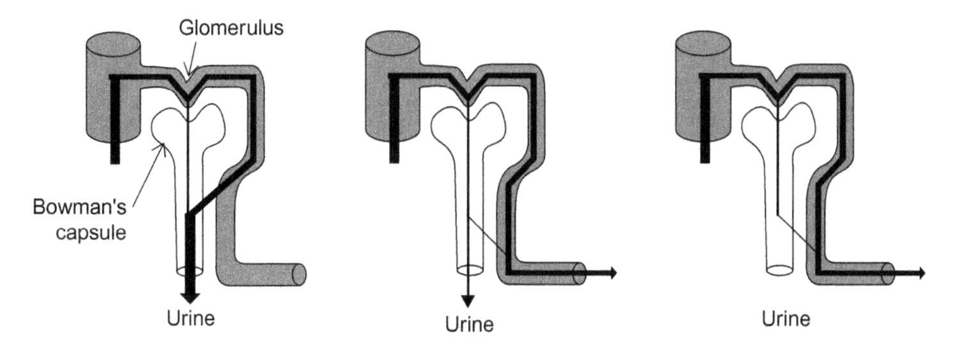

FIGURE 12.5 Three possible scenarios for glomerular filtration, tubule secretion, and tubule reabsorption. In the first case, a small amount of substance is filtered and a large amount is secreted, so that there is a minimal concentration left in the peritubular capillaries. In the second scenario, some of the substance is filtered and reabsorbed. In the last case, all of the filtered substance is reabsorbed. Urea would fall under the first case (although it is not completely secreted), bicarbonate would be an example for the second case, and glucose would be an example for the third case. *Adapted from Widmaier et al. (2007).*

water and all of the dissolved components can move freely between the vascular space and Bowman's capsule. Proteins do not move with this filtrate process because most of the pores within the glomerular capillaries are negatively charged and the majority of plasma proteins also have a partial negative charge. Also, most high molecular weight compounds cannot move through the pore space themselves due to a size restriction. As stated above, the majority of low molecular weight compounds can move freely into Bowman's capsule except for those molecules that are associated with plasma proteins. For instance, the majority of plasma calcium ions is associated with plasma proteins and is therefore not filtered by the glomerulus. Calcium ions can be actively secreted into the nephron along the peritubular capillary system, if necessary.

The effective flow rate of blood into the glomerular capillaries from the afferent arteriole is in the range of 600 mL/min. The effective glomerular filtration rate is on the order of 120 mL/min, thus approximately 20% of the plasma is removed within the glomerulus to form the glomerular filtrate that enters the nephron tubule system. The efferent arteriole flow rate is in the range of 480 mL/min. The question remains, how do the glomerular capillaries accomplish this task and how do the glomerular capillaries restrict the movement of species into the filtrate. The barrier to diffusion out of the glomerular capillaries consists of three layers: the endothelial cells that comprise the capillaries, a basement membrane that surrounds the capillaries, and podocytes that surround the basement membrane (Figure 12.6).

Endothelial cells that comprise the glomerular capillaries are fenestrated capillaries, which are capillaries that have pores throughout the entire cell. These pores are designed to have negligible resistance to fluid flow, but due to the pore size they restrict the

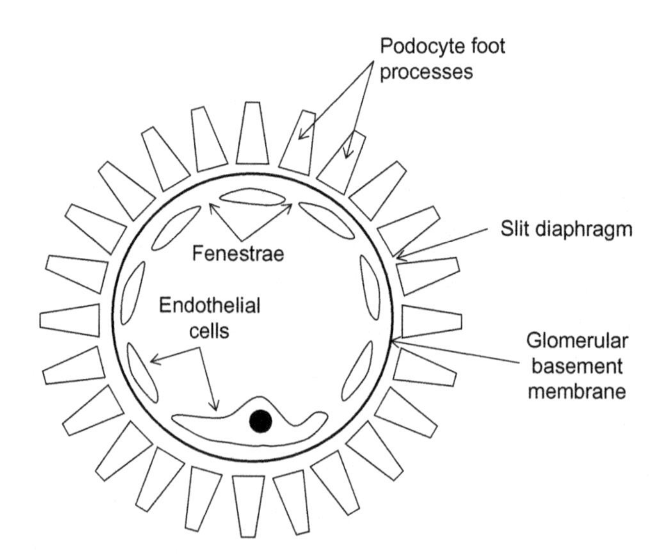

FIGURE 12.6 Cross-section of a glomerular capillary surrounded by a basement membrane and foot processes of podocytes. The integrity of this barrier is critical to the processing of plasma and the production of the glomerular filtrate. For each particular molecule, a sieving coefficient can be defined that can be used to describe the filtration of that compound.

movement of all cellular matter. The diameter of the pores is in the range of 100 nm, which allows the passage of all plasma constituents including plasma proteins, and the fenestrae cover nearly 20–25% of the entire capillary surface area. The basement membrane is primarily composed of collagen, laminin, and proteoglycans (heparin sulfate in the case of the kidney basement membrane), forming a layer that is approximately 200 nm thick. The net negative charge of the proteins that comprise the basement membrane restricts the movement of main proteins further through the filtration barrier.

Podocytes make up the remaining portion of the filtration barrier. Podocytes extend projections, termed foot processes, onto the glomerular basement membrane. The space between neighboring foot processes forms specialized pore structures termed slit diaphragms, which contain a variety of proteins, such as podocin and nephrin. The diameter of these pore structures is in the range of 30 nm. The net charge on these proteins is negative restricting the motion of negatively charged molecules through the slit diaphragm. Additionally, the density of these proteins within the slit diaphragm is relatively high, which allows this barrier to also restrict molecules based on their size. The sieving coefficient is one method to quantify the effectiveness of a filtration barrier against particular solutes. In the case of the glomerular filtration barrier, the sieving coefficient can be defined as

$$S = \frac{C_{Bowman's\ Space}}{C_{Glomerular\ Capillaries}} \tag{12.1}$$

where S is the sieving coefficient and C is the concentration of a particular species within either Bowman's space or the glomerular capillaries. Note that when defining a sieving coefficient the denominator always represents the source term and the numerator always represents the sink term. For all molecule types, the sieving coefficient of the glomerular filtration barrier decreases as the size of the molecule increases.

The glomerular filtrate is formed as a balance between the hydrostatic pressure within the glomerular capillary, the hydrostatic pressure within Bowman's capsule, and the osmotic pressure within the glomerular capillary. The glomerular capillary hydrostatic pressure favors filtration while the hydrostatic pressure of Bowman's capsule and the osmotic pressure of the glomerular capillaries resist filtration. Under normal conditions, the glomerular capillary hydrostatic pressure is approximately 60 mmHg, the Bowman's space hydrostatic pressure is approximately 15 mmHg, and the capillary osmotic pressure is approximately 30 mmHg; therefore, there is a net movement of fluid from the glomerular capillaries into Bowman's capsule. Under normal conditions, the net filtration pressure (approximately 15 mmHg into Bowman's capsule) is always positive, and this initiates urine formation by forcing the filtrate into and through the tubule system.

As briefly mentioned earlier, this barrier effectively restricts how much plasma moves from the glomerular capillaries into Bowman's space. One way to quantify the glomerulus function is through the glomerular filtration rate. This in essence is the amount of fluid that passes from the glomerular capillaries into Bowman's space per unit time. The glomerular filtration rate is determined by the net filtration pressure (which is on average 10 mmHg), the permeability of the glomerular capillary endothelial cell membrane, and the area available for filtration. In an average person, the glomerular filtration rate is nearly 120 mL/min or approximately 180 L/day. This is nearly 50 times more fluid being filtered out of the glomerular capillaries than across all other capillaries within the systemic circulation. Remember that in an average adult, the total blood volume is 5 L and that the plasma composes approximately 60% of this or 3 L. That suggests that the kidneys filter the entire plasma volume approximately 60 times per day. This allows the kidneys to very tightly regulate the composition of plasma constituents.

As discussed above, the glomerular filtration rate is not constant (e.g., the positive feedback from the juxtaglomerular system). In fact, neural and hormonal input on the kidneys can affect the filtration rate (we will discuss this more in Section 12.8). It is important to also recall that the glomerulus is located in between two arterioles, the afferent arteriole and the efferent arteriole, which both can regulate the blood flow through the glomerulus. Constriction of the afferent arteriole reduces the hydrostatic pressure of the glomerulus and thus reduces the glomerular filtration rate. This occurs because there is an increase in the resistance to flow through the afferent arteriole into the glomerular capillaries. Constriction of the efferent arteriole increases the hydrostatic pressure of the glomerulus and this increases the glomerular filtration rate. This occurs because there is an increase in the resistance to flow through the efferent arteriole out of the glomerulus. A third case is a constriction of both the afferent arteriole and efferent arteriole, which tends to have no effect on the glomerular filtration rate. This occurs because of the opposing effects of the increase in flow resistance. Dilation of the two arterioles has the reverse effect of those described here.

It is possible to quantify the amount of a plasma molecule that enters the nephron each day. This quantity is termed the filtered load and is obtained by multiplying the plasma concentration of a molecule by the glomerular filtration rate. For instance, the plasma concentration of sodium is approximately 3.5 g per liter of blood. This makes the filtered load for sodium equal to $3.5\ \text{g/L} \times 180\ \text{L/day} = 630\ \text{g}$ of sodium filtered per day. By comparing this value to the concentration of the same molecule within urine, one can determine if that molecule experiences a net reabsorption or a net secretion within the nephron. If the urine concentration of a molecule is lower than the filtered load, this suggests that there was a net reabsorption of the molecule into the peritubular capillaries. However, if the urine concentration of a molecule is higher than the filtered load, this suggests that there was a net secretion of the molecule into the nephron. An example of a molecule that experiences net reabsorption is glucose, whereas a molecule that experiences net secretion is para-amino hippuric acid. Therefore, urine is not a function of only the glomerular filtration properties.

The dynamics of the glomerular capillaries determines the net filtration pressure, which helps to drive fluid into the nephron tubule. Using Starling's relationship, the net filtration pressure is the difference in the hydrostatic pressure of the glomerular capillaries (which averages to ~60 mmHg but varies from ~62 mmHg at the afferent arteriole end to ~58 mmHg at the efferent arteriole end), the hydrostatic pressure of Bowman's space (which is relatively constant at approximately 20 mmHg) and the glomerular capillary oncotic pressure (which averages to ~30 mmHg but varies from ~25 mmHg at the afferent arteriole end to ~35 mmHg at the efferent arteriole end) (Figure 12.7). The oncotic pressure of Bowman's space is approximately zero, since negligible proteins enter Bowman's space. Although we highlighted the dependency of filtration pressure on the

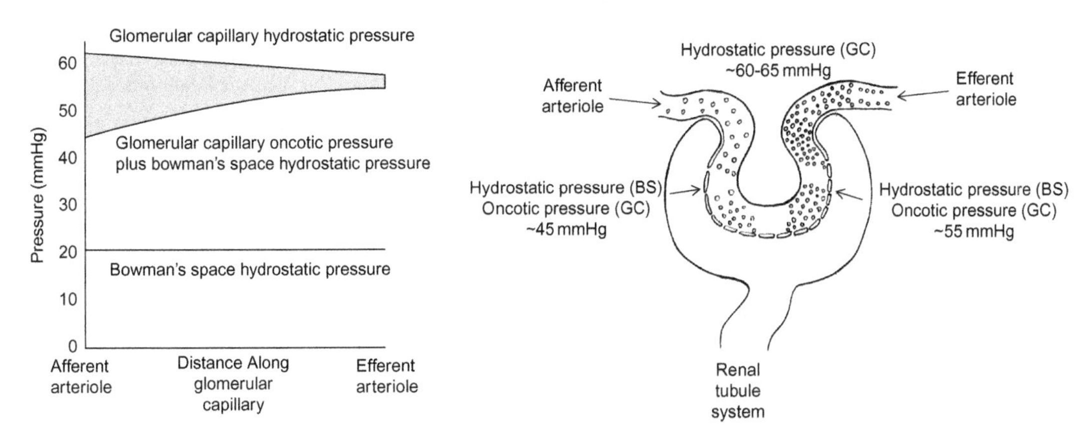

FIGURE 12.7 Variation in the hydrostatic pressure of the glomerular capillaries and Bowman's Space and the glomerular capillary oncotic pressure as a function of distance along the glomerular capillary. There is a subtle drop in glomerular capillary hydrostatic pressure and an increase in the glomerular capillary oncotic pressure along the length of the glomerular capillaries. The oncotic pressure of the glomerular capillaries increases due to a net movement of protein-free materials out of the glomerular capillaries into the nephron tubule system. The hydrostatic pressure of Bowman's space (BS) is relatively constant over this length. Notice that the net filtration (the gray-shaded region) always favors movement of materials out of the glomerular capillaries. This is in contrast to "normal" systemic capillaries, where materials move back into the capillary along the venous end. Note that this variation is a function of distance along the glomerular capillary (GC).

diameter of the afferent and efferent arteriole in the previous section, the flow of fluid through the nephron also alters the glomerular filtration pressure. As the renal flow increases, the hydrostatic pressure of Bowman's space decreases, and thus the net driving force into the nephron tubule system also increases. Similar effects would be observed when the flow into the afferent arteriole increases.

A significant amount of work has been conducted on the overall hydraulic conductivity of the glomerular capillaries. Note that these values would take into account the sieving coefficient. What has been observed is that as the surface area for exchange increases, there is a nearly linear decrease in the hydraulic conductivity. Although these values are relatively stable across different species, they are altered due to vasoactive compounds. For instance, exposure to histamine or various prostaglandins will decrease the overall permeability of glomerular capillaries.

12.4 TUBULE REABSORPTION/SECRETION

Tubule reabsorption is the process by which molecules from the glomerular filtrate are returned back to the plasma. This occurs along the entire nephron unit. Metabolically important molecules are nearly completely reabsorbed, whereas wastes are typically reabsorbed to some extent, with the majority of waste molecules making it into the urine. By looking at the relative percent reabsorbed of various molecules, it is possible to draw some

TABLE 12.1 Average Reabsorption Values for Various Metabolically Important Compounds

Compound	Filtered Load	Quantity Excreted	Percent Reabsorbed
Water	180 L/day	1.5 L/day	99%
Glucose	180 g/day	0 g/day	100%
Lipids	1080 g/day	3.6 mg/day	99.99%
Sodium	630 g/day	3.3 g/day	99.5%
Bicarbonate	110 g/day	8.5 g/day	92.3%
Urea	55 g/day	35 g/day	36%

conclusions about tubule reabsorption (Table 12.1). Metabolically important molecules (water, ions, and organic molecules) are completely reabsorbed so that there is no need to constantly ingest (or produce) these molecules. Waste products are not fully reabsorbed so that they can be removed from the body. It is important to note that the reabsorption of most organic compounds (e.g., glucose) is not regulated and is typically very high. Therefore, under normal conditions, none of these compounds are found within the urine. For these compounds, it can be considered that the kidneys do not exist, because the kidneys have no effect on their plasma concentration. However, the reabsorption of most non-organic metabolically important molecules (e.g., water and ions) is tightly regulated but is also very high under normal conditions. You can prove this by drinking a few 64-oz double-big sodas from your local gas station in under 10 min (hopefully this is not a normal condition). It is guaranteed that you will be visiting the bathroom soon to remove the excess water from your system. However, if you eat a few candy bars with a lot of sugars in them, the amount of sugar in your urine does not increase, but is stored for later use.

There are two main processes that account for the reabsorption of compounds into the peritubular capillaries. Some substances can be reabsorbed by diffusion and others involve some receptor-mediated transport. Diffusion typically occurs across the tight junctions of the tubule epithelial cells or through specific channels, whereas the receptor-mediated transport occurs through the epithelial cells themselves. For example, the reabsorption of urea occurs by diffusion. However, because the composition of the glomerular filtrate is the same as the plasma composition, there should be no concentration gradient driving force for the movement of urea into or out of the plasma/urine. Early on within the proximal tubule system, water is removed from the filtrate (via receptor-mediated transport). With the removal of water, the effective concentration of urea increases within the tubule lumen, and therefore a concentration gradient is formed between the fluid within the nephron and the plasma within the peritubular capillaries. Urea can then diffuse down its concentration gradient from the filtrate into the plasma at a distal location along the nephron. The reabsorption of the majority of lipid soluble compounds occurs in this manner and is therefore dependent on the early reabsorption of water, to effectively increase the molecular concentration within the nephron.

For a material to be reabsorbed via receptor-mediated transport, the molecule must first diffuse to the wall of the nephron tubule. The molecule must then cross the luminal wall into the tubule epithelial cells. The molecule could then diffuse across the tubule epithelial cell to the basolateral cell membrane. The molecule then crosses this cell membrane into the peritubular capillaries. It is not necessary for the molecule to be actively transported across both of the cell membranes, and typically the molecule would move down its concentration gradient when crossing one of the barriers. For example, sodium can diffuse into the tubule epithelial cells across the apical membrane, but it is then actively transported across the epithelial cells basolateral membrane to enter the bloodstream. If the transport of a molecule is active across at least one barrier, then it falls into the category of receptor-mediated transport.

It is interesting to note that the reabsorption of many molecules is coupled to sodium movement across the tubule epithelial cells. This type of movement is mediated by a cotransporter, which in this case utilizes the energy derived from the movement of sodium in the direction of its electrochemical gradient to drive the movement of another molecule (e.g., glucose, many organic compounds, and some inorganic ions) against its electrochemical gradient. The activity of cotransporters is classified by the amount of molecules that can be transported in a unit time. Under most normal conditions, the maximum rate of transport is never reached by the cotransporters. However, if the nephron concentration of a particular compound becomes so large that all of the binding sites on every transporter are occupied, then the maximum transportation rate is reached (i.e., the transporters are saturated) and this compound may enter the urine. As an example, under diabetic conditions, it is possible for the glomerular filtrate glucose concentration to exceed the maximum transportation rate, and then glucose enters the urine and is excreted. At this point, the glucose concentration in the nephron fluid has saturated all of the transport mechanisms for glucose to be reabsorbed. There is actually an old legend that before the age of modern medicine, diabetes mellitus would be diagnosed by determining how "sweet" a patient's urine was. How much truth is in this legend is up for debate, but regardless, diabetic patients can excrete glucose in urine, whereas under normal conditions, all of the glucose within the glomerular filtrate is reabsorbed into the plasma.

Tubule secretion is the process by which molecules from the peritubular capillaries move into the nephron tubule lumen. Similar to reabsorption, secretion can occur via diffusion or receptor-mediated transport. You may be wondering why the peritubular capillaries secrete compounds into the lumen. There are a variety of reasons based on the particular compound. Many toxins or foreign compounds are secreted to be fully removed from the plasma. Hydrogen ions are secreted to regulate the pH of the blood. Secretion mediated by diffusion occurs similarly to the diffusion associated with tubule reabsorption, except that it occurs in the opposite direction. Interestingly, the receptor-mediated secretion of molecules is typically coupled to sodium reabsorption. Therefore, the electrochemical gradient of sodium drives the movement of other compounds against their electrochemical gradients. These types of transports are typically called antiporters because the two molecules move in opposite directions. Hydrogen ion secretion makes use of an antiporter coupled to sodium.

It is important to note before we discuss specific examples of reabsorption and secretion, what components of the nephron perform what functions during urine formation. The primary function of the proximal tubule is to reabsorb large quantities of water and other solutes within the glomerular filtrate. This helps to form the concentration gradient which will be used in later segments of the nephron to drive the reabsorption and/or secretion of particular compounds. The Loop of Henle also functions to reabsorb large quantities of solutes and small quantities of water. The proximal tubule system is also responsible for the secretion of the majority of compounds, except for potassium. This early movement of solutes and water within the proximal tubule and the Loop of Henle are by bulk processes, where the major goal of these processes is to return the plasma/urine solute concentration close to its acceptable level. The distal convoluted tubule and the collecting duct system are primarily responsible for fine-tuning the concentrations of the solutes and determining the final excreted concentration and the final plasma concentration. Therefore, it should be intuitive that the majority of the mechanisms that exert control over nephron reabsorption/secretion (e.g., hormones) to control the final urine concentration of a molecule, act on the distal convoluted tubule and the collecting duct. As a summary, the amount of a compound that is excreted can be calculated by measuring the amount filtered, secreted, and reabsorbed, as follows:

$$\text{Amount Excreted} = \text{Amount Filtered} + \text{Amount Secreted} - \text{Amount Reabsorbed} \quad (12.2)$$

It is possible to quantify the difference between the amount of a solute that is reabsorbed versus the amount of a solute that is secreted, by modifying Eq. (12.2). The filtered load is a measure of the amount of solute that enters the nephron tubule system. In general this is defined as

$$FL_x = GFR \cdot S_x \cdot [x]_{\text{plasma}} \quad (12.3)$$

where FL represents the filtered load for a solute x, GFR is the glomerular filtration rate (~120 mL/min), S is the sieving coefficient for solute x, and $[x]_{\text{plasma}}$ is the plasma concentration for x. The filtered load of a compound represents the rate of filtration for that particular molecule and is used in a modified equation (12.2), which looks at rates

$$\text{Rate of Excreted} = \text{Rate of Filtration} + \text{Rate Secreted} - \text{Rate Reabsorbed} \quad (12.4)$$

The rate of excretion is typically represented as the urine flow rate multiplied by the urine concentration of the particular species. The kidney reabsorption and secretion kinetics would need to be investigated in order to determine the functions that represent these quantities. In the case of glucose, it is known that there is no mechanism for glucose secretion to occur, thus the rate of glucose secretion is equal to zero. Following this analysis the rate of reabsorption for glucose can be represented as

$$\text{Rate of reabsorption}_{\text{glucose}} = GFR \cdot S_{\text{glucose}} \cdot [\text{glucose}]_{\text{plasma}} - Q_{\text{urine}}[\text{glucose}]_{\text{urine}}$$

One way that this is observed *in vivo* is to quantify the clearance of the particular solute over some time period. The clearance of a molecule provides a measure of how effective the kidney is clearing plasma of a particular solute. Clearance is defined as

$$C_x = \frac{Q_{\text{urine}}[x]_{\text{urine}}}{[x]_{\text{plasma}}} \quad (12.5)$$

Under normal plasma glucose concentrations (e.g., when the transport mechanisms are not saturated) the clearance of glucose is zero and therefore, the kidneys do not alter the plasma glucose concentration. However, if the saturation limits are met for glucose transport, then the clearance can exceed a value of zero. It was mentioned earlier that para-amino hippuirc acid (PAH) is secreted by the peritubular capillaries into the nephron. Under these cases, saturation can limit the transport of PAH into the nephron, however, the clearance of PAH is very high for low plasma concentrations (because the dominator in Eq. (12.5) is very low) and decreases as the plasma concentration increases. A second way to quantify how effective the tubule sections of the nephron are reabsorbing/secreting a compound is to quantify the double ratio. The double ratio relates the relative concentration of a species in the tubule fluid versus the concentration of the same species in plasma and then relates this to the same concentrations of inulin. The double ratio is defined as

$$\text{Double Ratio} = \frac{\left(\frac{[x]_{\text{tubule fluid}}}{[x]_{\text{plasma}}}\right)}{\left(\frac{[\text{inulin}]_{\text{tubule fluid}}}{[\text{inulin}]_{\text{plasma}}}\right)} \tag{12.6}$$

The denominator of the double ratio determines where and how much water has been removed or added to the tubule fluid because the nephron can filter but cannot reabsorb or secrete inulin. Therefore, the quantity of inulin remains the same along the entire tubule length but the concentration may be altered based on water movement. The numerator of the double ratio provides similar information about a particular species x. The issue is that if the ratio that makes up the numerator is 1, this can suggest that either species x and water did not move, or that species x and water both moved at a value that maintained the ratio as 1. Thus, the double ratio relates this ambiguous information to the ratio of inulin which cannot be secreted or reabsorbed and thus inulin concentration changes are dictated by water movement only.

12.5 SINGLE NEPHRON FILTRATION RATE

Single nephron filtration rates are similar to the total glomerular filtration rates that were discussed earlier. In general, the glomerular filtration rate can be defined as

$$GFR = Q_{\text{urine}}\left(\frac{[\text{inulin}]_{\text{urine}}}{[\text{inulin}]_{\text{plasma}}}\right) \tag{12.7}$$

In a similar fashion the single nephron filtration rate (SNFR) can be defined as

$$SNFR = Q_{\text{tubule fluid}}\left(\frac{[\text{inulin}]_{\text{tubule fluid}}}{[\text{inulin}]_{\text{plasma}}}\right) \tag{12.8}$$

Tubule fluid information would be obtained by periodic measurements of the tubule fluid contents and flow rate using micropuncture techniques. In fact, many of the data points and localization information that we have discussed previously were obtained from these types of experiments. In general, under physiological conditions, single nephron filtration rates average to approximately 100 nL/min, but can range from as low as 20 nL/min to as high as 200 nL/min. Using the single nephron filtration rate and the glomerular filtration rate, one can approximate the number of nephrons as 1.2 million.

Many groups have investigated the effects of glomerular capillary pressure on single nephron filtration rates. By applying a negative pressure within the glomerular capillaries, a decrease in the overall pressure within Bowman's space was observed, however, the pressure within the proximal convoluted tubule was not altered, suggesting that this location partially acts as a nonlinear resistor to prevent the transmission of altered pressure from the glomerular capillaries throughout the nephron tubule system. Under these reduced pressure conditions, single nephron filtration rates were not altered significantly suggesting the presence of regulatory mechanisms that aim to maintain the overall function of the kidneys (Section 12.8 will discuss some of these mechanisms in more detail). The single nephron filtration rate varies somewhat depending on where the values are recorded along the nephron tubule system. In general, the values vary by no more than 5–10 nL/min along a single nephron tubule length. As mentioned in an earlier section, the flow through cortical nephrons varies as compared with juxtamedullary nephrons. In general, the difference is on the order of 20–30 nL/min, with the juxtamedullary nephrons have a higher flow rate as compared with the cortical nephrons. If one is comparing the single nephron filtration rates over a long-time interval (e.g., years), the number of glomeruli is an important factor to consider. In general, with the loss of function of nephrons as one increases in age, the glomerular filtration rate does not decrease, but instead remains relatively constant. To accommodate this, the single nephron filtration rates increase with age, but if these filtration values are normalized by the number of active nephron units, then the overall normalized single nephron filtration rate remains the same.

There has been significant amount of work that aimed at mathematically modeling the resistance of the renal microvasculature at various locations throughout the vascular network. The resistance of the afferent arteriole is an important value, which can regulate the flow through the entire glomerular capillary bed and the peritubular capillary bed (or vasa recta capillaries). The resistance of the afferent arteriole (R_{AA}) has been shown to be a function of the mean arterial pressure (P_{MA}), glomerular capillary pressure (P_{GC}), and the glomerular blood flow rate (Q_G), as follows:

$$R_{AA} = \frac{P_{MA} - P_{GC}}{Q_G} \tag{12.9}$$

The glomerular blood flow rate is a function of the hematocrit of the afferent arteriole, which should be equivalent to the systemic arteriole hematocrit, the single nephron filtration rate and the single nephron filtration fraction (*SNFF*), as follows:

$$Q_G = \frac{SNFR}{SNFF(1 - Hct_{AA})} \tag{12.10}$$

The single nephron filtration fraction is a means to quantify the ability of the glomerular capillaries to filter the blood and is calculated from the proportional difference between the efferent arteriole protein concentration and efferent arteriole hematocrit as compared with the systolic protein concentration and hematocrit.

Using a similar approach, the resistance of each efferent arteriole can be approximated by the following formulation:

$$R_{EA} = \frac{P_{GC} - P_{PC}}{Q_{EA}} \tag{12.11}$$

where P_{PC} is the hydrostatic pressure within the peritubular capillaries (or vasa recta) and Q_{EA} is the efferent arteriole flow rate, which can be defined as

$$Q_{EA} = Q_G - SNFR \tag{12.12}$$

The total arteriolar resistance of the preglomerular to postglomerular vascular tree can be defined as

$$R_T = R_{AA} + R_{EA} \tag{12.13}$$

In most animals the resistance of the afferent arteriole accounts for approximately 60% of the total arteriole resistance of the renal circulation, whereas the efferent arteriole accounts for approximately 40% of the total renal arteriolar resistance. Combining the arteriolar and venule resistance in a global renal vascular resistance term, the arteriolar side of the vascular tree accounts for approximately 90% of the vascular resistance, whereas the venous side accounts for approximately 10% of the vascular resistance. As with all vascular beds, changes to vascular resistance can occur as a function of changes to other fluid parameters. For instance, with an increase in the blood volume, the resistance of the afferent arteriole and the efferent arteriole reduces at a similar rate, allowing the glomerular capillary hydrostatic pressure to remain the same. A reduction in the mean arterial pressure tends to decrease the afferent arteriole pressure, however, this is accompanied by an increase in the efferent arteriole pressure, which acts to maintain the glomerular capillary hydrostatic pressure. Increases in hematocrit tend to increase the afferent and efferent arteriole resistance, again at a similar rate to maintain the glomerular capillary pressure. With a decrease in hematocrit, the resistances of the afferent and the efferent arteriole tend to reduce at similar rates as well. It is important to observe that while these changes can have a net effect on the flow through the glomerulus and the peritubular capillaries, there tends to be no net effect on the filtration, reabsorption, and secretion of materials from the glomerular capillaries, the peritubular capillaries, or the nephron tubule system.

12.6 PERITUBULAR CAPILLARY FLOW

As we have highlighted above, the reabsorption of filtrate into the peritubular capillaries is accounted for by two major processes; the transport of solutes through the nephron epithelial tissue (passive or active transport mechanisms) and the uptake of solutes into the peritubular capillaries. Many groups have investigated whether or not the hydrostatic pressure gradient or the osmotic pressure gradient is the driving force for this motion (e.g., which forces in the Starling equation dictate the motion of the solutes out of the nephron system). Early work has shown that only approximately 5% of the net reabsorption within the proximal tubule system can be accounted for by the hydrostatic and osmotic pressure gradients that exist between that tubule system and the peritubular capillaries. Similar work has also attributed the vast majority of this 5% to the hydrostatic pressure gradient, by disrupting the oncotic pressure gradient and observing no changes to the net movement of materials that are reabsorbed within the proximal tubule system. Starling's equation only holds for the passive movement of solutes through the nephron system, thus it would appear that the active transport of materials across the tubule wall is the more important transport mechanism. Many studies have shown that the flow of sodium (and other ions) induces a net movement of water out of the proximal tubule system. This is accompanied by an increase in the hydrostatic pressure of the interstitial space that surrounds the nephron tubule epithelial cells and the peritubular capillary endothelial cells. The increase in hydrostatic pressure aids in movement of solutes into the peritubular capillaries, however, if the permeability of the peritubular capillaries increases, this driving force may act to force materials back into the proximal tubule system. This suggests that the uptake of materials by the peritubular capillaries is the rate limiting step in the reabsorption processes that occurs within the nephron tubule system.

The uptake of material from the capillaries is equal to the single nephron filtration rate minus the rate of fluid flow within the nephron segment of interest. When quantifying these variables, one must take care in order to accurately assess the location of the peritubular capillary network in relation to the nephron tubule system, so that the fluid transport is accounted for at the appropriate level. Additionally, the peritubular capillary network does not surround the entire nephron unit and there are some locations that have extensive lymph flow. All of these variables should be accounted for within the mathematical approximation. The reabsorptive coefficient for fluid materials within the nephron unit has been represented as

$$K_r = \frac{SNFR - Q_{TS}}{\Pi_{PC} - \Pi_{IS} - P_{PC} - P_{IS}} \tag{12.14}$$

where $SNFR$ is the single nephron filtration rate, Q_{TS} is the flow rate through the nephron tubule segment of interest, Π is the oncotic pressure of the peritubular capillaries (PC) or the interstitial space (IS), and P is the hydrostatic pressure of the peritubular capillaries or the interstitial space. Capillary uptake in these instances averages to approximately 50 nL/min under physiological conditions. The Starling forces vary based on a number of parameters, as discussed in previous chapters. The oncotic pressure of the peritubular capillary is approximately 30 mmHg, the oncotic pressure of the interstitial space is approximately 7 mmHg, the hydrostatic pressure of the capillary is approximately 15 mmHg and the hydrostatic pressure of the interstitial space is approximately 6 mmHg. Using these values, one can calculate the reabsorptive coefficient and obtain a value of approximately 400 nL mmHg/s.

A number of mathematical models exist that describe the movement of solutes through the nephron unit as a function of distance along the nephron tubule. Since, the net pressure for reabsorption to occur is relatively low (e.g., the summation of the Starling forces), small changes in any of these values may have drastic effects on the solute transport. These models have shown that small changes in the protein concentration (either in the interstitial space or the peritubular capillaries) have a drastic effect on the movement of solutes, whereas large changes within the flow properties (e.g., the flow through the nephron unit or the flow through the peritubular capillaries) have a much lower effect on the transport of materials. These findings suggest that even though much of the early transport occur through receptor-mediated processes the reuptake of solutes from the interstitial space is very highly dependent on the maintenance of the pressures across the peritubular capillaries.

The lymphatic system plays an important role in the reuptake of solutes from the interstitial space between the nephron epithelial cells and the peritubular endothelial cells. In the kidney, as we discussed before, significant changes in the flow properties within the vascular system or the nephron tubule system have very small effects on flow of lymph. In fact, the flow of lymph appears to be unaltered under most pathological conditions and this has been attributed to the overall small changes to the flow within the nephron unit, with changes to the vascular flow conditions. With that said, since the kidney obtains and filters a relatively large percentage of blood, the lymph flow of the kidney is greater than the lymph flow of many other organs, but remains relatively stable throughout all possible altered conditions.

12.7 SODIUM BALANCE AND TRANSPORT OF IMPORTANT MOLECULES

As discussed in previous sections, sodium is reabsorbed throughout the majority of the entire nephron tubule system. Recall that the concentration of sodium is high in both the tubule lumen and the interstitial fluid surrounding the nephron, but importantly the movement of sodium appears to drive the movement of many of the salient molecules through the nephron system. The movement of sodium initiates from the passive diffusion of sodium, through a sodium channel, from the tubule lumen, into the epithelial cells lining the tubule. To maintain the electrochemical neutrality of luminal epithelial cells, potassium efflux (from the epithelial cell into the tubule lumen) is generally coupled to the process of sodium movement. This movement of potassium is by bulk diffusion as well. The movement of sodium out of the epithelial cell into the interstitial fluid is achieved by an active mechanism: the sodium–potassium ATPase pump. This pump moves three sodium ions out of the epithelial cell and two potassium ions into the epithelial cell for every ATP that is hydrolyzed into ADP. Sodium then moves into the peritubular capillaries.

Sodium reabsorption is regulated via the amount and type of channels present on the luminal surface of the epithelial cell. The entry of sodium into the epithelial cells that compose the cortical collecting duct is by diffusion through sodium channels. Along the proximal collecting tubule, the movement of sodium into the epithelial cells is typically by cotransport with organic molecules such as glucose or by countertransport with hydrogen ions. Therefore, the movement of sodium down its electrochemical gradient provides the energy for the reabsorption of glucose and the secretion of hydrogen ions. The transport of sodium across the basolateral membrane, via the sodium–potassium ATPase pump, is typically constant along the entire length of the tubule lumen. The quantity of these pumps present can be regulated by other mechanisms to determine the amount of sodium reabsorbed into the peritubular capillaries, although this does not normally occur.

Many cotransporters and countertransporters are present along the proximal convoluted tubule. For instance, nearly all glucose is reabsorbed within the proximal convoluted tubule system via a sodium-glucose cotransporter (SGLT). There are two main types of these transporters, the SGLT1 moves two sodium down its electrochemical gradient to drive the motion of one glucose molecule up its concentration gradient, whereas the SGLT2 moves one sodium down its electrochemical gradient to drive the motion of one glucose molecule into the tubule epithelial cell. These transporters reside on the apical membrane epithelial cell membrane. On the basolateral membrane a glucose transporter (GLUT1 or GLUT2) is present to facilitate the diffusion of glucose out of the epithelial cell into the peritubular capillaries. Amino acids are reabsorbed within the proximal convoluted tubule using specific transport proteins, many of which require sodium to be cotransported along with the amino acid (some transporters do not require any molecule to be transported in parallel to the amino acid). On the basolateral side of the epithelial cell, many of these same transporters exist to facilitate the movement of the amino acids into the peritubular capillaries.

Urea is an example of a molecule that does not need sodium present to transport into the tubule epithelial cell. Urea can passively cross through biological membranes; however, the permeability of these membranes to urea is relatively low. Urea transporters are present, which can facilitate the motion of urea through both the apical and basolateral sides of the epithelial membrane. Water movement is similar to urea, in that it typically moves by passive diffusion through aquaporin channels on the apical and basolateral membrane of the epithelial cells. There are various aquaporin channels that are found in the kidney but all of them have the same overall function. We will discuss more details regarding water movement at the end of this section.

It is interesting to point out that some proteins can enter the filtrate because the sieving coefficient is not zero for proteins. The proximal tubule has pit regions that attract proteins and then take the proteins up by endocytosis mechanisms. Interestingly, the majority of the proteins that are reabsorbed in this way are digested into the amino acid components because the endosome containing the protein typically fuses with a lysosome. The amino acids are then transported into the peritubular capillaries by specific transporters discussed above. This typically occurs very early within the proximal tubule system.

We will now move our discussion to the control of sodium reabsorption. In essence, no sodium is secreted into the nephron. Therefore, the total amount of sodium that is excreted is equal to the amount reabsorbed subtracted from the amount filtered. Changes in the glomerular filtration rate or the reabsorption rate of sodium will affect the excretion rate of sodium, but how are changes in sodium concentration sensed in the body? Currently, there are no known sensors in the body for sodium concentration. Instead, sodium concentration is regulated by changes in plasma volume (e.g., water volume), which will directly change the arterial blood pressure. Baroreceptors, located in various places throughout the body, can sense changes in blood pressure, and they have a direct effect on the kidneys as well as an indirect effect on the kidneys. The direct effect of a reduction in blood pressure is that the glomerular filtration rate will reduce because the hydrostatic pressure balance across the glomerular capillaries and Bowman's capsule decreases (see Section 12.3). The indirect effect of a decreased arterial blood pressure on sodium excretion is via activation of a sympathetic nervous signaling pathway. The renal sympathetic nerve directly innervates the juxtaglomerular cells and causes them to increase the production and secretion of renin. Renin enters the bloodstream and converts angiotensinogen into angiotensin I. Angiotensinogen is produced by the liver and under normal conditions is always present in the plasma. Angiotensin I is then converted into angiotensin II via an enzyme termed angiotensin-converting enzyme (or ACE). Angiotensin-converting enzyme is bound to the luminal surface of capillary endothelial cells and is always expressed. Angiotensin II has various effects within the body, but the most important for sodium reabsorption is the stimulation of the adrenal glands to produce and secrete aldosterone into the blood. Aldosterone stimulates the cortical collecting duct epithelial cells to produce more sodium channels and sodium–potassium ATPase pumps. This effectively increases the reabsorption of sodium (or more accurately, decreases the excretion of sodium). An increased plasma volume has the opposite effect as described above, so that there is an increased excretion of sodium.

As mentioned earlier, sodium drives the reabsorption of water within the nephron. As sodium ions begin to move out of the nephron, the osmolarity of the solution within the nephron decreases. A decrease in osmolarity can also be considered as an increase in water concentration. As the solute moves out of the tubule lumen, the osmolarity of the interstitial fluid and/or epithelial intercellular fluid increases, and this can also be considered as a decrease in the water concentration. Therefore, there now exists a concentration gradient and an osmotic gradient for water that will cause water molecules to diffuse out of the tubule lumen into the interstitial space. In general, the transporters that were discussed in this section only work in one direction and thus water will move out of the nephron unit to reduce the concentration of sodium.

Water movement across the tubule epithelial cells occurs via bulk diffusion across the tight junctions or through aquaporin channels on the apical epithelial membrane. The proximal tubule is very highly permeable to water, and large quantities of water are always reabsorbed within this segment of the nephron. The Loop of Henle experiences both water reabsorption (descending limb) and water secretion (ascending limb) to maintain a countercurrent flow of other nutrients that are important, which will not be discussed here. The water movement through the distal tubule and the collecting duct is variable and is subject to physiological control. In a similar process to the regulation of sodium reabsorption, changes in the water reabsorption rates of the collecting ducts are controlled by the blood volume. Decreases in blood volume will reduce the arterial pressure, which is again sensed by baroreceptors within the cardiovascular system. This reduction causes an increase in the production and secretion of vasopressin (also known as antidiuretic hormone). Vasopressin is released into the blood and eventually makes its way to the collecting ducts. Vasopressin then stimulates the production of new aquaporin channels which are placed into the collecting duct epithelial membrane. With a higher quantity of aquaporin channels within the collecting duct epithelial cell, permeability to water increases, and therefore, the reabsorption of water increases. With an increase in blood volume, the opposite response will occur, causing the permeability of the collecting ducts toward water to decrease.

12.8 AUTOREGULATION OF KIDNEY BLOOD FLOW

The control of blood flow to and through the kidneys is central to the function of the kidneys. With significant changes in blood flow, we can imagine that the clearance of metabolic wastes would be altered, which would likely affect many other processes within the body. For approximately 100 years, many groups have investigated conditions that can alter the renal blood flow. It has been observed that changes in systemic arterial blood pressure (either increase or decrease) elicit negligible effects on the blood flow to and through the glomerular and peritubular capillaries. In fact, later work has shown that even a direct occlusion to the renal artery only elicits a short-term decrease in the renal perfusion that quickly returns to control levels. Combined this suggests that mechanisms are present within the kidney to regulate and maintain the blood flow to and through the glomerular and peritubular capillaries at near constant levels. To take this one step farther, groups have removed all kidney innervation and have observed that the altered flow through the renal artery is normalized out by the time that the blood reaches the glomerular capillaries. This self-regulating property, in the face of changes to the arterial flow properties has been termed autoregulation.

We have already discussed that the nephrons are perfused differently based on their location within the kidney. Early work suggested that the limited changes to the kidney perfusion rate may be accounted for by autoregulation events that only affect certain regions of the kidney. This would suggest that with a decreased inflow pressure, certain regions of the kidney would experience a decreased flow, whereas other regions of the kidney would autoregulate to normal flow conditions. However, it has become apparent that all regions of the kidney appear to have autoregulation procedures but they may act different to adjust particular regions of the kidney at different rates.

Many groups have tried to identify the regions of the kidney circulation that are responsible for the autoregulation events. It has been observed that the glomerular filtration rate and the filtration fraction of the kidneys remain relatively constant under altered pressure conditions and this suggests that the afferent arteriole or a vessel in close proximity to the afferent arteriole is primarily responsible for the autoregulation events that are observed. Changes to the afferent arteriole vascular resistance have been observed under altered inflow blood pressure conditions and this can account for the return to control conditions for the glomerular blood flow rates. Robertson et al. altered the mean arterial pressure of rats from 60 mmHg to approximately 120 mmHg. Under these conditions, the glomerular perfusion rate and single nephron perfusion rates increased by approximately 30–40%, however, the single nephron filtration rates remained unchanged. Similarly, the hydrostatic pressure of the glomerular capillaries and the peritubular capillaries and the oncotic pressure of the afferent and efferent arteriole remained largely unaltered under the changes in systolic pressure. However, there was an approximate 200% increase in the resistance of the afferent arteriole accompanied by an approximate 20% reduction in the resistance of the efferent arteriole. Thus, autoregulation processes appear to be restricted to the afferent arteriole.

There are a number of theories that try to describe the mechanisms that are responsible for autoregulation changes. Many of these have been difficult to verify experimentally, but we will highlight them to illustrate possible mechanisms of action. The first theory is the cell separation theory, which states that a change in systemic pressure alters the degree and location of red blood cells within the vasculature. By altering the red blood cell concentration within the kidney microcirculation, the hematocrit is altered and thus viscosity and resistance are altered. In a similar argument, changes to the systemic pressure will alter the interstitial pressure and this may act to regulate the perfusion rate. These two theories have some experimental support but do not adequately describe all of the observations made during autoregulation conditions.

The myogenic theory, which was described in a different context in Chapter 6, states that changes to the vascular wall tension instigate changes to the tone of the vessel to minimize the changes to the vascular wall tension. Recall that the myogenic theory predicts that with an increased arterial pressure, there would be an accompanying increase in the wall tension. This would initiate a constriction of smooth muscle cells to inhibit the stretch-induced expansion of the vascular wall. Support for this mechanism stems from studies where agents were added to the kidneys that inhibit the activity of vascular smooth muscle cells. Under these conditions, a significant portion of the autoregulation response was abolished; however, the kidney still retained some of its ability to regulate the filtration rates.

The metabolic theory is the second major mechanism that aims to explain the autoregulation response of the kidneys. This theory suggests that with a decreased perfusion there would be an increase in the concentration of particular vasoactive compounds that would tend to induce vasodilation (the opposing argument would be made for an increase in perfusion pressure). A significant supporting argument for the metabolic theory is the close association of the distal tubule and the afferent/efferent arteriole at the juxtaglomerular apparatus. Although there are some clear observations of an altered tubule concentration effecting the flow rate by juxtaglomerular signaling, the exact mechanisms of this potential feedback loop and the extent of its role during autoregulation responses remain to be identified. This feedback mechanism has been termed tubuloglomerular feedback and is a source of much debate. Again it is likely that both the myogenic theory and the metabolic theory play different but parallel roles in order to regulate the nephron filtration rates. Please note that we have mentioned the renin–angiotensin system as a regulating agent in previous discussions. This would appear to be a supporter of the metabolic theory, however, it is still unclear how this system initiates and what controls the release of renin from the juxtaglomerular apparatus.

A third mechanism for autoregulation is through the sympathetic nervous system. The renal sympathetic nervous system innervates the nephron tubules, the glomerular and peritubular capillaries, and the juxtaglomerular cells. In general, the activation of the renal sympathetic nervous system increases the reabsorption of sodium and water along the tubular system, decreases the renal blood flow and the glomerular filtration rate, and stimulates the release of renin from the juxtaglomerular apparatus. However, work that has been conducted on denervated kidneys has shown that the kidneys can still self-regulate the blood flow along the nephron unit and the composition of the glomerular filtrate. Thus, it is unlikely that the sympathetic nervous plays a major role in the autoregulation of the kidney function.

12.9 COMPARTMENTAL ANALYSIS FOR URINE FORMATION

This section will describe one of the mathematical techniques that can be used to describe the movement of solutes through the nephron. In general, this technique can be extended to describe the movement of solutes through any regions divided by some permeable membrane. When analyzing the transfer of substrates within the body, it is typically convenient to describe the regions that are involved in the transfer as compartments. For instance, the tubule lumen, the tubule epithelial cells, and the peritubular capillary lumen can all be individual compartments used to describe solute movement. With this analysis method, it is typical to assume that the boundary has a spatially uniform permeability (although the permeability can change with time; including this would only increase the complexity of the solution by including a time varying component), and that the compartments have some known volume as a function of time. In general, compartmental analysis problems are solved using mass conservation laws.

We will begin our analysis by solving for the transfer of a substance between two compartments separated by a thin permeable membrane. To solve for the change in solute concentration with time across a thin membrane, Fick's law of diffusion can be used:

$$\frac{dc}{dt} = -\frac{DA}{V}\frac{dc}{dx} \tag{12.15}$$

In Eq. (12.15), c is the concentration of the solute, D is the diffusion coefficient, A is the surface area of the membrane separating the compartments, V is the compartment volume (constant in this case), and dx is the thickness of the membrane separating the compartments. The simplest system is composed of two compartments (Figure 12.8), and each quantity of interest would be denoted with a subscript to denote which compartment is being referenced. Equation (12.15) can be simplified to

$$\frac{dc_1}{dt} = -\frac{DA}{V_1}\frac{c_1 - c_2}{\Delta x} \tag{12.16}$$

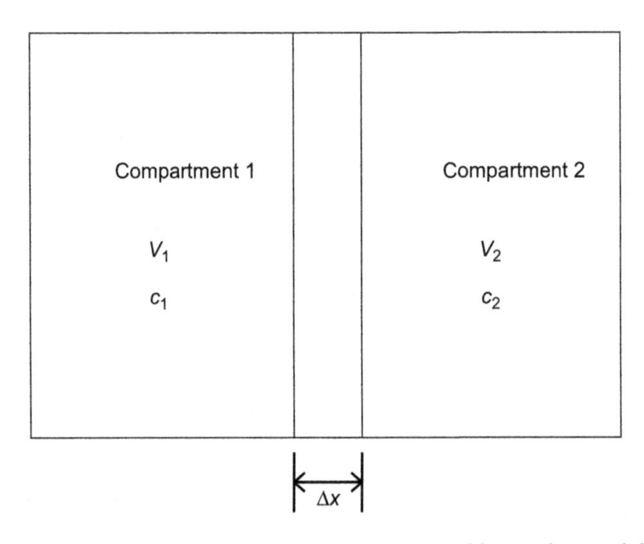

FIGURE 12.8 Two-compartment model, separated by a semipermeable membrane of thickness Δx. This type of model can be used to analyze the solute transport through the nephron. To increase the complexity of the model, multiple compartments can be included separated by different semipermeable membranes.

assuming uniformity across the semipermeable membrane, homogeneous distribution within each compartment and that we are referencing compartment 1. If at time 0, there is an initial concentration (c_0) of a solute in compartment 1 (with no solute in compartment 2 at time 0), then mass conservation states that

$$V_1c_1 + V_2c_2 = V_1c_0 \tag{12.17}$$

The multiplication of volume by a concentration is "quantity of solute" and that is what can be measured and must be conserved within the two-compartment system (e.g., concentrations are a function of the volume and the quantity and do not need to be conserved). If Eq. (12.17) is solved for the concentration within compartment 2 and substituted into Eq. (12.16), we get

$$c_2 = \frac{V_1c_0 - V_1c_1}{V_2}$$

$$\frac{dc_1}{dt} = -\frac{DA}{V_1}\frac{c_1 - \frac{V_1c_0 - V_1c_1}{V_2}}{\Delta x} = -\frac{DA}{\Delta x}\frac{c_1(V_1+V_2) - V_1c_0}{V_1V_2}$$

$$\frac{dc_1}{dt} + \frac{DA}{\Delta x}\frac{c_1(V_1+V_2)}{V_1V_2} = \frac{DA}{\Delta x}\frac{c_0}{V_2} \tag{12.18}$$

which is a first-order linear ordinary differential equation. Equation (12.18) can be solved in a variety of ways, depending on the conditions of the problem. For instance, if V_1 is equal to V_2, it will significantly simplify Eq. (12.18). The use of Laplace Transforms to solve the differential equation may prove to be easy, or other mathematical methods can be used. Regardless of the solution method, it will be typical to see that the solute concentration change with respect to time in one compartment will be described by an exponential decay, while the solute concentration in the other compartment will be described by an exponential rise (for a two-compartment system).

To use compartmental analysis to model the transfer of solutes between two compartments, the following assumptions are useful. First, assume that the volume of each compartment remains the same with time. This makes the time derivative only a function of concentration and not of volume (as shown in Eq. (12.18)). Also, we make the assumption that the solute is homogeneously mixed throughout the compartment immediately after it crosses the semipermeable membrane. The only factors that affect the transfer of the solute across the membrane are accounted for within the diffusion coefficient, the cross-sectional area of the membrane, the membrane thickness, and the concentration gradient across the membrane.

Example

Consider the reabsorption/metabolism of glucose within the nephron as modeled by a two-compartment system. Assume that the input of glucose into the nephron is constant and defined by K_1. The output of glucose from the nephron can be described by the metabolism of glucose by the nephron epithelial cells (K_2) and the reabsorption of glucose into the peritubular capillaries (K_3), where the Ks are defined as the transfer rate, which account for the diffusion coefficient and the membrane parameters. The initial quantity of glucose within the nephron is defined as q_0. Solve for the time rate of change of the quantity of glucose within the nephron.

Solution

The mass balance of glucose quantity can be defined as the accumulation of glucose within the nephron is equal to the input of glucose minus the output of glucose. Therefore, the mass balance equation is

$$\frac{dq}{dt} = q_{in} - q_{out}$$

The input quantity is defined as

$$K_1 q$$

The output quantity is defined as

$$(K_2 + K_3)q$$

Using this definition within the differential equation

$$\frac{dq}{dt} = K_1 q - (K_2 + K_3)q$$

$$\frac{dq}{q} = (K_1 - K_2 - K_3)dt$$

$$\ln(q) = (K_1 - K_2 - K_3)t + c$$

$$q(t) = Ce^{(K_1 - K_2 - K_3)t} = q_0 e^{(K_1 - K_2 - K_3)t}$$

where C is the initial concentration of glucose within the nephron. With increasing complexity, these methods can still be applied, but it is necessary to quantify the time rate of change of each substance, within each compartment. This can lead to multiple differential equations that must be solved simultaneously.

12.10 EXTRACORPOREAL FLOWS: DIALYSIS

Dialysis is a method to separate solutes within a solution by means of diffusion through a semipermeable membrane. Blood dialysis is typically needed when the kidneys can no longer remove toxins from blood. When the kidneys fail, the patients must be maintained on dialysis; otherwise, they will die, because there will be a buildup of toxins within the body. As with any dialysis, there are three major components of the system (Figure 12.9). The first is a compartment for the solution which will have substances removed. In our case, this will be blood. The second is a compartment containing a fluid with which the solutes will enter. The fluid in this compartment is termed the dialysate. The last is the membrane which is permeable to the substances of interest (e.g., toxins) and not permeable to other substances (e.g., plasma proteins). The design of these membranes is still of interest to many research groups. There are two common types of dialysis: the

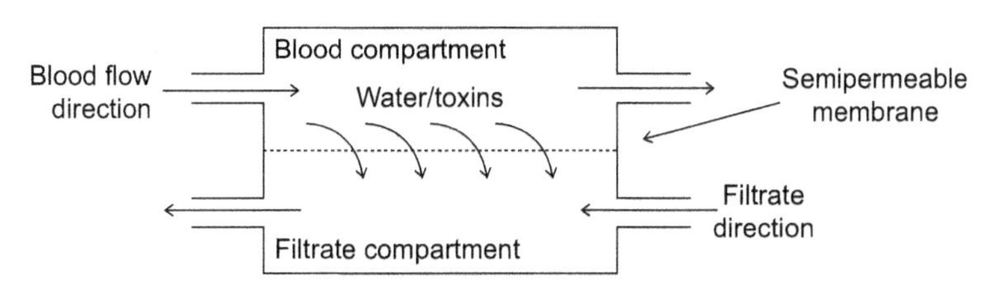

FIGURE 12.9 Typical arrangement of a dialyzer to remove toxins from blood. A dialyzer would contain a semipermeable membrane that would be able to filter toxins from the blood and would leave all proteins, cells, and other necessary organic molecules within the blood. A countercurrent flow is used to maintain a concentration gradient along the entire membrane.

arteriovenous fistula method and the peritoneal dialysis method. The characteristics of these methods will be described briefly below.

Blood dialysis occurs in more than 300,000 patients in the United States. Prior to dialysis, the patient's blood is anticoagulated and remains so during the entire dialysis period. Blood is continually drawn from the body and enters the extracorporeal device. Water, toxins, and potentially some ions are removed from the blood and enter water (which is the dialysate) that contains ions in a similar concentration as the plasma. The ions are in a similar concentration as the patient's blood, to minimize the concentration gradient of these molecules between the patient's blood and the dialysate. The semipermeable membrane used differs based on the patient's characteristics. Typically, membranes are composed of poly(methyl methacrylate), polyacrylonitrile, and polysulfone. These membranes allow the movement of molecules that are relatively small but block the movement of larger molecules. Importantly, these materials are biocompatible and do not activate inflammatory and/or thrombotic reactions.

To initiate the flow of water and solutes between the two compartments of the dialyzer, a hydrostatic pressure gradient and a concentration gradient (of the molecules of interest) are put into place across the semipermeable membrane. Also, the flow through the dialyzer is typically slow, so that a countercurrent gradient is established. A countercurrent gradient means that the blood inflow is at the same location as the dialysate outflow. This location would have the largest concentration gradient between the blood and the dialysate. At the blood outflow location, which is aligned with the dialysate inflow, a relatively large concentration gradient exists as well. Using this approach, all of the toxins can be removed from the blood without multiple passes through the dialyzer. In fact, the nephron uses a countercurrent multiplier system to efficiently transfer solutes from the nephron into the peritubular capillaries and vice versa (although this was not discussed). If the blood and the dialysate flow in the same direction, the concentration gradient reduces over the length of the membrane.

To mathematically model these types of systems, a modified compartmental analysis method can be used to describe the convection of blood and dialysate through the extracorporeal device and the diffusion of solutes out of the blood. We would first need to consider the rate at which the solute enters the compartment and the rate at which the solute exits the compartment. These can be represented as

$$\frac{dq_{\text{in}}}{dt} = \dot{q}_{\text{in}}$$
$$\frac{dq_{\text{out}}}{dt} = \dot{q}_{\text{out}} \tag{12.19}$$

These rates would also be equal to the volumetric flow rate of fluid into the compartment multiplied by the inflow or outflow concentration of the solute:

$$\dot{q}_{\text{in}} = \dot{Q}c_{\text{in}}$$
$$\dot{q}_{\text{out}} = \dot{Q}c_{\text{out}} \tag{12.20}$$

We can then describe the time rate of change of any solute within the compartment as

$$\frac{dq}{dt} = \dot{q}_{\text{in}} - \dot{q}_{\text{out}} = \dot{Q}(c_{\text{in}} - c_{\text{out}})$$

or

$$V_{\text{compartment}}\frac{dc}{dt} = \dot{Q}(c_{\text{in}} - c_{\text{out}}) \tag{12.21}$$

assuming that the volume of the compartment does not change with time, which is typically a good assumption for dialysis compartments. To completely model an extracorporeal device, a system of first-order linear ordinary differential equations would need to be coupled, such that the solute concentration in each compartment would affect the net diffusion across the semipermeable barrier and this would alter the concentration in the second compartment. Such a system would take the form of

$$\frac{dc_1}{dt} = -\frac{\dot{Q}_{1\to 2}}{V_1}c_1 + \frac{\dot{Q}_{2\to 1}}{V_2}c_2 + \dot{Q}_1$$
$$\frac{dc_2}{dt} = \frac{\dot{Q}_{1\to 2}}{V_1}c_1 - \frac{\dot{Q}_{2\to 1}}{V_2}c_2 + \dot{Q}_2 \tag{12.22}$$

where $\dot{Q}_{1\to 2}$ is the flow rate from compartment 1 into compartment 2 and $\dot{Q}_{2\to 1}$ is the flow rate from compartment 2 into compartment 1.

Example

A patient is currently connected to an extracorporeal device to remove the excess salt from his or her blood (Figure 12.10). Imagine that the inflow blood contains 5 g of salt and the blood inflow flow rate is 5 mL/min. The blood outflow rate is also 5 mL/min. The inflow rate of the dialysate is 2 mL/min, and the outflow flow rate is also 2 mL/min. There is a flow between the two compartments of 1 mL/min in each direction. The initial concentration of salt within the dialysate is 0 g, and the initial concentration of salt within the blood is 150 g. The volume of the blood compartment is 100 mL, and the volume of the dialysate compartment is 100 mL. Determine the salt quantity as a function of time for both compartments of this system.

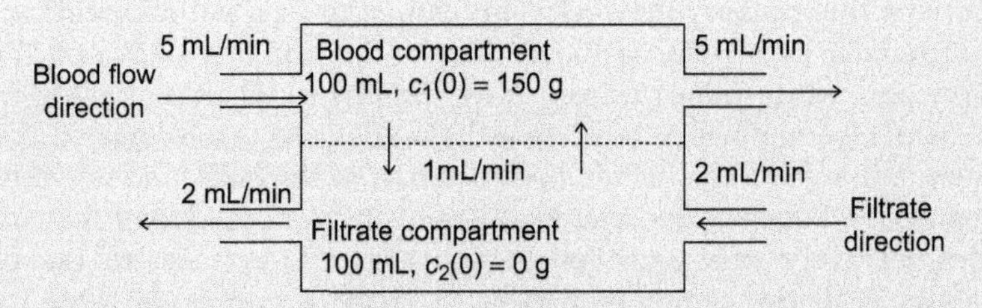

FIGURE 12.10 Flow through an extracorporeal device for the in-text example problem.

Solution

First, set up the differential equations and the initial conditions:

$$\frac{dc_1}{dt} = -\frac{1\ \text{mL/min}}{100\ \text{mL}}c_1 + \frac{1\ \text{mL/min}}{100\ \text{mL}}c_2 + 5\,g/\text{min}$$

$$\frac{dc_2}{dt} = \frac{1\ \text{mL/min}}{100\ \text{mL}}c_1 - \frac{1\ \text{mL/min}}{100\ \text{mL}}c_2$$

$$c_1(0) = 150g$$

$$c_2(0) = 0g$$

Using the Laplace Transform to solve this system of differential equations, this system becomes

$$sC_1 - c_1(0) = -0.01C_1 + 0.01C_2 + \frac{5}{s}$$

$$sC_2 - c_2(0) = 0.01C_1 - 0.01C_2$$

$$C_1(s + 0.01) - 0.01C_2 = \frac{5}{s} + 150 = \frac{150s + 5}{s}$$

$$0.01C_1 = C_2(s + 0.01)$$

$$C_1 = 100C_2(s + 0.01)$$

$$[100C_2(s + 0.01)](s + 0.01) - 0.01C_2 = \frac{150s + 5}{s}$$

$$100C_2(s^2 + 0.02s) = \frac{150s + 5}{s}$$

$$C_2 = \frac{1.5s + 0.05}{s^2(s + 0.02)} = \frac{-5}{s} + \frac{2.5}{s^2} + \frac{5}{s + 0.02}$$

$$C_1 = \frac{150s^2 + 6.5s + 0.05}{s^2(s + 0.02)} = \frac{200}{s} - \frac{2.5}{s^2} - \frac{50}{s + 0.02}$$

$$c_1(t) = 200 - 2.5t - 50e^{-0.02t}$$
$$c_2(t) = -5 + 2.5t + 5e^{-0.02t}$$

An arteriovenous (AV) fistula is a general term in which an artery is connected directly to a vein, bypassing the capillary network. This can occur as a pathological condition and it would be characterized by a reduced blood flow to the capillary beds and thus a reduced nutrient delivery and waste removal from these capillaries. During dialysis procedures an artificially AV fistula (sometimes referred to as hemodialysis) can be created by directly connecting an artery and a vein. This fistula becomes part of the body and is typically a reasonably robust connection between the arterial side and the venous side of the circulation. This fistula provides relatively easy sites for the physician to connect to the blood supply. Peritoneal dialysis is a less common dialysis method as compared with the AV fistula approach. During peritoneal dialysis the dialysate is perfused into the peritoneal space. The capillaries within the peritoneum serve as surrogates for the kidneys allowing for the wastes to be filtered into the dialysate. The dialysate is collected through a drain line. An advantage of peritoneal dialysis is that it can be conducted at home by the patient, since the peritoneal catheter is surgically implanted into the patient. However, there are some risks of infection with this approach and should only be used in certain patients.

12.11 DISEASE CONDITIONS

12.11.1 Renal Calculi

Renal calculi are commonly known as kidney stones. Most of the time, calculi appear in the kidneys due to urine becoming supersaturated with salts. These salts become crystallized and become associated with an organic matrix and therefore are no longer in solution. As the solid crystals pass through the kidney tubule system, they may become lodged in a region or simply form in the larger tubules (commonly the ureter). When this happens, urine cannot be processed and the calculi must be removed by natural means (e.g., passing a kidney stone) or through the use of a laser or ultrasound to break apart the salt deposits. Approximately 75% of calculi are formed from calcium because there is a high daily calcium intake that cannot be handled by the kidney when it also experiences a high salt environment. The easiest treatment for calculi is to decrease the intake of salts, which will reduce the load on the kidney. However, once a stone has formed, it must pass through the urinary system or be removed by some other means.

Renal calculi are found in about 1% of the population; however, the reoccurrence rate in afflicted individuals is normally in the range of 70–80%. Kidney stones that originate from abnormal calcium concentrations typically take the form of calcium oxalate (CaC_2O_4) and calcium phosphate (e.g., $Ca(H_2PO_4)_2$). Calculi that are not composed of calcium are typically composed of uric acid, cysteine, or magnesium ammonium phosphate. Interestingly, inhibitors of crystal growth can hinder the formation of calculi and these molecules include pyrophosphate, citrate, and glycosaminoglycans.

There are five major abnormalities that promote the formation of calculi. This include absorptive hypercalciuria, which is classified by an increase in the absorption of calcium within the intestine. Renal hypercalciuria occurs when there is an impaired nephron tubule reabsorption of calcium. Resorptive hypercalciuria is characterized by an excessive bone resorption, usually associated with an increased secretion of parathyroid hormone. Hyperoxaluria is associated with an enhanced formation of oxalate due to a disruption of enzyme kinetics in the synthesis of oxalate. Finally, hyperuricosuria is normally associated with an increase in the production of uric acid, but it may also be associated with altered dietary intake of purines. Regardless of the mechanism that initiates the calculi, the end product is similar.

12.11.2 Kidney Disease

Kidney disease is a general term that can apply to many different diseases that affect different functions of the kidney. Here we will discuss some of the common diseases of the kidneys that may lead to dialysis. Typically, kidney diseases are identified through abnormal urine analysis in asymptomatic healthy patients. One of the most common findings of an abnormal urine analysis is proteinuria. Normally kidneys excrete approximately 150 mg of proteins per day and this is regulated fairly tightly by the glomerular filtration barrier. An increased protein concentration is indicative of a breakdown of this filtration barrier or a decrease in the reabsorption of proteins from the filtrate. Regardless of the cause, there are serious repercussions associated with an increased protein excretion. A second common finding during kidney diseases is hematuria. Hematuria is characterized by an excretion of red blood cells within the urine. Under physiological conditions, the fenestrated endothelial cells do not allow for the passage of blood cells in the nephron filtrate. Thus, any blood cells in the urine (typically the urine turns reddish) indicate a significant breakdown of the glomerular filtration barrier. Similarly, leukocyturia is the presence of leukocytes within the urine. Less common findings associated with kidney disease is chyluria, lipiduria, and pneumaturia. Chyluria is associated with lymph entering the urine, which occurs when there is a breakdown of the lymph flow within the kidneys. Urine typically appears to be milky under these conditions. Lipiduria is associated with an increased concentration of fats (typically in the form of droplets) in the urine. Lipiduria is common with some poisonings (such as heavy metal or carbon monoxide poisoning). Finally, pneumaturia is characterized as gas bubbles in urine. This is commonly found during urinary tract infections, when the infectious agents (e.g., bacteria) produce gas.

Diseases of the kidney are typically coupled with cardiovascular diseases. In many instances, with a reduced filtration, there would be an increased water retention (edema), circulatory volume overload, and hypertension. Acute renal failure describes any event where the function of the kidneys is suddenly but briefly impaired, whereas chronic renal failure would typically be associated with a slower but longer lasting impairment of renal function. It is typically quantified under normal ingestion conditions as urine volume of <400 mL per day. Under either of these conditions it may be necessary for dialysis to be prescribed.

Other diseases of the kidney include nephrotic syndrome, which is the consequence of sustained and heavy proteinuria. Typically, nephrotic syndrome is associated with edema, proteinuria, and hyperlipidemia. There are also common defects of the nephron tubule system. The majority of these are hereditary and can be characterized by a progressively reducing filtration rate coupled with altered plasma or urine concentrations of particular solutes. Some of these nephron tubule diseases are also associated with anatomical defects, which alter the overall function of the nephron system. Depending on the particular disease, a different treatment course may be prescribed.

END OF CHAPTER SUMMARY

12.1 The kidneys remove metabolic waste products and foreign particles from the body, as well as maintain the water volume and the concentration of various ions within the body. Urine formed within the kidneys passes through the ureters into the bladder, where it is held until it is excreted from the body. Blood is supplied to the kidneys via the renal arteries. Interestingly, the renal circulation is composed of two capillary beds that are separated by an arteriole. The efferent arteriole predominantly regulates the pressure in both arterioles. The first set of capillaries forms the glomerulus, which is the site for blood filtration. The second set of capillaries, the peritubular capillaries, is involved in the reabsorption and secretion of compounds. The juxtaglomerular apparatus forms a direct feedback mechanism within each nephron to control the filtration rate based on the concentration of sodium ions within the nephron.

12.2 The renal artery, which bifurcates from the aorta, passes through the hilum, and branches into the interlobar arteries. The interlobar arteries bifurcate and bend and are termed arcuate arteries, which give rise to the radial arteries. The afferent arterioles that enter the nephron units branch off of the arcuate arteries. The venous circulation through the kidney follows a similar path, but in reverse to the artery path. The principle differences between the peritubular capillaries and the vasa recta is that the peritubular capillaries do not penetrate as deep into the medulla. Blood flow to different capillary beds within the renal circulation is perfused differently.

12.3 Glomerular filtration is the process by which compounds are removed from the blood and enter the nephron. Filtration is a bulk flow process which is dependent on the hydrostatic pressure within the glomerulus, the hydrostatic pressure within Bowman's capsule, and the osmotic pressure within the glomerulus. Under normal conditions, the total blood volume is filtered approximately 60 times per day. By knowing the plasma concentration of a compound, it is possible to calculate the nephron's filtered load for that ion.

12.4 Tubule reabsorption is a process by which compounds within the filtrate can be brought back into the blood. Secretion is a process by which compounds that were not filtered can be brought into the nephron. Most compounds are reabsorbed or secreted by both passive and active transport mechanisms. The reabsorption of many molecules is coupled to sodium movement within the nephron. The amount of any substance that is secreted can be calculated from

$$\text{Amount Excreted} = \text{Amount Filtered} + \text{Amount Secreted} - \text{Amount Reabsorbed}$$

12.5 Single nephron filtration rates are similar to the total glomerular filtration rates that were discussed earlier, but are used to describe the actions of single nephrons. This can be quantified from

$$SNFR = Q_{\text{tubule fluid}}\left(\frac{[\text{inulin}]_{\text{tubule fluid}}}{[\text{inulin}]_{\text{plasma}}}\right)$$

In general, under physiological conditions, single nephron filtration rates average to approximately 100 nL/min, but can range from as low as 20 nL/min to as high as 200 nL/min. Using the single nephron filtration rate and the glomerular filtration rate, one can approximate the number of functional nephrons as 1.2 million/kidney. The resistance of the renal microcirculation is critical to the flows through the nephron units, these can be quantified from

$$R_{AA} = \frac{P_{MA} - P_{GC}}{Q_G}$$

and

$$R_{EA} = \frac{P_{GC} - P_{PC}}{Q_{EA}}$$

to give a total renal resistance of

$$R_T = R_{AA} + R_{EA}$$

12.6 Work has shown that only approximately 5% of the net reabsorption within the proximal tubule system can be accounted for by the hydrostatic and osmotic pressure gradients that exist between that tubule system and the peritubular capillaries. Many studies have shown that the flow of sodium (and other ions) induces a net movement of water out of the proximal tubule system. This is accompanied by an increase in the hydrostatic pressure of the interstitial space. The reabsorptive coefficient, which is a measure of the nephron function can be represented as

$$K_r = \frac{SNFR - Q_{TS}}{\Pi_{PC} - \Pi_{IS} - P_{PC} - P_{IS}}$$

12.7 Sodium reabsorption is regulated by the amount and type of channels present within the luminal surface of the nephron epithelial cells. Sodium movement into the epithelial cells is coupled to glucose or hydrogen movement. Sodium movement out of the epithelial cells is regulated by the sodium–potassium ATPase. The balance between sodium reabsorption/secretion is coupled to both the juxtaglomerular apparatus and the renin–angiotensin system. Water movement out of the nephron is either by bulk diffusion or is facilitated through the presence of aquaporin channels. Reuptake of proteins (or amino acids) and other metabolically important molecules are typically coupled to sodium transport with highly specific transporters that are specific to the renal epithelial tissue.

12.8 Autoregulation of the kidneys is important to maintain the filtration conditions under altered inflow conditions to the kidney. A significant amount of work has shown that mechanisms are present within the kidney to regulate and maintain the blood flow to and through the glomerular and peritubular capillaries at near constant levels. There are a number of theories that try to describe the mechanisms that are responsible for autoregulation changes, which include the cell separation theory, the myogenic theory, and the metabolic theory.

12.9 Compartmental analysis is a technique that is used to describe the movement of various solutes throughout different systems. Fick's law of diffusion is the governing equation for this type of analysis. It states that

$$\frac{dc}{dt} = -\frac{DA}{V}\frac{dc}{dx}$$

Solving this for a simple two-compartment system, the change in concentration of a single species would be

$$\frac{dc_1}{dt} + \frac{DA}{\Delta x}\frac{c_1(V_1 + V_2)}{V_1 V_2} = \frac{DA}{\Delta x}\frac{c_0}{V_2}$$

12.10 Dialysis is a method used to remove wastes from the blood, when the kidneys are no longer functioning properly or efficiently. An extracorporeal device is designed with both a blood compartment and a filtrate compartment. The direction of blood flow and filtrate flow is typically opposing to maintain a concentration gradient along the entire device. There are different mechanisms that can be used to initiate the dialysis; these include the AV fistula method or peritoneal dialysis. Each has their advantages and shortcomings.

12.11 Renal calculi appear when the urine becomes supersaturated with salts. These salts can crystallize, and therefore, they are no longer in solution. As these crystals pass through the nephron (or kidney) they can become lodged within the tube, thus preventing flow through that tube. The majority of renal calculi are formed from calcium salts. There are many kidney diseases which are characterized by altered urine production and/or altered concentration of compounds within urine. Some of these diseases are hereditary in nature.

HOMEWORK PROBLEMS

12.1 The functional unit of the kidney is the nephron. Discuss some of the important functions that the nephron performs and how the nephron performs those functions.

12.2 Urine formation involves all of the following process except which of the following and why?

a. reabsorption of water
b. reabsorption of particular solutes
c. hydrostatic pressure within the glomerular capillaries
d. secretion of excess nutrients
e. filtration of blood

12.3 What is the effect of having two capillary beds within the renal circulation? How is the blood flow regulated within each of these capillary beds?

12.4 By increasing the diameter of the afferent arteriole and decreasing the diameter of the efferent arteriole, what effect can one predict to occur for the glomerular filtration rate?

12.5 Discuss the juxtaglomerular complex functions and locations.

12.6 Calculate the expected filtered load for potassium ions, chlorine ions, and calcium ions. What would you anticipate the percent reabsorption for each of these ions to be? Why?

12.7 There are some compounds that are filtered by the kidneys but not reabsorbed (such as mannitol). Would a sudden increase in the mannitol concentration have an effect on the glomerular filtration rate and the amount of urine produced? If it has an effect, why?

12.8 The daily elimination of magnesium is approximately 0.15 g/day at a concentration of 14 mg/dL. Calculate the amount filtered and amount reabsorbed if we assume that the nephron does not secrete any magnesium.

12.9 Cellular matter enters the nephron at a rate of approximately 600 cells/mL of urine produced (combined red blood cells and white blood cells). What are some of the reasons that cellular matter can enter the nephron? If there was an increase in the red blood cell concentration in the urine, what would this suggest? How about an increase in the white blood cell urine concentration?

12.10 For the system shown in Figure 12.11, determine a single differential equation that depicts the movement of the solutes between the various compartments as a function of time, rate constants, and solute. Assume that the concentration of solute in compartment 1 at time zero is $q_1(0)$, in compartment 2 at time 0 is 0, and in compartment 3 is $q_3(0)$.

12.11 Find the maximum concentration of solute in the plasma if $K_1 + K_2 = 0.012\ \text{min}^{-1}$ and K_3 is equal to $0.034\ \text{min}^{-1}$. At time 0, 75 g of the solute was ingested into the digestive system compartment and there was no solute in the plasma. See Figure 12.12 for the compartmental setup.

12.12 Renal arteriosclerosis is a risk factor for hypertension. Why and what are some of the treatments that you would recommend to lower the blood pressure?

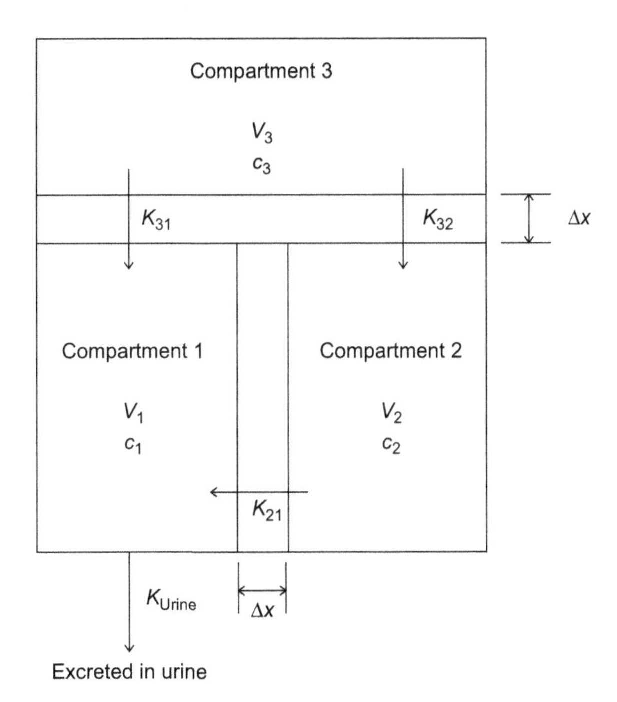

FIGURE 12.11 Three-compartment model for Homework Problem 12.10.

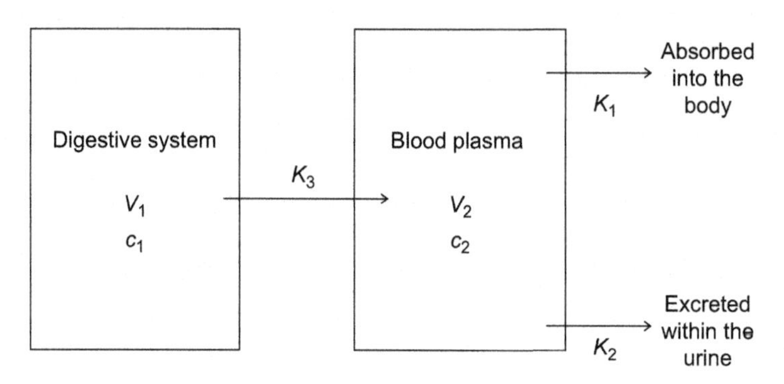

FIGURE 12.12 Two-compartment model for Homework Problem 12.11.

12.13 Knowing one single nephron filtration rate, would you be able to accurately predict the glomerular filtration rate? What information would you need to make a relationship between these variables? What assumptions would you need to make regarding this relationship? Do you predict that these would be valid assumptions?

12.14 We defined autoregulation in terms of vascular function. The kidney has similar autoregulation features. Do you suppose that the autoregulation features of the vasculature and the kidney interact with each other or do they act autonomously? Would it be beneficial to have some overlap over these two systems, if you believe that there is no cross-talk?

12.15 From your knowledge of inulin concentration in the plasma (which is approximately 3 mg %) and within the tubule fluid (you may need to approximate this), calculate the flow rate of tubule fluid if the single nephron filtration rate is 125 nL/min.

12.16 Given that the single nephron filtration rate averages to 125 nL/min, that the single nephron filtration fraction is 0.35 and that the hematocrit in the afferent arteriole is 0.52, calculate the flow rate within the efferent arteriole. Additionally, if the pressure difference across the glomerular capillaries and the vasa recta is 35 mmHg, determine the resistance of the efferent arteriole.

12.17 Discuss the reabsorptive coefficient for different segments of the nephron tubule system. Would there be differences in the reabsorptive coefficient along these different segments? If so, where would the differences manifest (e.g., in the net difference in the flow rates or in the net difference of the pressures acting across the segments of interest or both)?

References

[1] C.A. Berry, H.E. Ives, F.C. Rector, Renal transport of glucose, amino acids, sodium, chloride and water, in: B. M. Brenner (Ed.), Brenner and Rector's the Kidney, fifth ed., W. B. Saunders, Philadelphia, 1996.

[2] W. Bowman, On the structure and use of the Malpighian Bodies of the Kidney with Observations on the Circulation through the Gland, Philos. Trans. R Soc. London Ser. B 132 (1842) 57.

[3] B. Braam, K.D. Mitchell, J. Fox, L.G. Navar, Proximal tubular secretion of angiotensin II in rats, Am. J. Physiol. 264 (1993) F891.

[4] B. Braam, K.D. Mitchell, H.A. Koomans, L.G. Navar, Relevance of the tubuloglomerular feedback mechanism in pathophysiology, J. Am. Soc. Nephrol. 4 (1993) 1257.

[5] B. Braam, L.G. Navar, K.D. Mitchell, Modulation of tubuloglomerular feedback by angiotensin II type 1 receptors during the development of Goldblatt hypertension, Hypertension 25 (1995) 1232.

[6] B.M. Brenner, R. Beeuwkes, The renal circulations, Hosp. Pract. 13 (1978) 35.
[7] J.O. Davis, R.H. Freeman, J.A. Johnson, W.S. Spielman, Agents which block the action of the renin–angiotensin system, Circ. Res. 34 (1974) 279.
[8] J.O. Davis, R.H. Freeman, Mechanisms regulating renin release, Physiol. Rev. 56 (1976) 1.
[9] J.M. Gonzalez-Campoy, J. Kachelski, J.C. Burnett Jr., J.C. Romero, J.P. Granger, F.G. Knox, Proximal tubule response in aldosterone escape, Am. J. Physiol. 256 (1989) R86.
[10] A.C. Guyton, D.B. Young, J.W. DeClue, N. Trippodo, J.E. Hall, Fluid balance, renal function, and blood pressure, Clin. Nephrol. 4 (1975) 122.
[11] A.C. Guyton, J.E. Hall, J.P. Montani, Kidney function and hypertension, Acta Physiol. Scand. 571 (Suppl) (1988) 163.
[12] J.A. Haas, J.P. Granger, F.G. Knox, Effect of renal perfusion pressure on sodium reabsorption from proximal tubules of superficial and deep nephrons, Am. J. Physiol. 250 (1986) F425.
[13] J.E. Hall, A.C. Guyton, T.E. Jackson, T.G. Coleman, T.E. Lohmeier, N.C. Trippodo, Control of glomerular filtration rate by renin–angiotensin system, Am. J. Physiol. 233 (1977) F366.
[14] J.E. Hall, T.G. Coleman, A.C. Guyton, The renin–angiotensin system. Normal physiology and changes in older hypertensives, J. Am. Geriatr. Soc. 37 (1989) 801.
[15] J.E. Hall, M.W. Brands, J.R. Henegar, Angiotensin II and long-term arterial pressure regulation: the overriding dominance of the kidney, J. Am. Soc. Nephrol 10 (Suppl. 12) (1999) S258.
[16] A.A. Khraibi, J.P. Granger, J.A. Haas, J.C. Burnett Jr., F.G. Knox, Intrarenal pressures during direct inhibition of sodium transport, Am. J. Physiol. 263 (1992) R1182.
[17] M.A Knepper, J.B. Wade, J. Terris, C.A. Ecelbarger, D. Marples, B. Mandon, et al., Renal aquaporins, Kidney Int. 49 (1996) 1712.
[18] J.P. Kokko, The role of the collecting duct in urinary concentration, Kidney Int. 31 (1987) 606.
[19] I. Ledebo, Principles and practice of hemofiltration and hemodiafiltration, Aritf. Organs. 22 (1998) 20.
[20] P.S. Malchesky, Membrane processes for plasma separation and plasma fractionation: guiding principles for clinical use, Ther. Apher. 5 (2001) 270.
[21] C.R. Robertson, W.M. Deen, J.L. Troy, B.M. Brenner, Dynamics of glomerular ultrafiltration in the rat III. Hemodynamics and autoregulation, Am. J. Physiol. 223 (1972) 1191.
[22] J.M. Sands, Regulation of renal urea transporters, J. Am. Soc. Nephrol. 10 (1999) 635.
[23] V.M. Sanjana, P.M. Johnston, W.M. Deen, C.R. Robertson, B.M. Brenner, R.L. Jamison, Hydraulic and oncotic pressure measurements in inner medulla of mammalian kidney, Am. J. Physiol. 228 (1975) 1921.
[24] A.M. Scher, Focal blood flow measurements in cortex and medulla of kidney, Am. J. Physiol. 167 (1951) 539.
[25] H.A. Schroeder, A.E. Cohn, Reaction of renal blood flow to partial constriction of the renal artery, J. Clin. Invest. 17 (1938) 515.
[26] A.E. Taylor, Capillary fluid filtration. Starling forces and lymph flow, Circ. Res. 49 (1981) 557.
[27] J.A. Trueta, A.E. Barclay, P.M. Daniel, K.J. Franklin, M.M.L. Prichard, Studies on the Renal Circulation, Blackwell, Oxford, UK, 1948.
[28] E.J. Weinman, M. Kashgarian, J.P. Hayslett, Role of peritubular protein concentration in sodium reabsorption, Am. J. Physiol. 221 (1971) 1521.
[29] D.B. Young, Y.J. Pan, A.C. Guyton, Control of extracellular sodium concentration by antidiuretic hormonethirst feedback mechanism, Am. J. Physiol. 232 (1977) R145.
[30] D.B. Young, T.E. Lohmeier, J.E. Hall, J.E. Declue, R.G. Bengis, T.G. Coleman, et al., The role of the renal effects of angiotensin II in hypertension, Cor. Vasa. 22 (1980) 49.
[31] M.L. Zeidel, T.G. Hammond, J.B. Wade, J. Tucker, H.W. Harris, Fate of antidiuretic hormone water channel proteins after retrieval from apical membrane, Am. J. Physiol. 265 (1993) C822.
[32] M.L. Zeidel, Hormonal regulation of inner medullary collecting duct sodium transport, Am. J. Physiol. 265 (1993) F159.
[33] M.L. Zeidel, K. Strange, F. Emma, H.W. Harris Jr., Mechanisms and regulation of water transport in the kidney, Semin. Nephrol. 13 (1993) 155.

CHAPTER

13

Splanchnic Circulation: Liver and Spleen

LEARNING OUTCOMES

1. Describe the physiology and the anatomy of the liver
2. Identify the functional unit of the liver
3. Examine the filtration of salient species within the liver and spleen
4. Describe the physiology and the anatomy of the spleen
5. Discuss the salient aspects of the splanchnic circulation
6. Compare the blood flow through the liver and spleen under various physiological and pathological conditions
7. Describe the flow of blood through the microcirculatory units within the liver and the spleen
8. Discuss how the liver can store and release blood under various physiological conditions
9. Compare the active and passive components of the splanchnic circulation
10. Examine the role of sympathetic nervous system activity on the spleen
11. Discuss disease conditions that commonly affect the liver and the spleen

13.1 LIVER AND SPLEEN PHYSIOLOGY

The splanchnic circulation is a specialty circulation that is associated with the gastrointestinal system and is composed of the gastric (stomach), small intestinal, colonic (large intestine), pancreatic, hepatic (liver), and splenic (spleen) circulations. What makes this circulation especially unique is that these circulations are largely in parallel with each other, supplied by three major arteries. These arteries are the celiac artery, the superior mesenteric artery, and the inferior mesenteric artery, which branch extensively into anastomosing arterioles and capillaries. Additionally, the importance of the splanchnic circulation is

© 2015 Elsevier Inc. All rights reserved.

illustrated by the distribution of the cardiac output to this circulation. Under normal physiological conditions, the splanchnic circulation receives approximately 20–25% of the cardiac output and this system extracts approximately 20% of the oxygen available in blood. Changes to this blood distribution and usage have been used as markers of stress to the cardiovascular system. We will focus our discussion on the liver circulation and the spleen circulation, which are of particular interest to biofluids mechanics. Each of these circulations has unique characteristics that relate to inflammation, thrombosis, absorption of nutrients, etc.

The liver is the largest visceral organ in the human body (with a mass of approximately 1.5 kg), is at the center of many metabolic processes, and regulates many other metabolic processes that occur in the body. The liver is surrounded by a tough and fibrous capsule that functions to protect the organ from injury. This capsule is divided by the falciform ligament that marks the division between the left and right lobes of the liver. The falciform ligament thickens into the round ligament toward the posterior section of the liver. This ligament is a remnant of the umbilical vein, which is the outflow location of blood during fetal development. There is a small region on the posterior section of the liver, which is termed the porta hepatis. This is the location where the hepatic artery proper, the hepatic portal vein, and the common bile duct enter/exit the liver.

The blood flow to the liver is quite unique. Nearly one-third of the blood that enters the liver arises from the hepatic artery proper. The hepatic artery proper is a branch of the common hepatic artery, which is a branch from the celiac trunk. The celiac trunk directly branches off from the abdominal aorta (Figure 13.1). Thus, this portion of the blood is typical arterial blood. The remaining two-thirds of the blood that enters the liver comes from the hepatic portal vein. The blood within the hepatic portal vein is venous blood that comes from the capillaries of the esophagus, stomach, small intestines, and the vast majority of the large intestines. Any blood vessel that connects to capillary beds is termed a portal vessel. The purpose of the hepatic portal system is to deliver nutrients absorbed from the gastrointestinal system directly to the liver for storage, excretion, or modification for later use. The hepatic portal vein obtains blood from the inferior mesenteric vein (which collects blood from the descending colon, the sigmoid colon, and the rectum), the splenic vein (which collects blood from the spleen, portions of the orad stomach, and the pancreas), and the superior mesenteric vein (which collects blood from the caudad region of the stomach, the small intestines, the ascending colon, the transverse colon, and a small portion of the pancreas) (Figure 13.1). The majority of the flow within the hepatic portal vein is contributed by the superior mesenteric vein. The blood entering the liver eventually makes it into the liver sinusoids, which is the major location in which the concentration of nutrients in the blood is modified. After passing through the liver sinusoids, the venous blood is collected into the hepatic veins, which converge directly with the inferior vena cava. Under normal physiological conditions, the concentration of nutrients within the blood is stable, even after ingestion, because of the actions of the liver.

Each main lobe of the liver is divided into approximately 100,000 lobules, which are the most basic functional unit of the liver (Figure 13.2). Each lobule is separated by an interlobular septum that helps to prevent transfer of materials between the different lobules. Lobules are relatively large with an average diameter in the range of 1 mm and have a hexagonal shape. At the center of each lobule is the central vein, which collects and mixes

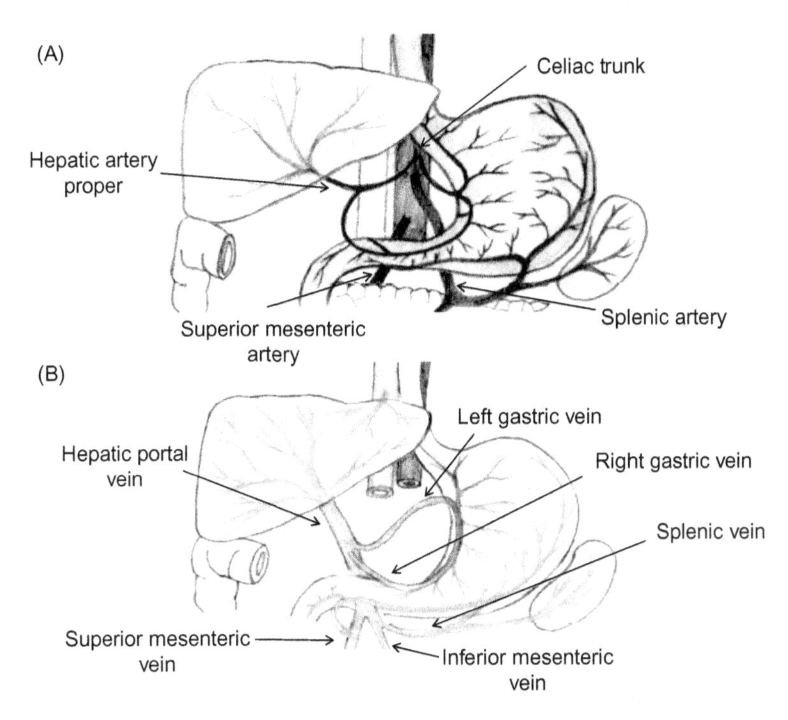

FIGURE 13.1 Arteries supplying the superior abdominal organs (A) and the hepatic portal system (B). These vascular systems illustrate that the liver obtains blood from the main arterial systemic circulation (as a branch of the celiac trunk) and from the venous circulation from the majority of the gastrointestinal system. *Source: Adapted from Martini and Nath (2009).*

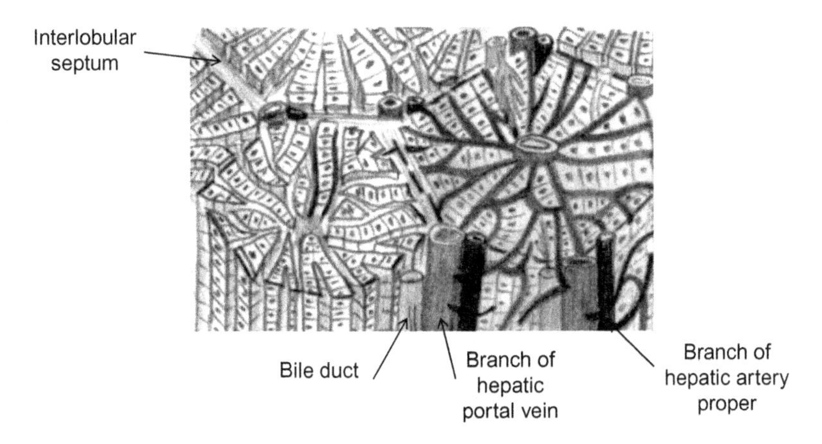

FIGURE 13.2 A schematic view of the liver lobule structure illustrating that branches of the hepatic artery proper and the hepatic portal vein reach the lobule level at each vertex of a hexagon-like shape. Capillaries from the vessels penetrate toward the central vein region, which is illustrated in Figure 13.4. Each lobule is approximately 1 mm in length. *Source: Adapted from Martini and Nath (2009).*

blood from both the hepatic artery proper and the hepatic portal vein. Each lobule is composed of numerous liver sinusoids, which are separated by a sandwich of endothelial cells (on the two blood sides, e.g., the bread of the sandwich) and hepatocytes (in between the endothelial cells, e.g., the meat of the sandwich). Blood flows through the sinusoids and collects in the central vein.

Sinusoids are a specialized type of fenestrated capillary beds that are found in various locations of the body. It is typical for sinusoids, including those found in the liver, to have a more flattened, rectangular shape instead of the more classical circular shape that is associated with typical capillaries. The endothelial cells that compose sinusoid capillaries are both fenestrated (e.g., have pores that pass through the cell from the apical side to the basolateral side) and have gaps between neighboring endothelial cells (e.g., the cell membranes of neighboring cells are not stitched together) (Figure 13.3). In many cases, the basement membrane surrounding the epithelial cells is very porous, thin, or even absent, which is the case in the liver sinusoids (compare this to the fenestrated renal filtration barrier discussed in Chapter 12). The consequence of these large openings in sinusoids, including those of the liver, is that all constituents of plasma, including the largest plasma proteins, can diffuse out of the capillary network into the space adjacent to the hepatocytes (Figure 13.4). A specialized macrophage, termed the Kupffer cell, resides in the liver sinusoids. Their role is to screen and engulf any foreign particle or damaged blood cell that enters the liver.

Blood enters the liver sinusoids from branches of both the hepatic artery proper and the hepatic portal vein. Recall that each lobule is a hexagonal shape made up of numerous sinusoids. A small branch from the hepatic artery proper, a small branch from the hepatic portal vein, and a small bile duct form a triad at each of the six corners of a liver lobule. Branches from the two blood vessels converge to form the sinusoids, thus one side of the sinusoids contains venous blood from the gastrointestinal tract, whereas the other side of the sinusoid contains arterial blood from the aorta. Since, there is a very well-defined organization of the liver sinusoids, each triad delivers blood to three liver lobules. The

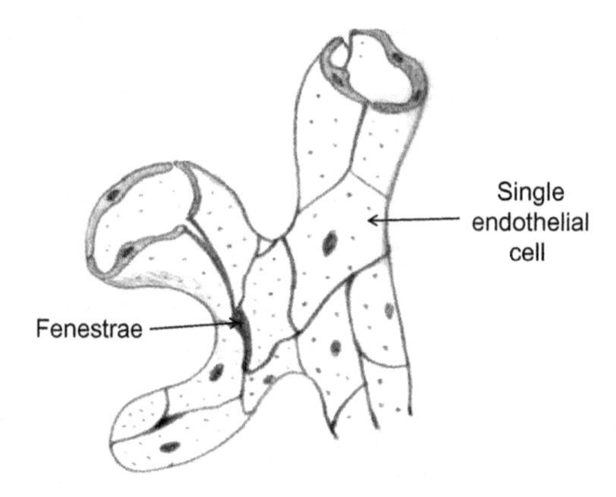

FIGURE 13.3 Schematic of a fenestrated capillary system illustrating that there are large pores between neighboring endothelial cells. These large pores allow for the rapid exchange of water and solutes between blood and interstitial fluid and penetrate the entire endothelial wall. *Source: Adapted from Martini and Nath (2009).*

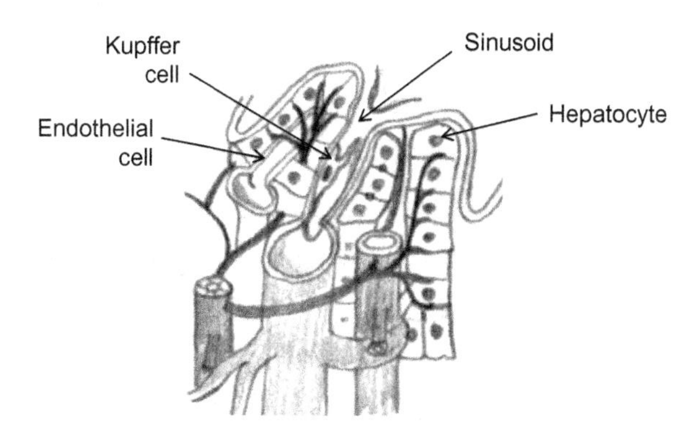

FIGURE 13.4 Schematic of a single liver lobule illustrating both the cells that comprise the liver and sinusoid subunit. Blood from the hepatic artery proper and the hepatic portal vein converge at the sinusoid level, are filtered by hepatocytes and Kupffer cell, and then pass to the central vein location. *Source: Adapted from Martini and Nath (2009).*

hepatocytes absorb solutes from the plasma throughout the length of the sinusoid and secrete plasma proteins into the plasma for later use. This modified blood enters the central vein, which eventually converges into the hepatic vein, which converges with the inferior vena cava.

The liver conducts hundreds of functions, which fall into three general categories. The first of these functions is the regulation of metabolism. Blood composition is primarily regulated by the actions of the liver. As mentioned earlier, all blood from the gastrointestinal tract enters the liver first prior to entering the general systemic circulation. Since the nutrient concentration in blood is fairly stable, regardless of ingestion periods, this suggests that the liver extracts all of the excess nutrients from the hepatic portal system and corrects deficiencies in the same nutrients. Similarly, any toxin that enters the blood from the gastrointestinal tract, whose absorptive surface is technical outside the body, is cleared by the liver.

In terms of carbohydrates, the liver maintains the blood glucose level at approximately 90 mg/dL. If the glucose concentration exceeds this value, hepatocytes absorb glucose from the bloodstream and either store excess glucose as glycogen or synthesize lipids from the excess glucose. If the glucose concentration falls below this level, glycogen is broken down into glucose and the new glucose is secreted into the bloodstream. Similar mechanisms exist for lipids and amino acids. Excess of either of these metabolites can be removed from the circulation and stored for later use. When lipid levels drop, the liver secretes lipids into the bloodstream. The liver also stores minerals, such as iron, and fat-soluble vitamins. Many drugs can be metabolized within the liver and waste products can be removed by actions taken by the liver.

The second major function of the liver is the regulation of blood components that are not related to metabolic needs of the body. For instance, the liver produces many plasma proteins including albumin, clotting proteins, transport proteins, and complement proteins. Any damaged blood cell or a blood cell that is presenting a pathogen can be removed by the liver. The liver plays a prominent role in the recycling of released hormones, including insulin, epinephrine, and many of the sex hormones (e.g., estrogen), and the removal of antibodies.

Finally, the liver produces and secretes bile into the gastrointestinal tract via the gallbladder. Bile is a watery mixture, which contains some ions, bilirubin, and various bile salts (which are lipids). The bile salts help in the digestion of non-water soluble lipids, by emulsifying the lipids. If the bile salts are not present, large ingested lipids coalesce to form large lipid droplets, which cannot be absorbed or broken down by digestive enzymes. The bile salts act to stabilize small droplets of lipids, thereby increasing the surface area for enzyme activities and allowing lipids to eventually be absorbed along the gastrointestinal tract.

The spleen plays a salient role in the lymphatic system and immune reactions that occur throughout the body. The largest lymphoid tissue in the adult body is the spleen. The adult spleen is approximately 12 cm long and has a mass of approximately 150 g. The spleen is located along the orad section of the greater curvature of the stomach and is in close proximity to the stomach, the left kidney, the diaphragm, and the pancreas. The posterior portion of the spleen toward the inferior portion contains a hilum, where the splenic artery, the splenic vein, and the splenic lymphatic vessels enter/exit the spleen (Figure 13.5). The hilum is located at the junction between the gastric (stomach) region and the renal region.

The human spleen is surrounded by a thick fibrous tissue composed of mostly collagen and elastin (this is in contrast to other mammals, that have a layer of smooth muscle cells within this fibrous tissue, allowing for the spleen to "eject" blood components). The interior portion of the spleen is composed of the red pulp, which contains blood, and the white pulp, which is similar to lymph nodes.

Immediately after the splenic artery enters the hilum, it branches into a number of small arteries, termed trabecular arteries, which spread throughout the red and white pulp. The

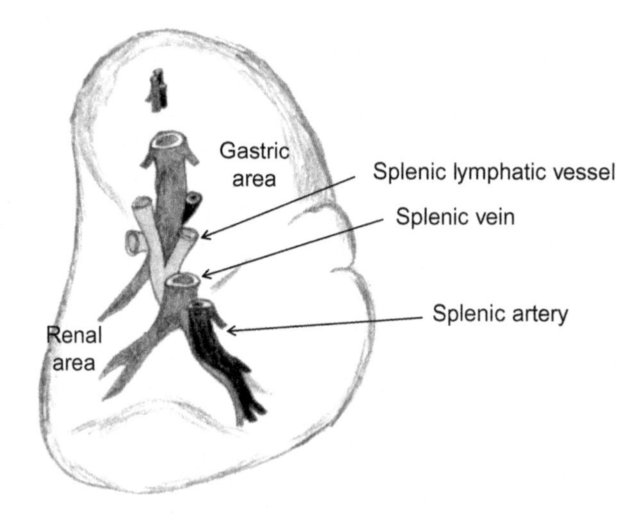

FIGURE 13.5 Posterior view of the spleen illustrating the hilum where the splenic artery, splenic vein, and the splenic lymphatic vessel enter/leave the spleen. *Source: Adapted from Martini and Nath (2009).*

trabecular arteries continually branch into small arterioles and capillaries that eventually enter into the red pulp regions. In general, the white pulp surrounds medium to small size arterioles. The small capillaries empty into sinusoids that are lined by macrophages and that contain a mesh-like framework of proteins. The role of the macrophages and the mesh-like structure are to identify, filter, and engulf foreign particles. Additionally, since the white pulp, which acts as a lymph node, is in close proximity to this sinusoid, any inflammatory response is quickly sensed and cleared by lymphocytes in the white pulp. After flowing through the splenic sinusoids, blood enters small venules that converge to the trabecular veins and eventually the splenic vein.

In a general sense, the spleen has a very similar function as lymph nodes, which were described in Chapter 8 and some additional functions that are related to the regulation of blood properties. First, the spleen can remove abnormal/damaged blood cells, especially platelets, by phagocytosis mechanisms. Some materials from these cells may be recycled for use in other locations of the body. The spleen also has a relatively large store of red blood cells, primarily from the clearance and recycling of any damaged red blood cells. Finally, as with most lymph nodes, the spleen can initiate an immune response that involves B- and T-cells. Thus, the spleen plays a large role in the adaptive immunity reactions.

13.2 HEPATIC/SPLENIC BLOOD FLOW

As we have mentioned earlier, the liver obtains approximately 25% of the cardiac output, over which approximately 70–75% of this comes from the hepatic portal system. The distribution of hepatic portal blood and hepatic artery proper blood is homogenous throughout the liver, and the inflow pressure of either branch has been shown to have little effect on this distribution and perfusion of liver tissue. However, it has been shown that when vessels prior to the lobule regions become obstructed there is a shift of blood from the vessel that is occluded (for sake of argument the hepatic portal branch) to the vessel that is not occluded (using the same argument the hepatic artery proper). This has been described fully in many experimental conditions where a reduction in the blood flow (or pressure) within a branch of the hepatic portal blood flow elicited an increased flow through the neighboring branch of the hepatic artery proper. With a reduced flow in branches of the hepatic artery proper, a decrease in the vascular resistance of the neighboring branch of the hepatic portal vein was observed, but the flow through this vessel was not altered significantly. There are many theories explaining these mechanisms, but it appears that since the majority of the oxygenated blood comes from the hepatic artery proper, the system is designed as to either maintain this flow and/or to maintain the concentration of nutrients throughout the liver tissue. Much of these responses appear to be related to myogenic processes, however, there are some passive effects that have been observed in this system. The myogenic responses appear to be the more likely effects, since the pressures are very different between the hepatic artery proper system (in the range of 70 mmHg) and the hepatic portal system (in the range of 10 mmHg). Thus, small changes in pressure may be able to elicit significant and differing effects between the two systems.

Splenic blood flow appears to be more variable than other organs and averages approximately 100 mL/(min 100 g). It appears that splenic flow can be altered based on the activity of other organs and there may be some rhythmic activity that can describe the filling of the spleen. The splenic flow is somewhat dependent on the activity of the splenic nerve and presence/absence of hormones within the blood. However, along with this marginal splenic activity, the human spleen does not have a large role as a blood reservoir (remember that other animals have smooth muscle cells surrounding their spleen to induce blood ejection, in fact rhythmic activity of the spleen of these animals has been linked to the cardiac cycle and the intra-arterial pressure). In the case of the human spleen, it appears that the activity is restricted to the resistance of the splenic artery but the total volume of the spleen does not change significantly, thus there is little ejection/absorption of blood based on the activity of other regions of the human body. It is possible that this activity is related to the storage of red blood cells within the spleen because the spleen tends to collect red blood cells that are toward the end of their useful time. If needed, these cells may be released prior to recycling processes based on the activity of the spleen.

13.3 HEPATIC/SPLENIC MICROCIRCULATION

We have previously discussed the anatomical organization of the hepatic microcirculation but the physiological relevance of this organization and the overall control of blood perfusion have not been described. Recall, that the hepatic microcirculation is defined within the hepatic sinusoids. At this region, the circulation from the hepatic artery proper and the hepatic portal vein converge into a small capillary. The sinusoids have large fenestrae and gaps between neighboring endothelial cells to allow plasma and the constituents of plasma to have free access to the spaces of Disse (an anatomical region between endothelial cells and hepatocytes) and the hepatocytes that line the sinusoids region. Recall, that it is the hepatocytes that have a large role in modifying the plasma protein concentration and the plasma solute concentration. The endothelial cells near the entrance to the sinusoid region and those that line the central vein have been shown to modify the intra-sinusoidal pressure as a means of regulating the flow/pressure throughout the liver sinusoid. Under most conditions, the pressure throughout the sinusoids varies from approximately 12 mmHg at the inflow to 4 mmHg at the central vein. Note that the precapillary sphincters of the hepatic microcirculation are regulated by the nervous stimuli, hormones, and typical vasoactive substances.

There is an interesting segregation of activity within the hepatic microcirculation. Cells close to the inflow region of the liver sinusoids, where the hepatic artery proper and the hepatic portal vein blood mixes, are geared to absorption/secretion of sugars and amino acids. Cells closer to the central vein region produce/metabolize steroids and pigment. Thus, changes in the flow through the circulatory unit can drastically alter the composition of blood, since there is little to no sharing of functions between hepatocytes along the sinusoid.

The anatomical organization of the splenic microcirculation has not been fully investigated at the time of writing. The splenic microcirculation is composed of sinuses, similar to those described for the liver in previous sections, however, unlike all other sinuses in the human body, the splenic sinuses are not the equivalent of capillaries. Splenic sinuses initiate within the venous microcirculation and continue through the venules. This suggests that the splenic sinuses are in addition to well-defined typical capillary beds. In addition to this, the spleen contains cavities where red blood cells, white blood cells, and platelets can reside not surrounded by endothelial cells. These regions are surrounded by fibroblasts and a well-defined extracellular matrix zone, forming an open mesh-like circulation.

There has been significant doubt in the splenic microcirculation field as to whether two microcirculation organization can be found in the spleen. The first is termed an open microcirculation where terminal arterioles drain into these well-defined open cavities that then collect into the venule sinuses of the spleen. This has been observed multiple times, however, an anastomosing closed loop system cannot be excluded based on the experiments conducted so far. This closed system would act as a shunt around a capillary bed, where the terminal arterioles would be connected directly to the venule sinusoids. While, the majority of studies have identified and suggest that the vast majority of the red pulp follows an open capillary structure, some of closed loop microcirculation has been observed in patients. It is unclear at this point, if closed microcirculations are associated with pathological events that have or will happen. Interestingly, it is also unclear how the physiology has been designed that can maintain platelets and white blood cells in a quiescent state when flowing through these spaces that are defined by extracellular matrix. If a platelet observes extracellular matrix, it would typically adhere and/or aggregate within this space. Also, and possibly most importantly, this open splenic microcirculation is the only known space in the human body where under normal conditions blood flows in a space that is not surrounded by endothelial cells.

13.4 STORAGE AND RELEASE OF BLOOD IN THE LIVER

As we have discussed, a significant portion of the cardiac output is delivered to the liver. Therefore, the liver can act as a blood reservoir and it is important to understand how the liver regulates these storage properties. One of the primary regulators of the blood volume within the liver is the hepatic sympathetic nerve. An increase in the activity of the hepatic sympathetic nerve, in the absence of any vagus nerve activity (due to surgical removal), causes a reflexive increase in the hepatic arterial resistance. This is due to both the increase in the hepatic nerve activity and an increase in catecholamine release from the adrenal gland. The net effect of this activity is a reduction in the hepatic volume by approximately 5–10%. It has also been observed that a decrease in carotid artery pressure causes a net decrease in the hepatic volume by approximately 20%. This is due to a baroreceptor reflex that activates the hepatic sympathetic nervous system. Similarly, an increase in the carotid artery pressure induces an increase in the uptake of blood by the liver; thus the liver in conjunction with pressure sensing locations in the vascular system can act to regulate the blood pressure by altering the blood volume. Under similar conditions, the cardiopulmonary receptors can also regulate the hepatic storage/release of blood. Decreased pressure at the cardiopulmonary receptors induced a decrease in the hepatic blood volume. The opposing effects were also observed; with an increased pressure at the cardiopulmonary receptor an increase in the hepatic blood volume was observed.

Interestingly, the majority of the baroreceptor reflex responses occur very rapidly; generally within 1 min of the receptor sensing a change. This suggests, (1) the importance of blood pressure to these locations (e.g., the brain and the lungs) and (2) the importance of the liver as being a source for blood and a regulator of blood volume/pressure. Under these rapid responses, it is common to see upward of 30–40% of the hepatic blood volume be mobilized or an increase in hepatic blood volume of around 25%. These changes can produce substantial changes to the systemic blood pressure.

Other mechanisms that can regulate hepatic blood volume are the hepatic venous pressure, significant blood volume changes (e.g., hemorrhage), and vasoactive compounds. As expected, with an increased hepatic venous pressure there was a parallel, rapid increase in hepatic blood volume. The relative change in hepatic volume as a function of change in hepatic venous pressure differs based on the species, but in general with an increased hepatic venous pressure in the range of 10 mmHg, an increase in hepatic blood volume was approximately 40 mL/100 g of tissue. Some species exhibit a nonlinear rate between an increase in hepatic blood volume and hepatic venous pressure and in most instances as the pressure increases, the rate of storage of blood in the liver also increases. Thus, this is an important mechanism to regulate rapid changes in blood volume. It is also interesting to note, for a general tissue an increase in pressure disturbs the balance of the Starling forces for 10–15 min. After this time, either a return to steady state is observed, so that the exchange of nutrients and wastes is maintained. However, in the liver, it was observed that an increase in the hepatic venous pressure is coupled with an increase in the filtration/retention of fluid within the liver extravascular space. This is maintained for hours. Note that the liver is especially susceptible to changes in pressure, because the osmotic pressure gradient across the liver sinusoids is effectively zero (since the liver pores are so large, there is no difference in the constituents, even including proteins, across the liver endothelial cells). The liver lymphatics and the peritoneal space can accommodate large changes in volume and thus do not hinder the net movement of fluid across the liver sinusoid wall.

In moderate to severe hemorrhage, a significant decrease in hepatic blood volume is observed, to approximately 80% of the normal liver blood volume. While there was a decrease in hepatic blood volume, this was accompanied by an active constriction of the venous circulation. In anesthetized animals, it has been observed that a hemorrhage of up to 20% loss of systemic blood volume induces a similar release of hepatic volume (approximately 25% of hepatic blood was released with a 20% hemorrhage). Interestingly, this release of hepatic blood volume after severe hemorrhage has been shown to be independent of signaling from the kidneys, the pituitary gland, the adrenal gland, and the nervous system, thus local changes can regulate the hepatic blood volume in times of significant demand. However, under these conditions, the total demand for increased blood volume could not be met by the liver alone.

The liver is also responsive to various vasoactive compounds. For instance, when subjected to epinephrine, at either the hepatic artery or the portal vein location, a reduction in hepatic arterial flow was observed, however, there was no net effect on the venous outflow. Thus, a net decrease in hepatic blood volume would be observed in response to epinephrine. Histamine infusion was observed to decrease the hepatic arterial resistance and increase the portal vein resistance. Under these conditions, a net increase in the hepatic blood volume was observed. Exposure to vasopressin had similar effects as histamine on the hepatic circulation and the hepatic blood volume.

13.5 ACTIVE AND PASSIVE COMPONENTS OF THE SPLANCHNIC CIRCULATION

We have already described some of the active components of the splanchnic circulation but it is important to reiterate the role of these active components at this point. The importance of the splanchnic circulation is not solely restricted to times of ingestion or times of nutrient need. During exercise, the splanchnic blood volume has been shown to decrease rather rapidly to accommodate the increase in oxygen consumption from other tissues. These changes to the splanchnic blood volume have been linked to both active components of the circulation and passive components of the circulation. This will be reviewed briefly here.

The passive components of the splanchnic circulation are directly related to an increase or decrease in the flow rates into or out of the splanchnic circulation. With an increased inflow flow rate, there was almost an immediate linear increase in the splanchnic blood pressure and the splanchnic blood volume. Decreased inflow flow rate was associated with a decreased splanchnic blood pressure and a decreased splanchnic blood volume. Opposing effects are typically observed with alterations to the outflow flow rate (which is modulated at the hepatic portal vein). A decreased outflow flow rate would increase the splanchnic pressure and the splanchnic blood volume whereas increased outflow flow rates would decrease the splanchnic pressure and blood volume.

Active components of the splanchnic circulation include the nervous innervation via the splanchnic sympathetic nerves. With an activation of this sympathetic nerve pathway, there is a relatively rapid decrease in splanchnic blood flow coupled with an increase in the splanchnic outflow. These changes were directly coupled to diameter changes of the splanchnic resistance and capacitance vessels. The degree of diameter change was correlated with the frequency of sympathetic nervous activity. We have already discussed the effects of the carotid baroreceptor reflex on the hepatic circulation and this can be extended to the entire splanchnic circulation. A decrease in pressure along the carotid sinuses induced a mobilization of blood from the splanchnic circulation. Thus, an increase in splanchnic blood flow and a decrease in splanchnic blood volume would be observed under these conditions. Most of these changes were observed within splanchnic compliance vessels and not the resistance vessels. Many of the remaining active components of the splanchnic circulation mimic those already discussed for the hepatic circulation.

13.6 INNERVATION OF THE SPLEEN

The spleen has significant innervation that is specific for the arterial resistance vessels, the venous capacitance vessels, and the splenic capsule. Upon activation of the splenic sympathetic nervous system, there is a significant release of blood from the spleen. This is induced by a decrease in the diameter of resistance vessels, an increase in the diameter of capacitance vessels, and a contraction of smooth muscle cells located with the splenic capsule. The magnitude of the response of these changes is dependent on the frequency of firing and the amplitude of the changes observed. In the human spleen, the maximal responses were observed to occur when the stimulation frequency was in the range of 10 Hz; however, the change in splenic blood volume was relatively small since the capsule does not contain a significant smooth muscle cell population. Interestingly, in animals that contain a significant smooth muscle cell region of the spleen released blood at a very high hematocrit (in the range of 80%) after sympathetic nerve activity. This shows the importance of the spleen red pulp as a source for red blood cells in a time of need.

As we discussed for the entire splanchnic circulation, the changes to the spleen blood volume and the spleen blood flow rate should exhibit both passive and active components. However, in the case of the spleen, it appears that either the passive components do not contribute to overall changes in the blood volume/blood flow rate or the active components are so significant that they overwhelm any all other possible mechanisms. Work has been conducted on both dogs and cats to tease out this possible difference and it was observed that splenic blood volume and blood flow are largely independent of the inflow pressure within physiological ranges. This suggests that the spleen can somewhat function autonomously, as long as it receives some blood supply. However, similar to the other organs within the splanchnic circulation, when baroreceptor activity increases, the spleen blood volume and outflow flow rate increase to increase the overall systemic blood volume and blood pressure. In the case of humans, it appears that sympathetic nerve activity and/or infusion of vasoactive compounds (the same compounds that were discussed earlier) altered the resistance of the resistance and capacitance vessels but did not have a significant effect on the spleen volume.

13.7 DISEASE CONDITIONS

13.7.1 Hepatitis

Hepatitis is any sustained inflammation of the liver that lasts for more than 1 year. Most cases of hepatitis are accompanied by hepatocyte death and in the most severe cases of hepatitis a deterioration of the liver lobule/sinus anatomy is observed. Hepatitis can initiate by the hepatitis B or C virus, drugs, toxins, and other metabolic diseases. Different pathophysiologies of hepatitis have been observed but possibly the most common is the periportal hepatitis, wherein many different regions of the liver experience necrosis. These regions of inflammation tend to be very difficult to treat and are encapsulated by scar tissue that may be hard to penetrate by treatment modalities. The second most common type of hepatitis is the subacute hepatic necrosis, in which different necrotic regions tend to bridge together to form large masses of necrotic tissue. It is more typical for this form of hepatitis to be associated with a cirrhosis.

Most forms of hepatitis tend to be asymptomatic and many patients with hepatitis believe that they are suffering from the common cold, flu, or similar disease. The "symptoms" associated with hepatitis are loss of appetite, fatigue, muscle pain, nausea, and abdominal pain. If these symptoms persist and antibiotics do not take care of the infection, it may be common to check for hepatitis. In most instances, hepatitis B can be treated like any other viral infection; rest eating healthy and drinking plenty of water. Alcohol should be avoided due to the negative effects of alcohol on the liver and patients with hepatitis B who have consumed significant quantities of alcohol have done more damage to their liver than just the disease. Patients with hepatitis C are generally treated with a combination of drugs.

13.7.2 Alcoholic and Fatty Liver Disease

Alcoholic liver disease can manifest in many ways but is associated with the overconsumption of alcohol. It is possible for hepatitis, cirrhosis, or other diseases of the liver to manifest after significant overconsumption of alcohol. Constant chronic consumption of alcohol, which is primarily detoxified in the liver, leads to increased pro-inflammatory responses in the liver, including the release of cytokines and oxidative stresses. If these pro-inflammatory responses persist, it leads to hepatocyte inflammation and apoptosis. This disease is found in approximately one-third of the heavy drinkers, but it does not seem to be related to the type, quantity, and rate of drinking among heavy drinkers, but instead appears to manifest due to a complex interaction of environmental and genetic factors. If alcoholic liver disease manifests, it is typically treated by a cessation of alcohol consumption. Liver damage may be reduced or even reversed after ending excessive alcohol consumption. However, medications and antioxidant treatments may be needed to correct some of the damage to the liver.

Fatty liver disease can be induced by alcohol consumption or other activities; here we will restrict our discussion to nonalcoholic fatty liver disease. Typically, this is classified as a fat content of the liver greater than 10% of the weight of the liver. For reference, the vast majority of heavy drinkers, upward of 90%, develop fatty liver disease. Nonalcoholic fatty liver disease is the most common cause of chronic liver diseases in the United States. Over 20% of the United States adult population have a form of nonalcoholic fatty liver disease. In some of these, the disease may never fully manifest, however, with poor diet and consumption of alcohol, the incidence of pathologies associated with this disease increases. Causes for nonalcoholic fatty liver disease are typically associated with diet and heredity factors. Nonalcoholic fatty liver disease tends to be asymptomatic, but you may experience fatigue and a loss of appetite. Unfortunately, no medications or other standard treatments exist for nonalcoholic fatty liver disease. In most cases, a physician would prescribe lifestyle changes.

13.7.3 Splenomegaly

Splenomegaly is an observed enlargement of the spleen and is associated with an increased workload of the spleen. Thus any disease where there is an enhanced deterioration of red blood cells, white blood cells, or platelets may lead to splenomegaly. Splenomegaly can be classified as moderate (largest spleen dimension in the range of 11–20 cm) or severe (largest spleen dimension over 20 cm). Symptoms for splenomegaly tend to be associated with abdominal pain but sometimes may extend to back pain. In many instances, splenomegaly is associated with an infection such as syphilis or endocarditis, although there are many instances of metabolic diseases that are associated with splenomegaly. The treatment for splenomegaly lies with the cause of the disease. In many instances, if the workload on the spleen is reduced, it is possible to see a reduction in the spleen size back to normal levels. However, in some instances, it is more common to remove the spleen from the patient. The side effects of a splenectomy are the possibility for an increased infection rate and the possibility for increased platelet counts. Since the spleen acts to filter pathogens from blood it is possible that some of these pathogens may not be cleared by the blood. Likewise, the platelet count is somewhat controlled by the activity of the spleen and thus with removal of the spleen, the count may increase to abnormal levels. An increased platelet count can lead to thrombosis at other locations with the circulation. Thus, patients tend to be maintained on anti-platelets, anti-thrombotic, and antibody treatments.

END OF CHAPTER SUMMARY

13.1 The splanchnic circulation is a highly unique and special circulation that is involved with the absorption of materials from the gastrointestinal system and the removal of risk factors and pathogens that enter the circulation from the gastrointestinal system. The majority of these clearance functions are accomplished by the liver (which removes both pathogens and nutrients from the blood prior to the blood entering the systemic circulation) and the spleen (which removes pathogens and damaged blood cells). The splanchnic circulation obtains approximately 20–25% of the cardiac output under physiological conditions, suggesting its importance to the overall composition of blood. The liver has two distinct circulations that converge within the sinusoids of the liver. The first is the hepatic artery proper, which is typically arterial blood, and the second is the hepatic portal vein, which is blood from the stomach, intestines, and spleen. The sinusoids of the liver are highly porous capillaries that allow the exchange of plasma proteins. The liver has many salient functions related to homeostasis. The spleen plays a large role in inflammatory responses and is composed of two distinct sections; the red pulp, which contains blood, and the white pulp, which resembles a lymph node. Small capillaries of the splenic circulation enter the spleen sinusoids, which are an open circulation. This is currently the only known circulation that is not surrounded by a layer of endothelial cells.

13.2 The hepatic blood flow is homogenous under most conditions, showing that the liver regulates blood from the hepatic artery proper and the hepatic portal vein to a level that equilibrates their flow in the sinusoids. Splenic blood flow is somewhat variable and appears to be based on the activity of the other organs.

13.3 The microcirculatory units of both the liver and spleen are somewhat unique. The microcirculation of the liver is designed to modify all of the plasma constituents. The endothelial cells are fenestrated and provide blood with access to the hepatocytes. Also, there are specialized regions within the hepatic microcirculation by which the cells absorb or secrete different compounds into the plasma. The splenic microcirculation is somewhat unique in that it provides an open mesh-like system for blood to flow through. Many properties/functions of this mesh are still under investigation.

13.4 One of the important functions of the liver is to act as a reservoir for blood. Under pathological conditions, the liver can either store additional blood or release additional blood into the splenic circulation. The alteration in these flow conditions appears to be regulated at both the inflow arterial side of the circulation and the outflow conditions.

13.5 In general, all organs that feed the splanchnic circulation are regulated by both passive and active components. The passive components include the changes in blood flow rates and organ blood volume upon alterations to the inflow pressure. The active components include the nervous innervation and the response to vasoactive compounds. The text highlights a number of these responses within the splanchnic circulation.

13.6 The importance of the splenic nervous system is somewhat up for debate, since in humans the spleen does not provide a large source of blood and the spleen cannot "eject" blood into the systemic circulation. In other mammals, these properties of the spleen are observed and are regulated by the autonomic nervous system. The spleen does show some passive components, similar to the other organs involved with the splanchnic circulation.

13.7 Hepatitis, alcoholic and/or fatty liver disease, and splenomegaly are common diseases that affect the splanchnic circulation. Hepatitis is an inflammation of the liver that can degrade the sinusoid units. This reduces the functionality of the liver. Alcoholic liver disease is a degradation of the liver and a reduction of the liver function based on excessive alcohol consumption. Fatty liver disease is somewhat common and is associated with a large retention of fats within the liver, which may lead to liver diseases. Splenomegaly is an increase in the size of spleen, which is likely due to a primary disease that alters the properties of red blood cells, white blood cells, or platelets. With an increased destruction of these cells, the spleen enlarges to accommodate the removal processes. The spleen can be removed, however, there are significant risks associated with this surgical procedure.

HOMEWORK PROBLEMS

13.1 The splanchnic circulation has some unique features that are not found in many other vascular beds. What is it about the splanchnic circulation that makes it so unique and why are these anatomical features not found in many other vascular locations?

13.2 The blood inflow to the liver is also quite unique. What are the features of the arteries that feed the liver and why is this vascular bed so unique?

13.3 The functional units of the liver have a very specialized organization that is very important to the function of the liver. Describe the organization, key features of this organization, and what role these features have on the circulation in general.

13.4 The spleen is also composed of units termed sinusoids. Do these sinusoids have similar functions as the liver sinusoids? Compare and contrast these functions.

13.5 The blood flow into many visceral organs is kept "constant" during normal physiological activities (e.g., do not consider the blood shift that occurs directly after eating). Why is blood flow to the spleen so variable during normal physiological conditions?

13.6 Fenestrated endothelial cells are found in particular locations within the liver. What is the purpose of these specialized endothelial cells within the liver?

13.7 What are the different microcirculation organizations that can be found in the liver and what would they be used for?

13.8 The liver can release significant quantities of blood. What would be the overall physiological effects of this blood release? Would the systemic blood pressure and/or peripheral resistance change due to the efflux of blood from the liver?

13.9 Why are there different controls on the splenic activity. Does one control effect the other control and do they act at different times for different purposes?

13.10 How would disease conditions of the liver and the spleen alter the systemic circulation? You can choose to discuss the disease conditions that we highlighted, or others.

13.11 Develop a mathematical relationship, using any (or all) of the principles that we have discussed in the previous sections, to describe the flow of blood through the liver sinusoids. Make sure to consider mixing, flow convergence, and transport when devising your relationship.

References

[1] R.W. Brauer, O.S. Shill, J.S. Krebs, Studies concerning functional differences between liver regions supplied by the hepatic artery and the hepatic vein, J. Clin. Invest. 38 (1959) 2202.
[2] C.V. Greenway, W.W. Lautt, Effect of infusions of catecholamines, angiotensin, vasopressin and histamine on hepatic blood volume in the anesthetized cat, Br. J. Pharamacol. 44 (1972) 177.
[3] C.V. Greenway, G.E. Lister, Capacitance effects and blood reservoir function in the splanchnic vascular bed during non-hypotensive hemorrhage and blood volume expansion in anesthetized cats, J. Physiol. Lond. 237 (1974) 279.
[4] A.M. Rappaport, The microcirculatory unit, Microvasc. Res. 6 (1973) 212.
[5] B. Steiniger, M Bette, H. Schwarzbach, The open microcirculation in human spleens, J. Histochem. Cyotchem. 59 (2011) 639.

PART V

MODELING AND EXPERIMENTAL TECHNIQUES

CHAPTER

14

In Silico Biofluid Mechanics

LEARNING OUTCOMES

1. Use computational methods to describe the flow through various physiological geometries
2. Describe the mathematics behind computational fluid dynamics
3. Explain the need for fluid structure interaction modeling
4. Use the Buckingham Pi Theorem to develop dimensionless numbers
5. Analyze conditions for dynamic similarity
6. Evaluate salient dimensionless numbers
7. Describe some relevant areas of current *in silico* biofluid mechanics research
8. Discuss some possible avenues for future *in silico* biofluid mechanics research

14.1 COMPUTATIONAL FLUID DYNAMICS

In earlier chapters, we were able to solve the fluid mechanics governing equations for various fluid properties of interest by making assumptions that significantly simplify the Navier-Stokes equations, the Conservation of Momentum equations, the Conservation of Energy equations, and the Conservation of Mass equations. An exact solution could only be reached if these assumptions were made and the boundary conditions as well as the initial conditions were known or if "valid" reasonable assumptions were made regarding these parameters. Note that our statement does not touch on the accuracy of any of the solutions that we obtained, only that the solution can be obtained if a well-defined set of conditions was applied to the problem. However, in many situations, the Navier-Stokes equations cannot be easily simplified such as when the flow is not steady, certain complex body forces need to be considered, or when all three dimensions must be considered (e.g., if there are forces or flow in all of the directions) in the solution. Under these and other conditions, it becomes very difficult to solve the Navier-Stokes equations as a

© 2015 Elsevier Inc. All rights reserved.

set of partial differential equations analytically along with any of the conservation laws that are of interest to the particular flow problem. Computational fluid dynamics is a tool often used to solve the coupled Navier-Stokes equations simultaneously with the conservation equations, omitting many but not all of the previous assumptions made in earlier chapters. Computational fluid dynamics techniques typically make use of numerical methods to define the flow conditions at discrete points within the volume of interest and thus computational fluid dynamics does not obtain a continuous solution for the entire flow field.

As we know, the Reynolds number can be defined as

$$\mathrm{Re} = \frac{\rho v d}{\mu} \tag{14.1}$$

where ρ is the density of the fluid, v is the mean fluid velocity, d is the diameter of the vessel (or some characteristic length to describe the vessel), and μ is the viscosity of the fluid. The equation for the Reynolds number can be rearranged to the following form:

$$\mathrm{Re} = \frac{\rho v^2}{\mu \cdot \frac{v}{d}} = \frac{\text{inertial forces}}{\text{viscous forces}} \tag{14.2}$$

Indeed, the Reynolds number reveals an important relationship between the inertial forces and the viscous forces applied to the fluid elements. For slow flows or flows with small Reynolds numbers (less than 1), the viscous terms dominate; therefore, we may be able to ignore the inertial terms and state that the viscous forces are more important than the fluid momentum (flows within the microcirculation may use this assumption). For faster flows or flows with Reynolds numbers greater than 1, we may potentially ignore the viscous resistance, because the inertial forces (e.g., momentum of the fluid) dominate the flow conditions. By using the Bernoulli equations under these conditions, we may obtain an approximate solution to these types of flows (see Section 3.8). However, for flows where the viscous terms have a similar weight to the inertial terms, we cannot ignore either of the terms in the governing fluid dynamics equations, and it typically becomes very difficult to solve the Navier-Stokes equations by hand. For example, when the fluid is relatively compressible (e.g., gases), typically the inertial terms and the viscous terms have a similar weight; therefore, not only do the Navier-Stokes equations need to be solved, but the energy and thermodynamic relationships are needed as well to completely characterize the flow conditions. However, most of the fluids we discussed in this textbook (e.g., blood and interstitial fluid, among others) are usually considered incompressible within the range of applied pressures. We also know that the Reynolds number of most flows within the human body is close to 1, thus suggesting that computational methods may be needed to more accurately characterize the flow conditions.

When it becomes very difficult to solve partial differential equations analytically, the Finite Element Method (FEM) or Finite Difference Method (FDM) is often used. FEM is most often used to solve solid mechanics problems, due to its advantage in handling complex geometries. FDM and its related Finite Volume Method (FVM) are often used to solve fluid mechanics problems. By using FDM, differential equations can be solved with numerical methods by using a finite value for Δx rather than allowing Δx to approach 0.

Again this would only provide us with the flow conditions at discrete points within the flow field Δx apart and would not provide us with a continuous solution throughout the entire flow field. Note that Δx can be very small such that you essentially obtain a continuous solution and that Δx does not need to be a fixed value throughout the volume of interest. You can decrease the spacing at regions of interest within the volume of interest (this would provide you with more information about that location) and increase the spacing at regions of less interest (which would provide less information at those locations). The mathematical equations behind the FDM can be developed from the Taylor series expansion for multiple independent variables. This method is derived by gridding (or meshing) the volume of interest with equal spacing in each direction (more complex solution methods relax the constraint of having even spacing as mentioned above). In two dimensions, the rectangular grid in the x/y directions corresponds to i and j steps, respectively (Figure 14.1). The spacing between the mesh are denoted as Δx and Δy, in the i and j directions, respectively. Therefore, the location of subsequent points within the mesh can be identified by

$$\begin{aligned} x_n &= x_0 + n\Delta x \\ y_n &= y_0 + n\Delta y \end{aligned} \tag{14.3}$$

where x_0 and y_0 are the initial points, at a predefined origin, within the mesh. Note that with this approach Δx and Δy do not need to be equal, although for our illustration we will set them to be equal for convenience. For any given function (f) that can be described at each of these grid points, the Taylor expansion states that the function at neighboring points can be defined using the spatial partial derivative of that function, provided that the function is continuous. All flow properties are continuous at steady-state conditions and this approach would try to identify those steady-state conditions. In one dimension, the Taylor expansion for a continuous function is defined by

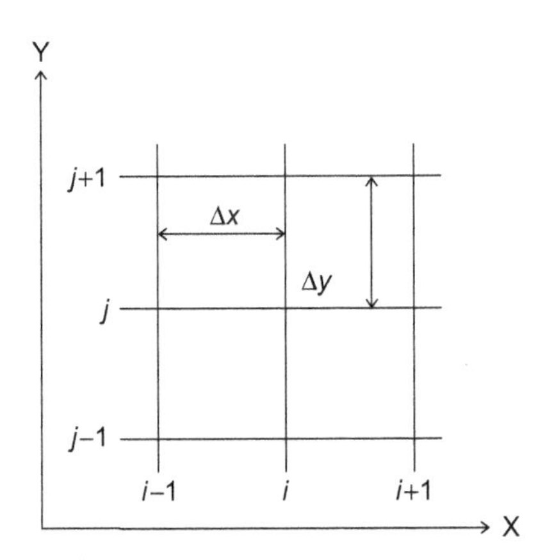

FIGURE 14.1 Two-dimensional (2D) grid for the FDM, where i denotes grid spaces in the x-direction and j denotes grid spaces in the y-direction. Note that the spacing does not have to be uniform across the entire geometry. By using either the forward difference method or the backward difference method, parameters of interest can be obtained from known boundary conditions and known initial conditions.

$$f_{i+1} = f_i + \left.\frac{\partial f}{\partial x}\right|_i \Delta x + \frac{1}{2}\left.\frac{\partial^2 f}{\partial x^2}\right|_i \Delta x^2 + \frac{1}{6}\left.\frac{\partial^3 f}{\partial x^3}\right|_i \Delta x^3 + \ldots$$
$$f_{i-1} = f_i - \left.\frac{\partial f}{\partial x}\right|_i \Delta x + \frac{1}{2}\left.\frac{\partial^2 f}{\partial x^2}\right|_i \Delta x^2 - \frac{1}{6}\left.\frac{\partial^3 f}{\partial x^3}\right|_i \Delta x^3 + \ldots \quad (14.4)$$

To make use of Eq. (14.4), the first spatial partial derivative of f at i can be solved using a forward difference (Eq. (14.5)) or a backward difference approach (Eq. (14.6)):

$$\left.\frac{\partial f}{\partial x}\right|_i = \frac{f_{i+1} - f_i}{\Delta x} + H.O.T. \quad (14.5)$$

$$\left.\frac{\partial f}{\partial x}\right|_i = \frac{f_i - f_{i-1}}{\Delta x} + H.O.T. \quad (14.6)$$

Note that the forward difference method makes use of f_{i+1} (using the trend ahead of the point of interest) whereas the backward difference method makes use of f_{i-1} (trend behind the point of interest). *H.O.T.* denotes all of the higher-order terms of order Δx^2 or more. These terms are typically neglected, which gives a truncation error in the FDM of the first degree. The *H.O.T.* can be neglected because they will typically be relatively small as compared to the Δx term (if and only if Δx approaches zero, if Δx is small then Δx^2 is much smaller, and so on). If Eqs. (14.5) and (14.6) are added together and averaged, a formulation for the central difference method is obtained, which has a second-order accuracy (if the *H.O.T.* are ignored).

$$\left.\frac{\partial f}{\partial x}\right|_i = \frac{f_{i+1} - f_{i-1}}{2\Delta x} + H.O.T. \quad (14.7)$$

The central difference method considers the trend ahead of and behind the current point of interest but it does not consider the trend at the point of interest.

Using any of these methods discussed, it is possible to calculate the velocity distribution through continuous geometries by relating neighboring points. For instance, let us begin our analysis with a 2D geometry in a sudden expansion (Figure 14.2). In biofluids, we can idealize the flow of blood from the microcirculation (small diameter) into the venous circulation (larger diameter) or the flow through the aortic valve into the aorta with this geometry (note that this geometry is clearly not a very accurate representation of the physiological conditions, but for illustration purposes we will make use of it). Let us assume that the diameter of the vessel increases by 2.5 times at the step and the length of interest is 4 times that of the smaller diameter (the length of interest is denoted by the dashed lines in Figure 14.2).

Taking this geometry and meshing it with a total of 5120 square elements, the FDM can be used to solve for the velocity field throughout the expansion (Figure 14.3, solved with FEMLab, COMSOL software). To obtain this solution, it was assumed that the velocity at the inlet had a known parabolic profile. This simplified geometry can also be analyzed by hand (using Eq. (14.7)), if the input boundary conditions and the wall boundary conditions are known. In general, the geometry and the grids do not need to be idealized as shown here; we are just making use of this simplified example to illustrate the concept. Later in this section, we will discuss some work that is being conducted using computational fluid dynamics, and we will see that the geometries and grids can be quite complex.

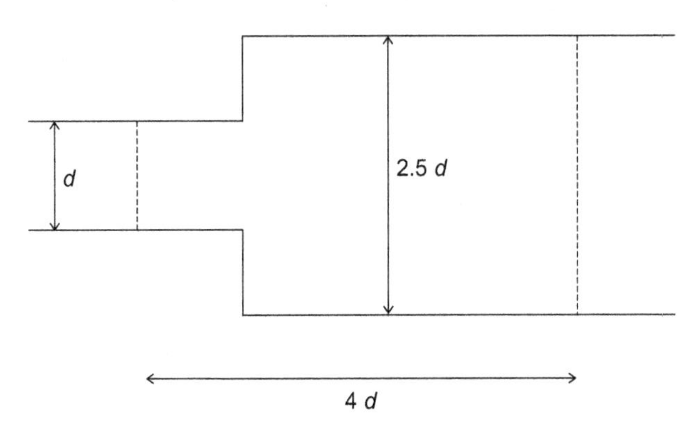

FIGURE 14.2 Idealized flow in a sudden expansion from the microcirculation to the venous system. Clearly, based on the geometry, the computational results will differ, but with simple geometries an idea of the flow field can be established.

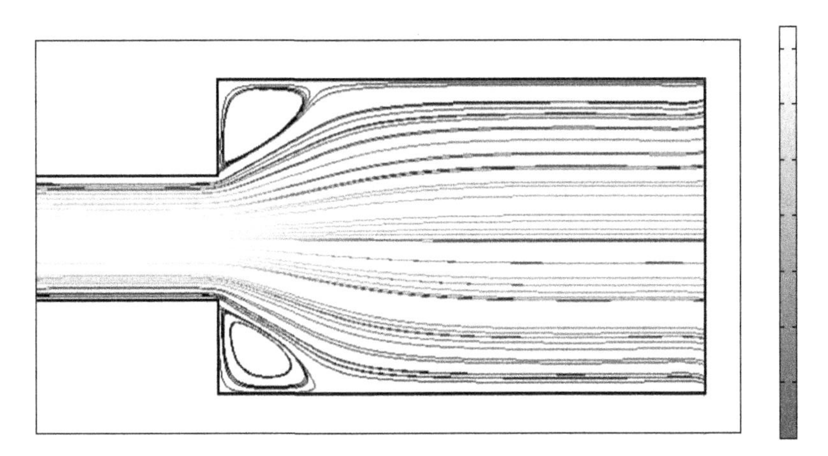

FIGURE 14.3 Computational fluid dynamic solution for the flow conditions in Figure 13.2. This illustrates that in a sudden expansion, there will be a stagnation zone in which the fluid becomes trapped.

When turbulence develops in the flow field, we would have to consider the turbulent Reynolds stress, which includes time-dependent fluctuations that are difficult to solve by hand, while incorporating them into the fluid dynamics equations. Recall that statistical methods were required for these formulations because turbulence cannot be accurately predicted but only observed. The solution to the flow profile does not only depend on the Navier-Stokes equations and the Conservation of Mass, but also the turbulent kinetic energy (k), which is defined as

$$k = \frac{1}{2}\overline{u_i u_i} \tag{14.8}$$

where u_i is the velocity of the fluid. The other variable required in turbulent flows is the viscous dissipation rate of the turbulent kinetic energy, denoted as ε. The dissipation rate is defined as

$$\varepsilon = \nu \cdot \overline{u_{i,j}u_{i,j}} = \nu \frac{1}{\Delta t} \int_t^{t+\Delta t} \widehat{u_{i,j}}\widehat{u_{i,j}} dt \tag{14.9}$$

where $u_{i,j}$ is the velocity and ν is the kinematic viscosity. The turbulent viscosity μ_t is directly proportional to the turbulent quantities k and ε as

$$u_t \propto \frac{\rho_0 k^2}{\varepsilon} \tag{14.10}$$

In the macrocirculation, blood vessels are relatively large and the fluid velocity is relatively high, and therefore, the standard k-ε turbulence model is commonly used to solve for the flow fields under these conditions (such as in the aorta). By using this model, one assumes an isotropic turbulence throughout the entire flow cycle with a high Reynolds number. However, when using the k-ε model, it could become challenging to estimate the velocity and the stress distribution in the near-wall regions, where the turbulent viscosity has a large effect and varies very significantly as a function of time and space. This would require a large amount of computational steps and a large processing time to accurately predict these spatial/temporal fluctuations. Also, the Reynolds number in the sublayer of the turbulent field close to the vessel wall is usually low because the viscosity is large, making the standard k-ε model inadequate. Therefore, a low Reynolds number Wilcox turbulent k-ω model is often used instead, which can more accurately predict the velocity and stress distribution at the near-wall regions when the effects due to the viscous forces are large. The k-ω model uses the turbulent frequency, ω, rather than turbulent viscous dissipation rate to characterize turbulence. In this model, the turbulent velocity u_t is related to k and ω in the form of

$$u_t \propto \frac{\rho k}{\omega} \tag{14.11}$$

As the turbulent frequency ω is related to k and ε by

$$\varepsilon = k\omega \tag{14.12}$$

The governing partial differential equations for the k-ω turbulent model are

$$\begin{gathered}
\frac{\partial}{\partial t}(\rho k) + \frac{\partial}{\partial x_j}(\rho U_j k) = \rho P - \rho \omega k + \frac{\partial}{\partial x_j}\left[\left(\mu + \frac{\mu_t}{\sigma_k}\right)\frac{\partial k}{\partial x_j}\right] \\
\frac{\partial}{\partial t}(\rho \omega) + \frac{\partial}{\partial x_j}(\rho U_j \omega) = \alpha \rho P - \beta \rho \omega^2 + \frac{\partial}{\partial x_j}\left[\left(\mu + \frac{\mu_t}{\sigma_k}\right)\frac{\partial \omega}{\partial x_j}\right]
\end{gathered} \tag{14.13}$$

where α and β are closure coefficients, which are known constants.

In the near-wall regions, a majority of the turbulent kinetic energy is dissipated due to the heightened shearing forces. Therefore, by including the dissipation rate within the solution, the k-ω turbulent model delivers a more accurate near-wall solution. Also, by averaging the Navier-Stokes equations, changes in the turbulent properties are smoothed. A method to calculate the irregular changes within turbulent flows is the large eddy simulation, which follows the development of small turbulent vortices within the fluid. Another method of turbulent modeling is through the direct coupling of the Navier-Stokes equations and the continuity equations, which is termed the direct numerical simulation of flow properties.

Many biofluid mechanics labs use computational fluid dynamics to predict regions within the flow fields that are likely to facilitate cardiovascular disease development (aneurysm and atherosclerotic lesions are very common), model the performance of implantable cardiovascular devices, or to just model and predict the flow profiles within the body. These studies have significantly helped the biofluid community gain a better understanding of many of the biofluid mechanics principles discussed within this textbook. Also, the design of implantable cardiovascular devices can be significantly improved with the help of computational fluid dynamics. For instance, mechanical heart valves tend to disrupt the flow profiles distal to the valve as compared to the native valves. A few research groups have built numerical models with accurate valve geometries to investigate how the implantation of an artificial heart valve affects the local flow field under transient and turbulent flow conditions. Also, these groups have asked, why do the mechanical heart valves disrupt blood flow, and how can we minimize the disruption to the fluid properties and biological properties during and after the implantation of a mechanical heart valve? With this knowledge, valve designs have been modified and performances of the artificial valves have been improved.

Studies have also been conducted by many research groups to identify areas in specific blood vessels (such as coronary arteries) that may induce large-magnitude transient shear stresses, oscillatory shear stresses, as well as large shear stress gradients. Altered shear stress in these regions has a higher potential to antagonize endothelial cells, platelets, and red blood cells to initiate activation and/or inflammatory processes. This tends to lead to an increased likelihood of cardiovascular disease formation at these areas. With early identification of these locations, it may be possible to target disease interventions toward these regions. Computation fluid dynamics modeling can be used to predict disease initiation and development.

For example, a model of a left coronary artery has been built using ANSYS CFX (Version 12.0) (Figure 14.4), and a stenosis can be introduced downstream of the bifurcation where the left main coronary artery divides into the left anterior descending artery and the left circumflex artery. Using the Wilcox k-ω turbulent model, flow fields in the left coronary artery under normal and stenosis conditions can be characterized relatively easily, using a normal cardiac output as the inflow boundary conditions (these waveforms are therefore transient). Under normal flow conditions, due to the large bifurcation angle (about 75° between the left anterior artery and the left circumflex artery), the flow skewed toward the left circumflex artery and separated from the blood vessel wall, causing a recirculation zone to develop near the bifurcation region (Figure 14.5). With the presence of an 80% stenosis approximately 8 mm downstream of the bifurcation in the left descending artery, the flow field changes significantly. With the recirculation zone near the bifurcation, new recirculation zones developed immediately distal to the stenosis, due to the jet flow induced by the vessel narrowing. Figure 14.6A depicts the velocity vectors in the left coronary artery near the stenosis, with a zoom-in view of the velocity vectors in the recirculation zones shown in Figure 14.6B.

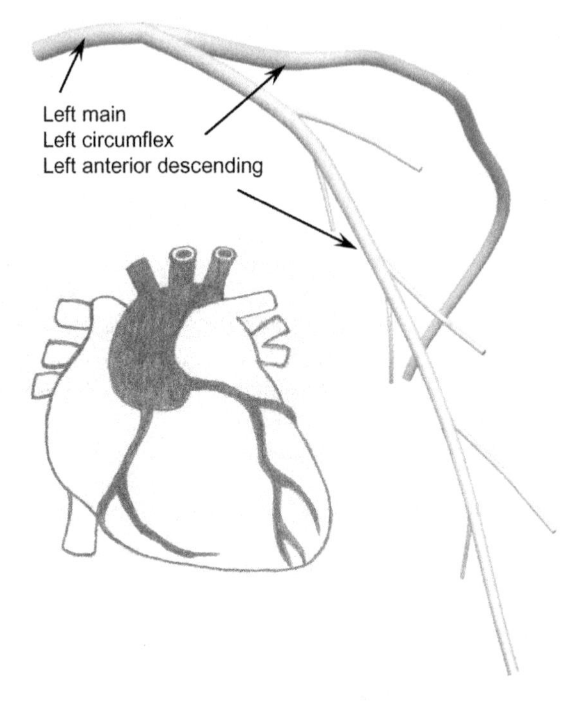

FIGURE 14.4 A three-dimensional (3D) model of the left coronary artery with branches, generated by Pro-Engineering software. This geometry can be imported into most CFD software for flow field (velocity vectors, pressure, among others) computation.

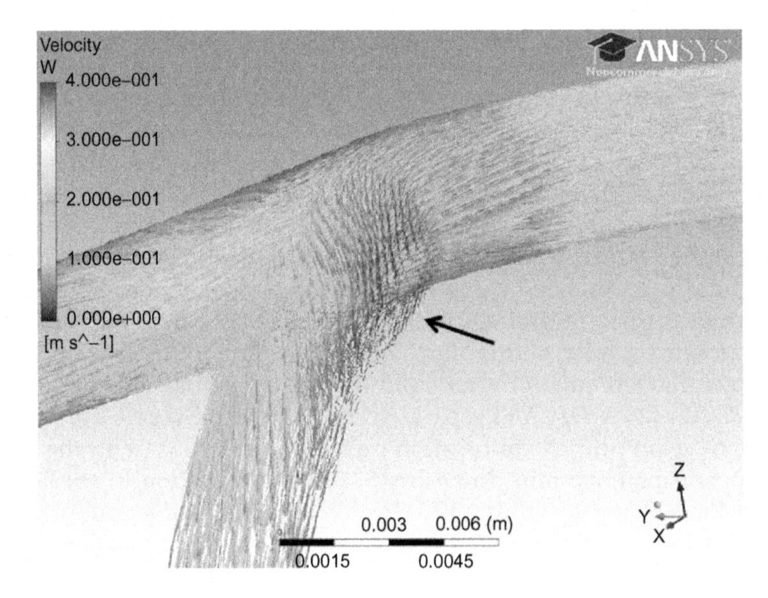

FIGURE 14.5 Velocity vectors in the left coronary artery near the bifurcation. Due to flow separation at the bifurcation, a recirculation zone developed (arrow), which can potentially trap cells and induce abnormal shear stresses.

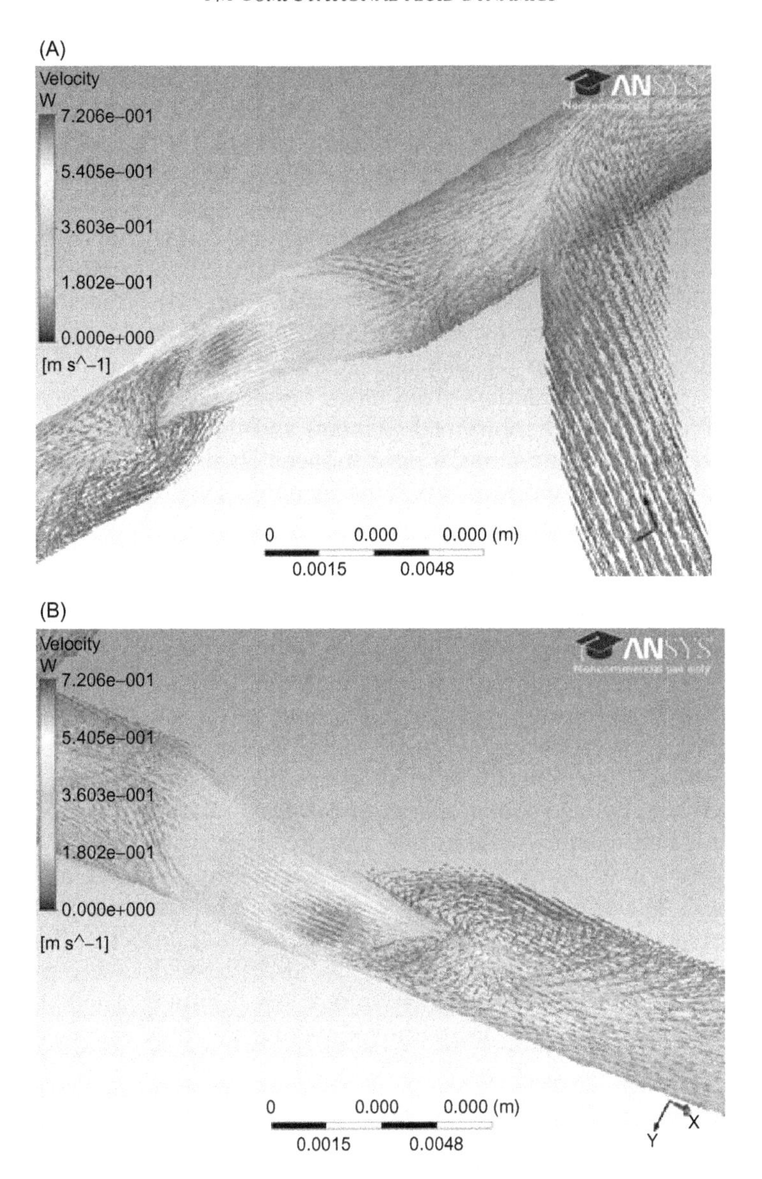

FIGURE 14.6 Velocity vectors in the left descending artery with an 80% stenosis (A). Jet flow developed at the stenosis throat, inducing a large recirculation zone distal to the stenosis. The recirculation zone by the bifurcation upstream is still visible. A close-up view of the stenosis and the recirculation zone is shown in (B).

With this model, one was able to calculate the velocity, velocity gradient, kinetic energy (k), energy dissipation rate (ε), and many other parameters for every single fluid element (there are approximately 2.2 million elements in the model). For this particular flow condition, a no-slip wall boundary condition was used and the inlet velocity waveform mimicked the temporal changes in the cardiac output waveform during one cardiac cycle. The outflow was considered to be a pressure outlet. Based on the information obtained from the simulation, shear stress and pressure distribution in the left coronary artery under normal and stenosis conditions can be estimated directly and then used as an input during *in vitro* experiments to confirm results on altered stresses. It is also important to note that there are many other commercial CFD software packages available, such as ADINA, COMSOL, FLOW-3D, and many others.

However, complex fluid mechanics equations or models cannot provide all the information about biofluid behavior, since not only fluid mechanics, but also biochemical responses are involved in physiological or pathological processes. The best scenario would be when we want to simulate the behavior of blood within a blood vessel. We know that blood is composed of plasma and formed cellular elements, including red blood cells, white blood cells, and platelets. In large or medium-sized blood vessels, we usually do not need to consider the interactions between these formed elements and the fluid phase, as long as we are not studying the cellular responses to the fluid induced stress. Additionally, under physiological conditions, these cells tend to have little activity in large vessels; however, under pathological conditions these cells may have more significant roles in these large vessels and we may want to consider how the flow alters the function of these cells. However, in small blood vessels, especially in the microcirculation, the interaction between the particle phase and the fluid phase becomes relatively important; therefore, the particle phase has to be considered when modeling both physiological and pathological flow conditions. Additionally, the functions of cells (such as red blood cell nutrient delivery and white blood cell extravasation) may be more important to consider at the smaller level. To model the cells as a second phase, discrete-phase or even multi-phase models can be used.

Many CFD software (including ANSYS) are capable of simulating the movement of discrete-phase particles, by calculating energy and momentum exchange between discrete-phase particles and fluid elements. From previous chapters, we have learned that platelets are extremely sensitive to blood flow–induced shear stress and platelet biochemical responses are closely related to the local flow conditions and shear stress distribution. Using the coronary artery example again, we can simply define platelets as a discrete phase and provide information on particle size, shape, density, viscosity, and release profile in ANSYS (note that this can be accomplished in other software packages as well). Thus, blood, the fluid phase, can be treated as a continuum by solving the time-averaged Navier-Stokes equations (as before). The behavior of platelets, the particle phase, can be calculated through the flow field at each time step. Forces on platelets can be determined by

$$\frac{du_p}{dt} = F_D(u - u_p) + F_x \tag{14.14}$$

where u_p is platelet velocity. F_x is an additional acceleration term, which is the force required to accelerate the fluid surrounding the particle. $F_D(u - u_p)$ is a drag force per unit mass term and it is dependent on the shape, size, density, and velocity of the particle, as well as density and velocity of surrounding fluid elements. Based on the velocity of a particle, its trajectory can be determined using the kinematic equation as

$$\frac{dx_p^i}{dt} = u_p^i \tag{14.15}$$

where index i refers to the coordinates directions. To account for the effects of turbulence, a stochastic model developed by Gosman and Ioannides can be used to calculate particle phase velocity by adding random fluctuations obtained from the k-ω simulation to carrier phase velocity. Furthermore, the stress that one platelet experiences along its trajectory can also be estimated. For example, we can use the Boussinesq approximation to compute the total stress (laminar plus turbulent stresses) applied to a single platelet (or any other particle immersed within a fluid), as

$$\sum_{t=t_0}^{t_{\max}} \left(\frac{(\varepsilon_i - \varepsilon_{i+1})}{2} \times (\mu_i + \mu_i^t) + KE_i \times \rho \right) \times \Delta t_i \tag{14.16}$$

where ε_i is the strain rate, μ_i is the viscosity, μ_i^t is the turbulent viscosity, ρ is the density, KE_i is the turbulent kinetic energy, and Δt is the time step.

Figure 14.7 shows platelet trajectories calculated using the methods discussed above. In Figure 14.8, the shear stress history of single platelets with "interesting" trajectories was exported from the simulation along the trajectory for five cardiac cycles (hence the time varying waveform), calculated using the Boussinesq approximation.

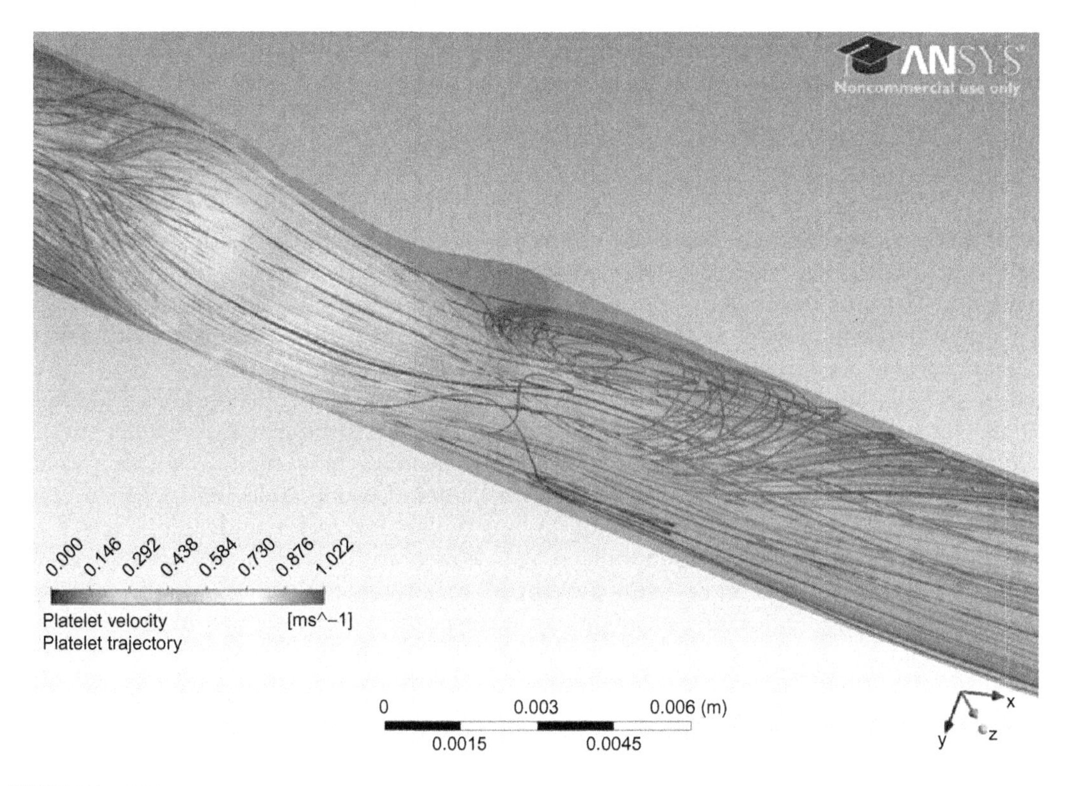

FIGURE 14.7 A discrete-phase model can be used to estimate platelet trajectories after they pass an 80% stenosis in the left descending artery. This image shows the trajectories of every particle seeded at the inlet and we can see that a significant proportion of these become trapped in the recirculation zone downstream of the stenosis.

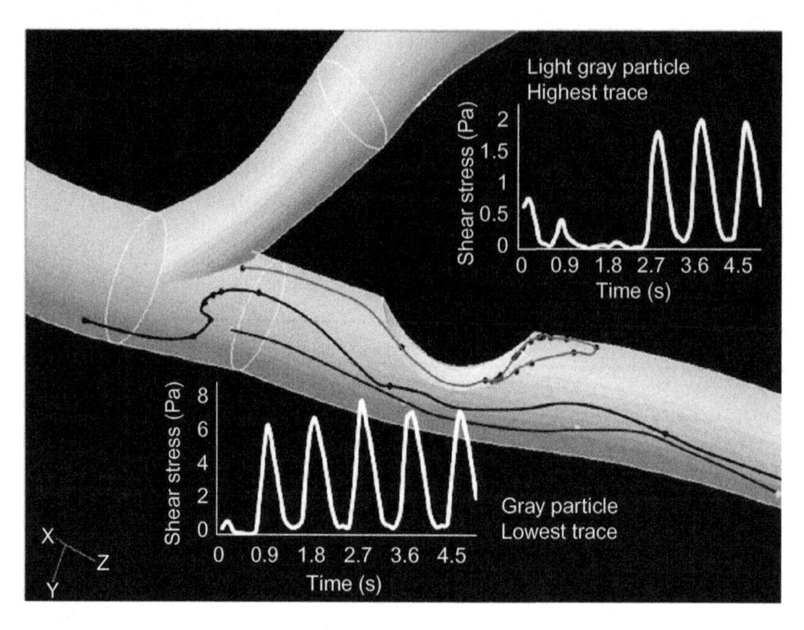

FIGURE 14.8 The shear stress history of platelets within the left coronary model estimated using the Boussinesq approximation, following their trajectories for five cardiac cycles. The two exported shear stress histories are for the lowest trace (gray particle) and the highest trace (light gray particle).

Using the approaches discussed above, more information about the discrete phase in blood (e.g., platelets in this case) can be gained, which, however, is still very limited. Huang and Hellums developed a population balance equation model more than two decades ago to describe platelet aggregation kinetics under shear flow, where platelet behavior was majorly described using collision/splitting formulations. A few years later, Tandon and Diamond developed a more physiologically relevant collision model to simulate platelet aggregation under shear flow, by incorporating fibrinogen and platelet surface glycoprotein GPIIb/IIIa bond formation and dissociation (this model would include kinetic principles). A 2D analytical platelet flipping model was established by Mody et al. more recently to describe platelet adhesion to von Willebrand factor under flow, based on images of platelet tethering to immobilized von Willebrand factor. In the past 10 years, various numerical models have been developed to simulate platelet motion, collision, adhesion, and aggregation. Different mathematical models were employed in these studies to describe platelet translocation, slow translations close to the wall, tethering, and protein bond formation, providing detailed information on the physical forces applied to platelets, and elucidating how platelet shape and force change affect platelet trajectory and associated biological responses.

Besides platelets, red blood cells and white blood cells, as well as proteins and many molecules in the blood stream can all be treated as discrete-phase particles, and their motion and behavior can be simulated using discrete (or multi-phase) models. To simulate leukocyte adhesion to endothelium, Khismatullin and Truskey defined a leukocyte as a viscouselastic drop, and the receptor–ligand interaction (between the leukocyte and the endothelium) was described using a spring-peeling kinetic model.

Overall, to achieve more accurate information on the discrete phase, such as cell deformational change, microparticle shedding, surface protein expression, more complicated models and modeling techniques (including kinetic reactions), as well as sufficient physiological knowledge are required (this will be discussed in Section 14.5). No commercial software is available to conduct these types of simulations, and therefore, sophisticated and elegant programming will be required by the research group interested in these questions. User defined programs need to be incorporated into commercially available code, or an entire new fluid dynamics simulation would need to be developed. Please note that the methods described above may not be applicable for each cell type, based on the assumptions made (e.g., spring-peeling kinetic model for leukocytes is a good model, but it may not work as well for red blood cells).

There is one important concept that should be introduced regarding numerical modeling of physical events. We have illustrated two models in this section (Figure 14.2, with approximately 5000 elements (2D) and Figure 14.4, with approximately 2.2 million elements (3D)) that used vastly different meshes. How do we know that the mesh that we made is sufficient to accurately predict the flow conditions? There is a concept termed mesh independence that should be considered for every numerical model that is attempted. Mesh independence states that the solution to the numerical model is independent of the mesh size (e.g., number of elements) and mesh shape (e.g., square vs. triangular elements in 2D). If it is found that the solution of the model varies with the mesh chosen, then it would not be clear which solution is accurately predicting the "real" flow conditions better. So how does one determine if the solution to the model is mesh independent? A simple way would be to run the model with two different meshes. For instance, the model shown in Figure 14.2 could have been run with a much coarser mesh (maybe 250 elements) and a much finer mesh (maybe 20,000 elements). For the sake of argument, let's assume that the solution between the 250 element model and the 5000 element model varied significantly, whereas the solution between the 5000 element model and the 20,000 element model did not vary significantly. This would suggest that the 5000 element model was sufficient to depict the flow conditions (typically, but not always, the more elements that you add the more accurate your model is, so the 20,000 element model would be the most accurate). However, why would we run the 5000 element model in lieu of the 20,000 element model? The reason is that, the computation time associated with a 20,000 element model would be significantly greater than a 5000 element model. This does not mean that we should choose models with small quantities of elements to save us computational time. Instead it says that we can choose to use a model with a lower number of elements that is considered to be mesh independent in comparison to a model with larger number of elements. To ensure mesh independence, a general practice is to change the mesh size gradually until the solution difference is small (such as 1–5%). Therefore, when one reports findings obtained from numerical modeling, it is always important to report (1) if the model is mesh independent and (2) how many elements are used in the model, to ensure the readers that a sufficient number of elements are used and the predicted results are accurate. Note that it also may be important to change the mesh geometry, we are just using the mesh size as an example of calculating mesh independence.

14.2 FLUID STRUCTURE INTERACTION MODELING

Traditional computational fluid dynamics simulations can provide a great deal of information regarding the flow conditions. However, in some instances, assumptions regarding physiological aspects of the flow conditions may be made, which reduce the overall relevance of the numerical solution. One modeling technique that is commonly used to make a more physiologically relevant model is termed fluid structure interaction (FSI) modeling. The necessity for FSI modeling arises when a flexible structure either has a fluid flowing through it or the flexible structure is immersed within a fluid. One of the most common examples of FSI modeling is of blood flow through the cardiovascular system. As we have learned in previous sections, blood vessels are not rigid, and they distend in response to the heart's pressure pulse (this is especially prevalent in the large arteries, which are common sites for cardiovascular disease development). Under cardiovascular disease conditions such as aneurysm, the distensibility of blood vessels increases, and the need for FSI modeling to more accurately describe these conditions is increased. During atherosclerotic conditions, the distensibility of the blood vessels tends to decrease; again these altered parameters and the effects of these parameters on the flow conditions should be considered. As another example, blood cells are not rigid and the deformations of the cells may cause protein expression changes that can lead to pathological events. To accurately predict and understand the flow through these types of structures, the motion of the structure (or the cells) must be depicted. The mechanical properties of the vessel must be known or assumptions must be made about these properties so that the movement of the vessel and the relationship of this movement with the fluid flow can be accurately assessed. There are a number of different models that can be used to solve these types of flow conditions, which include time-dependent motions, linear/nonlinear models, and time-linearized models. We will highlight some of the equations within these methods and discuss some of the current biofluid research that uses FSI modeling as one of the tools to characterize the flow through a system. More details about current research can be found in Section 14.4.

The easiest method to model a solid boundary movement is based on the pressure that the fluid exerts on that boundary at a particular instant in time and space. This can be quantified as

$$p = \rho u \left[\frac{\partial v}{\partial t} + u \nabla v \right] \tag{14.17}$$

where ρ is the bulk fluid density, u is the average fluid velocity, and v is the fluid velocity as a function of time and space. This model is fairly useful in most biofluid applications, but it is not necessarily accurate when the average fluid velocity is relatively large (e.g., close to the speed of sound, which we typically do not approach in biofluids). Also, this model is simple enough that it can be used as an initial approximation to determine if the model parameters are accurate and the flow through the vessel looks reasonable. As the simulation includes more parameters, such as the fluid viscosity and possibility for turbulence to develop, the equations become much more complex (this is termed the full-potential theory/equations). Using this method, a nonlinear wave equation would be coupled to the structure movement. The wave equation is nonlinear due to the elastic and viscous mechanical properties of the vessel wall. If one generalizes this formulation, considering that the motion of the body is relatively small and the pressure is uniform throughout the wall (with an inviscid, irrotational fluid), a small displacement form of the full-potential theory can be developed. Making these assumptions, the convected-wave equation becomes linear and can be solved for any fluid property as

$$\nabla^2 \Phi - \frac{\partial^2 \Phi}{\partial t^2} - u\left[\frac{\partial^2 \Phi}{\partial x^2} + \frac{\partial^2 \Phi}{\partial y^2} + \frac{\partial^2 \Phi}{\partial z^2}\right] = 0 \tag{14.18}$$

where Φ is the fluid property of interest (e.g., velocity, pressure, among others). There are various approaches to make the solution of these equations simpler. The first would be to solve a quasi-linear set of fluid equations that would govern the fluid flow through a system that is completely rigid. This solution would be able to determine the flow properties at each spatial location within the volume of interest. From this solution, the boundary is given a small temporal disturbance and then the changes in the fluid properties are calculated. This gives a direct coupling between the fluid properties and the solid boundary; however, with this simplification the fluid is not directly causing boundary motion and only the boundary is affecting the fluid properties.

The direct coupling of the fluid properties to the solid boundary properties is the challenge of FSI modeling. For each time step within the computational solution, the fluid properties must be considered in relation to the motion of the boundary and changes in the boundary position induced by the fluid must also be considered. This coupling is what makes the complexity of FSI so great, and there are many examples of simplifications that have been used to obtain a solution. To give a brief idea of the formulations that are needed to be considered to model these types of flow conditions, a mechanical model would be needed to describe the induced motion, relaxation, elasticity, etc. of the fluid container. In parallel with this motion, the fluid properties would need to be updated to consider the change in the geometry (either increase or decrease in the available area). As these parameters are solved for one time step, they are used as the initial boundary conditions for the next time step. A complexity of this technique is that within each time step, the mesh would need to be regenerated to account for the changes in area. It may be necessary to gain or lose mesh nodes to account for the changes in area. However, meshing complex geometries is not a trivial problem, especially if the research group makes use of varying density meshes to focus calculation points to an area(s) of interest. With the consideration of cell dynamics and vessel wall dynamics, these types of simulations can become quite complex. As with computational fluid dynamics, these simulations are useful in describing cardiovascular disease conditions and the flow around cardiovascular devices among many other biological flow conditions.

Let's use the example presented in Figure 14.4 again to demonstrate how an FSI model can be developed and how it affects the final solution. As we mentioned earlier, for a transient FSI model, at each time step during the simulation, the mesh needs to be regenerated due to the geometry change, which would add a tremendous load to the computation, especially when the number of elements is large. Therefore, for exercise, only a small portion of the artery ($\sim$50 mm) was simulated using an FSI approach. The solid domain (the blood vessel wall) and the fluid domain (blood) were meshed separately, and mesh independence was checked individually. The basic continuity equations and Navier-Stokes equations were used to solve the fluid domain as described before, and arbitrary Lagrangian–Eularian methods were used to describe the blood vessel wall movement. Furthermore, the arterial wall was modeled as a nonlinear elastic solid using a Mooney–Rivlin hyper-elastic model, when the displacement of the wall was prescribed based on reported *in vivo* studies. During simulation, the fluid domain solver (ANSYS CFX) and the solid domain solve (ANSYS Mechanical) were coupled iteratively, allowing total mesh displacement and total force to be transferred and shared between the two domains. The results demonstrated that using the FSI model, flow velocity was slightly lower than that estimated using a rigid-wall model, as the extensible blood vessel wall would consume kinetic energy. Blood shear stress within the blood vessel and at the vessel wall was hardly affected by the arterial wall motion. However, stresses within the vessel wall (due to axial, circumferential, and radial stretch of the blood vessel) were heavily dependent on the vessel motion. These results were consistent with that obtained by others using different FSI models. As we discussed before, vascular wall endothelial cells are sensitive to their mechanical environment and contribute greatly to the development of cardiovascular diseases. Therefore, employing an FSI model can improve the physiological relevance of our numerical model and help us to better understand the mechanical stress/strain imposed on vascular wall endothelial cells, which will help us to predict their biological responses to altered mechanical environment more accurately.

Besides simulating blood flow interaction with the blood vessel wall under disease conditions such as stenosis and abdominal aortic aneurysm, FSI models are also commonly used to describe heart valve movement, since hemodynamics associated with a heart valve could significantly affect the valve performance. FSI models can also be used to investigate the interaction of blood cells with the blood vessel wall. As discussed previously, platelets and white blood cells can adhere to endothelial cells under stress conditions. It is thought that this adhesion can mediate many cardiovascular disease conditions. To accurately portray this interaction, the movement of the blood vessel wall must be included because the wall is not rigid and there may be some rebound/recoil after impact. From these research projects, many fundamental biological events have been described and have been validated with experimental methods.

Example

Figure 14.9 demonstrates the drainage route of aqueous human based on John and Kamm's model. Trabecular mesh was modeled as a porous elastic body, shown in the figure as springs. Schlemm's canal is modeled in two dimensions, with the outer wall (top) and the inner wall (bottom) being parallel to each other. The inner wall is permeable (so aqueous humor leaks through), and it deforms as the trabecular mesh work stretches. The local height of the canal is a function of position x and intraocular pressure (IOP). Under certain disease conditions (such as glaucoma), the size of Schlemm's canal may decrease due to the increase in IOP; therefore, aqueous humor drainage resistance increases. We can use FSI to model the interface of the moving inner wall and aqueous humor.

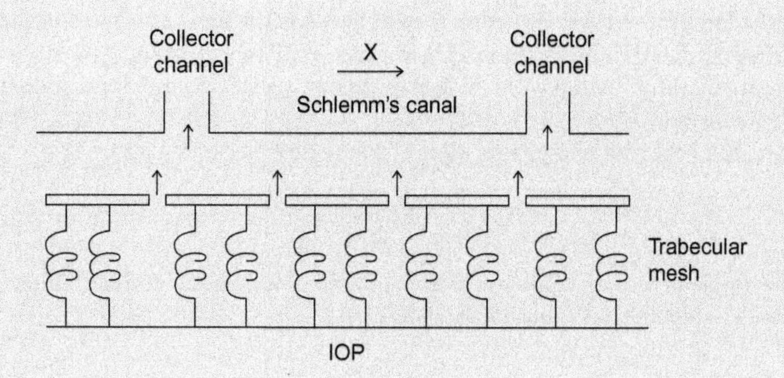

FIGURE 14.9 Model of Schlemm's canal for in-text example. *Source: Adapted from John and Kamm.*

Solution

First of all, we need to develop the mesh motion for our model. According to Newton's second law, the acceleration of the inner wall is proportional to the net force acting on it. In differential form, the equation to describe the motion of the inner wall is

$$m\frac{dv}{dt} = F_{flow} - F_{spring}$$

where m is the mass of the inner wall, v is the velocity of the inner wall on y direction, F_{flow} is the flow of the aqueous humor applied to the wall, and F_{spring} is the spring force applied on the wall through the trabecular mesh.

The velocity of the wall can be determined by the displacement of the wall as

$$v = \frac{y - y_0}{\Delta t}$$

and:

$$\frac{v - v_0}{\Delta t} = \frac{dv}{dt}$$

The new displacement of the inner wall determines the spring force in the form of

$$F_{spring} = k_{spring}(y - y_0)$$

Therefore, the discrete form of the equation for the inner wall motion can be rearranged, and we can get the expression for the inner wall displacement as

$$y = \frac{F_{flow} + \frac{m}{\Delta t}(v_0) + \frac{m}{\Delta t^2}(y_0)}{\frac{m}{\Delta t^2} + k_{spring}}$$

$$F_{flow} = p(x)L$$

$$\frac{\text{IOP}(t) - p(x)}{E} = -\frac{y - y_0}{y_0}$$

where $p(x)$ is the local pressure in Schlemm's canal and E is the spring stiffness of the canal.

For each time step (Δt), the location of the inner wall (mesh changes) can be determined based on the equation derived above, which depends on the IOP(t), the local pressure $p(x)$, and the size of the time step.

Furthermore, the flow rate of aqueous humor can be determined by $p(x)$:

$$\frac{dp}{dx} = 12\frac{\mu Q(x)}{\omega h(x)^3}$$

where μ is the aqueous humor viscosity, w is the depth of the canal, and $h(x)$ is the local height of the canal, which depends on y and initial conditions.

14.3 BUCKINGHAM PI THEOREM AND DYNAMIC SIMILARITY

Many real flows cannot be solved by analytic methods alone, but even for those analytic models, the use of full-sized prototypes/simulation geometries may not be practical. In order to relate the actual flow conditions with the simulation, they must be linked by some scaling factors. When the simulations and the real conditions match, they are said to be similar based on dimensional scaling. The Buckingham Pi Theorem is a mathematical approach that allows the formation of a relationship between a quantity of interest between the model and the real scenario. As an example, let us consider the drag force on a white blood cell. If we wanted to derive a relationship between the drag force and fluid properties, we would first make a list of all of the properties that we felt were important in determining the drag force. For instance, we would probably include the fluid's velocity, the viscosity, and the density as well as some geometric properties of the white blood cell. In symbolic form, we would state that the drag force F is a function of our fluid properties of interest:

$$F = f(v, \mu, \rho, d) \tag{14.19}$$

In Eq. (14.19), d is the characteristic diameter of the white blood cell. Now in the relationship that we derived, we may have unknowingly omitted parameters that are important (e.g., the material properties of the white blood cell membrane), and we may have included parameters which are not important (for sake of argument, maybe for this example we would have thought that the ratio of the cell diameter to the tube diameter would have been important). If we wanted to experimentally determine the relationship between the drag force and the four parameters of interest that we have chosen, let us consider the time it would take to complete this experiment. First, a suitable facility, including all experimental apparatus, would be needed to conduct these tests. Once this facility is built, we can start our experiments. For these experiments, we would need to hold the fluid viscosity, density, and white blood cell diameter at fixed values and vary the velocity (imagine the difficulty of finding white blood cells with the exact same diameter for the entire duration of the experiment). We would then vary viscosity and hold the other three parameters fixed. This would then need to be repeated for the remaining two variables. Even if we only conducted these experiments for five different fixed velocity, viscosity, density, and diameter values, we would need to complete 625 experiments (with no repetition for statistical purposes). For 10 different values, we would conduct 10,000 experiments. For fun, calculate how much time this would consume if each experiment took 1 h to complete and you worked 40 h/week at 100% efficiency (for 10 different values, you would be working for just shy of 5 years for this one relationship, without considering the time it takes to build the facility). Also note, that by conducting all of these experiments, you may learn that some of the parameters are not as important and that others are important (thus this would cause an increase in your experimentation time). The Buckingham Pi Theorem is a relatively easy approach to minimize this work and obtain a potentially meaningful relationship between all of the fluid properties of interest.

Using the same example as above, we would assume that some physical property of interest is dependent on n independent variables. This can be represented as a function, h, which is dependent on any (or all) of the parameters of interest. For instance, we stated that under our conditions drag force (F) was a function of velocity (v), viscosity (μ), density (ρ), and diameter (d) (Eq. (14.19)). We would therefore build a function, h, such that some combination of all of the properties equate to zero:

$$h(F, v, \mu, \rho, d) = 0 \tag{14.20}$$

The Buckingham Pi Theorem states that for any grouping of n parameters, they can be arranged into n-m independent dimensionless ratios (termed Π parameters). The number m is normally equal to the minimum number of independent dimensions represented by the quantities of interest. By knowing all of the possible Π ratios of important parameters, it is then possible to simplify the results by testing which of the Π ratios match the data obtained (thus you would not need to conduct all 10,000 experiments described earlier, only a subset to characterize the data). The Buckingham Pi Theorem begins by listing all of the dimensional parameters involved in the particular problem (this is equal to n). A consistent fundamental dimension system must be chosen (for SI units it is mass, length, time represented as MLT; for English units it is typically force, length, and time or FLT). We then list the fundamental dimensions for each dimensional parameter and quantify the minimum number of independent dimensions which is equal to m. We then select any set of m dimensional parameters which include all of the primary dimensions, and solve for a relationship between each of these parameters and the remaining parameters that is dimensionless. See the example for a more detailed discussion of this procedure.

Example

Determine all of the dimensionless groups possible to describe the drag force on a white blood cell, using Eq. (14.20).

Solution

List the dimensional parameters:

$$F \quad v \quad \mu \quad \rho \quad d$$

Select MLT as the fundamental dimension set.

List the fundamental dimensions of each parameter of interest (this is simply writing the units of each parameter in terms of the fundamental dimensions, e.g., we do not write Newton or kgm/s^2 for force, but instead write $\frac{mass*length}{time^2}$.

$$\begin{array}{ccccc} F & v & \mu & \rho & d \\ \frac{ML}{t^2} & \frac{L}{t} & \frac{M}{Lt} & \frac{M}{L^3} & L \end{array}$$

The minimum number of independent parameters is 3 (M, L, and t).

The number of Pi groups is 2 ($5-3=2$).

Pick v, d, and ρ as the subset of dimensional parameters and relate this to F (they represent M, L, and t). These values are related to F since this is the parameter that we are trying to find a relationship for. At this point, any grouping of three parameters that have M, L, and t represented would suffice.

The first Π group will be set up as

$$\Pi_1 = Fv^a d^b \rho^c$$

$$\left(\frac{ML}{t^2}\right)\left(\frac{L}{t}\right)^a (L)^b \left(\frac{M}{L^3}\right)^c = M^0L^0t^0$$

$$\left(\frac{ML}{t^2}\right)\left(\frac{L^a}{t^a}\right)(L^b)\left(\frac{M^c}{L^{3c}}\right) = M^0L^0t^0$$

$$ML(t^{-2})(L^a)(t^{-a})(L^b)(M^c)(L^{-3c}) = M^0L^0t^0$$

The dimensions must equate to a zero power, since we are trying to find a dimensionless number. Separate each dimension and relate the exponents:

$$\begin{gathered} M{:}1+c=0 \\ L{:}1+a+b-3c=0 \\ t{:}-2-a=0 \end{gathered}$$

From the mass equation,

$$c=-1$$

From the time equation,

$$a=-2$$

From the length equation,

$$b=3(-1)-1+2=-2$$

Therefore, the first Π group is

$$\Pi_1 = \frac{F}{v^2 d^2 \rho}$$

Pick v, d, and ρ as the subset of dimensional parameters and relate this to μ (they represent M, L, and t). We choose μ here since we already investigated F. Again, we could try to make a relationship for any of the variables that we believe are important.

The second Π group will be set up as

$$\Pi_2 = \mu v^a d^b \rho^c$$

$$\left(\frac{M}{Lt}\right)\left(\frac{L}{t}\right)^a (L)^b \left(\frac{M}{L^3}\right)^c = M^0 L^0 t^0$$

$$\left(\frac{M}{Lt}\right)\left(\frac{L^a}{t^a}\right)(L^b)\left(\frac{M^c}{L^{3c}}\right) = M^0 L^0 t^0$$

$$M(L^{-1})(t^{-1})(L^a)(t^{-a})(L^b)(M^c)(L^{-3c}) = M^0 L^0 t^0$$

Separate each variable and relate the exponents:

$$M{:}1 + c = 0$$
$$L{:} -1 + a + b - 3c = 0$$
$$t{:} -1 - a = 0$$

From the mass equation,

$$c = -1$$

From the time equation,

$$a = -1$$

From the length equation,

$$b = 3(-1) + 1 + 1 = -1$$

Therefore, the second Π group is

$$\Pi_2 = \frac{\mu}{\rho v d}$$

From these two Π groups, we can state that

$$\Pi_1 = f(\Pi_2)$$

$$\frac{F}{v^2 d^2 \rho} = f\left(\frac{\mu}{\rho v d}\right)$$

but the actual relationship between the Π groups would need to be determined experimentally (e.g., are there constants of proportionality?). However, the usefulness of this analysis technique is that we can obtain dimensionless parameters that can relate fluid properties. In our example, drag force was related to the fluid's velocity, density, and a characteristic length, and the viscosity was related to the same three parameters. The relationship is different in each case, it may not be correct, but at least now we have a starting point to analyze any experimental data that we collect.

There are many different dimensionless parameters that are important in biofluid flows, and we will take the time to describe some of these parameters here. This is not an all-inclusive list, but instead is a list that comprises many of the commonly used/discussed dimensionless parameters. The first parameter is the Reynolds number (Re), which is defined as

$$\mathrm{Re} = \frac{\rho v d}{\mu} = \frac{v d}{\nu} \tag{14.21}$$

where d is some characteristic length which is typically the blood vessel diameter in biofluid parameter (see previous example for Reynolds number formulation). The Reynolds number relates the inertial forces to the viscous forces and also is a criterion to describe the flow regime as either laminar or turbulent. As the Reynolds number becomes large, the flow is more likely to be turbulent. As the Reynolds number approaches 1, the inertial forces and the viscous forces balance and as the Reynolds number approaches zero, the viscous forces dominate and the flow is normally slow and laminar.

Another important dimensionless number is the Womersley number (α), which relates the pulsatility of the flow to the viscous effects. The Womersley number is defined as

$$\alpha = d\left(\frac{\omega}{\nu}\right)^{\frac{1}{2}} \tag{14.22}$$

where ω is the angular frequency. When α is less than 1, the fluid can become fully developed during each flow cycle (a flow cycle is defined as the period of the pulse frequency within the flow field). As α increases (>10), the velocity profile is typically represented as a plug-flow scenario and the flow will never fully develop. The Womersley number is important in biofluid mechanics because nearly all flow in larger blood vessels has some pulsatility to it. It is important to be able to determine whether the flow can fully develop in between each pressure pulse or whether the flow never develops fully.

The Strouhal number is another dimensionless parameter that is important in pulsatile flows. The Strouhal number is defined as

$$\mathrm{St} = \frac{fd}{v} \tag{14.23}$$

where f is the frequency of vortex shedding, d is a characteristic length, and v is the fluid velocity. For a Strouhal number greater than 1, the fluid will move as a plug following the oscillating frequency of the driving force. As Strouhal numbers decrease past 10^{-4}, the velocity of the fluid dominates the oscillation and little to no vortices are shed. For intermediate Strouhal numbers, there is a rapid vortex formation and many vortices can be shed into the mainstream fluid. Interestingly, the Womersley parameter can be defined using the Reynolds number and the Strouhal number as

$$\alpha = \sqrt{2\pi \, \mathrm{Re} \, \mathrm{St}} \tag{14.24}$$

The cavitation number (Ca) relates the local pressure to the vapor pressure and the fluid kinetic energy and provides a criterion for how likely it is for cavitation to occur. The cavitation number is defined as

$$\mathrm{Ca} = \frac{p - p_v}{\frac{1}{2}\rho v^2} \tag{14.25}$$

where p_v is the vapor pressure of the fluid at a particular temperature. As the cavitation number decreases, it is more likely for cavitation to arise in the fluid. Cavitation is a phenomena where gases within the fluid can either coalesce into large "bubbles" or gases surrounding the fluid enter and remain in the fluid in large quantities. Cavitation is typically a phenomena that incurs damage on the surrounding tissue but has been used to break apart kidney stones.

The Prandtl number (Pr) is the ratio of the momentum diffusion to the thermal diffusion and provides criteria for whether the flow will deliver a significant amount of heat to the surrounding container or if it retains its heat. The Prandtl number is defined as

$$\mathrm{Pr} = \frac{\mu C_P}{k} \tag{13.26}$$

where C_P is the specific heat of the fluid and k is the thermal conduction coefficient. As the Prandtl number reduces below 1, the fluid will not retain its heat very well, for example, the fluid dissipates large quantities of heat to the surrounding tissue. Flows characterized by a larger Prandtl number can retain their heat better than lower Pr flows.

The Weber number (We) and the capillary number (C) are important for flows with a free surface. The Weber number relates the inertial forces to the surface tension forces, and the capillary number relates the viscous forces to the surface tension forces. They are defined as

$$\mathrm{We} = \frac{\rho v^2 L}{\sigma} \tag{13.27}$$

$$C = \frac{\mu v}{\sigma} \tag{13.28}$$

where v is the average fluid velocity and σ is the surface tension of the fluid. There are many more dimensionless numbers that can be used to describe fluid properties, but we will restrict ourselves to the ones listed here.

Depending on the exact conditions that describe the characteristics of a particular flow, some of these dimensionless parameters may not be applicable. In general, in order to model a particular flow condition accurately, as many of these dimensionless numbers should be matched as possible. This leads us to a discussion of similarity within the experimental or numerical model. For a model to be geometrically similar, the model has to be the same shape as the real flow scenario and all lengths have to be related to the real scenario by the same scaling factor. For a model to be kinetically similar, the velocity at each location within the model has to be oriented in the same direction as the velocity under the real flow conditions. Also, the magnitude of each and every velocity vector must be related to the real flow scenario by a constant scaling factor. The hardest criterion to meet is that all of the forces within the model have to be identical (in magnitude by a constant scaling factor and direction). If these three conditions are met, then the flows are dynamically similar. To maintain dynamic similarity all of the viscous forces, the buoyancy forces, the inertial forces, the pressure forces, and the surface tension forces, among others, must be matched. When dynamic similarity is achieved, the data obtained in the model can be related back to the real flow condition.

From the previous example, which obtained the dimensionless group that related the drag force to the flow properties, we were able to end up with a mathematical relationship that states

$$\frac{F}{v^2 d^2 \rho} = f\left(\frac{\rho v d}{\mu}\right)$$

Therefore, if we were considering a flow where drag forces were important, it would be a necessary requirement that the Reynolds numbers equate between the actual scenario and the model. By satisfying this criterion, the forces acting on the immersed objects would be scalable between the model and the realistic case. This same analysis can be conducted for any of the dimensionless parameters (or as many as necessary) in order to satisfy similarity between the flow that is being modeled and the model itself. While it is important to maintain dynamic similarity between all fabricated models, it may be unreasonable or not feasible to actually fabricate a dynamically similar model. Under these cases, it would be important to identify the key properties that should be similar and other parameters that should be as close to similar as possible.

Example

Imagine that we are making a model of blood flow through an artery; however, we could only use water as our fluid flowing through the model artery. Calculate what the angular frequency of the pulsatile waveform and the initial inlet velocity should be if the characteristic length (diameter) of the blood vessel and the model of the blood vessel is 10 cm, the heartbeat is 72 beats/min (angular frequency is 5.24 rad/s), and the inlet velocity is 50 cm/s.

Solution

Using the Reynolds number, we can calculate the initial inlet velocity of the model:

$$\mathrm{Re}_{blood} = \mathrm{Re}_{model}$$

$$\frac{\rho_b v_b d_b}{\mu_b} = \frac{\rho_m v_m d_m}{\mu_m}$$

$$v_m = \frac{\rho_b v_b d_b \mu_m}{\mu_b \rho_m d_m} = \frac{1050\ kg/m^3(50\ cm/s)(10\ cm)(1\ cP)}{1000\ kg/m^3(10\ cm)(3.5\ cP)} = 15\ cm/s$$

Using the Womersley number, we can calculate the angular frequency of the model:

$$\alpha_{blood} = \alpha_{model}$$

$$d_b\left(\frac{\omega_b}{\nu_b}\right)^{\frac{1}{2}} = d_m\left(\frac{\omega_m}{\nu_m}\right)^{\frac{1}{2}}$$

$$d_b\left(\frac{\rho_b \omega_b}{\mu_b}\right)^{\frac{1}{2}} = d_m\left(\frac{\rho_m \omega_m}{\mu_m}\right)^{\frac{1}{2}}$$

$$\omega_m = \frac{\mu_m}{\rho_m}\left[\frac{d_b}{d_m}\left(\frac{\rho_b \omega_b}{\mu_b}\right)^{\frac{1}{2}}\right]^2 = \frac{1\ cP}{1000\ kg/m^3}\left[\left(\frac{10\ cm}{10\ cm}\right)\left(\frac{(1050\ kg/m^3)(72\ beats/min)}{3.5\ cP}\right)^{\frac{1}{2}}\right]^2$$
$$= 21.6\ beats/min$$

What happens if complete dynamic similarity is not achievable, because there are too many parameters to match between the real scenario and the model? In some instances, by solving the dimensionless groups simultaneously, it can be found that no fluid exists that can match the properties needed. For instance, if one was modeling the flow around an airplane wing, you may want to make a 1/10 or 1/50 model, which would then dictate the other properties of the fluid and the experimental conditions. By matching as many parameters as possible (velocity, wave frequency), it is typical that the fluid viscosity or density is what needs to be modified to make the solution dynamically similar. What happens when there is no known fluid that exists that can match the required parameters? Should some of the other parameters be altered to find a suitable fluid? If there is an incomplete dynamic similarity, then the only way to achieve a similar model is to use a full-scaled model, which may not be reasonable (what if it is a nuclear submarine or a single cell?). However, studies have shown that some of the data that are obtained from incomplete dynamically similar solutions are still useful. Therefore, it is best to match as many parameters as possible when modeling a real scenario, but if all of the parameters cannot be matched, then this must be accounted for when analyzing the data.

14.4 CURRENT STATE OF THE ART FOR BIOFLUID MECHANICS *IN SILICO* RESEARCH

In the last three chapters of this textbook, we will discuss some areas of open research that are being pursued by current biofluid mechanics research groups. We do not intend for this discussion to encompass all of the areas of research that are being explored, but instead we use this as an area to describe current trends and provide the reader with an avenue of exploration on their own. As a data point only, at the time of writing, conducting a PubMed search (http://www.ncbi.nlm.nih.gov/pubmed) with the search string of "computational fluid dynamics" returns over 3300 publications. The same search string in ScienceDirect.com (http://www.sciencedirect.com/) or Google Scholar (http://scholar.google.com/) retrieves over 100,000 and 1.68 million publications, respectively. Clearly, computational fluid dynamics is a very active research field that could not be adequately described in this small section.

There are two very common fields in current *in silico* biofluid mechanics research that can branch into multiple avenues. The first is the evaluation of cardiovascular implantable devices (e.g., stents, heart valves) and the second is the understanding of the development or flow properties that instigate/propagate cardiovascular diseases. The first area of research has proven to be an invaluable tool in designing new devices that minimize damage to blood cells or disruptions to the flow field. The importance of this area of research is that many of the original implantable cardiovascular devices failed due to damage induced to blood cells or a disruption to the fluid flowing around the device. For instance, it was common for patients with mechanical heart valves to be placed on an anticoagulant treatment to minimize the negative reactions that the heart valves induced on the hemostatic system. While the implantation of these valves would save the lives of the patients, there was an increased burden on the patient lifestyle that could potentially lead to valve failure and/or death.

Computational fluid dynamics approaches have focused on identifying critical geometries within the heart valve structure that elicit potentially adverse flow properties. Engineering optimization/design procedures have been implemented to redesign these critical geometries and then computational studies of the effects on the flow field have been conducted. The importance of this work is that the cost of conducting a computational fluid dynamics study to optimize particular properties of the flow conditions would be significantly less than building a prototype that can be used within *in vitro* or *in vivo* studies. "Good" designs can be refined and "bad" designs can be excluded relatively easily with these techniques.

For the second area of research, atherosclerosis and aneurysm development are probably the two most common research areas. Groups can make use of patient specific geometries to predict the onset of the disease or research groups can make use of these geometries to determine how the flow field is impacted around the developing disease. Many fundamental fluid dynamics issues related to cardiovascular disease development have been elucidated through the use of these models. In more recent work, models with greater physiological relevance have been developed to more accurately predict disease onset or development. For instance, cells within the flow field and the mechanical properties of the blood vessel have been considered to more accurately predict the flow conditions.

There are many other avenues of research that make use of computational fluid dynamics approaches. For instance, recent work has investigated the air flow through the branching lung conducting pathways under normal and pathological conditions. One of the major purposes of this work was to determine the effective forces that act on the respiratory epithelial cells in order to determine possible disease conditions. Models have been made that approximate the flow through the nephron tubule system to observe how altered cardiovascular conditions or altered regulatory mechanisms can alter the flow through nephron units. Similarly, models of the flow through the gastrointestinal tract have been developed to determine the effects of altered pressure or chemical regulation on digestion and other processes.

It is important to note that regardless of the area of biofluid mechanics research that the particular group is pursuing, these models need to be validated by some other measures. Some of this validation can come from previous reports, but it is important to conduct experiments that have similar constraints as the model that is being developed. Without this validation, it would be difficult to determine how accurate the model represents the physiological flow conditions. It would also be difficult for readers external to the research group to interpret the findings.

Finally, all of the research areas that we have discussed can be investigated using FSI modeling and/or multiscale modeling approaches. We did not highlight the different studies in these areas but be aware that each of these more thorough techniques can be applied to the general areas that we have discussed.

14.5 FUTURE DIRECTIONS OF BIOFLUID MECHANICS *IN SILICO* RESEARCH

In the last three chapters in this textbook, we will include a section about possible future directions of research that may be pursued to advance the field of biofluid mechanics research. As with the previous section of this chapter, this is not an all-inclusive list of possible future research endeavors, but instead our discussion is intended to highlight areas of research that the authors believe can significantly advance our knowledge of biofluid mechanics research.

In the author's opinion, it is crucial to make models that are more physiologically relevant and to incorporate more biological processes into the modeling approach. This would allow the model to more accurately predict the behavior of the system and this would also provide the community with more relevant information that can be used for the design of therapeutics and/or medical devices. More accurate FSI modeling and multiscale modeling approaches are two areas that may be developed in the near future to address these issues that we are describing.

In Section 14.2, we described the need for FSI modeling. This area, while it is currently in use in some biofluid mechanics labs, still needs significant development to become more widely used. For instance, to make these models better, it would be important to incorporate more physiologically relevant/pathophysiologically relevant parameters within the model. This is principally related to the velocity/pressure waveforms and/or the mechanical properties of the tissue/cells. As we have described, the interaction of fluid and solid elements is very important to the progression of cardiovascular diseases and thus more relevant information regarding these interactions needs to be incorporated into models. These types of models may be able to elucidate how critical mechanical forces are to the generation/progression of cardiovascular diseases.

Multiscale modeling is an approach in which the model incorporates multiple length (or time) scales within each iteration. In the case of biofluid mechanics problems, the largest scale would typically consider the bulk flow of the fluid through the channel of interest. A scale below this would consider the motion of cells and/or the deformation of the containing volume or the cells themselves. A smaller scale would consider the motion of proteins or other molecular elements within the flow field (e.g., do coagulation or inflammatory proteins come into contact with each other during the flow scenario). Finally, the kinetics of these protein interactions can be modeled (e.g., coagulation cascade modeling or adhesion mechanics/dynamics between cells and the vessel wall). We have described some of these modeling events in previous section, but the difficulty in multiscale modeling arises principally from the coupling between the different scales. It is relatively easy to model the fluid flow through a channel, but to include the motion of particular discrete-phase elements would be more difficult. If the discrete phase is given a true mass (and thus inertia/drag), the motion of the fluid alters the motion of the particles and the motion of the particle alters the motion of neighboring fluid elements. The motion of proteins or other dissolved elements would be even more difficult to include because most proteins typically do not have well-defined mechanical properties and the relationship between applied mechanical forces and geometric considerations would be difficult to validate. Finally, if the dynamics of the proteins could be modeled, then their interactions would need to be modeled. The protein–protein interaction would elicit some product that would then need to be tracked. Additionally, the kinetics of these interactions would need to be included within the time steps. Overall, the coupling of all of these processes would be extremely difficult to accomplish. However, these models would provide us with the most relevant information since many of the processes that are of concern to the researcher would be investigated at once. This would allow for the generation of a predictive model that could potentially look into many different physiological aspects of the flow conditions. At this point, it is typical for many simplifications to be included within the multi-scale model, so that some data can be obtained.

The authors also would like to see the combination of models with different scales be combined with FSI models, so that solid and fluid parameters can be included within the one simulation. This would provide us with significant information regarding the process that is under investigation. With that said, one has to consider what parameters are "important" to the flow and whether or not those parameters should be included within the flow conditions. For instance, if one is looking at blood flow through the coronary artery, then the heart motion and the dynamics of the vessel opening/closing would be critical to include. This would necessitate FSI modeling. Discrete flow through capillary beds is less susceptible to motion within the vessel and thus it may not be an important parameter to include within the flow conditions. Kinetics of cell adhesion can occur anywhere within the cardiovascular system and it may be important to include, however, if adhesion is atypical under the particular conditions investigating, it may not be important to include this parameter within the simulation. Thus, while more physiologically relevant models should be developed, the relevance of these additions to the model needs to be weighed against the difference in the data that is obtained from the simulation.

END OF CHAPTER SUMMARY

14.1 Computational fluid dynamics is a method that can be used to solve the coupled governing equations of biofluid flow, without making many simplifying assumptions. The most common method to solve CFD simulations is to mesh a relevant geometry and then by using either the forward difference method or the backward difference method, the fluid properties can be obtained at subsequent grid points. The equation takes the form of

$$\left.\frac{\partial f}{\partial x}\right|_i = \frac{f_{i+1} - f_{i-1}}{2\Delta x} + H.O.T.$$

Under these simulations, it is also possible to incorporate turbulent flow properties into the numerical methods. The most commonly used models are either the k-ε or the k-ω turbulent models, which are defined by

$$k = \frac{1}{2}\overline{u_i u_i}$$

$$\varepsilon = v\overline{u_{i,j}u_{i,j}}$$

$$\varepsilon = k\omega$$

A second useful tool that is incorporated into many CFD commercial software packages is the ability to solve for the discrete movement of particles within the fluid. By coupling Newton's laws to the fluid dynamics laws, equations can be composed that determine the forces that act on particles within the fluid. A common solution for the force and the velocity of a particle is

$$\frac{du_p}{dt} = F_D(u - u_p) + F_x$$

and

$$\frac{dx_p^i}{dt} = u_p^i$$

If the shear stress is a critical parameter in the flow scenario, then the accumulation of shear stress with time can be calculated from

$$\sum_{t=t_0}^{t_{\max}} \left(\frac{(\varepsilon_i - \varepsilon_{i+1})}{2} \times (\mu_i + \mu_i^t) + KE_i \times \rho\right) \times \Delta t_i$$

14.2 A more accurate model can incorporate the movement of the blood vessel wall and the deformability of particles within the fluid. This type of modeling is termed FSI model, and in these types of models, mass, momentum, and energy can be transferred between the wall, the particles, and the fluid. The solution of these types of simulations makes use of the partial differential equation:

$$\nabla^2 \Phi - \frac{\partial^2 \Phi}{\partial t^2} - u\left[\frac{\partial^2 \Phi}{\partial x^2} + \frac{\partial^2 \Phi}{\partial y^2} + \frac{\partial^2 \Phi}{\partial z^2}\right] = 0$$

which can be solved for any fluid parameter of interest. An important feature of FSI modeling is that the mechanical properties of the solid elements should be known under the conditions that are being observed. For instance, under disease conditions, it is likely that localized mechanical properties of the blood vessel wall are altered based on the deposition and/or removal of material. These changes should need to be addressed within the model.

14.3 To make the numerical model the most accurate possible, the models should be dynamically similar, which means that the length scales and the forces that act within the model are the same. A method to ensure that a model is similar is the Buckingham Pi Theorem, which devises a number of functions between a parameter of interest and various fluid properties. In biofluid mechanics, there are a number of dimensionless parameters that can be considered when making models. These are the Reynolds number, the Womersley number, the Strouhal number, the cavitation number, the Prandtl number, the Weber number, and the capillary number, among others. These are defined, respectively, as

$$\mathrm{Re} = \frac{\rho v d}{\mu} = \frac{vd}{\nu}$$

$$\alpha = d\left(\frac{\omega}{\nu}\right)^{\frac{1}{2}}$$

$$\mathrm{St} = \frac{fL}{\nu}$$

$$\mathrm{Ca} = \frac{p - p_v}{\frac{1}{2}\rho v^2}$$

$$\mathrm{Pr} = \frac{\mu C_P}{k}$$

$$\mathrm{We} = \frac{\rho v^2 L}{\sigma}$$

$$\mathrm{C} = \frac{\mu u}{\sigma}$$

14.4 Current state of the art of *in silico* biofluid mechanics research areas are very broad and encompass many different physiological and pathophysiological events. Many groups are making use of traditional techniques but are supplementing these techniques with customized codes that can depict particular aspects of the physiology that the group wants to investigate. These techniques have provided the community with a significant amount of information regarding disease progression and the development/optimization of new implantable therapeutic devices.

14.5 In the future, *in silico* biofluid mechanics research groups should focus on developing models that are more relevant and incorporate more fundamental processes to more accurately depict the pathological event (or whatever is being modeled). Some areas that are currently underdeveloped is the incorporation of multiple length scales into the model to depict the flow, cellular deformations, kinetics, etc. These types of models, if the initial conditions, assumptions, and linking functions can be assumed accurately, can provide a great deal of information to the research community.

HOMEWORK PROBLEMS

14.1 Using the central difference method in 2D space, where each point is the average of its four nearest neighbors, calculate the velocity through the specified channel with the known boundary conditions and the known inlet conditions (Figure 14.10). The velocity along the wall grid points (1, 2, 3, 19, 20, and 21) is 0 mm/s. The velocity at the inlet grid points (4, 9, and 14) is 40 mm/s. The velocity at the outflow grid points 8 and 18 is 20 mm/s, and the velocity at the grid point 13 is 35 mm/s. It may be easiest to use linear algebra to solve the system of equations that will be generated. Assume that the grid is uniformly spaced in the x- and y-directions.

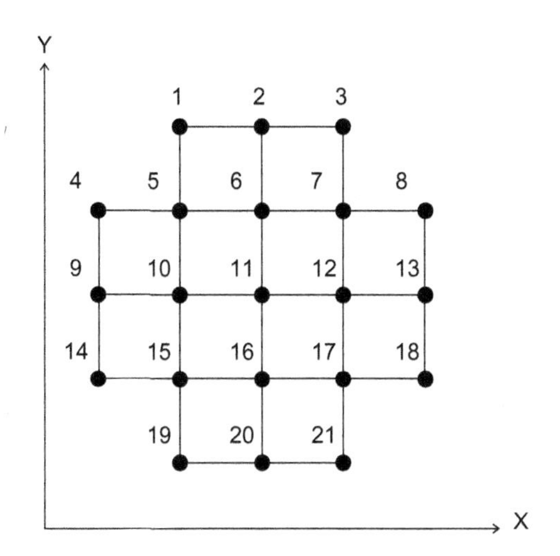

FIGURE 14.10 Grid for homework problem 14.1.

14.2 Using an available computational fluid dynamics software package, make a simple geometry for a rapid constriction (similar to Figure 14.3). Calculate the flow through this constriction. If no software is available, make a prediction for the flow through the constriction.

14.3 The spin on particulate matter within blood may play an important role in the activation or recycling of the cell. Therefore, the spin dissipation rate may be an important parameter to know. We believe that the torque on the particle, T, is dependent on the particle's speed (v), the blood density (ρ), the blood viscosity (μ), the cell diameter (d), and the spin rate (angular velocity, ω). Determine the dimensionless parameters that can be formed from this grouping.

14.4 When a mechanical heart valve closes rapidly, a hammer pressure wave is initiated in the fluid. This high-pressure wave may damage the blood vessel wall distal to the mechanical heart valve (potentially shearing endothelial cells off the inner lining). The maximum hammer pressure generated (p_{max}) is a function of the blood density (ρ), the blood velocity at time of closure (v), and the blood deformation modulus (E, which has units of a general elastic modulus). Determine the relationship between hammer pressure and the variables provided.

14.5 To match the Reynolds number between blood flow and a water flow model, using a twofold larger model, which flow will require the higher flow speed? How much higher will it be?

14.6 Develop an argument for the use of computational fluid dynamics methods in biofluid research. Consider and delineate the pros and cons of implementing a method such as this. Are there better models that can be employed; what are the shortcomings of those models?

14.7 We described a handful of dimensionless parameters in this section that are important to biofluid mechanics computational research. Choose 2–3 of these parameters and describe situations in which they would be important in biofluid research. Additionally, are there other parameters that you can think of that should be included within this discussion (e.g., are there a grouping of variables that we did not list that should be included).

14.8 Prepare a research report on the current state of the art for a field of your choosing within computational biofluid mechanics research. Describe the history of the field, how it has progressed to where it is now, and where you believe future efforts should be focused toward.

References

[1] Y. Alemu, D. Bluestein, Flow-induced platelet activation and damage accumulation in a mechanical heart valve: numerical studies, Artif. Organs 31 (2007) 677.
[2] F. Ali, E. Waller, Design of a hybrid computational fluid dynamics-Monte Carlo radiation transport methodology for radioactive particulate resuspension studies, Health Phys. 107 (2014) 311.
[3] T. AlMomani, H.S. Udaykumar, J.S. Marshall, K.B. Chandran, Micro-scale dynamic simulation of erythrocyte-platelet interaction in blood flow, Ann. Biomed. Eng. 36 (2008) 905.
[4] M. Bathe, R.D. Kamm, A fluid–structure interaction finite element analysis of pulsatile blood flow through a compliant stenotic artery, J. Biomech. Eng. 121 (1999) 361.
[5] D. Bluestein, K. Dumont, M. De Beule, J. Ricotta, P. Impellizzeri, B. Verhegghe, et al., Intraluminal thrombus and risk of rupture in patient specific abdominal aortic aneurysm – FSI modelling, Comput. Methods Biomech. Biomed. Eng. 12 (2009) 73.
[6] E. Buckingham, On physically similar systems: illustrations of the use of dimensional equations, Phys. Rev. 4 (1914) 345.
[7] M.A. Castro, M.C. Olivares, C.M. Putman, J.R. Cebral, Unsteady wall shear stress analysis from image-based computational fluid dynamic aneurysm models under Newtoniam and Casson rheological models, Med. Biol. Eng. Comput. 52 (2014) 827.
[8] D.I. Cattoni, C. Ravazzola, V. Tüngler, D.E. Wainstein, O. Chara, Effect of intestinal pressure on fistula closure during vacuum assisted treatment: a computational approach, Int. J. Surg. 9 (2011) 662.
[9] L.M. Crowl, A.L. Fogelson, Computational model of whole blood exhibiting lateral platelet motion induced by red blood cells, Int. J. Numer. Method Biomed. Eng. 26 (2010) 471.
[10] L.M. Crowl, A.L. Fogelson, Analysis of mechanisms for platelet near-wall excess under arterial blood flow conditions, J. Fluid Mech. 676 (2011) 348.
[11] S. De Gruttola, K. Boomsma, D. Poulikakos, Y. Ventikos, Computational simulation of the blood separation process, Artif. Organs 29 (2005) 665.
[12] G. Dubini, R. Pietrabissa, F.M. Montevecchi, Fluid–structure interaction problems in bio-fluid mechanics: a numerical study of the motion of an isolated particle freely suspended in channel flow, Med. Eng. Phys. 17 (1995) 609.
[13] H.A. Dwyer, P.B. Matthews, A. Azadani, N. Jaussaud, L. Ge, T.S. Guy, et al., Computational fluid dynamics simulation of transcatheter aortic valve degeneration, Interact. Cardiovasc. Thorac. Surg. 9 (2009) 301.
[14] W. Dzwinel, K. Boryczko, D.A. Yuen, A discrete-particle model of blood dynamics in capillary vessels, J. Colloid Interface Sci. 258 (2003) 163.
[15] D.J. Evans, A.S. Green, N.K. Thomas, Wall shear stress distributions in a model of normal and constricted small airways, Proc. Inst. Mech. Eng. H 228 (2014) 362.
[16] D.A. Fedosov, H. Noguchi, G. Gompper, Multiscale modeling of blood flow: from single cells to blood rheology, Biomech. Model Mechanobiol. 13 (2014) 239.
[17] M.J. Ferrua, R.P. Singh, Modeling the fluid dynamics in a human stomach to gain insight of food digestion, J. Food Sci. 75 (2010) R151.
[18] FLUENT 6.3 User's Guide. 2006. <https://www.sharcnet.ca/Software/Fluent6/html/ug/main_pre.htm>.
[19] M.D. Ford, H.N. Nikolov, J.S. Milner, S.P. Lownie, E.M. Demont, W. Kalata, et al., PIV-measured versus CFD-predicted flow dynamics in anatomically realistic cerebral aneurysm models, J. Biomech. Eng. 130 (2008) 021015.
[20] K.H. Fraser, M.X. Li, W.T. Lee, W.J. Easson, P.R. Hoskins, Fluid–structure interaction in axially symmetric models of abdominal aortic aneurysms, Proc. Inst. Mech. Eng. H 223 (2009) 195.
[21] L. Ge, S.C. Jones, F. Sotiropoulos, T.M. Healy, A.P. Yoganathan, Numerical simulation of flow in mechanical heart valves: grid resolution and the assumption of flow symmetry, J. Biomech. Eng. 125 (2003) 709.

[22] A. Gosman, L. Ioannides, Aspects of computer simulation of liquid-fueled combustors, J. of Energy 7 (1983) 482.
[23] L. Goubergrits, U. Kertzscher, B. Schoneberg, E. Wellnhofer, C. Petz, H.C. Hege, CFD analysis in an anatomically realistic coronary artery model based on non-invasive 3D imaging: comparison of magnetic resonance imaging with computed tomography, Int. J. Cardiovasc. Imaging 24 (2008) 411.
[24] L. Goubergrits, E. Wellnhofer, U. Kertzscher, K. Affeld, C. Petz, H.C. Hege, Coronary artery WSS profiling using a geometry reconstruction based on biplane angiography, Ann. Biomed. Eng. 37 (2009) 682.
[25] B. Gunther, Dimensional analysis and theory of biological similarity, Physiol. Rev. 55 (1975) 659.
[26] M. Hasan, D.A. Rubenstein, W. Yin, Effects of cyclic motion on coronary blood flow, J. Biomech. Eng. 135 (2013) 121002.
[27] K. Hassani, M. Navidbakhsh, M. Rostami, Modeling of the aorta artery aneurysms and renal artery stenosis using cardiovascular electronic system, Biomed. Eng. Online 6 (2007) 22.
[28] J.O. Hinze, Turbulence, McGraw-Hill, New York, NY, 1987.
[29] P.L. Hsu, R. Graefe, F. Boehnung, C. Wu, J. Parker, R. Autschbach, et al., Hydraulic and hemodynamic performance of a minimally invasive intra-arterial right ventricular assist device, Int. J. Artif. Organs 37 (2014) 697.
[30] P.Y. Huang, J.D. Hellums, Aggregation and disaggregation kinetics of human blood platelets: Part I. Development and validation of a population balance method, Biophys. J. 65 (1993) 334.
[31] D.C. Ipsen, Units, Dimensions and Dimensionless Numbers, McGraw-Hill, New York, NY, 1960.
[32] J.M. Jimenez, P.F. Davies, Hemodynamically driven stent strut design, Ann. Biomed. Eng. 37 (2009) 1483.
[33] M.C. Johnson, R.D. Kamm, The role of Schlemm's canal in aqueous outflow from the human eye, Invest. Ophthalmol. Vis. Sci. 24 (1983) 320.
[34] J. Jung, A. Hassanein, Three-phase CFD analytical modeling of blood flow, Med. Eng. Phys. 30 (2008) 91.
[35] D.B. Khismatullin, G.A. Truskey, Three-dimensional numerical simulation of receptor-mediated leukocyte adhesion to surfaces: effects of cell deformability and viscoelasticity, Phys. Fluids 17 (1994) 031505.
[36] M.J. Kraus, E.F. Strasser, R. Eckstein, A new method for measuring the dynamic shape change of platelets, Transfus. Med. Hemother 37 (2010) 306.
[37] S.A. Kock, J.V. Nygaard, N. Eldrup, E.T. Frund, A. Klaerke, W.P. Paaske, et al., Mechanical stresses in carotid plaques using MRI-based fluid–structure interaction models, J. Biomech. 41 (2008) 1651.
[38] Y.G. Lai, K.B. Chandran, J. Lemmon, A numerical simulation of mechanical heart valve closure fluid dynamics, J. Biomech. 35 (2002) 881.
[39] C.A. Leguy, E.M. Bosboom, A.P. Hoeks, F.N. van de Vosse, Assessment of blood volume flow in slightly curved arteries from a single velocity profile, J. Biomech. 42 (2009) 1664.
[40] Z. Li, C. Kleinstreuer, Fluid–structure interaction effects on sac-blood pressure and wall stress in a stented aneurysm, J. Biomech. Eng. 127 (2005) 662.
[41] N.A. Mody, O. Lomakin, T.A. Doggett, T.G. Diacovo, M.R. King, Mechanics of transient platelet adhesion to von Willebrand factor under flow, Biophys. J. 88 (2005) 1432.
[42] N.A. Mody, M.R. King, Platelet adhesive dynamics. Part I: characterization of platelet hydrodynamic collisions and wall effects, Biophys. J. 95 (2008) 2539.
[43] S.M. Moore, K.T. Moorhead, J.G. Chase, T. David, J. Fink, One-dimensional and three-dimensional models of cerebrovascular flow, J. Biomech. Eng. 127 (2005) 440.
[44] J.L. Pelerin, C. Kulik, C. Goksu, J.L. Coatrieux, M. Rochette, Fluid/structure interaction applied to the simulation of abdominal aortic aneurysms, Conf. Proc. IEEE Eng. Med. Biol. Soc. 1 (2006) 1754.
[45] A.K. Politis, G.P. Stavropoulos, M.N. Christolis, F.G. Panagopoulos, N.S. Vlachos, N.C. Markatos, Numerical modeling of simulated blood flow in idealized composite arterial coronary grafts: steady state simulations, J. Biomech. 40 (2007) 1125.
[46] C. Pozrikidis, Flipping of an adherent blood platelet over a substrate, J. Fluid Mech. 568 (2006) 161.
[47] V.L. Rayz, L. Boussel, G. Acevedo-Bolton, A.J. Martin, W.L. Young, M.T. Lawton, et al., Numerical simulations of flow in cerebral aneurysms: comparison of CFD results and *in vivo* MRI measurements, J. Biomech. Eng. 130 (2008) 051011.
[48] L.I. Sedov, Similarity and Dimensional Methods in Mechanics, Academic Press, New York, NY, 1959.
[49] I. Sgouralis, A.T. Layton, Autoregulation and conduction of vasomotor responses in a mathematical model of the rat afferent arteriole, Am. J. Physiol. Renal Physiol. 303 (2012) F229.
[50] I. Sgouralis, A.T. Layton, Control and modulation of fluid flow in the rat kidney, Bull. Math. Biol. 75 (2013) 2551.

[51] S.K. Shanmugavelayudam, D.A. Rubenstein, W. Yin, Effect of geometrical assumptions on numerical modeling of coronary blood flow under normal and disease conditions, J. Biomech. Eng. 132 (2010) 061004.
[52] Y. Shi, Y. Zhao, T.J. Yeo, N.H. Hwang, Numerical simulation of opening process in a bileaflet mechanical heart valve under pulsatile flow condition, J. Heart Valve Dis. 12 (2003) 245.
[53] S. Smith, S. Austin, G.D. Wesson, C.A. Moore, Calculation of wall shear stress in left coronary artery bifurcation for pulsatile flow using two-dimensional computational fluid dynamics, Conf. Proc. IEEE Eng. Med. Biol. Soc. 1 (2006) 871.
[54] B. Sul, A. Wallqvist, M.J. Morris, J. Reifman, V. Rakesh, A computational study of the respiratory airflow characteristics in normal and obstructed human airways, Comput. Biol. Med. 52 (2014) 130.
[55] P. Tandon, S.L. Diamond, Hydrodynamic effects and receptor interactions of platelets and their aggregates in linear shear flow, Biophys. J. 73 (1997) 2819.
[56] R. Torii, M. Oshima, T. Kobayashi, K. Takagi, T. Tezduyar, Fluid–structure interaction modeling of aneurysmal conditions with high and normal blood pressures, Comput. Mech. 38 (2006) 482.
[57] R. Torii, M. Oshima, T. Kobayashi, K. Takagi, T.E. Tezduyar, Influence of wall elasticity in patient-specific hemodynamic simulations, Comput. Fluids 36 (2007) 160.
[58] W. Wang, N.A. Mody, M.R. King, Multiscale model of platelet translocation and collision, J. Comput. Phys. 244 (2013) 223.
[59] J. Wen, K. Liu, K. Khoshmanesh, W. Jiang, T. Zheng, Numerical investigation of haemodynamics in a helical-type artery bypass graft using non-Newtonian multiphase model, Comput. Methods Biomech. Biomed. Eng. 18 (2015) 760.
[60] D.C. Wilcox, Re-assessment of the scale-determining equation for advanced turbulence models, AIAA (1988) 1414.
[61] C. Xu, D.M. Wootton, Platelet near-wall excess in porcine whole blood in artery-sized tubes under steady and pulsatile flow conditions, Biorheology 41 (2004) 113.
[62] W. Yin, Y. Alemu, K. Affeld, J. Jesty, D. Bluestein, Flow-induced platelet activation in bileaflet and monoleaflet mechanical heart valves, Ann. Biomed. Eng. 32 (2004) 1058.
[63] W. Yin, S.K. Shanmugavelayudam, D.A. Rubenstein, 3D numerical simulation of coronary blood flow and its effect on endothelial cell activation, Conf. Proc. IEEE Eng. Med. Biol. Soc. 2009 (2009) 4003.
[64] P. Zhang, A. Sun, F. Zhan, J. Luan, X. Deng, Hemodynamic study of overlapping bare-metal stents intervention to aortic aneurysm, J. Biomech. S0021 (2014) 00461.
[65] P. Zhang, C. Gao, N. Zhang, M.J. Slepian, D. Yuefan, D. Bluestein, Multiscale particle-based modeling of flowing platelets in blood plasma using dissipative particle dynamics and coarse grained molecular dynamics, Cell Mol. Bioeng. 7 (2014) 552–574.
[66] S.Z. Zhao, X.Y. Xu, A.D. Hughes, S.A. Thom, A.V. Stanton, B. Ariff, et al., Blood flow and vessel mechanics in a physiologically realistic model of a human carotid arterial bifurcation, J. Biomech. 33 (2000) 975.

CHAPTER

15

In Vitro Biofluid Mechanics

LEARNING OUTCOMES

1. Discuss common experimental techniques that are used for *in vitro* biofluid mechanics studies
2. Compare particle imaging velocimetry techniques and the data that can be obtained from these methods
3. Describe the equipment necessary to conduct particle imaging velocimetry measurements
4. Describe parallel plate and other viscometry techniques
5. Mathematically represent the flow field within a parallel plate viscometer and a cone-and-plate viscometer
6. Discuss current and future work that makes use of *in vitro* biofluid mechanics techniques.

15.1 PARTICLE IMAGING VELOCIMETRY

Particle imaging velocimetry (PIV) is a computer-based technique that tracks a sequence of reflective particles through a flow chamber. In the most basic PIV technique, one camera is used to obtain flow information about one cross section through the flow chamber. However, it is more common to use more than one camera in order to collect multiple views so that a more complete three-dimensional velocity flow profile can be obtained. Also, flow patterns are more easily obtained from multiple cameras collecting information about the same particles within the flow field. There are multiple techniques that can fit into PIV, and we will discuss some of them here. The most popular technique is particle tracking. In this technique, one or more cameras are used to image a three-dimensional velocity profile that is projected onto a two-dimensional plane. If only one camera is used, information about the distance from the camera will be lost, since a single camera can only project a three-dimensional image onto a two-dimensional plane (e.g., imagine taking a picture of a crowded train terminal, the exact position of every person cannot be obtained

© 2015 Elsevier Inc. All rights reserved.

accurately from the one image). However, by adding more cameras, this information can be obtained through a cross-correlation of the data (also termed triangulation), and therefore, a full three-dimensional image of the flow field can be generated (although you may not have information regarding the entire flow field, since this is dependent on the optics of the imaging system). To obtain these images of flow scenarios, it is preferred to fabricate an accurate (possibly scaled up or down) model, which can depict the geometry of the real flow conditions. Dynamic similarity, which was described in Chapter 14, is used to make sure the scaling and the forces throughout the model are matched. The basis for collecting data with PIV is through the use of reflective beads, which are placed within the fluid that will be circulated throughout the model. A high-intensity light source (commonly a laser source) is focused onto a region of interest within the flow chamber. A digital camera with a high acquisition frame rate is also focused at this region of interest to collect data on the particle motion over time. An external synchronization trigger is needed to control the illumination of the laser (which is typically pulsed) and the camera is needed. Particle motion is obtained through the light reflection from the seeded particles, which is captured by the digital camera. Using a computer algorithm, each subsequent image is correlated to the previous image, temporally, so that each particle's velocity can be calculated from a known change in position divided by the known frame rate of the camera. The position information can be obtained by scaling the camera images to the physical dimensions of the model. By collecting multiple data sets at each region of interest, it is possible to obtain mean velocities (magnitude and direction) throughout the flow field. A typical PIV setup is schematized in Figure 15.1. It should be noted that it is

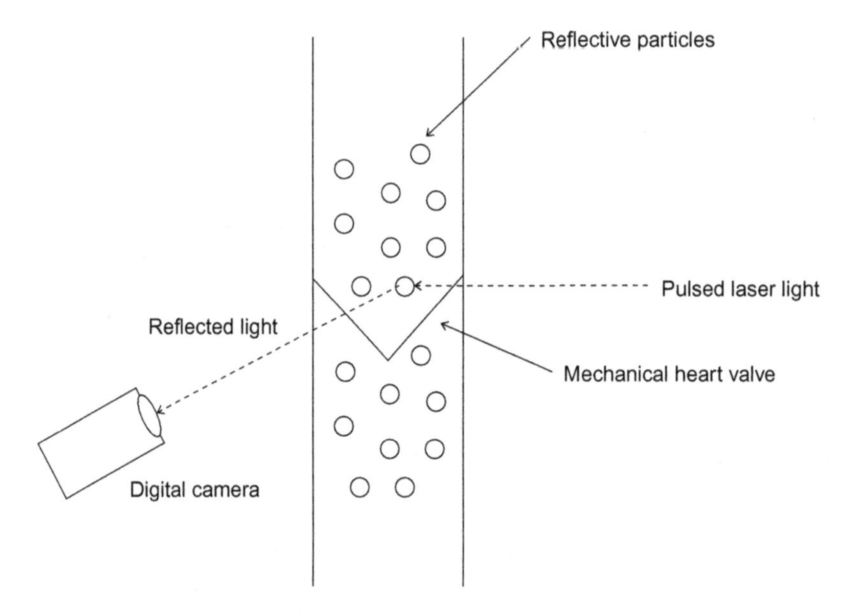

FIGURE 15.1 Schematic of a typical PIV system, which consists of a flow chamber (shown with a bileaflet mechanical heart valve), a fluid seeded with reflective particles, a high-intensity laser light source, and a digital camera. Not shown is the computer system that would be coupled to both the laser and the camera to calculate particle velocities.

possible for the particle tracking algorithm to not correlate the correct particles in subsequent images. For instance, if the particles are too concentrated within the flow field, then the reflected light could be outside the cameras resolution limit and thus the algorithm will be unable to discriminant between near particles. Also, if the particle does not follow a "normal" trajectory, it may be difficult for the algorithm to determine particle motion. However, if the particles are too sparsely populated, then not enough information about the flow field can be obtained. Therefore, it is necessary to optimize the particle seeding density within the flow chamber. Besides the particle density, many of the issues regarding correlating of neighboring particles can be addressed by the optics and the optical control systems.

The usefulness of PIV systems in biofluid mechanics research is that new designs for cardiovascular implantable devices can be tested to determine if they disrupt the flow field and by how much the flow field is altered. Computational methods provided a limited amount of insight, depending on the accuracy of the geometry of the model, the assumptions built into the model and the algorithm used to solve the governing equations. Using PIV systems, an actual device (such as a mechanical heart valve, a total artificial heart, or a stent) can be investigated within a dynamically similar environment. This means that if a physical model can be fabricated with properties relevant to the actual flow conditions, a real device can be investigated within an environment that mimics the intended environment. For most cardiovascular *in vitro* simulations, it would be most appropriate to match the Reynolds number and the Womersley number to cardiovascular flow as well as the physical properties of the container to reach a dynamically similar solution (as discussed in Chapter 14).

There are a number of variations in PIV systems based on the incident light and/or the collection method. One of these methods, laser speckle velocimetry (LSV), makes use of the speckle pattern of particles suspended within the fluid instead of the reflective pattern as discussed above for PIV. LSV is typically used when the exact velocity distribution is wanted at all locations within the fluid. The fluid is seeded with a high concentration of particles, whose reflected light would not be able to be distinguished by a PIV computer algorithm (and therefore the particles themselves could not be differentiated) coupled with a high-speed camera system and one particular bandwidth of incident light. For LSV, a coherent light source is used which generates a speckle pattern based on the interference of the particles. A coherent light source is needed so that the interference pattern is largely stable both over short time durations and spatial dimensions. This speckle pattern is collected by a digital camera, and more powerful computer processing techniques are needed to determine the flow profile based on this interference speckle pattern. This is typically used to model the movement of large solids, such as thromboemboli, through a dynamically similar system.

Another PIV technique, holographic PIV (or HPIV), is used to obtain a three-dimensional velocity profile by collecting the particle location information on a hologram and then computationally reconstructing this image. The advantage of HPIV systems is that a true three-dimensional flow profile image can be obtained instead of projecting the three-dimensional velocities onto a two-dimensional plane. A recent PIV technique can be used to investigate microflows, with the use of an epi-fluorescent microscope. Instead of using reflective particles, fluorescent particles are placed in the flow stream, and data are recorded through a camera coupled to an epi-fluorescent microscope. Fluorescent particles are useful because they excite at one particular wavelength of light and emit at a second wavelength of light. Through the use of appropriate filters, these wavelengths can be discerned precisely and all other wavelengths of light can be ignored. Also, common reflective particles are typically too large for studies on the microcirculation, but through the use of fluorescent particles, images on a smaller scale can be obtained. A disadvantage of using a fluorescent source is that the fluorescence may overpower small features within the flow field.

15.2 LASER DOPPLER VELOCIMETRY

Laser Doppler velocimetry is primarily concerned with quantifying the microstructures of flows that are subjected to obstacles within the flow field (similar to PIV techniques). Direct measurements of the kinematics of the fluid motion can be obtained with this method. The benefit is that one can observe the smallest of flows surrounding these obstructions; which may not appear when imaging particles within a flow field. For laser Doppler velocimetry, the measurement of the fluid velocity is made at the intersection of two laser beams that are focused at a point of interest within the flow field. Laser light sources are required for this technique because the two beams must be monochromatic (one wavelength) and be coherent, so that the intersecting beams can create an interference pattern. The interference pattern occurs within the fluid and is documented with a digital camera that is focused at the region of interest. Based on the wavelength of the incident beams and the angle between the two beams, fringe patterns would be formed with a spacing of

$$d = \frac{\lambda}{2\sin\left(\frac{\theta}{2}\right)} \tag{15.1}$$

where d is the fringe pattern spacing, λ is the wavelength of the laser light, and θ is the angle between the two laser beams (Figure 15.2). As a particle moves through the laser beam area of intersection, scattered light will have a different intensity and frequency than the normal fringe patterns. This new frequency (ω) is defined by the velocity of the particle and the fringe spacing as

$$\omega = \frac{v}{d} = \frac{2v\sin\left(\frac{\theta}{2}\right)}{\lambda} \tag{15.2}$$

where v is the velocity of the particle in the direction perpendicular to the fringe patterns. This frequency can be obtained by the Fourier transformation of the intensity signal acquired by a camera focused at the region of interest. With a high frame rate camera, multiple particle velocities can be obtained and correlated by measuring the distortion at different subsequent fringes as the particle moves through the region of interest. Similar to PIV techniques, particles can be tracked through the flow field as they continually distort the flow pattern. In this way, various properties of the flow field can be obtained from the single experiments.

The only component of a laser Doppler system that may hinder the usage of this technique is the size of the particles within the fluid. Small air bubbles and dust (present

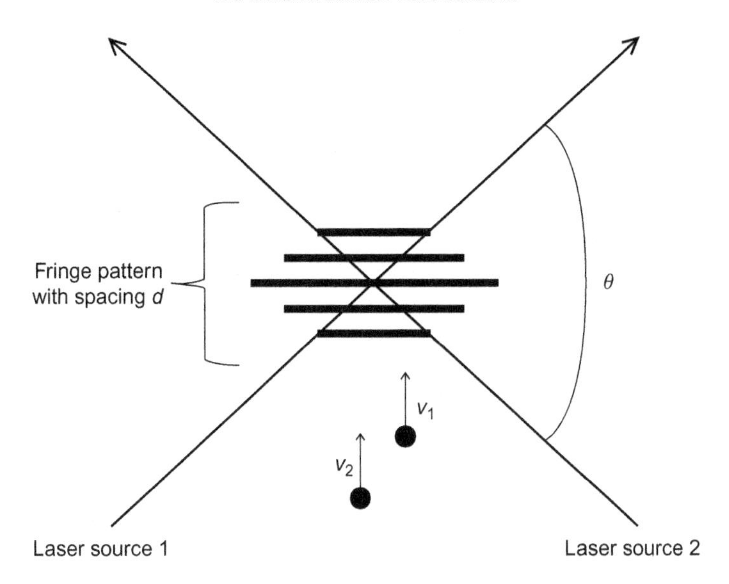

FIGURE 15.2 Schematic of a laser Doppler system, with fringe patterns that are distorted by particles within the fluid. A digital camera (not shown) would be focused at a region that would contain the fringe patterns so that as particles distort the fringes, information can be obtained about the particle and fluid velocity.

within the fluid) can be used to generate distortions within the fringe pattern, but they might not be present in a large enough quantity to obtain meaningful data. By overseeding particles within the fluid, it is difficult for the computer program to determine where the distortion is coming from because the particle distortions can be very large and may overlap. An ideal system would have 10–20 particles passing through the various portions of the fringe pattern at any instant in time (e.g., you do not want all of the particles at the center of the pattern, but distributed throughout the pattern). Also, the use of forward scattering can collect the most data from the system, but this may not always be attainable based on the geometry of the device of interest. Back- or side-scattered fringe patterns typically lose most of the information from the pattern because the intensity of these scatterings is not as large as forward scattering. However, with careful design one may be able to limit all of the shortcomings of the laser Doppler system and obtain meaningful data for the particular flow conditions.

An improvement over laser Doppler systems is a dynamic laser Doppler light-scattering velocimetry technique. Here, the total intensity of the scattered light is measured as well as the fringe pattern distortion. The intensity is typically averaged over time so that some of the time-dependent fluctuations induced by differences in particle size, location, or scattering properties can be removed. Without going through all of the mathematical equations that are required to smooth the intensity signals of a distorted fringe pattern, this technique can provide a direct measurement of fluid velocities based on time-average values of scattered intensities. This is a better collection method than laser Doppler velocimetry, but it is more difficult to implement because of the cross-correlation that is needed between particles.

15.3 FLOW CHAMBERS: PARALLEL PLATE/CONE-AND-PLATE VISCOMETRY

Many cells that we have discussed so far in this textbook are subjected to shear stress in their natural environment (endothelial cells, epithelial cells, and all blood cells, among others). Therefore, there is a need for an accurate experimental technique that can be used to expose cells to precise physiological shear stresses. There are two types of flow chambers that are typically used in labs, and they are the parallel plate–based flow chambers and the viscometer-based flow chambers.

Parallel plate–based systems use a pressure gradient to drive the fluid motion through the system. The most commonly used is termed the parallel plate flow chamber (Figure 15.3), which uses a driving force to push fluid through a rectangular channel. Cells can be cultured on the bottom surface of the flow chamber or can be seeded within the fluid itself. Typically, these devices are designed so that the velocity is fully developed (a parabolic profile) at the location where the cells will be subjected to shear. The wall shear stress can be approximated by the Poiseuille equation and is therefore only dependent on the inflow flow rate, the fluid viscosity, and the channel dimensions (thickness and potentially length). The advantage of this type of system is that cells can experience a particular shear stress (if adhered to the bottom plate) or can be subjected to varying shear stresses if they are in the fluid solution (similar to the *in vivo* conditions for blood cells). At the same time, the inflow flow rate can be temporally varying and thus, the cardiac waveforms, respiratory waveforms, or others can be simulated under these conditions.

A slight modification of the parallel plate system is termed the radial parallel plate flow chamber (Figure 15.3). In this case, fluid flows in from an opening within the center of the plate, and it exits out in all radial directions. The shear stress is dependent on the radial position and therefore is not constant for all cells that may be adhered to the bottom surface of the plate. The shear stress gradients need to be accounted for during data analysis techniques, and therefore, the data may not be as easy to report (e.g., one shear stress value) for cells adhered to the bottom surface in the parallel plate studies. It is typical in a

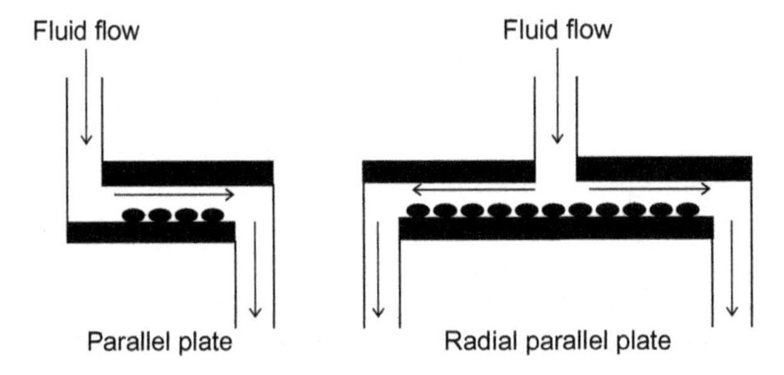

FIGURE 15.3 Schematic of a parallel plate and a radial parallel plate system. In both of these systems, cells (shown as black ovals) can be adhered to the bottom of the plate or can be seeded within the fluid. The shear stress can be modulated based on the fluid properties and the channel dimensions, and therefore, many different physiological conditions can be investigated.

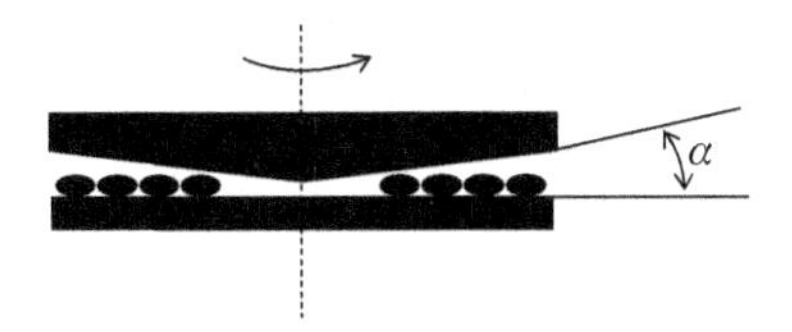

FIGURE 15.4 Schematic of a cone-and-plate viscometer. In these systems, cells within the fluid would be subjected to a uniform shear stress, which is only dependent on the fluid viscosity and the angular rotation of the cone. This is the case if the angle between the cone and the plate remains small, so that the velocity profile is linear.

radial flow chamber to use steady flow (instead of pulsatile), which simplifies the analysis slightly, but definitely makes the waveform easier to report.

Viscometer-based systems are most commonly used to expose cells in solution (e.g., blood cells) to a uniform shear stress, meaning that the shear stress throughout the fluid has the same value. Cells can be grown on the bottom surface, but there is no advantage of this type of system as compared to the parallel plate system for adherent cells. In these systems, the velocity profile is linear (not parabolic) because one of the constraining walls is moving, while the other is held fixed. In cone-and-plate viscometry, a cone (the top plate) is rotated to generate the flow profile (Figure 15.4). The cone has to have a small incident angle so that the velocity profile does not become disrupted (e.g., is no longer linear or turbulent) when the distance between the cone and the plate increases. The design of these systems is somewhat difficult since the angle between the flat plate and the cone (α in Figure 15.4) is on the order of $0.5^\circ-1^\circ$ and the spacing between the walls of the cone and the constraining cylinder need to be equal to the maximum spacing between the cone and the plate. In these systems, the fluid velocity is dependent on the angular velocity of the cone, the cone angle, and the distance between the plate and the cone. The shear stress in these systems is only dependent on the viscosity, the angular velocity, and the cone angle, which are all typically held constant in experiments. The cone's angular velocity can be modulated to mimic transient waveforms. This is an advantage over parallel plate experiments where it is normally more difficult to control the fluid flow rate. Again, there could be a slight variation to this design, where the top disc would be parallel to the bottom plate. This is termed the parallel disc viscometer. The velocity is then dependent on the radial location, and the shear stress is no longer constant throughout the flow field. This arrangement is less common than the previous three methods discussed.

One advantage of the plate designs over the viscometer designs is that it is typically easier to design a plate system that is optically transparent. This would give one the advantage of imaging the effects of the shear flow conditions on your cells of interest in real time. In fact, many groups have fabricated a system, where the bottom portion of the chamber is composed of a glass cover slip, so that (1) cells can be cultured on the "disposable" glass under normal cell culture conditions and (2) an objective of a microscope can be brought into close proximity to the coverslip. A viscometer-based system typically cannot include this because the cone is directly attached to a motor or belt system which would interfere with the optical path.

15.4 CURRENT STATE OF THE ART FOR BIOFLUID MECHANICS *IN VITRO* RESEARCH

As with the previous chapter, we focused on a very small subset of *in vitro* research techniques that are commonly used by biofluid mechanics research teams. Likewise, in this section we will highlight some of the current research topics that are being investigated using these techniques.

As we discussed above, the benefits of PIV are that scaled models can be fabricated and using dynamically similar flow conditions, the flow field about these models can be investigated and documented. In parallel to these studies, blood can be used in the models and the effects of the documented flow fields on blood functions (inflammation, thrombosis, hemolysis, etc.) can be investigated. One of the current major areas of research for velocimetry techniques is understanding the flow patterns around obstructions or disease conditions using patient specific geometries. The advantage of patient specific models is that the researcher can focus on the effects of flow in one patient and try to predict specific outcomes for that patient. Little to no assumptions about the geometric profile need to be made when using these models (non-patient specific models would need to use averaged or assumed values for the geometry and many geometries of interest to biofluid mechanics researchers are quite complex). A downside of patient specific models is that translating these predictive models to a clinical setting may be difficult (with the current state of the technology) because each patient would need to go through the imaging procedures and then the three-dimensional geometric model would need to be constructed and used for velocimetry experiments. The modeling, the fabrication of the model, and conducting the experiments may take a significant amount of time. In some instances, patients are in a critical condition and would need immediate attention, whereas under other conditions, a patient that is presenting with mild conditions may be able to wait for the experiment time.

Regardless of the limitations of these models, many researchers are using these techniques to evaluate existing biofluid implantable devices (especially cardiovascular devices) in assumed or patient specific geometries. For instance, recent work has evaluated the effects of carotid artery stenosis severity on plaque formation and shear stress levels. Other groups have evaluated the micro-rheological changes of the extracellular matrix when a tumor is present. In addition, flow through the normal and dysfunctional heart, un-stented and stented aneurysms, and the ability of cells to pass between different cellular aggregates have all been investigated recently with velocimetry techniques. The purpose of highlighting these varying fields is to illustrate that many of the biofluid mechanics topics that we have discussed can be evaluated with velocimetry techniques. These different biofluid mechanics topics that have been evaluated with PIV techniques illustrate the ability for PIV to evaluate different length and time scales, when one carefully considers the model parameters.

In recent years, laser Doppler velocimetry has become more popular than many of the other PIV techniques because this technique can be applied to living patients/animals. Although we will not discuss these techniques at this time, one can imagine the usefulness of monitoring flow conditions under their innate conditions. Similar to PIV techniques, recent uses for Doppler velocimetry techniques have included patient specific geometries and the implantation of biofluid mechanics devices. As with PIV techniques, the goal would be to understand the flow conditions near these high-risk areas and potentially predict the occurrence of future events. Similar shortcomings of fabricating a model would be applicable to Doppler techniques; however, since this technique can be used with a living patient, some of the intrinsic challenges may be overcome. The computations associated with the prediction ability of the measurements may still reduce the overall applicability of this technique. Many areas of research that are currently under investigation with PIV techniques are also being investigated with Doppler velocimetry techniques. Thus, we will not reiterate these areas but include some recent references that the reader can review at their leisure.

Both parallel plate flow chamber systems and cone-and-plate viscometry systems have allowed biofluid mechanics researchers to evaluate the behavior of cells to precisely controlled flow fields. Major advantages of this system are that the flow field can be precisely controlled and the behavior of cells can be monitored in real time (or near real time). However, disadvantages stem from the ability to machine precise parts (e.g., with slight variations in the cone geometry the flow field can be altered) and the limitations on the motor response times. Also, in some instances, the programmed waveforms may have assumptions built into them. For instance, the waveforms may come from a numerical model or some other experimentally obtained scenario, which will have its own intrinsic limitations.

Even though there are some intrinsic limitations to the use of flow chamber–based systems, the information that they provide on basic biology of cells under flow conditions is invaluable. Many groups use these systems to determine the effects of flow condition on many cellular responses or flow conditions coupled with other risk factors on these same cellular responses (for instance, our group has added tobacco smoke to the perfusate). Under these conditions, it is possible to evaluate how these systems alter the physiological response of cells. It will be quite difficult to list all of these research areas succinctly, since there are many wide varying avenues of research. A quick internet search (with particular keywords of interest) can help the reader find topics of interest. Some areas of research include the genetic profile of a particular cell under shear flow, the proteomic profile of a particular cell under shear flow, how multiple cells interact under shear flow, kinetic reaction rates under flow, adhesion/migration of particular cells under shear flow, or biocompatibility of new devices under flow conditions. In parallel, disease progression can be monitored, disease conditions can be simulated, or new therapeutics can be evaluated under shear flow. A major benefit of these techniques is that users can evaluate a system from many different standpoints while using the same exposure conditions. Cells can be sampled over-time (up to days of shearing) to observe how they respond and adapt to altered loading conditions.

15.5 FUTURE DIRECTIONS OF BIOFLUID MECHANICS *IN VITRO* RESEARCH

As with the previous section, we will discuss here some possible future directions in biofluid mechanics research using *in vitro* techniques. We are not trying to suggest that these are the only avenues of research or the only important avenues of research but we are attempting to generalize some salient areas of research that may be or should be addressed in the near future. In the author's opinions, one of the major current limitations of *in vitro* biofluid mechanics research is in the availability of models that mimic multiple realistic physiological conditions. For instance, we have focused our discussion in Section 15.4 mostly on current models that use patient specific geometries and/or patient specific flow waveforms. This has provided the biofluid mechanics community with a significant amount of insight into the pathophysiology of diseases and the overall effect of shear flow on cellular responses, however, other physical, biochemical, and/or mechanical factors can be included within the experiments to more accurately simulate the disease conditions.

For instance, most of the techniques that we have discussed in this chapter make use of "rigid" wall containers or containing vessels with wall mechanical properties that are drastically different than biological tissue. For groups that are investigating the progression of atherosclerotic plaques or aneurysms, it would be beneficial for them to simulate the blood vessel wall mechanical properties within their measurements. It is especially important to simulate wall mechanical properties for these two diseases, because the blood vessel wall gets more rigid during atherosclerotic processes and less rigid during the progression of an aneurysm. These mechanical changes are likely to elicit changes in the flow conditions around these pathologies and may even alter the cellular responses of vascular smooth muscle cells and endothelial cells that reside within/along the vessel wall. If these mechanical properties can be incorporated into PIV techniques, it would be possible to obtain more realistic flow patterns and waveforms around these pathologies. These more realistic waveforms could then be used in other experiments, such as viscometry techniques, so that a better understanding of the basic biology can be obtained.

As for parallel plate and viscometry techniques, a major hurdle to these experiments is not only the mechanical properties of the channel wall, but also the relevance of the fluid that is perfused throughout the device. In many instances, biofluid mechanics research groups make use of a standard cell culture media or a physiological salt solution as the perfusate. While this is beneficial from the standpoint of controlling experiments and/or consistency within experimental conditions, this type of perfusate does not match the physiology well. It may be more relevant to include blood cells within the perfusate (if simulating cardiovascular conditions), air particles (if simulating respiratory conditions), cytokines, risk factors, hormones, among many other biological relevant chemicals. This may complicate the experimental conditions and make the data analysis more challenging, but the information garnered from studies such as these may be more accurate when modeling biofluid mechanics conditions.

Parallel plate and other viscometry techniques would also benefit from mechanical/flow conditions that are more physiologically relevant, as discussed for velocimetry techniques. Studies on the effects of altered fluid mechanics on flow would be more relevant under these simulated conditions instead of under steady flow or rigid conditions. However, the challenge with altering these properties would be in the design of a system that can maintain the precisely controlled flow fields while allowing for some flexibility in the container. There are some systems that are being developed that allow this flexibility and the use of these systems in future biomedical work will significantly progress the field of biofluid mechanics research forward.

END OF CHAPTER SUMMARY

15.1 PIV is a computer-based technique that tracks a sequence of reflective particles through a flow chamber. A computer program is used to control a high-intensity light source (normally a laser) and correlate the data obtained regarding the movement of particles through the fluid. Based on the known particle displacement and the camera frame rate, it is possible to obtain the velocity of the particles seeded within the fluid. Through the use of multiple cameras, true three-dimensional velocity fields can be obtained. LSV and holographic PIV are more specialized PIV systems.

15.2 Laser Doppler velocimetry is used to quantify the microstructures of flows. Two monochromatic light sources are focused at a region of interest and this forms a fringe pattern that is based on the fluid velocity. By seeding particles within the fluid, the fringe pattern changes, based on

$$\omega = \frac{v}{d} = \frac{2v\sin\left(\frac{\theta}{2}\right)}{\lambda}$$

and this is what is collected by the digital camera. Most laser Doppler systems have the problem of transient fluctuations in the fluid properties, caused by imperfections in the experimental system, which can confound the results. By using a dynamic time-averaging system, these transient changes can be smoothed.

15.3 Flow chambers are used to subject cells to particular shear stresses. Through the particular design of the chamber, cells can either be subjected to uniform stresses along the walls of the chamber or to varying stresses. It is most common to use either a parallel plate system or a cone-and-plate viscometer system to model physiological conditions *in vitro*.

15.4 The current state of the art for biofluid mechanics research is primarily focused on obtaining realistic geometries for velocimetry techniques. These "patient specific" geometries would be helpful in obtaining a predictive model; however, the applicability of these solutions to the clinic needs to be addressed prior to their use. Viscometry techniques are primarily used to determine the basic biology that underlies pathological events and/or how shear flow alters cell behavior. One of the shortcomings of each of these techniques is that while groups are aiming to make increase the relevance of these models, technological limitations still exist that prevent physiologically relevant model fabrication, with considerations on multiple biochemical and biophysical properties.

15.5 In the author's opinion, future efforts in *in vitro* biofluid mechanics research should focus on fabricating models that are more relevant to the physiological conditions. For instance, incorporating materials that simulate the mechanical properties of the biological tissue would help to increase the accuracy of the models. Similarly, the addition of biologically relevant and active compounds to the perfusate that are similar to the particular flow conditions would help researchers to understand the basic biology of the particular process.

HOMEWORK PROBLEMS

15.1 Compose a small algorithm that can be used to quantify approximately 10 particle velocities, within subsequent digital images, for a PIV system.

15.2 In a laser Doppler system, two monochromatic laser light sources are placed 80° apart. The wavelength of light emitted from these two lasers is 540 nm. Calculate the normal fringe spacing for these laser light sources and the anticipated change in frequency if a particle intersects the laser at a velocity of 20 mm/sec. If the light sources are changed to a laser that emits at 650 nm, how does this affect the calculations?

15.3 Using the Navier-Stokes equations, model the flow through a parallel plate flow chamber and a cone-and-plate viscometer. Assume that the parallel plate is only under pressure-driven flow $\left(\frac{\partial p}{\partial x}\right)$, the channel height is h, and that the fluid viscosity is μ. The cone-and-plate has a radius of r, an angle of α (assumed to be small), and an angular velocity of ω. Gravity can be neglected for both systems.

15.4 Discuss a possible re-design of a flow chamber that can be used to model physiological flows within the microcirculation.

References

[1] L. Agati, S. Cimino, G. Tonti, F. Cicogna, V. Petronilli, L. De Luca, et al., Quantitative analysis of intraventricular blood flow dynamics by echocardiographic particle image velocimetry in patients with acute myocardial infarction at different stages of left ventricular dysfunction, Eur. Heart J. Cardiovasc. Imaging 15 (2014) 1203.

[2] K.M. Ainslie, J.S. Garanich, R.O. Dull, J.M. Tarbell, Vascular smooth muscle cell glycocalyx influences shear stress-mediated contractile response, J. Appl. Physiol. 98 (2005) 242.

[3] E. Akagawa, K. Ookawa, N. Ohshima, Endovascular stent configuration affects intraluminal flow dynamics and *in vitro* endothelialization, Biorheology 41 (2004) 665.

[4] K.D. Barclay, G.A. Klassen, R.W. Wong, A.Y. Wong, A method for measuring systolic and diastolic microcirculatory red cell flux within the canine myocardium, Ital. Heart J. 2 (2001) 740.

[5] V. Bassaneze, V.G. Barauna, C. Lavini-Ramos, J. Kalil, I.T. Schettert, A.A. Miyakawa, et al., Shear stress induces nitric oxide-mediated vascular endothelial growth factor production in human adipose tissue mesenchymal stem cells, Stem Cells Dev. 19 (2010) 371.

[6] G. Beaune, T.V. Stirbat, N. Khalifat, O. Cochet-Escartin, S. Garcia, V.V. Gurchenkov, et al., How cells flow in the spreading of cellular aggregates, Proc. Natl. Acad. Sci. USA 111 (2014) 8055.

[7] B.R. Blackman, K.A. Barbee, L.E. Thibault, *In vitro* cell shearing device to investigate the dynamic response of cells in a controlled hydrodynamic environment, Ann. Biomed. Eng. 28 (2000) 363.

[8] P. Browne, A. Ramuzat, R. Saxena, A.P. Yoganathan, Experimental investigation of the steady flow downstream of the St. Jude bileaflet heart valve: a comparison between laser Doppler velocimetry and particle image velocimetry techniques, Ann. Biomed. Eng. 28 (2000) 39.

[9] L.P. Chua, K.S. Ong, G. Song, W. Ji, Measurements by laser Doppler velocimetry in the casing/impeller clearance gap of a biocentrifugal ventricular assist device model, Artif. Organs 33 (2009) 360.

[10] S. Cito, A.J. Geers, M.P. Arroyo, V.R. Palero, J. Pallarés, A. Vernet, et al., Accuracy and reproducibility of patient-specific hemodynamic models of stented intracranial aneurysms: report on the virtual intracranial stenting challenge 2011, Ann. Biomed. Eng. 43 (2014) 154.

[11] T.L. De Backer, M. De Buyzere, P. Segers, S. Carlier, J. De Sutter, C. Van de Wiele, et al., The role of whole blood viscosity in premature coronary artery disease in women, Atherosclerosis 165 (2002) 367.

[12] J. Dörler, M. Frick, M. Hilber, H. Breitfuss, M.N. Abdel-Hadi, O. Pachinger, et al., Coronary stents cause high velocity fluctuation with a flow acceleration and flow reduction in jailed branches: an *in vitro* study using laser-Doppler anemometry, Biorheology 49 (2012) 329.

[13] M.D. Ford, H.N. Nikolov, J.S. Milner, S.P. Lownie, E.M. Demont, W. Kalata, et al., PIV-measured versus CFD-predicted flow dynamics in anatomically realistic cerebral aneurysm models, J. Biomech. Eng. 130 (2008) 021015.

[14] R. Gerrah, A. Brill, S. Tshori, A. Lubetsky, G. Merin, D. Varon, Using cone and plate(let) analyzer to predict bleeding in cardiac surgery, Asian Cardiovasc. Thorac. Ann. 14 (2006) 310.

[15] N. Gomes, M. Berard, J. Vassy, N. Peyri, C. Legrand, F. Fauvel-Lafeve, Shear stress modulates tumour cell adhesion to the endothelium, Biorheology 40 (2003) 41.

[16] D.I. Hollnagel, P.E. Summers, D. Poulikakos, S.S. Kollias, Comparative velocity investigations in cerebral arteries and aneurysms: 3D phase-contrast MR angiography, laser Doppler velocimetry and computational fluid dynamics, NMR Biomed. 22 (2009) 795.

[17] D.P. Jones, W. Hanna, H. El-Hamidi, J.P. Celli, Longitudinal measurement of extracellular matrix rigidity in 3D tumor models using particle-tracking microrheology, J. Vis. Exp. 88 (2014) e51302.

[18] S. Kefayati, J.S. Milner, D.W. Holdsworth, T.L. Poepping, *In vitro* shear stress measurements using particle image velocimetry in a family of carotid artery models: effect of stenosis severity, plaque eccentricity, and ulceration, PLoS One 9 (2014) e98209.

[19] S. Khosla, S. Murugappan, R. Lakhamraju, E. Gutmark, Using particle imaging velocimetry to measure anterior-posterior velocity gradients in the excised canine larynx model, Ann. Otol. Rhinol. Laryngol. 117 (2008) 134.

[20] V. Kini, C. Bachmann, A. Fontaine, S. Deutsch, J.M. Tarbell, Integrating particle image velocimetry and laser Doppler velocimetry measurements of the regurgitant flow field past mechanical heart valves, Artif. Organs 25 (2001) 136.

[21] W.O. Lane, A.E. Jantzen, T.A. Carlon, R.M. Jamiolkowski, J.E. Grenet, M.M. Ley, et al., Parallel-plate flow chamber and continuous flow circuit to evaluate endothelial progenitor cells under laminar flow shear stress, J. Vis. Exp. 59 (2012) e3349.

[22] H.L. Leo, H.A. Simon, L.P. Dasi, A.P. Yoganathan, Effect of hinge gap width on the microflow structures in 27-mm bileaflet mechanical heart valves, J. Heart Valve Dis. 15 (2006) 800.

[23] K.B. Manning, T.M. Przybysz, A.A. Fontaine, J.M. Tarbell, S. Deutsch, Near field flow characteristics of the Bjork-Shiley Monostrut valve in a modified single shot valve chamber, ASAIO J. 51 (2005) 133.

[24] R.S. Meyer, S. Deutsch, C.B. Bachmann, J.M. Tarbell, Laser Doppler velocimetry and flow visualization studies in the regurgitant leakage flow region of three mechanical mitral valves, Artif. Organs 25 (2001) 292.

[25] M.J. Mitchell, K.S. Lin, M.R. King, Fluid shear stress increases neutrophil activation via platelet-activating factor, Biophys. J. 106 (2014) 2243.

[26] T. Moriguchi, B.E. Sumpio, PECAM-1 phosphorylation and tissue factor expression in HUVECs exposed to uniform and disturbed pulsatile flow and chemical stimuli, J. Vasc. Surg. 61 (2015) 481.

[27] J. Partridge, H. Carlsen, K. Enesa, H. Chaudhury, M. Zakkar, L. Luong, et al., Laminar shear stress acts as a switch to regulate divergent functions of NF-kappaB in endothelial cells, FASEB J. 21 (2007) 3553.

[28] M. Qian, J. Liu, M.S. Yan, Z.H. Shen, J. Lu, X.W. Ni, et al., Investigation on utilizing laser speckle velocimetry to measure the velocities of nanoparticles in nanofluids, Opt. Express. 14 (2006) 7559.

[29] R. Saxena, J. Lemmon, J. Ellis, A. Yoganathan, An *in vitro* assessment by means of laser Doppler velocimetry of the medtronic advantage bileaflet mechanical heart valve hinge flow, J. Thorac. Cardiovasc. Surg. 126 (2003) 90.

[30] P.R. Schuster, J.W. Wagner, Holographic velocimetry: an alternative approach for flow characterization of prosthetic heart valves, Med. Prog. Technol. 14 (1988) 177.

[31] T.W. Secomb, T.M. Fischer, R. Skalak, The motion of close-packed red blood cells in shear flow, Biorheology 20 (1983) 283.

[32] P.P. Sengupta, B.K. Khandheria, J. Korinek, A. Jahangir, S. Yoshifuku, I. Milosevic, et al., Left ventricular isovolumic flow sequence during sinus and paced rhythms: new insights from use of high-resolution Doppler and ultrasonic digital particle imaging velocimetry, J. Am. Coll. Cardiol. 49 (2007) 899.

[33] P. Sundd, X. Zou, D.J. Goetz, D.F. Tees, Leukocyte adhesion in capillary-sized, P-selectin-coated micropipettes, Microcirculation. 15 (2008) 109.

[34] S. Tateshima, F. Vinuela, J.P. Villablanca, Y. Murayama, T. Morino, K. Nomura, et al., Three-dimensional blood flow analysis in a wide-necked internal carotid artery-ophthalmic artery aneurysm, J. Neurosurg. 99 (2003) 526.

[35] W. Trasischker, R.M. Werkmeister, S. Zotter, B. Baumann, T. Torzicky, M. Pircher, et al., *In vitro* and *in vivo* three-dimensional velocity vector measurement by three-beam spectral-domain Doppler optical coherence tomography, J. Biomed. Opt. 18 (2013) 116010–116011.

[36] B.R. Travis, M.E. Andersen, E.T. Frund, The effect of gap width on viscous stresses within the leakage across a bileaflet valve pivot, J. Heart Valve Dis. 17 (2008) 309.

[37] H.Q. Wang, L.X. Huang, M.J. Qu, Z.Q. Yan, B. Liu, B.R. Shen, et al., Shear stress protects against endothelial regulation of vascular smooth muscle cell migration in a coculture system, Endothelium 13 (2006) 171.

[38] R. Yamaguchi, H. Ujiie, S. Haida, N. Nakazawa, T. Hori, Velocity profile and wall shear stress of saccular aneurysms at the anterior communicating artery, Heart Vessels 23 (2008) 60.

[39] C.H. Yap, N. Saikrishnan, G. Tamilselvan, A.P. Yoganathan, Experimental technique of measuring dynamic fluid shear stress on the aortic surface of the aortic valve leaflet, J. Biomech. Eng. 133 (2011)061007.

[40] Y. Zeng, Y. Qiao, Y. Zhang, X. Liu, Y. Wang, J. Hu, Effects of fluid shear stress on apoptosis of cultured human umbilical vein endothelial cells induced by LPS, Cell Biol. Int. 29 (2005) 932.

CHAPTER

16

In Vivo Biofluid Mechanics

LEARNING OUTCOMES

1. Discuss the need and use of live animal preparations in biofluid mechanics
2. Describe the necessary components of an intravital microscopy system
3. Explain the use of Doppler ultrasound technology
4. Calculate the apparent Doppler frequency shift
5. Discuss magnetic resonance imaging techniques
6. Understand the physics behind magnetic resonance imaging
7. Discuss current work and future trends related to *in vivo* biofluid mechanics research

16.1 LIVE ANIMAL PREPARATIONS

Although there is a great deal of information that can be learned from *in vitro* experiments, *in vivo* studies may provide the closest, most accurate, and most relevant models for researchers to understand what occurs under undisturbed physiological conditions. We make this distinction between disturbed conditions and undisturbed conditions because we will focus this section on intravital microscopy, which is a technique where a vascular bed is dissected out of the animal, but it remains connected to the feeding vessels, nerve cells, musculature, and other biological structures. Therefore, the vascular system under investigation (usually from muscle) is still under physiological regulation (nervous, hormonal, among others), but there may be some other unanticipated factors that alter the flow through the vascular bed that stems from the dissection. For these experiments, animals are placed under anesthesia to allow for the researcher to dissect the tissue, and then comprehensive observations of changes within the flow conditions can be made throughout the dissected tissue. Traditional intravital microscopy uses the mesenteric capillary beds or various skeletal muscle capillary beds to observe changes in flow under various

 © 2015 Elsevier Inc. All rights reserved.

conditions of interest. These capillary beds are used because it is relatively easy to access and the tissue is relatively transparent, so that optical microscopes in conjunction with recording equipment can be used to document the flow through the vascular bed. However, many other capillary beds can be used for intravital microscopy techniques, as long as the preparations can be made from a thin section of tissue to allow for enough light to pass through the tissue.

When these preparations are being observed on the microscope stage, various conditions can be investigated to determine the effects of these conditions on flow properties, vascular tone, or blood cell motion. For instance, various agonists can be added to the tissue perfusate, which is used to keep the tissue moist during the experiments, to determine how the addition of a compound to the entire vascular network affects that vascular bed (e.g., flow through it, communication along it, among others). For clarification, this perfusate bathes the entire dissected tissue so that it does not dry out. In some regards, this provides a measure of a global effect on the compound, but not all compounds are released in sufficient concentrations to alter the entire network physiology. Additionally, micropipette techniques can be used to localize the injection of agonists to determine the endothelial cell communication throughout the vascular network and how that communication affects flow conditions among many other properties (e.g., the function of vascular smooth muscle cells in slightly larger vessels). Micropipette techniques may provide more accurate measurements of physiological changes, but induce some technical difficulties when trying to ensure that the same location is instigated in multiple trials (e.g., with different concentrations, or with different agonists). Changes in the vascular network, such as vessel diameter, flow rate, and cell volume, can all be investigated with intravital microscopy. Therefore, there are many possible outcomes for this type of research and many different research avenues, as long as the tissue is prepared consistently.

In recent years, the information that has been collected from intravital microscopy has improved significantly with better optics, better cameras, and better antibodies/fluorescent probes (targeting, longevity, and optical properties). Better optics/cameras have allowed the researchers to target smaller capillary bed segments more accurately, as well as small blood vessels. The use of better antibodies/fluorescent markers has allowed researchers to target specific compounds, specific cell receptors, and other indicators to understand the changes that occur within the vascular bed down to the molecular level. As an example, many cellular processes are regulated by the movement of calcium through a single endothelial cell and by the calcium and other ion movement through gap junctions connecting neighboring endothelial cells. By using fluorescent markers that can target intracellular calcium compartments and/or intracellular calcium signaling molecules, intravital microscopy can investigate physiological and pathological events instead of just physical parameters of the vascular bed. Also, signal transduction pathways, gene regulatory proteins, and many other processes that occur during the regulation/control of gene expression can be observed. How the gene expression relates to physical changes can also be observed. For instance, changes in blood cell flux, hematocrit, and blood pressure can all be related to intracellular ion concentrations, and the expression of cell membrane-bound receptors. Also, the adhesion/rolling of white blood cells onto the endothelial cells membrane and how this relates to disease processes and the expression of cytokines can be investigated with intravital microscopy. Therefore, many physiological and pathological

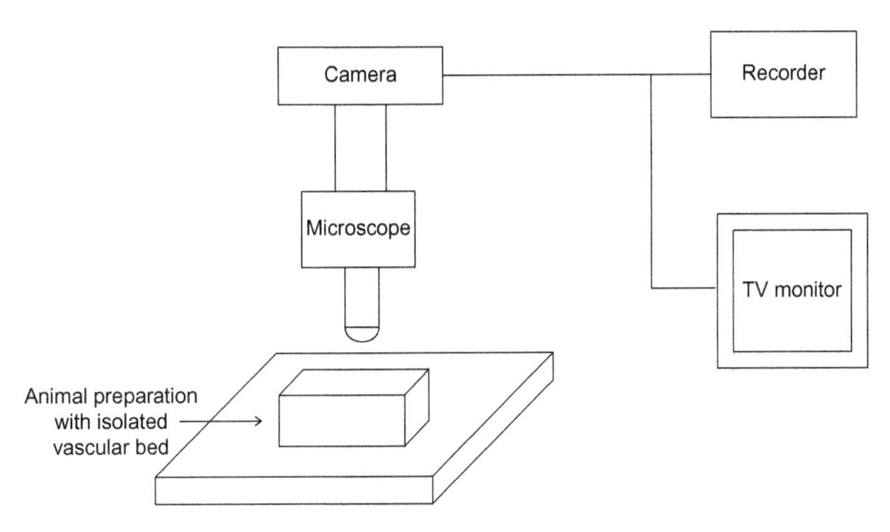

FIGURE 16.1 Schematic of a typical intravital microscopy system. The key parameters are the microscope, which is coupled to a live-feed imaging system, and a real-time data acquisition component. Intravital microscopy can be used to monitor disease progression, therapeutic efficacy, clearance, and distribution of blood cells within the microvasculature.

conditions such as ischemia/reperfusion, inflammatory responses, endothelial cell barrier integrity, graft success, and nutrient delivery can all be investigated with this technique.

A standard intravital microscopy apparatus (Figure 16.1) consists of an epi-fluorescence upright (or inverted) microscope, which is connected to a camera, and then feeds directly into a recording device and a live video monitor. A perfusion system is needed to keep the tissue moist during the experiment. An anesthesia controlling system is needed to make sure that the animal is kept under anesthetic during the entire experiment, which can last up to 6–8 h. When conducting intravital microscopy studies, it is critical to choose anesthetics that do not alter vascular properties. A micropipette system can be used to locally inject antagonists within the vascular bed or they can be injected via the perfusate. Data collected from this apparatus are typically fed into an image analysis system which can be used to accurately quantify the particular data of interest from the experiment. It is critical that these systems are integrated properly and the technician is well trained so that he/she can collect and analyze the data correctly. Remember that a poorly trained technician can alter the results during the dissection or the data collection.

Intravital microscopy is a very powerful research tool to investigate changes to vascular networks under multiple conditions. The importance of this method is that the vascular network that is under investigation remains connected to the animal and is under physiological regulation from the animal. Many groups use this technique to understand the development of diseases and fundamental vascular physiology because as described above, various conditions can be applied to the entire tissue or locally to individual cells. Other benefits of this approach are that relatively long *in vivo* effects can be monitored because the animal preparation can be maintained for at least 6 h. Therefore, clearance of new therapeutics can be investigated in tandem with the local effects of the therapeutic agent. Therefore, this technique has applications in not only basic biology of the microcirculation but also the design of drugs, implantable devices, and anything that may come into contact with or interact with the vascular system.

16.2 DOPPLER ULTRASOUND

Before we discuss the Doppler mode associated with ultrasound image acquisition, we will first discuss the general procedures involved with three-dimensional ultrasound technology. The first device necessary to acquire ultrasound images is the ultrasound scanner, which is used to emit and collect sound waves aimed at specific tissue locations. Standard probes are equipped with a transducer that emits a frequency in the range of 1–10 MHz. Ultrasound technology also requires the use of an electromagnetic position and an orientation measurement device, which normally consists of a component that is located within the examination table and a second component located within the ultrasound probe. This device records the spatial location of the probe in relation to the fixed examination table position. The angle of the probe in relation to the fixed axis is also recorded by this device, because an ultrasound probe has 6 degrees of freedom and an accurate distance and angle of incidence is needed to analyze the data properly. By sweeping the ultrasound probe over a tissue bed, multiple images can be obtained to observe the tissue from different locations. The importance of these locators cannot be understated; they are analogous to multiple satellites being put into use to triangulate the location of a device (e.g., a GPS).

There are multiple imaging modes that can be used in conjunction with the ultrasound probe. The one of most interest to this textbook is the Doppler mode, which can obtain a significant amount of information about the flow of blood through vascular networks that are relatively close to the skin. Doppler ultrasound technology takes advantage of the apparent change in the ultrasound frequency caused by the velocity of blood and the incident angle of the sound wave in relationship to the blood vessel (Figure 16.2). The apparent frequency shift can be calculated from

$$f_D = \frac{2vf_0}{c}\cos(\theta) \tag{16.1}$$

where f_D is the recorded Doppler frequency, v is the blood velocity, c is the speed of sound, θ is the angle between the ultrasound probe and the blood vessel, and f_0 is the emitted frequency from the probe. Because the blood velocity is variable based on the radial position within the blood vessel and the pulsatile nature of blood, the recorded frequency is variable. Therefore, a spatially and temporally varying velocity distribution can be obtained if the receiving equipment is sensitive enough to differentiate between these small changes in frequency within a small blood vessel. An additional software component to the general Doppler ultrasound technique is the use of the color flow imaging module. This add-on color codes the blood velocity, and therefore, one can obtain a real-time depiction of the relative blood flow speed across the radial direction of a blood vessel as well as the direction of the blood flow. Color Doppler scans are normally obtained in conjunction with the spectral Doppler mode (also known as the Duplex Doppler mode). Using the spectral Doppler mode, images can obtain a time-dependent mean velocity

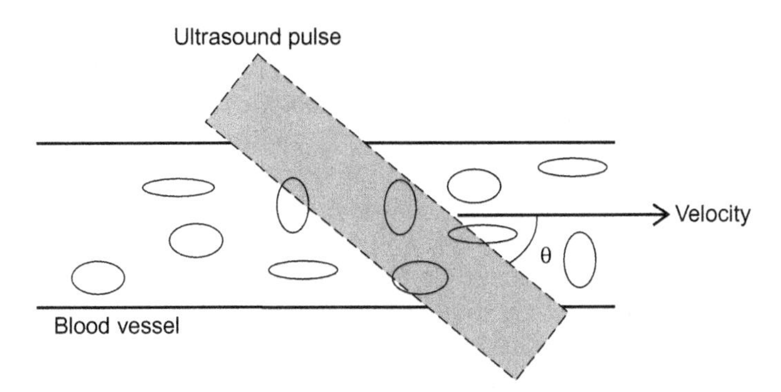

FIGURE 16.2 Doppler ultrasound wave intersecting with a blood vessel. Based on the apparent shift of the ultrasound frequency, one can obtain information about the blood velocity. This shift occurs due to the reflection of sound waves after coming into contact with blood particles.

waveform over a small section of the blood vessel. Therefore, one can determine if there are temporal shifts within the blood flow of that vessel. The usefulness of this technique is that the blood velocity can be compared between vessels of similar sizes. This would help to determine if regions are not obtaining enough blood, potentially due to a partially blocked vessel upstream of the imaging location or other existing pathologies. Additionally, these techniques can be applied to both humans and animals and thus, the effects of therapeutic agents on the flow through vascular networks among other changes can be observed. Lastly, Doppler ultrasound is useful to quantify the frequency of emboli or other large particulate matter by recording high-magnitude transient changes to the velocity profile. These high-magnitude changes only occur when there is a large shift in the velocity that is generated by a "rigid" material flowing through the velocity field.

Doppler ultrasound has proven to be a good research tool to characterize the potential for new cardiovascular devices to induce thromboemboli within the cardiovascular system. These new devices can be implanted into an animal, and after postoperative recovery the generation of high-intensity transient events can be compared to prior implantation in order to determine if this device induces new emboli formation. Because the ultrasound probe is noninvasive, it is also possible to investigate the changes in emboli formation months to years after implantation. A good example of this application is a Transcranial Doppler Ultrasound system, which is often used to record high-intensity transient Doppler signals (HITS) in human patients or large animals (such as sheep). HITS are usually induced by microemboli, which can generate a high-pitched sound, different from the sounds that the blood flow normally makes and different from the normal particle matter within the blood. The duration of a HITS is usually less than 50 ms, and the amplitude of a HITS is usually much higher than the background blood flow signal. Based on the velocity and duration of a HITS, the nature of the microemboli, for example, particle emboli (more dangerous to patients) or gaseous emboli, can be distinguished. Again, this is based on the relative size of the particle relative to the size of the blood vessel.

Another major use for ultrasound technology is to obtain vascular architectures. The advantage of ultrasound technology is that it has a low cost, is not harmful, and has the ability to carry out real-time imaging and the ability to obtain flow information at the same time as geometry information. This capability makes it possible to form a feedback loop for calculating the flow through real geometries and validating the flow calculations with the ultrasound recordings. The disadvantage of this technique is that most ultrasound probes have a limited scanning depth (at larger depths, the resolution decreases) which restricts the usage of this technique to superficial blood vessels, such as the carotid arteries, which are critical blood vessels related to brain perfusion, or capillaries in the skin, which are typically not considered to be critical blood vessels. However, if an ultrasound probe can be inserted into the body, deep blood vessels or tissue can be assessed. A good example is the transesophageal echocardiography. With this technique, an ultrasound probe is inserted into an animal (or human patient) through the mouth and passed into the esophagus. This places the probe in close proximity to the aortic root and importantly, the aortic valve. Short-axis and long-axis views of the aortic valve and the left ventricle can be obtained and therefore the diameter of the aortic root can be determined. Meanwhile, the pressure gradient across the aortic valve, blood flow velocity, and other fluid properties can be calculated. By adjusting the location and angle of the probe head, information about the left ventricle and right atrium can be obtained as well. Actually, transesophageal echocardiography is often used in patients before a heart valve replacement surgery, to determine the size of the original heart valve and local blood flow conditions. The risks of using a transesophageal probe are relatively low, however, the usage of this type of imaging modality for different locations and pathologies is somewhat limited.

Commonly, ultrasound measurements are carried out under the Brightness mode (B-mode), which renders a two-dimensional image of the blood vessel, where the intensity could be correlated to the blood velocity. Based on the Brightness mode imaging, three-dimensional ultrasound images can also be achieved, as the ultrasound probe is normally small and handheld. Three-dimensional images can be obtained by moving the probe in relation to the imaging location. This set of two-dimensional Brightness mode images can then be reconstructed into a three-dimensional image. A few groups have investigated the use of this technique to obtain information about blood vessel architecture and the flow through these blood vessels, and they all found that it was necessary to either have a precise computer-controlled mechanism to maneuver the probe or an accurate monitoring system for the probe location in relation to the imaging area. This has proved to be relatively difficult to carry out, but the information gained from obtaining ultrasound images from different orientations has proved to be beneficial. Also, the amount of noise associated with the image is reduced by overlaying and reconstructing images from multiple directions. Therefore, one clearer image is obtained from the same location. However, most three-dimensional work has not been conducted on imaging the vascular system. Instead, there has been a large population that has investigated the use of three-dimensional ultrasound techniques in imaging the heart motion, including heart valve motion.

16.3 PHASE CONTRAST MAGNETIC RESONANCE IMAGING

We will briefly discuss standard magnetic resonance imaging (MRI) to highlight the general technique employed for this image acquisition mode, before we move to phase contrast MRI. MRI has been successful in imaging the entire human body because the body has a very high water content (approximately 60% of the total body weight is from water). Hydrogen atoms have a single proton in their nucleus that spins continually. As this proton spins, a small magnetic field is created about the atom. The basis of MRI is to place the water ions in a large static magnetic field, which aligns all of the water molecules within the body. When the magnetic field is removed, how rapidly the water molecules return to their "normal" spin state is quantified and can provide a great deal of information about the tissue composition. Two magnetic fields are generated as the water molecules move back into their "normal" positions, and those two magnetic fields provide information on the thermal tissue properties (longitudinal field) and the tissue inhomogeneities (transverse field).

There are a number of ways that a magnetic field can be generated to affect the atomic structure of the water. Two easy ways to form a magnetic field are by either driving a current through a wire or by applying a voltage field across a magnet. No matter how the magnetic field is generated, there are a few important measurable effects on the atomic structure that concern MRI. The first is the spin state of the electrons. Electrons generate their own small magnetic field that can be measured and this relates to the energy required to change the electrons spin state. This transformation energy is

$$\Delta E = hv \tag{16.2}$$

where ΔE is the change in energy, h is Planck's constant, and ν is the orbital frequency of the electron, which is equal to

$$\nu = \frac{\gamma B_0}{\pi} \tag{16.3}$$

where γ is the magnetic ratio and B_0 is the applied magnetic field. The magnetic ratio for the hydrogen isotope most often used in MRI measurements is 42.58 MHz/T, which is reasonable to measure within a MRI electric field less than 4 T.

The second atomic parameter that is commonly measured during MRI is the precession of the atoms. In a stationary two-dimensional magnetic field (e.g., $x-y$ plane), it is common for an electron to spin along the z-axis. This is very difficult to measure, and therefore, a magnetic force is applied that pushes the electron to spin off the z-axis and into the $x-y$ plane (it is possible to have three rotational directions, but for illustration we use the two-dimensional example). This downward spiraling orbit is termed the precession of the electron. To induce electron precession, a temporal magnetic field is applied perpendicular to the z-axis. This causes the net electron magnetic field to precess at angle ϕ from the z-axis. By measuring the change in the direction of the magnetic field and the time for the magnetic field to return back to an aligned state, information about the tissue composition can be obtained.

There are a few major components to an MRI system, which include a super-conducting magnet, surface coils that can induce a transient magnetic field (to investigate precession), and data collection/analysis components. The super-conducting magnet is used to generate a static magnetic field within the interior of the MRI system. This static magnetic field is used to align all of the atoms within a certain axis. This static field is normally around 3 T, although both stronger and weaker magnetic fields do exist. With a higher magnetic field strength the resolution of the image increases. The surface coils, use transient magnetic fields, typically within the range of 5% of the static magnetic field to localize the spin, precession, and time constants for the atoms as they are pulled away from their alignment and return back to the aligned configuration. The data collection software can detect the magnetic fields generated by the electrons themselves. The data analysis software takes the collected frequency data and determines the average tissue homogeneity/composition. Image processing techniques can then be used to generate a two-dimensional image of the tissue composition. It is typical that the magnetic field alignment is changed during the image processing time (e.g., from the *z*-axis to the *y*-axis) to allow for the data acquisition at different slices/orientations throughout the body. Images can be acquired on the order of every 50 ms with the current technology.

MRI technology has been applied to many different functional applications to detect changes within the tissue composition. Changes in brain activity can be quantified by changes in cerebral metabolism, blood flow, or oxygenation, in response to various stimuli. Changes in these parameters are all linked, because increased metabolism generally decreases tissue oxygenation, while increased blood flow increases tissue oxygenation levels. This is typically quantified as differences in the alignment and time constants of hemoglobin. Phase contrast MRI has been used to obtain pictures of arteries and entire vascular networks. This has been used to diagnose the likelihood of cardiovascular disease onset (such as stenosis and/or aneurysm). There are two common techniques that have been used to image the vasculature with phase contrast MRI. The first is through the administration of a paramagnetic contrast agent, most commonly gadolinium, into the bloodstream. That contrast agent (similar to standard angiography) is responsive to changes in the magnetic field and emits a strong signal that can be recorded by the MRI data acquisition equipment. A second technique, termed FLASH MRI, saturates the tissue that neighbors a vascular region of interest. As new blood enters the vascular bed, the blood itself will not be saturated; therefore, the MRI signal from the blood is much higher than the MRI signal from the surrounding tissue. This provides a direct measurement of velocity changes, as well as the vascular architecture.

16.4 REVIEW OF OTHER TECHNIQUES

Angiography or arteriography is a technique to visualize blood vessels *in vivo*. Usually, a catheter will be inserted through the femoral artery of a patient, with a contrast dye being injected into the vasculature. X-ray images will then be taken and the structures of the blood vessels can be studied. Digital subtraction angiography (DSA) is most commonly used because it can subtract the surrounding tissues and bones, leaving only the vessels filled with the dye visible. This enables cardiologists to identify stenosed or blocked vessels. Coronary angiography is performed to examine blood in the coronary arteries; cerebral angiography is used to visualize arteries and veins in the brain; peripheral angiography is commonly performed to identify stenosis in peripheral circulation, such as in the legs and feet.

Besides the techniques discussed above, there are many much simpler methods that can be used to conduct *in vivo* studies and collect information on blood or other biofluids. For example, a simple pressure cuff can be used to measure blood pressure and air-puff tonometry can be used measure intraocular pressure. Even though compared to Doppler ultrasound or MRI, information acquired from these methods is very limited, these simple methods may suit the need of one's study and provide the information desired. Also, associated experimental procedures are simple for these techniques and the equipment/devices involved are relatively inexpensive. But if more detailed information is required, one would certainly turn to the more complex techniques mentioned above.

16.5 CURRENT STATE OF THE ART FOR BIOFLUID MECHANICS *IN VIVO* RESEARCH

As with the previous chapters in this section, we will turn our focus onto some of the current areas of biofluid mechanics *in vivo* research that use some of the techniques that we discussed within this chapter. We highlighted very few techniques in this section, but we will illustrate here, how those techniques can be applied to salient research topics in biofluid mechanics.

Intravital microscopy is a very powerful technique for understanding the flow properties, cellular communication mechanisms, cell distributions among many other parameters as a function of cardiovascular antagonists, new therapeutics, new materials, etc. As such, it is quite difficult to summarize its current uses concisely, but we will try to highlight some major focus areas that make use of this technique. Groups that are interested in the microcirculation and changes to microcirculatory parameters commonly use intravital microscopy techniques, because this technique focuses on visualizing these small vessels. A significant portion of microcirculation research is focused on understanding the basic biology of the microcirculation in response to physiologically relevant stressors. For instance, the effects of various cardiovascular risk factors on disease progression have become a very important area of research. As with the previous techniques described in Chapters 14 and 15, it is important to understand the effects of these mediators on disease progression, so that new therapeutics and intervention techniques can be designed. Investigators that make use of intravital techniques can study the effects of these mediators on the microcirculation when it is under physiological control.

Other groups make use of ischemia-reperfusion models to determine the effects of localized hypoxic conditions followed by altered biomechanical factors (rapid increase in pressure and shear stress) on various microcirculatory parameters. One of the goals of recent work in this area is understanding how inflammatory reactions precede under these conditions and how these inflammatory reactions can interact with other cardiovascular mechanisms (e.g., thrombotic reactions) to alter the physiology of the microvascular bed. In parallel with this work, new drugs to minimize the inflammatory/thrombotic reactions induced by ischemia can be evaluated.

Some groups have used intravital microscopy to evaluate new blood vessel growth in response to various biochemical and biomechanical stimuli relevant to the microcirculation. Again, the importance of this work stems from the ability to observe new blood vessel growth under physiological control. This allows the research group to augment and control the innate response and try to determine what conditions are important during new vessel growth. Importantly, the ability for new materials to sustain new vessel growth can also be evaluated under these conditions. There are many other areas of research that make use of intravital microscopy; a quick search on pubmed.gov with a keyword of "intravital microscopy" finds approximately 5000 papers.

The other major techniques that we discussed in this section were Doppler ultrasound and MRI techniques. Again, there are many research avenues that can make use of these techniques and we will focus our discussion on a very brief subset of these techniques. The majority of work that makes use of ultrasound or MRI techniques is interested in understanding the flow patterns around pathologies and/or the flow patterns that instigate pathological events. In many instances, the subject will be treated with a risk factor, recreational drug or therapeutic drug and the effects of these treatments on flow properties will be evaluated. A great deal of information regarding the flow conditions can be identified from these investigations. Additionally, much of the information can be used within *in vitro* or *in silico* experiments as given parameters related to the flow conditions and/or the geometries. It is also common for these techniques to be used in case studies on particular clinical pathologies that present at hospitals.

16.6 FUTURE DIRECTIONS OF BIOFLUID MECHANICS *IN VIVO* RESEARCH

Similar to previous chapters, the authors would like to discuss some possible directions for future research interests that make use of some of the *in vivo* research techniques that we have described. Again, we are not suggesting that these are the only avenues of research or the only important avenues of research, instead we are attempting to highlight some avenues of research that we believe are important to develop. In the author's opinions, one of the major current limitations of *in vivo* biofluid mechanics research is the availability of equipment and model systems that can more fully characterize systems of interest. For instance, while intravital microscopy is a very powerful technique for the investigation of microcirculatory parameters, many cardiovascular disease processes are localized to the macrocirculation. Therefore, it would be important to understand and correlate occurrences within the macrocirculation and the microcirculation to fully understand disease progression.

To accomplish this type of integrated system, it would be important to couple imaging techniques that can be used in parallel. For instance, groups are working on combining positron emission tomography (PET) scanners with both traditional and more modern imaging techniques. PET scanners have a strong history of evaluating the functional activity of the heart and if PET imaging can be coupled with other techniques that can quantify blood flow dynamics or transient signaling molecules within the small blood vessels, a larger picture of hemodynamic changes under pathological conditions could be obtained. At the moment, groups that make use of intravital microscopy try to evaluate the macrocirculation through *ex vivo* investigations of disease progression. For instance, it is common for groups to dissect key organs or blood vessels from the macrocirculation after the intravital microscopy procedures have been completed and evaluate lesion growth, leukocyte infiltration, among other disease progression markers. However, a real-time correlation of these two events would be more beneficial.

Perhaps the most likely candidate for coupling two or more systems together is a system that includes a Doppler ultrasound imaging technique. In general, the instrumentation associated with ultrasound techniques is minimal and would not interfere with other imaging modalities, such as intravital microscopy. We envision that an intravital microscopy preparation can be made, a tissue can be antagonized with a specific factor of interest, and one can observe both the localized and the systemic change (if there are any). Clearly, collecting, segmenting, and processing all of the data that would be generated from such a system would be a challenge, since the viability of the data is largely assessed in real time (e.g., the imaging locations, the focus of the imaging locations), while the majority of the data quantification is completed afterward. With careful consideration on the locations of the probes and the associated equipment, the author's believe that information gained from this type of system would be invaluable in understanding the real-time effects of risk factors (or other antagonist) on both the microcirculation and the macrocirculation.

One of the other advantages of systems designed with multiple imaging modalities at multiple length scales is that information gained from this work can be used to develop more comprehensive models. Both *in silico* and *in vitro* models would benefit significantly from a more comprehensive understanding of possible interactions between the microcirculation and the macrocirculation. All of these improvements would be able to positively feedback on one another, since we may be able to gain an improved understanding of the basic biology of these reactions while correlating them to modeling techniques and other predictive techniques that can be used to diagnosis and characterize the systems of interest.

Again, there are many techniques, avenues of research (both current and future) and new technologies that can be developed to facilitate *in vivo* biofluid mechanics research. Our goal was to highlight one of these topics that we believe should be a focus in coming years.

In the second edition of this textbook, the authors choose to include sections regarding current research areas and possible avenues of future research in biofluid mechanics. It is our hope, that through this very brief overview of techniques, the reader understands the complexity of the biofluid system that they are interested in and that one particular experimental technique and/or focus will not be sufficient to fully characterize the system of interest. The benefit of multiple techniques that are investigating different properties of the same physiological event is that each of these techniques can focus on different aspects gaining new information that can help to refine and focus the other techniques. This feedback is one of the major positive aspects of conducting studies at these different levels.

END OF CHAPTER SUMMARY

16.1 A significant amount of information can be gained from *in silico* and *in vitro* work; however, *in vivo* models may provide the closest, most accurate, and most relevant conditions to study biofluid mechanics. In some of these studies, there is some manipulation of the *in vivo* tissue, and in others there is no manipulation. This may or may not affect the obtained results. Using intravital microscopy, blood flow through microvascular networks can be observed, in order to determine changes in blood cell populations, efficacy of new therapeutic agents, and cell–cell communication within the vascular network, among many other parameters. Research fields, such as wound healing, angiogenesis, or cardiovascular disease development/progression can all be documented with intravital techniques.

16.2 Doppler ultrasound techniques can be used to obtain three-dimensional images of biological tissue. For our interests, it can also be used to quantify blood flow through superficial blood vessels. This makes use of an apparent Doppler frequency shift which can be calculated from

$$f_D = \frac{2vf_0}{c}\cos(\theta)$$

By rotating the Doppler probe in space, in relation to the location of interest, it is possible to obtain a three-dimensional image of the blood vessel with a known velocity profile throughout the vessel.

16.3 MRI makes use of the spin and alignment of ions within the body. Under normal conditions, the magnetic field generated by ions within the body is randomly distributed, but when the body is placed within a large magnetic field, all of the ions (especially hydrogen) align with the magnetic direction. When the magnetic field is removed, then the time that it takes for the ions to move back into a random arrangement can be quantified. Individual tissues with different compositions will have multiple times for the hydrogen ions to return back to the respective normal random directions and therefore, pathological conditions (such as cancer within a tissue) can be identified from this technique.

16.4 Angiography or arteriography is a technique to visualize blood vessels *in vivo*. DSA is most commonly used because it can subtract the surrounding tissues and bones, leaving only the vessels filled with the dye visible. There are many much simpler methods that can be used to conduct *in vivo* studies and collect information on blood or other biofluids.

16.5 There are many current areas of *in vivo* biofluid mechanics research that cover a very broad range of topics. We choose to highlight some of those topics in our discussion and this was primarily related to the use of intravital microscopy for understanding the basic biology of various physiological and pathological events, drug diagnostics, and/or materials compatibility testing. We also highlighted how ultrasound and MRI techniques can be used to identify pathological locations and/or predict the occurrence of pathological events. These techniques may also be used to gather information that can be used in model development or initial model conditions.

16.6 The future of *in vivo* biofluid mechanics research is ever evolving. The author's believe that the future research efforts should be focused on combining length and time scales, so that real-time information can be gathered about disease progression in multiple locations. For instance, in the cardiovascular system, it would be interesting to uncover both local and systemic effects induced by agonist exposure. To accomplish this, new techniques need to be developed for simultaneous evaluation of these different length scales. As we have discussed throughout Part 5 of this textbook, we believe that the information gained from any experimental technique needs to be put into perspective with other techniques. For instance, information gained from *in vivo* techniques can be used to develop better *in vitro* models. These better *in vitro* models may provide us with a better understanding of the basic biology of the underlying processes, which may help the community to refine areas of interest for investigation and/or prediction tools/algorithms for disease diagnostic purposes. Combining all of the sections of Part 5, we would like the reader to understand that not one technique will be able to provide us with all of the information needed to understand these complex processes, but the combination of these techniques may help us to better understand the underlying physiology of the events that are being observed.

HOMEWORK PROBLEMS

16.1 Discuss some of the difficulties that may be encountered when conducting live animal preparations. Are there any ways to minimize these difficulties? Are these new methods associated with their own difficulties?

16.2 An ultrasound probe is being used to image a blood vessel. The probe is emitting a sound wave at a frequency of 2 MHz. Calculate the Doppler frequency if the angle between the blood vessel and the incident wave is 65° and blood is flowing at 50 mm/s.

16.3 The magnetic ratio being used in a MRI system is 42.5 MHz/T. Determine the orbital frequency for an electron within a 2 T magnetic field. How can we enhance the imaging with this machine?

References

[1] R. Borgquist, P.M. Nilsson, P. Gudmundsson, R. Winter, M. Leosdottir, R. Willenheimer, Coronary flow velocity reserve reduction is comparable in patients with erectile dysfunction and in patients with impaired fasting glucose or well-regulated diabetes mellitus, Eur. J. Cardiovasc. Prev. Rehabil. 14 (2007) 258.

[2] H.E. Cline, W.E. Lorensen, W.J. Schroeder, 3D phase contrast MRI of cerebral blood flow and surface anatomy, J. Comput. Assist. Tomogr. 17 (1993) 173.

[3] O. Goertz, L. von der Lohe, H. Lauer, T. Khosrawipour, A. Ring, A. Daigeler, et al., Repetitive extracorporeal shock wave applications are superior in inducing angiogenesis after full thickness burn compared to single application, Burns 14 (2014)000033-3.

[4] N. Hecht, U.C. Schneider, M. Czabanka, M. Vinci, A.K. Hatzopoulos, P. Vajkoczy, et al., Endothelial progenitor cells augment collateralization and hemodynamic rescue in a model of chronic cerebral ischemia, J. Cereb. Blood Flow Metab. 34 (2014) 1297.

[5] M. Hiratsuka, T. Katayama, K. Uematsu, M. Kiyomura, M. Ito, *In vivo* visualization of nitric oxide and interactions among platelets, leukocytes, and endothelium following hemorrhagic shock and reperfusion, Inflamm. Res. 58 (2009) 463.

[6] C.J. Jones, M.J. Lever, Y. Ogasawara, K.H. Parker, K. Tsujioka, O. Hiramatsu, et al., Blood velocity distributions within intact canine arterial bifurcations, Am. J. Physiol. 262 (1992) H1592.

[7] J.P. Ku, C.J. Elkins, C.A. Taylor, Comparison of CFD and MRI flow and velocities in an *in vitro* large artery bypass graft model, Ann. Biomed. Eng. 33 (2005) 257.

[8] H.A. Lehr, M. Krober, C. Hubner, P. Vajkoczy, M.D. Menger, D. Nolte, et al., Stimulation of leukocyte/endothelium interaction by oxidized low-density lipoprotein in hairless mice. Involvement of CD11b/CD18 adhesion receptor complex, Lab Invest. 68 (1993) 388.

[9] S. Lendemans, A. Peszko, R. Oberbeck, D. Schmitz, B. Husain, M. Burkhard, et al., Microcirculatory alterations of hepatic and mesenteric microcirculation in endotoxin tolerance, Shock 29 (2008) 223.

[10] Q. Long, X.Y. Xu, U. Kohler, M.B. Robertson, I. Marshall, P. Hoskins, Quantitative comparison of CFD predicted and MRI measured velocity fields in a carotid bifurcation phantom, Biorheology 39 (2002) 467.
[11] K. Mende, J. Reifart, D. Rosentreter, D. Manukyan, D. Mayr, F. Krombach, et al., Targeting platelet migration in the postischemic liver by blocking protease-activated receptor 4, Transplantation 97 (2014) 154.
[12] F. Nicolini, D. Corradi, A. Agostinelli, B. Borrello, T. Gherli, Aortic valve regurgitation secondary to ectopia and atresia of the left main coronary artery, J. Heart Valve Dis. 23 (2014) 158.
[13] A.R. Pries, T.W. Secomb, P. Gaehtgens, Structural adaptation and stability of microvascular networks: theory and simulations, Am. J. Physiol. 275 (1998) H349.
[14] A.R. Pries, T.W. Secomb, Microvascular blood viscosity *in vivo* and the endothelial surface layer, Am. J. Physiol. Heart Circ. Physiol. 289 (2005) H2657.
[15] S.Y. Sinkler, S. Segal, Aging alters reactivity of microvascular resistance networks in mouse gluteus maximus muscle, Am. J. Physiol. Heart Circ. Physiol. 307 (2014) H830.
[16] M.B. Srichai, R.P. Lim, S. Wong, V.S. Lee, Cardiovascular applications of phase-contrast MRI, AJR Am. J. Roentgenol. 192 (2009) 662.
[17] S. Stoquart-Elsankari, P. Lehmann, A. Villette, M. Czosnyka, M.E. Meyer, H. Deramond, et al., A phase-contrast MRI study of physiologic cerebral venous flow, J. Cereb. Blood Flow Metab. 29 (2009) 1208.
[18] J. Sugawara, H. Komine, K. Hayashi, M. Yoshizawa, T. Otsuki, N. Shimojo, et al., Systemic alpha-adrenergic and nitric oxide inhibition on basal limb blood flow: effects of endurance training in middle-aged and older adults, Am. J. Physiol. Heart Circ. Physiol. 293 (2007) H1466.
[19] R. Sumagin, C.W. Brown III, I.H. Sarelius, M.R. King, Microvascular endothelial cells exhibit optimal aspect ratio for minimizing flow resistance, Ann. Biomed. Eng. 36 (2008) 580.
[20] M. Takata, M. Nakatsuka, T. Kudo, Differential blood flow in uterine, ophthalmic, and brachial arteries of preeclamptic women, Obstet. Gynecol. 100 (2002) 931.
[21] Y.H. Tang, S. Vital, J. Russell, H. Seifert, E. Senchenkova, D.N. Granger, Transient ischemia elicits a sustained enhancement of thrombus development in the cerebral microvasculature: effects of anti-thrombotic therapy, Exp. Neurol. 261C (2014) 417.
[22] M.C. Wagner, J.R. Eckman, T.M. Wick, Sickle cell adhesion depends on hemodynamics and endothelial activation, J. Lab Clin. Med. 144 (2004) 260.
[23] E.A. Waters, S.D. Caruthers, S.A. Wickline, Correlation analysis of stenotic aortic valve flow patterns using phase contrast MRI, Ann. Biomed. Eng. 33 (2005) 878.
[24] L. Wigstrom, T. Ebbers, A. Fyrenius, M. Karlsson, J. Engvall, B. Wranne, et al., Particle trace visualization of intracardiac flow using time-resolved 3D phase contrast MRI, Magn. Reson. Med. 41 (1999) 793.
[25] R.G. Wise, A.I. Al-Shafei, T.A. Carpenter, L.D. Hall, C.L. Huang, Simultaneous measurement of blood and myocardial velocity in the rat heart by phase contrast MRI using sparse q-space sampling, J. Magn. Reson. Imaging 22 (2005) 614.
[26] F.G. Zollner, J.A. Monssen, J. Rorvik, A. Lundervold, L.R. Schad, Blood flow quantification from 2D phase contrast MRI in renal arteries using an unsupervised data driven approach, Z. Med. Phys. 19 (2009) 98.

Further Readings Section

Biomedical Engineering/Biomechanics

Biomechanics: Circulation by Y.C. Fung. Springer, 1997. ISBN: 0-387-94384-6.

Biomechanics: Mechanical Properties of Living Tissue by Y.C. Fung. Springer, 1993. ISBN: 0-387-97947-6.

Introduction to Biomedical Engineering by J. Enderle, S. Blanchard and J. Bronzino. Elsevier, 2005. ISBN: 0-12-238662-0.

Introductory Biomechanics: From Cells to Organisms by C.R. Ethier and C.A. Simmons. Cambridge, 2007. ISBN: 978-0-521-84112-2.

Transport Phenomena in Biological Systems by G.A. Truskey, F. Yuan and D.F. Katz. Pearson, 2004. ISBN: 0-13-042204-5.

Cell Biology/Anatomy and Physiology

Biology by N. Campbell and J. Reece. Benjamin Cummings, 2007. ISBN: 978-0-321-54325-7.

Fundamentals of Anatomy and Physiology by F.H. Martini and J.L. Nath. Pearson, 2009. ISBN: 0-321-50589-1.

Hurst's The Heart by V. Fuster, R.W. Alexander and R.A. O'Rourke. McGraw-Hill, 2004. ISBN: 0-07-142264-1.

Molecular Biology of the Cell by B. Alberts, A. Johnson, J. Lewis, M. Raff, K. Roberts, and P. Walter. Taylor and Francis, 2002. ISBN: 0-8153-3218-1.

Textbook of Medical Physiology by A.C. Guyton and J. E. Hall. Saunders, 2000. ISBN: 0-7216-8677-X.

Vander's Human Physiology: The Mechanisms of Body Function by E.P. Widmaier, H. Raff and K. T. Strang. McGraw-Hill, 2007. ISBN: 978-0-077-21609-2.

Williams Hematology by E. Beutler, M.A. Lichtman, B. S. Coller, and T.J. Kipps. McGraw-Hill, 1995. ISBN: 0-07-070386-8.

Fluid Mechanics/Heat Transfer

Heat and Mass Transfer: Fundamentals and Applications by Y. Cengel and A. Ghajar. McGraw-Hill, 2010. ISBN: 978-0-077-36664-3.

Introduction to Fluid Mechanics by A.T. McDonald, R. W. Fox, and P.J. Pritchard. Wiley, 2008. ISBN: 978-0-471-74299-9.

Introduction to Thermal and Fluid Engineering by D. A. Kaminski and M.K. Jensen. Wiley, 2005. ISBN: 0-471-26873-9.

Viscous Fluid Flow by F.M. White. McGraw-Hill, 1991. ISBN: 0-07-069712-4.

Solid Mechanics/Statics and Dynamics

Engineering Mechanics: Combined Statics and Dynamics by R.C. Hibbeler. Pearson, 2010. ISBN: 978-0-13-814929-1.

Mechanics of Materials by J.M. Gere. Thomson Learning, 2001. ISBN: 0-534-42167-9.

Vector Mechanics for Engineers: Statics and Dynamics by F. Beer, E. Johnston, E. Eisenberg, P. Cornwell, and D. Mazurek. McGraw-Hill, 2009. ISBN: 978-0-07-727555-1.

Index

Note: Page numbers followed by "*f*" and "*t*" refer to figures and tables, respectively.

C

D

E

F

G

H

I

J

K

L

M

Printed in the United States
By Bookmasters